AF342137

IUTAM SYMPOSIUM ON COMPUTATIONAL PHYSICS AND NEW PERSPECTIVES
IN TURBULENCE

IUTAM BOOKSERIES
Volume 4

Aims and Scope of the Series

The IUTAM Bookseries publishes the proceedings of IUTAM symposia under the auspices of the IUTAM Board.

Series Editors:

G.M.L. Gladwell, University of Waterloo, Waterloo, Ontario, Canada
R. Moreau, INPG, Grenoble, France

Editorial Board:

D. van Campen, Eindhoven University of Technology, Eindhoven, The Netherlands
L.B. Freund, Brown University, Providence, USA
H.K. Moffatt, University of Cambridge, Cambridge, UK
J. Engelbrecht, Institute of Cybernetics, Tallinn, Estonia
T. Kambe, IIDS, Tokyo, Japan
A. Kluwick, Technische Universität, Vienna, Austria
N. Olhoff, Aalborg University, Aalborg, Denmark
Z. Zheng, Chinese Academy of Sciences, Beijing, China

For a list of related mechanics titles, see final pages.

IUTAM Symposium on Computational Physics and New Perspectives in Turbulence

Proceedings of the IUTAM Symposium
on Computational Physics
and New Perspectives in Turbulence,
Nagoya University, Nagoya, Japan,
September, 11–14, 2006

Edited by

YUKIO KANEDA

Nagoya University, Nagoya, Japan

 Springer

A C.I.P. Catalogue record for this book is available from the Library of Congress.

ISBN 978-1-4020-6471-5 (HB)
ISBN 978-1-4020-6472-2 (e-book)

Published by Springer,
P.O. Box 17, 3300 AA Dordrecht, The Netherlands.

www.springer.com

Printed on acid-free paper

Scientific Committee

(Chair)
Y. Kaneda	Nagoya University, Japan

C. Cambon	École Centrale de Lyon, France
P. A. Davidson	University of Cambridge, UK
B. Eckhardt	Philipps-Universität Marburg, Germany
T. Gotoh	Nagoya Institute of Technology, Japan
J. Jiménez	Universidad Politécnica, Spain, and Stanford University, USA
A. Pouquet	National Center for Atmospheric Research, USA
K. R. Sreenivasan	International Center for Theoretical Physics, Italy

(IUTAM Representative)
R. Narasimha	Jawaharlal Nehru Centre for Advanced Scientific Research, Indian

Local Organizing Committee

T. Gotoh	Nagoya Institute of Technology, Japan
T. Ishihara	Nagoya University, Japan
Y. Kaneda	Nagoya University, Japan
Y. Tsuji	Nagoya University, Japan
K. Yoshimatsu	Nagoya University, Japan

Sponsors of Symposium

International Union of Theoretical and Applied Mechanics (IUTAM)
The 21st Century COE Program "Frontiers of Computational Science"
Daiko Foundation
The Kajima Foundation
The Kao Foundation for Arts and Sciences
Research Foundation for the Electrotechnology of Chubu
Springer Science and Business Media

Preface

Turbulence remains one of the most challenging problems in classical physics. The papers collected in this volume are the proceedings of an IUTAM Symposium on turbulence, entitled Computational Physics and New Perspectives in Turbulence. The symposium was held in September 2006, at Nagoya University in Japan.

The last few years have witnessed a rapid and dramatic rise in our ability to compute highly complex physical systems. As in other fields, this has had a major impact on the way in which we approach the problem of turbulence, opening up a new phase of research by providing an opportunity to study the nature of fully-developed turbulence in unprecedented detail. Leading experts in turbulence were brought together at this Symposium to exchange ideas and discuss, in the light of the recent progress in computational methods, new perspectives in our understanding of turbulence. The Symposium also fostered a vigorous interaction between those who pursue computations, and those concerned with developments in experiment and theory.

There were 104 participants representing 13 countries, and the presentations and consequent debate extended over a period of four days. Throughout, emphasis was placed on the fundamental physical interpretation of turbulent phenomenon. The topics covered included: (i) computational physics and the theory of canonical turbulent flows; (ii) experimental approaches to fundamental problems in turbulence; (iii) turbulence modeling and numerical methods; and (iv) geophysical and astrophysical turbulence.

We are grateful to all of the participants, to the authors of written contributions to this volume, and to the reviewers of those contributions who helped improve the quality of these proceedings. The success of the Symposium would not have been possible without the valuable advice of the members of Scientific Committee and the excellent work of the Local Organizing Committee. Sincere thanks are also extended to Professor Tsutomu Kambe, the representative of Japan in IUTAM, who encouraged and supported this Symposium from the early planning stages. I would also like to express my thanks for helpful editorial assistance to Ms. Mami Yamashita.

Generous support of the Symposium are gratefully acknowledged, including, IUTAM, The 21st Century COE Program "Frontiers of Computational Science", Daiko Foundation, The Kajima Foundation, The Kao Foundation for Arts and Sciences, and Research Foundation for the Electrotechnology of Chubu. Last but not least, the cooperation and financial assistance of Springer Science and Business Media, making possible the publication of these proceedings, is very much appreciated.

Nagoya

Yukio KANEDA
March 2007

Contents

Part II Experimental Approach to Fundamental Problems in Turbulence

Part III Turbulence Modeling and Numerical Methods

Part IV Geophysical, Astrophysical and Complex Turbulence

Part I

Computational Physics and Theory for Canonical Turbulence

Some Contributions and Challenges
of Computational Turbulence Research

Javier Jiménez

School of Aeronautics, Universidad Politécnica, 28040 Madrid, Spain
and Centre for Turbulence Research, Stanford Univ., USA
`jimenez@torroja.dmt.upm.es`

Abstract. The contributions of numerical simulations to the changes that have taken place during the past twenty years in our understanding of near-wall turbulence, and of the dissipative scales of isotropic flows, are briefly reviewed. It is argued that both problems have moved over this time from empirical observations to relatively coherent theoretical models, and that much of the reason is that they could be simulated cheaply enough to subject them to conceptual experiments. This required a lapse of ten to fifteen years after they were first computed, over which time the cost of simulations decreased by a factor of 100. Simulations of the logarithmic layer and of the inertial energy cascade and now beginning to be affordable. Both are still too expensive to experiment with them, but it is argued that, if history can be taken as a guide, both problems will become routinely computable in the next decade, and that we will then be able to attack their dynamics.

Keywords: turbulence, direct simulations, boundary layers, isotropic turbulence, computational methods

1 Introduction

There is little question that, if the elementary book of Tennekes and Lumley [1] were written today, it would be very different from the original, and that structure would at least in part replace statistics. Even if it is unclear whether such change should be considered advantageous, it is obvious that the intervening decades have changed our view of turbulence from that of the ancient maps of unknown lands, with their warnings of 'here be dragons', to the gentler one of a travel guide.

Both numerical simulations and experiments have contributed to this revolution, and both have been responsible for the consequent flourishing of turbulence research that began in the early 1960s and that continues today. But the enabling technology, the 'new thing' that made the last two decades different from the previous ones, has been numerics.

Much has been said about Colourful Fluid Dynamics, mostly unkindly, but pictures are useful, as experimentalists discovered even before simulations

Y. Kaneda (ed.), *IUTAM Symposium on Computational Physics and New Perspectives in Turbulence*, 3–10.

became practical [2]. Moreover, the power of numerical simulations is not primarily in the beautiful pictures, or even in the exceptionally complete data sets that they produce, but in their ability to go back to the original data and to ask questions that were unintended at the time that they were collected. Very specially in their ability to interrogate the flow, and even the constitutive equations, as if they were made of pieces to be taken apart and examined individually. To borrow a term that became common in philosophy a few decades ago, simulations allow us to 'deconstruct' turbulence.

Guidebooks are not good places to look for the essence of an alien country. That requires a good writer, preferably native, and it is unclear what it means to be native to the country of turbulence. But guidebooks are indispensable for touring, and touring is a necessary first step to understanding. Perhaps the main effect of numerical simulations has been to provide us with postcards from the country of turbulence, and to dispel the, at times convenient and even comfortable, feeling that dragons lurk behind the eddies.

The talk on which this paper is based was intended as a review of what simulations have achieved up to now, and, less credibly, of what could be expected from them in the next few years. Such a task was impossible even for a long talk, and it is even harder for a summary. In the next two sections I will give two examples of turbulence problems whose solutions are today much closer than twenty years ago, and to which numerics have made major contributions. The task of extrapolating the future is tackled in the conclusions. It should of course be stressed that neither the talk nor the paper are intended as summaries of the progress in turbulence over the past twenty years. That should also have included the contributions of theory and of experiments, and would lead to a much longer paper. Our emphasis is on the results obtained from a particular method, not on everything that has been achieved on a particular problem.

2 The Near-Wall Layer

The number of degrees of freedom in a turbulent flow increases rapidly with the Reynolds number, and our ability to pose questions about turbulence quickly exceeds the capacity of the available computers to answer them. Although computers improve rapidly, and this limitation keeps becoming less and less stringent, it has traditionally restricted numerical simulations either to flows at moderate Reynolds numbers, or to parts of the flow in which viscosity is important. One region that remains viscous even in high-Reynolds number flows is the near-wall layer of wall-bounded turbulence. If we admit that the velocity fluctuations scale approximately with the friction velocity u_τ [1], and that the smallest dimension of the eddies cannot be larger than their distance y to the wall, the distance y^+ expressed in wall units is a Reynolds number, and all the eddies below approximately $y^+ = 100$ are directly influenced by viscosity. Simulations were very soon able to resolve this region [3], and the result was an immediate change in what was previously considered a very complicated problem. It was known that this part of the flow was dominated by longitudinal streaks of the streamwise velocity, and that the wall-normal velocity was localized in 'sweeps' and 'bursts', but the relations between those features were unclear. It is seldom as easy as in this case to pinpoint the date of a revolution in our understanding of a phenomenon. It took a single set of colourful pictures, and specially the analysis of a computer movie, to clarify the geometry [4]. It turned out that the dominant structures of the buffer layer were the previously-mentioned velocity streaks, and that they were flanked by loose pairs of quasi-streamwise vortices. The latter were a surprise. Although the existence of streamwise vortices had been postulated before, to account for the formation of the streaks [5], they were widely believed to be about the same length as the latter. They turned out to be much shorter, with several vortices associated with each streak. In the buffer region they move faster than the mean flow, and it is probably because of this that they create the streaks as 'wakes' in the mean velocity profile. It also turned out that most of the sweeps and bursts that had been educed from the analysis of one-point hot-wire measurements were not fast unsteady phenomena when observed in the proper frame of reference, and could be explained by the passing of individual slowly-evolving vortices.

The knowledge gained from these early simulations was mostly kinematic. The catalogue of important structures was more or less complete, but their interactions were still unclear. It took a few more years, and several 'conceptual' experiments on intentionally unrealistic configurations [6, 7, 8] to get beyond kinematics. That part of the story involves many actors and has been reviewed before [9, 10]. It will not be repeated here. The end result was a simplified model of the viscous and buffer layers as a dynamical system organized around a family of fixed points that are numerically exact nonlinear traveling-wave solutions to the Navier-Stokes equations. Those solutions are unstable saddles in phase space, and as such are not expected to be seen in the flow, but the phase velocity of a system is low near fixed points, even if they are unstable, and the flow spends a disproportionately large fraction of its lifetime in their neighbourhood. In fact, fixed points account for about half of the energy production and dissipation in the near-wall layer. The rest can be modelled in terms of limit cycles and of homoclinic orbits loosely based around them [11, 12]. This incidentally reinstates the early observations of unsteady structural bursting in flow visualizations [5], which had been partly discredited by the association of hot-wire bursting with individual quiescent vortices. Many characteristics of the near-wall layer, such as the dominant spacing of the streaks, and the shape and intensity of the fluctuation profiles, are well predicted by those models.

We are today at the early stages of simulating wall-bounded flows that have a logarithmic layer, and, as at the comparable stage of near-wall simulations, the information that we are getting is essentially kinematic. The problem is one of cost, and was shared by the original low-Reynolds number simulations that eventually led to our present understanding of the buffer layer. The recent simulation in [13], at $h^+ = 2000$, took six months on 2100 supercomputer processors. It took a similar time, twenty five years ago, to run the simulation in [3] at $h^+ = 180$. As long as each simulation takes such long times, it is only possible to treat it as a better-instrumented laboratory experiment, and to observe the results.

As computers improve, other things become possible. When the low-Reynolds number simulations of the 1980s became roughly 100 times cheaper in the 1990s, it became possible to experiment with them in ways that were impossible in the laboratory. The 'conceptual' simulations that led to the results mentioned above were of this kind. We are only beginning to be able to do the same for the logarithmic layer. One of the first examples are the perturbed-wall simulations in [14], where the near-wall cycle described above is purposely destroyed, and the effects on the logarithmic layer are observed to study the interactions between the two regions. However, the Reynolds numbers of those simulations are still only marginal, and they do not really exploit the uniqueness of computing. They are conceptually similar to flows over rough walls.

3 The Dissipative Scales of Isotropic Turbulence

Another case in which we are interested in relatively low-Reynolds number structures, even if the Reynolds number of the flow as a whole is large, is the dissipative end of the isotropic energy cascade. The Kolmogorov length scale is defined as that at which viscosity becomes important for individual eddies [15], which can therefore be studied at moderate resolution. As in the case of the near-wall region, this is where the velocity gradients are largest and the time scales fastest, and one could expect that the dynamics of the dissipative structures should be relatively independent of the larger scales. It was very soon found experimentally that this was not true [16], and that the velocity gradients in a turbulent flow do not satisfy the simple dimensional scaling implied by [15]. It was also suggested by experiments that the dissipative scales were anisotropic [17], and did not therefore fit the cascade picture neatly, but it was not until the first numerical simulations become available that it was realized that the vorticity was organized at those scales in the form of long linear vortices [18, 19, 20]. The following years saw a concerted effort by many groups to determine the properties of those vortices, both numerically [21], and experimentally [22, 23].

It was shown that they are present in virtually all turbulent flows [24, 25], and that they have radii of the order of the Kolmogorov scale η, and much longer lengths. Their internal azimuthal velocities range from the Kolmogorov viscous velocity scale, to the root-mean-square velocity of the flow, u'. Persuasive evidence was obtained, mostly from simulations, that they originate from the roll-up of pre-existing vortex sheets [26], and that their lengths and high internal velocities are maintained by axial pressure waves [27].

It was also soon realized [21] that the strongest vortices have internal Reynolds numbers, $u'\eta/\nu$, of the order of $Re_\lambda^{1/2}$, where Re_λ is based on u' and on the Taylor microscale. For sufficiently high Re_λ, the vortex Reynolds number would be much larger than unity, and one could expect that the vortices would become unstable. It was thought for a while that such an instability had been observed experimentally around $Re_\lambda = 600$ [28], but the experiment could not be replicated in a different apparatus, and the effect does not seem to be present in numerical simulations of isotropic flows, which have for some time reached comparable Reynolds numbers [29]. It was concluded in [30] that the proposed instability was present at all Reynolds numbers, that the strongest vortices would themselves become turbulent, and that there would always be some stronger intermittent structures at scales $\eta Re_\lambda^{-1/2}$, much smaller than the Kolmogorov scale. In fact that smaller scale is proportional in a gas to the mean-free path for molecular collisions, divided by the Mach number of the fluctuations, calling into question the validity of the Navier–Stokes equations themselves. Numerical simulations do not generally capture such small structures, because their resolution is adjusted to the *average* Kolmogorov scale. Grid refinement studies in [30] suggested that their effect was negligible in most macroscopic quantities, and even in the mean properties of the small-scale vortices. Recently, however, the previous size estimates have been rediscovered [31], and numerical simulations are in progress to study the nature of these sub-Kolmogorov scales, and their effect on small-scale intermittency.

Although the kinematics, and to some extent the dynamics, of the dissipative vortices were, as in the case of the near-wall layer, essentially clarified in the 1990s, it is clear from the previous summary that the success has in this case been less compelling than in wall flows. While the structures that have been isolated in the near-wall layer provide good quantitative estimates for the bulk statistics of the flow in that region, it is still unknown, for example, whether the small-scale vortices explain the intermittency properties of isotropic turbulence. In fact, the early success of simulations in identifying structures in moderate-Reynolds number flows, led to attempts to explain turbulence in terms of forests or tangles of vortices, and to posit that the cascade somehow bypasses the intermediate scales to transport energy directly from the energy-containing eddies to the Kolmogorov vortices. Similar models were attempted for the momentum transfer across turbulent boundary layers.

The personal opinion of the present author is that those models have not contributed much to our understanding of turbulence, and they are intellectual artifacts created by the low Reynolds numbers of the simulations. In atmospheric turbulence, for example, where Re_λ can easily be several thousands, the inertial energy cascade spans a range of length scales of the order of 10^5, or from 1 cm to 1 Km. It is unlikely that any process can span such differences without intermediate stages. The same can be said of the models of the boundary layer in terms of packets of buffer-layer *vortices*, which assume that the outer-flow scales, which in the atmosphere can also be of the order of kilometres, are formed by the amalgamation of sublayer streaks and hairpin vortices, of the order of millimetres. The multiscale model of the turbulent cascade, which treats such processes in terms of relatively incoherent *eddies*, is still a much more likely explanation. Note that it is easy to confuse a multi-step cascade with a one-step one when describing flows with $Re_\lambda = O(100)$, in which the range of scales is a small number.

More interesting are recent efforts to explore the role of coherent structures in implementing the energy dissipation in turbulence [32]. The question of whether the small-scale vortices are important for the flow as a whole is controversial, because it depends on the method used to define them. Estimates of the total enstrophy contained in them run from a few percent to most of it. It is known however that artificially removing the highest-intensity vortices has little effect on the integrated energy dissipation [21], and that has often been interpreted to mean that the vortices themselves are unimportant. An alternative interpretation is that they are important, but only during a small fraction of their lifetimes, which any instantaneous observation would tend to miss. The question is interesting because the reason why (and if) the turbulent energy dissipation is independent of the Reynolds number, is still not understood. The new numerical experiments use highly symmetric flows [33] that are the equivalent in isotropic turbulence of the 'conceptual' systems that were used in wall-bounded flows to clarify the near-wall layer. What is interesting is that the authors find a numerically-exact limit cycle in the reduced system that shares with turbulence the property that its mean dissipation is (numerically) independent of the viscosity. The dissipation is intermittent, and the configuration of the vorticity at the moment of the dissipation maximum, an array of three symmetric vortex pairs, is strikingly reminiscent of the configuration suggested in [34] for the blow-up of an inviscid fluid. Viscosity avoids the singularity in this case, but the scale of the vorticity configuration at the moment in which the dissipation peaks, 50η, is very similar to the location of the spectral peak of the dissipation in isotropic turbulence.

4 Conclusions and Future Developments

We have briefly reviewed two cases in which numerical simulations have contributed decisively to our understanding of turbulent flows. In both of them, the dynamics of the near-wall layer over smooth walls, and the structure of the dissipative scales of the energy cascade, it is doubtful whether similar advances would have been possible without the aid of simulations. Both are subjects that, like most others in turbulence, are still not fully closed, but that have evolved in the last two decades from empirical observations to relatively coherent theoretical models. The reason why simulations have been useful is that, in both cases, the local Reynolds numbers of the structures being studied are low, even if the global Reynolds number of the flow is not.

This makes them accessible to computation, while experiments are difficult. For example the spanwise spacing of the streaks in the sublayer is of the order of $z^+ = 100$, which is less than a millimetre in most experiments, but that relatively low value can be interpreted as the Reynolds number of the streaks, based on their widths or heights, and we have seen that it is well predicted by the range of parameters in which the associated equilibrium solutions exist. The quasi-singularity of the near-wall layer, with its very high velocity gradients, makes the results obtained in this way relatively independent of the global Reynolds numbers [35], although some interactions are known to exist [36, 13].

This relative independence, that allows us to isolate part of the flow from the global problem, has been important for the analysis. We have emphasised several times that the power of numerics lies in this possibility of isolating individual aspects. In essence we have been able to isolate the near-wall layer and the dissipate scales because they lie at the ends of their respective cascades, the cascade of linear momentum in the case of wall turbulence, and the energy cascade in the isotropic case.

Moving beyond those limited domains will be harder, but we have noted that simulations with a logarithmic layer and with an inertial range are beginning to appear. At the moment they are little else than glorified experiments, in which it is impossible to do much more than to observe. Understanding the dynamics of the two cases reviewed above implied being able to move from such kinematic simulations to conceptual experiments, and that required the cost of simulations to decrease considerably. It took ten to fifteen years, and a jump of computer speeds of the order of a hundred, to take that step.

In will probably take about the same time to do the same thing for the present simulations. There is no reason to believe that computer improvements have stopped, and the next decade will bring the cost of the simulations of the logarithmic layer and of the inertial cascade to the level at which dynamical experiments should become commonplace. It is only then that we can expect a dynamical theory for these parts of the flow to emerge from simulations. The motivation for such research is both theoretical and technological. The cascade of momentum across the range of scales in the logarithmic layer will probably be the first three-dimensional self-similar cascade to become accessible to computational experiments. Its simplifying feature is the alignment of most of the net transfer along the direction normal to the wall. The main practical drive is probably large-eddy simulation, in which the momentum transfer across scales in the inertial range has to be modelled for the method to be practical [37]. Only by understanding the structures involved will we be sure of how to accomplish that.

The preparation of this paper was supported in part by the CICYT grant TRA2006–08226/TAIR.

References

1. Tennekes H, Lumley JL (1972) A first course in turbulence. MIT Press
2. Brown GL, Roshko A (1974) J Fluid Mech 64:775–816
3. Kim J, Moin P, Moser RD (1987) J Fluid Mech 177:133–166
4. Robinson SK (1991) Ann Rev Fluid Mech 23:601–639
5. Kim HT, Kline SJ, Reynolds WC (1971) J Fluid Mech 50:133–160
6. Jiménez J, Moin P (1991) J Fluid Mech 225:221–240
7. Hamilton JM, Kim J, Waleffe F (1995) J Fluid Mech 287:317–348
8. Jiménez J, Pinelli A (1999) J Fluid Mech 389:335–359
9. Jiménez J, Kawahara G, Simens MP, del Álamo JC (2004) The near-wall structures of turbulent wall flows. In: Kida S, (ed.) IUTAM Symp. on Elementary Vortices and Coherent Structures: Significance in Turbulence Dynamics, 53–70, Springer
10. Jiménez J, Kawahara G, Simens MP, Nagata M, Shiba M (2005) Phys Fluids 17:015105
11. Kawahara G, Kida S (2001) J Fluid Mech 449:291–300
12. Toh S, Itano T (2003) J Fluid Mech 481:67–76
13. Hoyas S, Jiménez J (2006) Phys Fluids 18:011702
14. Flores O, Jiménez J (2006) J Fluid Mech 566:357–376
15. Kolmogorov AN (1941) Dokl Akad Nauk SSSR 30:301–305. Reprinted in *Proc. R. Soc. London.* A **434**, 9–13 (1991)
16. Batchelor GK, Townsend AA (1949) Proc Roy Soc London A 199:238–255
17. Kuo AY, Corrsin S (1972) J Fluid Mech 56:447–479
18. Siggia ED (1994) J Fluid Mech 107:375–406
19. She Z, Jackson E, Orszag SA (1990) Nature 344:226–228
20. Vincent A, Meneguzzi M (1991) J Fluid Mech 225:1–25
21. Jiménez J, Wray AA, Saffman PG, Rogallo RS (1993) J Fluid Mech 255:65–90
22. Douady S, Couder Y, Brachet ME (1991) Phys Rev Lett 67:983–986
23. Belin F, Tabeling P, Willaime H (1996) Physica D 93:52–63
24. Tanahashi M, Iwase S, Miyauchi T (2001) J Turbul 2:1–17
25. Tanahashi M, Kang SJ, Miyamoto T, Shiokawa S, Miyauchi T (2004) Int J Heat Fluid Flow 25:331–340
26. Passot T, Politano H, Sulem PL, Angilella JR, Meneguzzi M (1995) J Fluid Mech 282:313–338
27. Verzicco R, Jiménez J, Orlandi P (1995) J Fluid Mech 299:367–388
28. Tabeling P, Willaime H (2002) Phys Rev E 65:066301
29. Kaneda Y, Ishihara T, Yokokawa M, Itakura K, Uno A (2003) Phys Fluids 15:L21–L24
30. Jiménez J, Wray AA (1998) J Fluid Mech 373:255–285
31. Yakhot V, Sreenivasan KR (2005) J Stat Phys 121:823–841
32. van Veen L, Kida S, Kawahara G (2006) Fluid Dyn Res 38:19–46
33. Kida S (1985) J Phys Soc Japan 54:2132–2136
34. Pelz RB (2001) J Fluid Mech 444:299–320
35. Klewicki JC, Metzger MM, Kelner E, Thurlow E (1995) Phys Fluids 7:857–863
36. DeGraaf DB, Eaton JK (2000) J Fluid Mech 422:319–346
37. Jiménez J, Moser RD (2000) AIAA J 38:605–612

Global Scaling Properties of Heat and Momentum Transport in Fluid Flow

Bruno Eckhardt, Siegfried Grossmann, and Detlef Lohse

[1] Fachbereich Physik, Philipps-Universität, Renthof 6, D-35032 Marburg, Germany, `bruno.eckhardt@physik.uni-marburg.de`, `grossmann@physik.uni-marburg.de`

[2] Department of Applied Physics, University of Twente, 7500 AE Enschede, The Netherlands, `d.lohse@utwente.nl`

Abstract. We identify corresponding pairs of currents and dissipation rates in Rayleigh-Bénard, Taylor-Couette and pipe flow which allow us to transfer the modelling arguments developed for thermal convection to shear flows.

Keywords: shear flows, boundary layers, Taylor-Couette, Rayleigh-Bénard, pipe flow

1 Introduction

Scaling relations are central to the description and modelling of hydrodynamic flows [1, 2]. Their uses range from relating flows on different scales and velocities if only their Reynolds numbers coincide to predicting the smallest scales and the energy distribution in turbulent flows within the Kolmogorov cascade picture. Self-similarity is also at the heart of momentum transport and the logarithmic profile in boundary layers. Barenblatt [1] gives intriguing examples of how scaling can be used to solve seemingly intractable problems, from Taylor's estimate of the power of the first atomic bomb to the variation of the speed of rowing boats with the number of oarsmen. On the mathematical side the ideas behind these scaling laws have been collected in the Buckminster theorem about the number of independent dimensionless combinations that can be formed [3]. However, this observation should not be misinterpreted as giving the relevant physics for free. In all cases additional physical assumptions are needed in order to arrive at the experimentally observed scaling relations. Furthermore, by their very nature, these relations are asymptotic at best, and the models usually do not provide an idea of where the asymptotics sets in. They also require additional assumptions if the dependence on different dimensionless quantities has to be disentangled. Hence, despite their successes, there is room for reevaluation and for reconsideration of scaling relations when additional experimental or new theoretical insights and relations become available.

11

Y. Kaneda (ed.), *IUTAM Symposium on Computational Physics and New Perspectives in Turbulence*, 11–18.
© 2008 *Springer*.

The example that will be discussed in the following deals with the relation between thermal convection (Rayleigh-Bénard, RB) on the one hand side and shear flows like Taylor-Couette (TC) and pipe or channel Poiseuille flow on the other. The two communities studying these systems have developed independent and different uses of scaling relations. Early studies in RB flow emphasized power-law dependencies and this has been carried over to the most recent theories (see [4] for a discussion of the various models). In shear flows, on the other hand, the dominance of the law of the wall has led to logarithmic representations of the global scaling of the force with Reynolds number [2]. Attempts to link the two classes of flows, as in the studies by Bradshaw [5] or Dubrulle and Hersant [6], take sides with the logarithmic law and derive torque variations that contain logarithms. However, they do not take these results back to RB and discuss how the logarithms impact the heat transport in RB . Our earlier efforts [7], motivated by the power law modelling in RB did result in a reasonable representation of the data using power laws, but did not allow a complete analogy.

The situation has changed with the discovery of the quantities for which corresponding pairs in the two classes can be given [8, 9, 10]. At the heart of these analogies are two key observations: one has to separate the momentum current from the dissipation, and one has to distinguish between the externally imposed parameters and the internal response of the fluid. We begin with a discussion of the latter point first.

2 External Parameters and Internal Response

The laminar state in RB flow is one of pure diffusion of heat in the absence of a flow. In the laminar state there is no dissipation. Once the temperature difference across the layer exceeds a critical value, a flow field develops. The advection of heat by this flow enhances the heat transport well above the laminar value. Externally imposed parameters then are the temperature difference, as well as all other fluid and geometry parameters (such as Prandtl number, aspect ratio, geometry, etc.). The internal response of the system to this externally imposed parameters is the increase in heat transport, the development of a flow field and the increase in dissipation. In the early studies the velocity field is strongest near the plates, and hence represents a 'wind' blowing along the plates. We will continue to refer to this kind of response as a 'wind', not only in the case of thermal convection but also in the shear flows. Note that once it is present it will change the mean temperature profile, producing the steeper gradients near the walls.

For TC between rotating cylinders the situation is not all that different: in the laminar case we have the Couette profile with the only non-vanishing velocity component pointing in the azimuthal direction and continuously varying between the angular velocities of the inner and outer cylinders. As this is a shear flow, there is energy dissipation already in the laminar flow, and a constant torque has to be applied to keep the cylinders moving. Once critical rotation speeds are exceeded, 3-d velocity fields develop, in the form of Taylor-vortices [11] or more complicated modulated cousins. Of the three velocity components, the one in radial and axial direction are new and not present in the laminar case. They reflect the response of the fluid to the externally imposed stress (the profile gradient or shear). Since they also flow along the plates and modify the mean azimuthal velocity profile, they correspond to the wind in the RB case and can be characterized by an appropriate Reynolds number R_w, which measures the wind amplitude. Note, however, that it is premature to conclude that this wind Reynolds number R_w and the externally imposed Reynolds numbers $R_{1,2}$ of the rotating cylinders 1 and 2 have to coincide. While this is perhaps suggested on dimensional grounds, it still requires a physical mechanism to connect the two fields, the azimuthal or longitudinal flow component and the radial and axial components as the transversal field. This mechanism cannot be derived by dimensional analysis and has to come out of the Navier-Stokes equation. One, therefore, should be prepared to find, in general, $R_w \neq R_i$! This is obvious for low Reynolds numbers, where the Taylor-vortices have an amplitude that is significantly smaller than the externally imposed one, and continues to higher R, as will be demonstrated below.

The third example we will consider is Poiseuille flow down a pipe. Again, one can identify the externally prescribed Reynolds number R from the laminar profile of the externally imposed pressure gradient. The laminar profile has a downstream component and radial variability only. The viscous drag gives rise to energy dissipation in the laminar flow already. The flow is linearly stable for all Reynolds numbers, but since it is connected with the appearance of finite amplitude travelling waves dominated by downstream vortices [12, 13, 14, 15, 16], the reasoning from TC flow can be transferred immediately. The transition to turbulence is then connected with the onset of some azimuthal and radial velocity components, again denoted as 'the wind'. This wind changes the mean profile from the parabolic shape to the familiar plug flow one. Again, there is an externally imposed Reynolds number R, but freedom in the internal response, reflected in the wind amplitude R_w, since the nonlinear waves are governed by their own mechanics.

3 Currents and Dissipation

In the case of RB flow there is an obvious current, the heat transport current. The heat is transported by molecular diffusion but in addition by convection, if the velocity field has appeared. The additional convective transport leads to additional dissipation as compared to the laminar state without flow. What are the analogies in other flows?

To see the analogies and similarities, consider the formula describing the heat current J_θ in RB flow. It is be obtained from the Oberbeck-Boussinesq equations of motion for the temperature field by averaging over surfaces parallel to the top and bottom plates, i.e., over surfaces perpendicular to the flux direction, and over time:

$$J_\theta = \langle u_z \theta \rangle_{A,t} - \kappa \partial_z \langle \theta \rangle_{A,t}. \tag{1}$$

The dimensionless Nusselt number is $N_\theta = J_\theta / \kappa \Delta d^{-1}$. The various physical parameters are the height d, the temperature difference Δ between the bottom and top plates, and the thermal conductivity κ. The temperature θ denotes the deviation from a reference temperature and u_z is the velocity field component in the direction of gravity, normal to the plates. By derivation from the equations of motion, this current is independent of the height z at which it is calculated.

A similiar expression exists for TC flow [8, 10]. The physical quantity, which corresponds to the temperature field in RB flow, is the azimuthal velocity u_φ, since u_φ is subject to an externally applied profile. Thus take the azimuthal (φ) component of the Navier-Stokes equation and average it over time and cylinder surfaces between the two rotating cylinders, i.e., perpendicular to the profile direction. Then

$$J_\omega = r^3 \left(\langle u_r \omega \rangle_{A,t} - \nu \partial_r \langle \omega \rangle_{A,t} \right) = \frac{d}{\kappa(\omega_1 - \omega_2) r_a^3} N_\omega. \tag{2}$$

is independent of the radius and hence the appropriate conserved current [10]. Here, $\omega = u_\varphi / r$ is the angular velocity, and ν the kinematic viscosity. The other quantities are the arithmetic mean of the radius ratios $r_a = (r_1 + r_2)/2$, the gap width d, and $\kappa = \nu r_1^2 r_2^2 / r_a^4$ is the equivalent of the thermal diffusivity. Unexpectedly, the quantity that appears is not the angular momentum, but the angular velocity. This appearance of $\omega(r)$ is dictated by the viscous term. Except for the factor r^3 which is primarily of geometrical origin, this expression (2) for the current has the same structure as the heat current (1) in the RB case: there is a Reynolds stress type term $\langle u_r \omega \rangle_{A,t}$ and a viscous derivative of the mean profile, $-\nu \partial \langle \omega \rangle / \partial r$. In the laminar case, the first part vanishes and conservation of J_ω gives the familiar profile $\omega(r) = A + B/r^2$, with constants A and B determined by the boundary conditions. The second equality in eq. (2) serves to define the dimensionless analog N_ω to the Nusselt number also for TC flow.

For Poiseuille flow through a pipe, the velocity component corresponding to the temperature in RB or the azimuthal component in TC flow is the axial component u_z, all having externally prescribed profiles. Averaging the equation about cylinders inside the pipe, one arrives (see [17]) at a conserved current

$$J_u = \frac{2}{r} \left(\langle u_r u_z \rangle_{A,t} - \nu \partial_r \langle u_z \rangle_{A,t} \right) = \frac{8\nu U}{a^2} N_u. \tag{3}$$

The last equality defines the corresponding dimensionless Nusselt number, with a is the radius of the pipe. Since the flow is pressure driven, the current equals the pressure gradient $\Delta p / \ell$ along the pipe of length ℓ. The Reynolds term now connects the fluctuations in the radial (u_r) and downstream (u_z) directions, while the viscous derivative contains the radial variation of the mean downstream velocity profile. The current J_u is constant, independent of the radial coordinate r, if the factor r^{-1} in front of the two contributions is respected.

As regards the dissipation, the situation in RB is straightforward, as the wind dissipation arises only when the flow sets in, so that

$$\epsilon = \nu \langle (\partial_i u_j)^2 \rangle = \frac{\nu^3}{d^4} Pr^{-2} Ra(N_\theta - 1) \,, \tag{4}$$

where Pr and Ra are the Prandtl and Rayleigh numbers, respectively. In the case of the shear flows, there is already shear dissipation in the laminar state, and this has to be subtracted from the total dissipation rate in order to measure the additional dissipation due to the additional (transverse) convective components, which give rise to the convective transport current. For the excess dissipation ϵ one then finds [10] for TC flow the expression

$$\epsilon = \epsilon_{tot} - \epsilon_{lam} = \frac{\nu^3}{d^4} \sigma^{-2} Ta(N_\omega - 1) \,, \tag{5}$$

where $\sigma = (r_a^2/(r_1 r_2))^2$ is a kind of Prandtl number and $Ta = d^2 r_a^2 (\omega_1 - \omega_2)^2/(\nu\kappa)$ the (appropriately defined) Taylor number. For pipe flow, the corresponding expression is (cf. [17])

$$\epsilon = \epsilon_{tot} - \epsilon_{lam} = \frac{\nu^3}{a^4} R^2 (N_u - 1) \,. \tag{6}$$

Given these similarities between (4), (5), and (6) for the excess dissipation rates ϵ, and the corresponding similarities (1), (2), and (3) for the respective transport currents J, one can now transfer the modelling assumptions from RB flow to the shear flows.

4 Modelling the Wind

As discussed before, there are two independent quantities for each flow, one conserved current J and one excess dissipation rate ϵ. Both are related to the wind, and hence should be modelled in terms of the wind Reynolds numbers.

For the dissipation, we appeal to the usual dimensional estimates which can also be made mathematically rigorous in terms of upper bounds:

$$\epsilon = c_1 R_w^{5/2} + c_2 R_w^3 \,. \tag{7}$$

The Reynolds number that appears must be the one of the wind and not the externally imposed one as ϵ only measures the excess dissipation introduced by the wind. If there is no wind, there will not be excess dissipation and vice versa. The second term then is the typical Kolmogorov dissipation rate in the volume $\propto U^3$ while the first one comes from the dissipation in the boundary layers (BL). It should be $\propto U^2$ and $\propto \delta^{-2}$, where δ denotes the boundary layer thickness. Since the BLs only contribute the relative fraction $\propto \delta$, the BL dissipation rate altogether is $\propto \delta^{-1}$, which according to Prandtl's boundary layer theory scales as $\sqrt{R_w}$. This in total implies the power $5/2$.

For the current, one might be tempted to use a combination of dominant scaling and Prandtl boundary layer theory to conclude $N \propto R_w$, perhaps with a square root correction. However, the case of RB flow shows that one obviously has to deal with *two* boundary layers, one for the temperature BL of thickness λ and another one for the wind BL of thickness δ [18]. The significance is that for $\delta \ll \lambda$, denoted as the lower case, the full advective velocity is reached within the temperature boundary layer, whereas for $\delta \gg \lambda$, called the upper case, only the fraction λ/δ of the wind velocity is relevant in the profile (temperature) boundary layer. As a consequence, the dimensionless current is modelled (cf. [4, 10, 17]) by

$$N_\theta = c_3 \sqrt{R_w\, Pr\, f(s)} + c_4 R_w\, Pr\, f(s), \tag{8}$$

with a switching function $f(s)$ of $s \propto \lambda/\delta$ that interpolates between a linear variation for small s and a constant behaviour for large s. The argument s itself is determined by the wind Reynolds number and the current, $s = \sqrt{R_w}/N_\theta$, so that (8) for given Pr is an implicit equation for the current.

The presence of two boundary layers in the case of thermal convection suggests to consider two boundary layers for shear flows as well. One is the boundary layer of the prescribed externally driven flow component, i.e., of the azimuthal velocity in TC and the downstream velocity in Poiseuille pipe flow, with thickness λ; the other boundary layer is that of the wind, i.e., that of the transverse flow field components, having the width δ. Continuing along those lines, one arrives [10] at modelling expressions for the angular velocity current in TC flow of the form

$$N_\omega = c_3 \sigma \sqrt{R_w\, g(s)} + c_4 \sigma R_w\, g(s) \tag{9}$$

and for the radial momentum transport in pipe flow of the form

$$N_u = c_3 \sqrt{R_w g(s)} + c_4 R_w\, g(s)\,. \tag{10}$$

However, in contrast to RB flow, the differences in the thicknesses of the two boundary layers are most likely only instantaneous, since all three velocity components are linked through the equations of motion in general and the incompressibility constraint in particular. Numerical studies to test the differences of the boundary layer thicknesses in the longitudinal and transversal components are under way. As a consequence of this coupling, it is plausible that while both boundary layers will fluctuate in their thicknesses, their mean thicknesses may be similar. Thus, while in RB flow the usual experimental protocol will cause a straight crossover from one region ($\delta \ll \lambda$) to the other

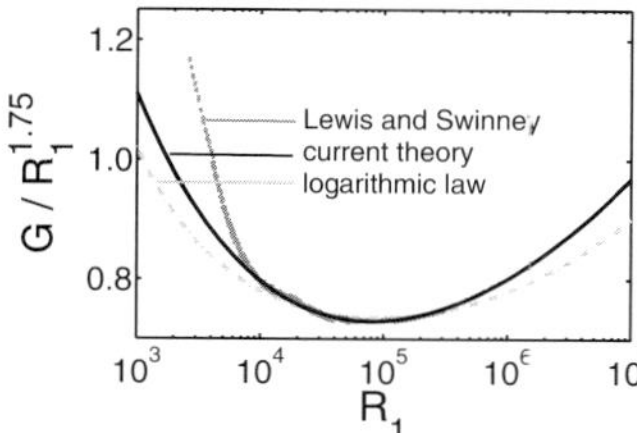 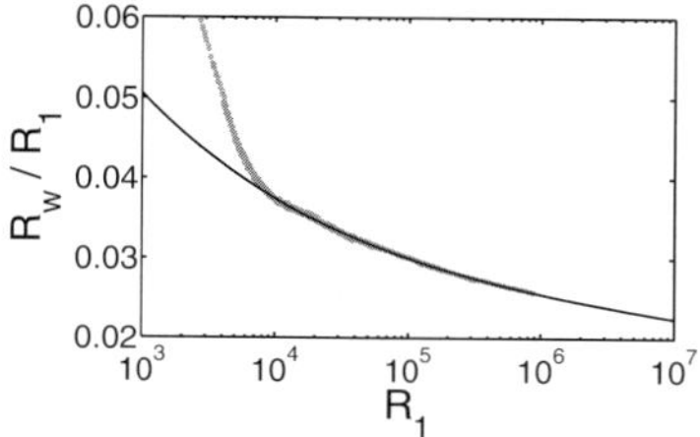

Fig. 1. Torque in TC with outer cylinder at rest. The data are from Lewis and Swinney [19] and shown as circles. The current theory is shown as a full line, a fit to a logarithmic profile as a dashed line. Left: rescaled torque vs. Reynolds number R_1 of inner cylinder. Right: Wind Reynolds number R_w relative to R_1 vs. R_1.

one ($\delta \gg \lambda$), the shear flows may live around the marginal situation where neither boundary layer dominates permanently. Since in such a situation fluctuations are usually important it was not yet possible to solve the implicit equation for the current. If the profile boundary layer is thicker than the wind boundary layer and thus $g \approx 1$, then for high Reynolds numbers $N_\omega \propto R_w$. If, on the other hand, the wind boundary layer is thicker, then the wind contributes only with a fraction of its value, thus $g(s) = s$ and $N_\omega \propto R_w^{3/4}$. Near the transition this implies large fluctuations, which can be modelled with an algebraic component in $g(s)$ (as discussed in [10]). Further investigations of this relation are ongoing, but the numerical results to be presented next are encouraging.

For the analysis of the model we take data from Lewis and Swinney on TC flow with the outer cylinder at rest and the inner one rotating [19]. Torque data for inner cylinder Reynolds numbers R_1 in the range $1.3 \cdot 10^4 \cdots 8.76 \cdot 10^5$ were used to determine the parameters in the model. The comparison between experiment and model results is shown in Fig. 1. The fit to the data is very good, relative errors are less than one percent and slightly better than with the logarithmic profile.

In Fig. 1b we display the wind amplitude R_w as a function of the external control parameter R_1, the inner cylinder rotation Reynolds number. Indeed, R_w differs from R_1. But both components, the longitudinal one ($\propto R_1$) as well as the transversal one ($\propto R_w$) increase with increasing rotation speed. The wind increases not as quickly as the externally controlled Reynolds number. It behaves like $R_w \propto R_1^{1-\xi}$, with a Reynolds number depending exponent $\xi(R_1)$, which is of order 0.05, consistent with earlier observations [7].

5 Final Remarks

The arguments and models presented here raise a couple of testable questions. It should be possible to measure and study the transverse velocity profiles, both in TC and pipe Poiseuille flow, so as to learn something about their strengths, i.e., amplitudes R_w, and the relation to the boundary layer thicknesses λ in the longitudinal and δ in the transversal velocity component direction. Secondly, it would be desirable if higher Reynolds numbers could be reached: not only does the modelling rest on the assumption of high Reynolds numbers, it is also there that different models begin to differ noticably in their predictions for the current and hence for the friction. The theory also has predictions for the transport properties for different radius ratios, in reasonable agreement with the data of Wendt [22], and for independently rotating cylinders, see [10].

It would also be of interest to go back to the RB case and to study the heat transport just along the transition line $\lambda \approx \delta$ in order to study the influence of the fluctuations on the heat transport there.

We thank DFG and FOM for their support.

References

1. Barenblatt GI (2003) Scaling Cambridge University Press Cambridge
2. Landau LD, Lifshitz IM (1987) Fluid Mechanics 2nd edition Butterworth-Heinemann
3. Buckingham E (1914) Phys Rev 4:345–376
4. Grossmann S, Lohse D (2000) J Fluid Mech 407:27–56
5. Bradshaw P (1969) J Fluid Mech 36:177–191
6. Dubrulle B, Hersant F (2002) Eur Phys J B 26:379–386
7. Eckhardt B, Grossmann S, Lohse D (2000) Eur Phys JB 18:541–545
8. Eckhardt B, Grossmann S, Lohse D (2005) Energy and Dissipation Balances in Rotating flows In: J. Peinke, et al. (eds) Progress in Turbulence. Springer, Berlin 47–50
9. Eckhardt B, Grossmann S, Lohse D (2006) Nonlinear Phenom Complex Sys 9:109–114
10. Eckhardt B, Grossmann S, Lohse D (2007) J Fluid Mech 581:221–250
11. Taylor GI (1936) Proc R Soc London Ser A157:546–564 and 565–578
12. Boberg L, Brosa U (1988) Z Naturforsch 43a: 697–726
13. Grossmann S (2000) Rev Mod Phys 72:603–618
14. Faisst H, Eckhardt B (2003) Phys Rev Lett 91:224502
15. Wedin H, Kerswell RR (2004) J Fluid Mech 508:333–371
16. Hof B et al. (2004) Science 305:1594–1598
17. Eckhardt B, Grossmann S, Lohse D (2006) Scaling of skin friction in turbulent pipe flow, Preprint
18. Grossmann S, Lohse D (2002) Phys Rev E 66:016305
19. Lewis GS, Swinney H S (1999) Phys Rev E 59:5457–5467
20. Lathrop DP, Fineberg J, Swinney H S (1992) Phys Rev Lett 68:1515–1518
21. Lathrop DP, Fineberg J, Swinney H S (1992) Phys Rev A 46:6390–6405
22. Wendt F (1933) Ingenieurs-Archiv 4:577–595

Linear and Angular Momentum Invariants in Homogeneous Turbulence

Peter Alan Davidson[1], Takaki Ishida[2], and Yukio Kaneda[2]

[1] Department of Engineering, University of Cambridge, Trumpington Street, Cambridge, CB2 1PZ. UK
`pad3@eng.cam.ac.uk`
[2] Department of Computational Science and Engineering, Graduate School of Engineering, Nagoya University, Chikusa-ku, Nagoya 464-8603, Japan
`ishida@fluid.cse.nagoya-u.ac.jp`, `kaneda@cse.nagoya-u.ac.jp`

Abstract. We discuss the constraints imposed on the evolution of freely-decaying turbulence by the laws of conservation of linear and angular momentum. In particular, we explain the results of recent numerical simulations in terms of angular momentum conservation. These simulations show that, once the turbulence reaches a mature state, with a fully-developed vorticity field, its kinetic energy decays as $t^{-10/7}$, a result which is consistent with the classical theories of Landau and Kolmogorov, and inconsistent with Markovianised closure models.

Keywords: invariants in turbulence, Loitsyansky's integral, angular momentum, decaying turbulence, Kolmogorov's decay law

1 Introduction

Perhaps the simplest problem in turbulence is the following. Suppose that we create a large cloud of turbulence, whose radius, R, is very much greater that the size of a typical large-scale eddy, ℓ. This turbulent cloud is, in effect, a seething tangle of vorticity, $\boldsymbol{\omega}(\boldsymbol{x}, t)$, which advects itself in a chaotic manner. Let us choose R so that the spherical surface $|\boldsymbol{r}| = R$ encloses all of the turbulence, except perhaps some exponentially weak far-field vorticity which is the product of viscous diffusion. What can we say about the evolution of this turbulent cloud? If we restrict ourselves to rigorous statements the answer is: 'very little'. In fact, there are probably only three rigorous but useful dynamical statements which can be made. They are:

(i) The linear momentum of the cloud, as measured by the linear impulse, $\boldsymbol{J} = (1/2) \int \boldsymbol{x} \times \boldsymbol{\omega} \mathrm{d}V$, is an invariant of the motion;
(ii) The angular momentum of the cloud, as measured by the angular impulse, $\boldsymbol{H} = (1/3) \int \boldsymbol{x} \times (\boldsymbol{x} \times \boldsymbol{\omega}) \mathrm{d}V$, is also an invariant of the motion;
(iii) The total kinetic energy of the motion declines according to

Y. Kaneda (ed.), *IUTAM Symposium on Computational Physics and New Perspectives in Turbulence*, 19–26.
© 2008 *Springer*.

$$\frac{\mathrm{d}}{\mathrm{d}t} \int \frac{1}{2} \boldsymbol{u}^2 \mathrm{d}V = -\nu \int \boldsymbol{\omega}^2 \mathrm{d}V. \tag{1}$$

One of the surprising features of most statistical theories turbulence is that, while they all agree on the importance of (iii), with almost equal unanimity they choose to ignore (i) and (ii). At first sight this is astonishing. After all, these conservation laws impose powerful constraints on the way in which the chaotic vorticity field can develop. We would expect such conservation principles to be particularly relevant to the behaviour of the large-scale vortices, as these contribute most to the net linear and angular momentum of the cloud.

The problem, perhaps, is that most two-point statistical theories are developed within the somewhat artificial (if convenient) framework of homogeneous turbulence. In homogeneous turbulence we have no well-defined 'cloud' of vorticity, and it is not immediately apparent how to restate these conservation laws for a sub-domain of the flow. Moreover, most theories are formulated in terms of two-point correlations, such as $\langle \boldsymbol{u} \cdot \boldsymbol{u}' \rangle$, and it is not obvious how to express integral quantities like $\boldsymbol{J}$ and $\boldsymbol{H}$ in terms of the more common currency of correlation functions.

In any event, it turns out that there *is* a statistical theory of the large scales based on these conservation principles, though it is somewhat controversial and little discussed. Broadly speaking, there are two theories and two camps. There are those who believe that the eddies which make up a field of homogeneous turbulence possess a significant linear impulse, in the sense that the impulse of a typical eddy, $\boldsymbol{J}_\mathrm{e} = (1/2) \int \boldsymbol{x} \times \boldsymbol{\omega}_\mathrm{e} \mathrm{d}V$, is of order $\sim |\boldsymbol{\omega}_\mathrm{e}| V_\mathrm{e} \ell_\mathrm{e}$. (Here the subscript e indicates an individual eddy, $|\boldsymbol{\omega}_\mathrm{e}|$ is a characteristic vorticity of the eddy, and V_e and ℓ_e are the characteristic volume and size of that eddy.) In such cases the global linear impulse, $\boldsymbol{J}$, of a cloud of such eddies will be non-zero, despite the random orientation of the vortices. That is, for a finite sized cloud of volume V_R there will always be some imperfect cancellation of linear impulse when we form the sum $\sum \boldsymbol{J}_\mathrm{e}$, with the central limit theorem suggesting $|\boldsymbol{J}| \sim V_R^{1/2}$. Saffman [1] has shown that turbulence of this type possesses the statistical invariant $\langle \boldsymbol{J}^2 \rangle / V_R = \int \langle \boldsymbol{u} \cdot \boldsymbol{u}' \rangle \mathrm{d}\boldsymbol{r}$, $\boldsymbol{r} = \boldsymbol{x}' - \boldsymbol{x}$, and that, as a consequence, its kinetic energy decays as $u^2 \sim t^{-6/5}$. It so happens that the energy spectrum in such cases grows as $E \sim k^2$ for small k.

There are others, however, who believe that typical eddies have negligible linear impulse, in the sense that

$$\frac{1}{2} \int_{V_\mathrm{e}} \boldsymbol{x} \times \boldsymbol{\omega}_\mathrm{e} \mathrm{d}V \ll |\boldsymbol{\omega}_\mathrm{e}| V_\mathrm{e} \ell_\mathrm{e}. \tag{2}$$

In such cases it can be shown that $\int \langle \boldsymbol{u} \cdot \boldsymbol{u}' \rangle \mathrm{d}\boldsymbol{r} = 0$, but that $\int r^2 \langle \boldsymbol{u} \cdot \boldsymbol{u}' \rangle \mathrm{d}\boldsymbol{r}$ is finite, negative and (as we shall see) more or less constant in fully-developed turbulence. We shall also see that the kinetic energy now decays approximately as $u^2 \sim t^{-10/7}$, while E grows as $E \sim k^4$ at small k. Moreover, it may be shown (see [2]) that the integral $-\int r^2 \langle \boldsymbol{u} \cdot \boldsymbol{u}' \rangle \mathrm{d}\boldsymbol{r}$ is an approximate measure of the angular momentum (squared) of a turbulent cloud,

$$\langle \boldsymbol{H}^2 \rangle / V_R \approx - \int \boldsymbol{r}^2 \langle \boldsymbol{u} \cdot \boldsymbol{u}' \rangle \mathrm{d}\boldsymbol{r}, \tag{3}$$

though the proof of this is controversial as it requires remote points in the turbulence to be statistically independent.

Now both $E \sim k^2$ and $E \sim k^4$ spectra are observed in the numerical simulations of turbulence. It is the initial conditions that dictate which form of turbulence emerges. If the initial linear momentum is large enough, then a Saffman spectrum is guaranteed. If it is small, an $E \sim k^4$ spectrum is obtained. The main controversy centres on the delicate issue of which form of turbulence is likely to be encountered in a real flow, such as grid turbulence. Here there is little agreement. A second controversy relates to the degree to which $- \int \boldsymbol{r}^2 \langle \boldsymbol{u} \cdot \boldsymbol{u}' \rangle \mathrm{d}\boldsymbol{r}$ is really conserved is cases where $\boldsymbol{J}_\mathrm{e} \approx \boldsymbol{0}$, and whether or not this integral is indeed proportional to $\langle \boldsymbol{H}^2 \rangle / V_R$. We focus on this second controversy here.

2 The Classical Theory of the Large Scales

Let us now try to justify the statements above by providing an overview of the classical theory of the large scales. For simplicity we restrict ourselves to freely-decaying, isotropic turbulence, though most of the statements generalize to anisotropic, homogeneous turbulence.

If we assume that there are no singularities in the two-point velocity spectral tensor, and we expand $E(k)$ about $k = 0$, then we obtain,

$$E(k) = Lk^2/4\pi^2 + Ik^4/24\pi^2 + \cdots \tag{4}$$

where,

$$L = \int \langle \boldsymbol{u} \cdot \boldsymbol{u}' \rangle \mathrm{d}\boldsymbol{r} \tag{5}$$

and

$$I = - \int \boldsymbol{r}^2 \langle \boldsymbol{u} \cdot \boldsymbol{u}' \rangle \mathrm{d}\boldsymbol{r}. \tag{6}$$

(Here $\boldsymbol{r} = \boldsymbol{x}' - \boldsymbol{x}$.) The integrals L and I are known as the Saffman and Loitsyansky integrals respectively, and it is readily confirmed that the Karman-Howarth equation demands

$$\frac{\mathrm{d}L}{\mathrm{d}t} = 4\pi \left[\frac{1}{r} \frac{\partial}{\partial r} \left(r^4 u^3 K \right) \right]_\infty \tag{7}$$

$$\frac{\mathrm{d}I}{\mathrm{d}t} = 8\pi \left[r^4 u^3 K \right]_\infty \tag{8}$$

where K is the usual triple longitudinal correlation function, u is defined by $u^2 = \langle \boldsymbol{u}^2 \rangle / 3$, and the subscript ∞ indicates a quantity evaluated at $r \to \infty$.

Now a fluctuation in the vorticity field at location $\boldsymbol{x}$ sends our pressure waves which decay as r^{-3} at large distance. This led Bachelor & Proudman [3] to suggest that, in general, we would expect turbulence to sustain long-range velocity-pressure correlations of the form, $\langle u^2 p' \rangle_\infty \sim r^{-3}$. It turns out that the gradients of just such correlations appear as source terms in the evolution equation for $\langle u_i u_j u'_k \rangle$, and so it is generally believed that the triple correlations, such as K, fall off algebraically as r^{-4}. This, in turn, tells us that L is an invariant, though I should be, in general, time dependent.

The invariance of L can be understood as follows. The equivalence of ensemble and volume averages tells us that

$$L = \left\langle \left[\int_V \boldsymbol{u} \mathrm{d}V \right]^2 \right\rangle \bigg/ V \tag{9}$$

for some large volume V, $V \gg \ell^3$. Thus L is a measure of the square of the linear momentum held in V. Now, in general, we would not expect this momentum to be conserved, because there will be fluxes of momentum across the surface of V, as well as pressure forces acting on that surface. However, it may be shown that, for a spherical volume, the pressure forces cannot change the square of the momentum in V. Moreover, as the radius R of the volume tends to infinity, the flux of momentum across the surface makes a negligible contribution to the net momentum budget [1] [4]. Thus the invariance of L is simply a manifestation of principle of conservation of linear momentum.

Saffman [1] also noted that a non-zero and constant value of L fixes the rate of decay of energy. That is, the large scales are usually observed to be self-similar, and so we have $L = c\, u^2 \ell^3$, where ℓ is the integral scale and the dimensionless pre-factor c is independent of time. Thus conservation of L requires $u^2 \ell^3 = $ constant. When combined with the energy equation

$$\frac{\mathrm{d}u^2}{\mathrm{d}t} = -\alpha \frac{u^3}{\ell}, \quad \alpha \sim \frac{1}{3}, \tag{10}$$

we obtain $u^2 \sim t^{-6/5}$.

Let us now consider a somewhat different scenario. Suppose that, when we initiate the turbulence, we endow the turbulent eddies with very little linear impulse. Then $L = 0$ at $t = 0$ and, because L is an invariant, it must stay zero for all time. This brings us to the second canonical case in decaying turbulence, in which

$$E(k) = I k^4 / 24\pi^2 + \cdots \tag{11}$$

Prior to the work of Saffman (see also [5]) all spectra were assumed to be of the form $E(k) \sim k^4 + \cdots$, largely because it is readily confirmed that

$$L = 4\pi u^2 \left[r^3 f \right]_\infty \tag{12}$$

where f is the usual longitudinal correlation function, and it was thought unlikely that such strong long-range correlations would exist in turbulence.

Now Landau noted that, for $E(k) \sim k^4 + \cdots$ spectra, Loitsyansky's integral has a property similar to (9). In particular he showed that, if the turbulence has no net linear momentum, and f_∞ decays faster than r^{-6}, then

$$I = \langle \boldsymbol{H}^2 \rangle / V, \quad \boldsymbol{H} = \int_V \boldsymbol{x} \times \boldsymbol{u} \mathrm{d}V \tag{13}$$

for some large spherical volume V. (Note that, to ensure that the turbulence has zero net linear momentum, Landau had to consider a closed spherical domain whose boundary is allowed to recede to infinity. One cannot use an open domain embedded within a field of homogeneous turbulence, because the net linear momentum in such a domain will be non-zero, and indeed it turns out that this residual linear momentum dominates $\boldsymbol{H}$, leading to $\langle \boldsymbol{H}^2 \rangle \sim \langle [\int \boldsymbol{u} \mathrm{d}V]^2 \rangle V^{2/3} \sim V^{4/3}$, and hence to a divergence of $\langle \boldsymbol{H}^2 \rangle / V$. See [4] for a detailed discussion.)

Considerable doubt has been expressed about (13), however. That is, we have just seen that, in general, we would expect $K_\infty \sim r^{-4}$ which, when combined with the Karman-Howarth equation, demands $f_\infty \sim r^{-6}$. At face value, then, it seems unlikely that the prerequisites for (13) to hold true are satisfied.

It is important to note, however, that there is no rigorous theory which predicts the magnitude of the pre-factor, a, in the expression $K_\infty = ar^{-4}$. It could, for example, be zero. If this happened to be the case then (8), combined with (13), would yield,

$$I = -\int r^2 \langle \boldsymbol{u} \cdot \boldsymbol{u}' \rangle \mathrm{d}\boldsymbol{r} = \langle \boldsymbol{H}^2 \rangle / V = \text{constant.} \tag{14}$$

In fact, prior to [3], it was generally assumed that remote points in a turbulence flow are statistically independent, in the sense that all two-point correlations decay rapidly (e.g. exponentially) with r. Thus (14) was indeed thought to be valid. In fact, Kolmogorov [6] exploited the alleged invariance of I to estimate the decay of $u^2(t)$ in $E(k) \sim k^4 + \cdots$ turbulence. He noted that self-similarity of the large scales, combined with (14), requires $u^2 \ell^5 = \text{constant}$, and this, in turn, may be combined with (10) to give

$$u^2(t) \sim t^{-10/7}, \tag{15}$$

which is known as Kolmogorov's decay law.

Now Markovianised two-point closure models, such as ENQNM and TFM, suggest that the pre-factor a in $K_\infty = ar^{-4}$ is finite. Indeed, in EDQNM the magnitude of a is simply set by one of the *ad hoc* model parameters in the closure, and as a result EDQNM predicts that I grows as a power law in t. If this were true, then it would invalidate both (14) and (15). However, such models are heuristic and it is of interest to put them to the test. The Markovianisation step, in particular, is hard to justify at the large scales [4].

3 The Result of 'Big-Box' Numerical Simulations

To test these ideas by direct numerical simulation is not easy because very large computational domains are required. (Typically, a minimum of 100 integral scales in each direction is needed.) However, just such simulations have been reported in [7]. The boundary conditions were periodic and the initial condition consisted of an isotropic energy spectrum $E \sim k^4 \exp[-2(k/k_\mathrm{p})^2]$ with random phases. The box size was 2π in any one direction and so the lowest wave-number was $k_\mathrm{min} = 1$. The wave-number at which $E(k, t = 0)$ peaks, k_p, was set to 40 or 80. The simulations lasted for 300 turnover times and the initial Reynolds number, based on the integral scale $\ell = k_\mathrm{p}^{-1}$, varied from 60 to 250. Note that, with this choice of integral scale, the initial value of L_box/ℓ is $2\pi k_\mathrm{p}$.

Some typical results are displayed in Figs. 1, 2 and 3. These show the temporal variation of Loitsyansky's integral, $I(t)$, the corresponding variation of the decay exponent, m, in the decay law $u^2 \sim t^{-m}$, and the evolution of a typical energy spectrum, $E(k, t)$. It is clear that, once the turbulence has matured, $\mathrm{d}I/\mathrm{d}t$ is very close to zero and we recover Kolmogorov's decay exponent of $m = 10/7$ (provided Re exceeds 100). This, in turn, tells us that the pre-factor, a, in $K_\infty = ar^{-4}$ is extremely small, so that (14) and (15) become good approximations. In short, the naïve, pre-1956 view, in which long-range interactions are ignored, provides a better approximation to mature turbulence than the later, more sophisticated, closure models, such as EDQNM.

In summary, then, the DNS of Ishida, Davidson & Kaneda [7] tentatively suggests that $\mathrm{d}I/\mathrm{d}t$, and by inference $[r^4 K]_\infty$, is very close to zero once the turbulence matures. That is to say, the long-range correlations, as measured by K, become extremely weak once the vorticity field has become teased out into fine-scale tubes. But why should these long-range correlations, induced by the pressure forces, or equivalently by the Biot-Savart law, progressively

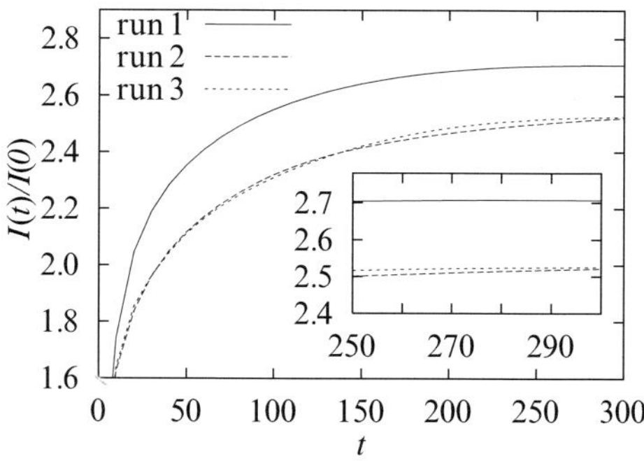

Fig. 1. $I(t)/I(0)$ versus time for run 1 ($k_\mathrm{p} = 80, \mathrm{Re} = 62.5$), run 2 ($k_\mathrm{p} = 40, \mathrm{Re} = 250$) and run 3 ($k_\mathrm{p} = 40, \mathrm{Re} = 125$).

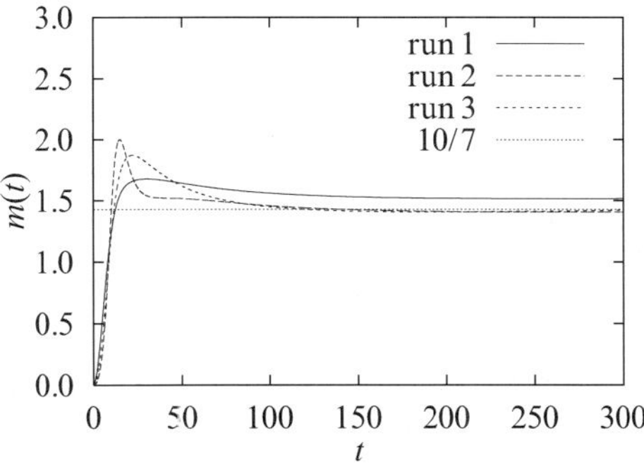

Fig. 2. $m(t)$ versus time for run 1 ($k_\mathrm{p} = 80, \mathrm{Re} = 62.5$), run 2 ($k_\mathrm{p} = 40, \mathrm{Re} = 250$) and run 3 ($k_\mathrm{p} = 40, \mathrm{Re} = 125$).

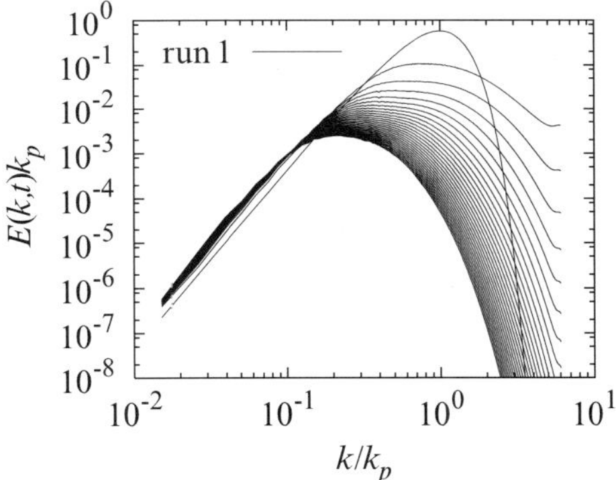

Fig. 3. $E(k,t)$ for run 1 ($k_\mathrm{p} = 80, \mathrm{Re} = 62.5$) from $t = 0$ up to $t = 300$ in steps of $\Delta t = 10$.

weaken as the turbulence matures ? This is an issue which, so far, has not been addressed in the literature. However, we might note that analogous behaviour is seen in other physical systems possessing many degrees of freedom. The most famous example is, perhaps, Debye-Huckel screening in plasmas and electrolytes, where the long-range Coulomb force between distant ions shuts down as a result of the clustering of oppositely signed charges. That is, the electrostatic potential, Φ, is governed by $\nabla^2 \Phi = -(\rho^+ - \rho^-)/\epsilon_0$ where ρ^+ and ρ^- are the charge densities of the positive and negative ions and ϵ_0 is the permittivity of free space. The long-range Coulomb force may be thought of as a consequence of the non-local nature of the inversion $\Phi \sim -\nabla^{-2}(\rho^+ - \rho^-)$ and this force is suppressed by a clustering of oppositely signed charges, which results in $\rho^+ - \rho^- \approx 0$. Moreover, such a clustering reduces the Gibbs free energy of the system, $E - TS$, as the electrostatic energy associated with the long-range forces is eliminated. (Here E is the electrostatic energy, T the temperature and S the entropy.) Such behaviour led Ruelle [8] to speculate that there should be an analogue of Debye screening in turbulence, in which the two-point vorticity correlation decays exponentially with separation, rather than the power law suggested by the Biot-Savart law. Interestingly, he suggested that turbulence whose vorticity field consists of thin, tube-like structures is most likely to favor such screening.

One tentative explanation for the observed suppression of long-range interactions in mature turbulence is the following. The long-range correlations are enforced by the pressure field. That is, the distortion or movement of an eddy can be detected at large distances because the eddy sets up a pressure field which pervades all space in accordance with $\nabla^2 p = (\omega^2/2 - S_{ij}S_{ij})/\rho$, S_{ij} being the strain-rate tensor. Just as a clustering of oppositely signed charges results in Debye-Huckel screening, so the long-range pressure forces will be suppressed if there is a strong degree of spatial overlap between the regions of intense enstrophy and intense dissipation, and indeed just such a spatial correlation may be seen in the DNS of mature, high-Re turbulence.

4 Extension of the Ideas to Other Homogeneous Turbulent Systems

We end by noting that the question of the conservation (or lack of conservation) of $I(t)$ in mature turbulence is not just relevant to isotropic turbulence. Many other systems, such as homogeneous MHD turbulence, and rotating, stratified turbulence, conserve one or more components of angular momentum. The fact that there is evidence of the suppression of long-range velocity correlations in isotropic turbulence gives us hope that these long-range interactions are also weak in these more complex flows. If so, they will possess Loitsyansky-like invariants of the form of (14), and these may be used to estimate decay laws in the spirit of Kolmogorov [6]. The MHD case, in particular, is discussed at length in [9], [4] and [10], where decay predictions based on an MHD-Loitsyansky integral are found to be consistent with the limited data available.

References

1. Saffman PG (1967) J Fluid Mech 27(3):581–593
2. Landau LD, Lifshitz EM (1959) Fluid mechanics 1st Ed, Pergamon
3. Bachelor GK, Proudman I (1956) Phil Trans R Soc Lond A 248:369–405
4. Davidson PA (2004) Turbulence: an introduction for scientists and engineers. Oxford University Press
5. Birkhoff G (1954) Comm Pure and Appl Math 7:19–44
6. Kolmogorov AN (1941) Dokl Akad Nauk SSSR 31(6):538–541
7. Ishida T, Davidson PA, Kaneda Y (2006) J Fluid Mech 564:455–475
8. Ruelle D (1990) J Stat Phys 61:865–868
9. Davidson PA (1997) J Fluid Mech 336:123–150
10. Davidson PA (2006) The role of angular momentum invariants in homogeneous turbulence In: Mathematical and Physical Theory of Turbulence, J Cannon & B K Shivamoggi eds, Chapman & Hall

Energy Dissipation and Pressure in 4d Turbulence

Toshiyuki Gotoh[1,2], Tohru Nakano[3], Yoshitaka Shiga[3],
and Yusaku Watanabe[1]

[1] Department of Engineering Physics, Graduate School of Engineering,
Nagoya Institute of Technology, Showa-ku, Nagoya 466-8555, Japan
[2] CREST, Japan Science and Technology Agency, 4-1-8 Honcho,
Kawaguchi, Saitama 332-0012, Japan,
`gotoh.toshiyuki@nitech.ac.jp`
[3] Department of Physics, Chuo University, Tokyo 112-8551, Japan
`nakano@phys.chuo-u.ac.jp`

Abstract. Energy transfer, intermittency of the energy dissipation and pressure in decaying turbulence in four dimensions are studied by direct numerical simulation in comparison with three dimensions. It is found that the energy transfer is more efficient in 4d than in 3d. The intermittency of the total energy dissipation and the associated velocity difference (spherical velocity difference) with separation distance r is weaker in 4d than in 3d, while the one dimensional surrogate dissipation rate and the associated longitudinal velocity difference show stronger intermittency in 4d. The statistics of the pressure and pressure gradient are examined and their dependence on the spatial dimension is discussed.

Keywords: 4d-turbulence, pressure, dissipation, transfer, intermittency

1 Introduction

Dynamics and statistics of turbulent flows depend strongly on space dimension d. In two dimensions, the energy cascades to large scales and the velocity statistics are very close to the Gaussian, while the enstrophy goes to small scales and the vorticity statistics deviate from the normal distribution [1, 2]. In three dimensions, the energy is transferred toward small scales, and the intermittency builds up with decrease of scale, which leads to the anomalous scaling of the velocity increments at very high Reynolds numbers. Recent studies on passive scalar of the Kraichnan model have found that the anomalous scaling exponents for the scalar increments tend to be normal as d increases [3, 4]. These facts motivate us to study effects of increasing the spatial dimension on incompressible turbulence, especially from three to four dimensions [5, 6, 7, 8, 9, 10, 11]. Relevant questions are whether the energy transfer toward small scales of motion is enhanced or diminished, whether intermittency becomes stronger or weaker, and what the statistics of the pressure field are.

27

Y. Kaneda (ed.), *IUTAM Symposium on Computational Physics and New Perspectives in Turbulence*, 27–34.

We have examined these issues by using direct numerical simulation (DNS) [12, 15].

2 Basic Equations and Direct Numerical Simulation

The Navier-Stokes equation for an incompressible fluid with unit density in any number of space dimensions is written as

$$\frac{\partial u_i}{\partial t} + u_j \Omega_{ji} = -\frac{\partial}{\partial x_i}\left[p + \frac{u^2}{2}\right] + \nu \Delta u_i, \tag{1}$$

where p is the pressure, and ν the kinematic viscosity, and $\Omega_{ij} = \frac{\partial u_j}{\partial x_i} - \frac{\partial u_i}{\partial x_j}$ is the 2-form of the velocity field, the counterpart of vorticity in 3d. Throughout this paper we employ the convention that the summation is taken over repeated indexes without stated otherwise.

The turbulence is assumed to be homogeneous and isotropic. Then the velocity field is expanded in the Fourier series. The total energy, the energy spectral density and the energy spectrum are defined as

$$E(t) = \frac{\langle \boldsymbol{u}^2(t)\rangle}{2} = \frac{d}{2}\bar{u}^2 = \int_0^\infty E(k,t)\,\mathrm{d}k, \tag{2}$$

$$E(k,t) = \frac{1}{2}S_d k^{d-1}Q(k,t), \quad S_d = \frac{2\pi^{d/2}}{\Gamma(d/2)}, \tag{3}$$

$$\langle u_i(\boldsymbol{k},t)u_j(-\boldsymbol{k},t)\rangle = \frac{1}{d-1}P_{ij}(\boldsymbol{k})Q(k,t), \tag{4}$$

where $\langle\cdot\rangle$ is the ensemble average, $k = |\boldsymbol{k}|$, S_d is the surface area of the d-dimensional sphere of a unit radius, $\Gamma(x)$ is the Gamma function, and $P_{ij}(\boldsymbol{k}) = \delta_{ij} - k_i k_j/k^2$. The energy equation is then written as $\left(\frac{\partial}{\partial t} + 2\nu k^2\right)E(k,t) = T(k,t)$ where $T(k,t)$ is the energy transfer function. Energy conservation by the nonlinear term means the property $\int_0^\infty T(k,t)dk = 0$, and then the energy transfer flux and the average energy dissipation rate per unit mass are defined as

$$\Pi(k,t) = \int_k^\infty T(k',t)\mathrm{d}k', \quad \bar{\varepsilon} = 2\nu \int_0^\infty k^2 E(k)\mathrm{d}k. \tag{5}$$

A set of the ordinary differential equations of the Fourier coefficients are integrated by the 4th order Runge-Kutta-Gill method in time and the pseudo spectral method is used for the nonlinear convolution term. The detail is found in Gotoh et al. [12, 15]. The initial random solenoidal velocity fields with given energy spectrum are generated by using the Gaussian random numbers. Three types of the initial energy spectra are used to examine the effects of the spatial dimensions on decay process, statistical invariants, spectrum evolution, and intermittency. They are

Table 1. DNS parameters and statistical quantities at initial time. M is the number of sampling runs. $k_{max}\eta$ is larger than unity for all runs, the minimum value being 1.04 for Run 3C.

Run	d	type	k_0	N	M	$\nu(\times 10^{-2})$	E	$\bar{\varepsilon}$	R_λ	$\eta(\times 10^{-2})$	λ/η	L/η
4A	4	I	4	64^4	2	2.00	1.97	1.59	24.7	3.99	12.5	15.6
4B	4	I	4	128^4	1	1.00	1.99	0.802	54.3	2.81	19.4	25.9
4C	4	I	4	128^4	1	0.500	2.01	0.402	110	1.99	27.6	35.8
4D	4	II	16	256^4	1	0.400	2.00	4.91	34.9	1.07	13.1	16.4
4E	4	II	16	256^4	1	0.300	2.00	3.68	46.5	0.926	15.1	18.9
4F	4	III	4	256^4	1	0.320	2.00	0.304	152	1.81	26.8	34.1
3A	3	I	4	64^3	18	2.00	1.45	1.21	24.1	5.07	9.83	12.1
3B	3	I	4	128^3	6	1.00	1.47	0.600	48.8	3.59	13.8	17.4
3C	3	I	4	128^3	6	0.500	1.48	0.300	98.6	2.54	19.5	24.5
3D	3	II	16	256^3	1	0.400	1.50	3.83	31.2	1.14	11.0	13.8
3E	3	II	16	256^3	1	0.253	1.50	2.42	49.4	0.904	13.8	17.4
3F	3	III	4	256^3	1	0.380	1.50	0.218	128	2.24	21.7	28.3
3G	3	II	32	1024^3	1	0.100	1.50	3.84	62.4	0.402	15.5	19.5

$$E_I(k,0) = Au_0^2 k_0^{-1}(k/k_0)^4 \exp\left(-2(k/k_0)^2\right), \tag{6}$$

$$E_{II}(k,0) = B_d u_0^2 k_0^{-1}(k/k_0)^{d+1} \exp\left(-\tfrac{d+1}{2}(k/k_0)^2\right), \tag{7}$$

$$E_{III}(k,0) = B'_d u_0^2 k_0^{-1}(k/k_0)^{d+1} \exp\left(-\tfrac{1}{2}(k/k_0)^2\right). \tag{8}$$

We choose $u_0 = 1$ for all runs and the constants A, B_d and B'_d are so chosen that the total energy is $d(u_0^2/2)$. Note that the energy per velocity component is the same in both dimensions. The spatial resolution is up to 256^3 or 256^4. Parameters of the direct numerical simulation (DNS) and the statistical quantities at initial time are given in Table 1, in which L, λ, and R_λ are the integral scale, the Taylor microscale, and the Taylor microscale Reynolds number, respectively. When it was needed, the ensemble average over several (M) random initial velocity fields was taken.

3 Energy Spectrum and Energy Transfer Flux

The energy spectra normalized in terms of the Kolmogorov variables are shown in Fig. 1. Collapse of the curves in both dimensions is satisfactory, which means that the Kolmogorov scaling holds in both 3d and 4d. R_λ is too low to see the inertial range in the present DNS's. The spectra in 4d look the same as in 3d, but the levels of the curves of 4d at wavenumbers of the inertial effect dominant are lower than those of 3d. The inserted guiding lines represent the curves $K_d(k\eta)^{-5/3}$ with the Kolmogorov constant $K_4 = 1.31$ in 4d and $K_3 = 1.72$ in 3d, respectively, which are computed by using a Lagrangian spectral theory of turbulence (Lagrangian Renormalized Approximation, LRA) [13, 14, 15]. It is also found from the theory that the Kolmogorov constant is a slowly

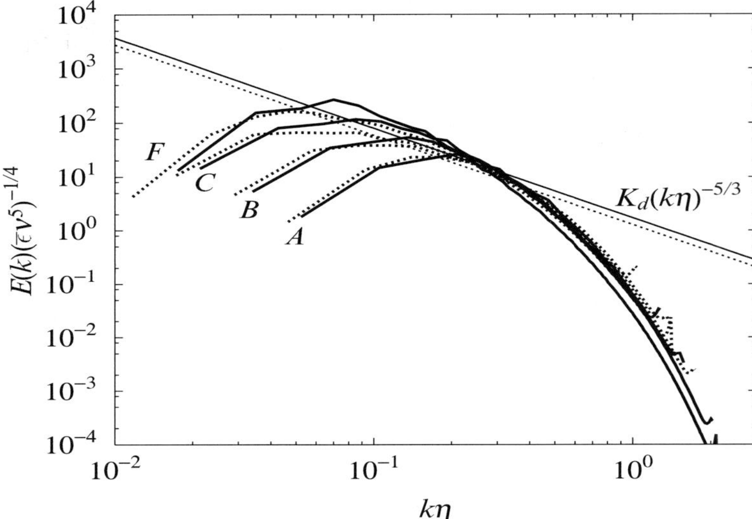

Fig. 1. Energy spectrum normalized by the Kolmogorov variables. Solid lines: 3d, dashed lines: 4d. Straight lines stand for $K_d(k\eta)^{-5/3}$, with the Kolmogorov constants $K_3 = 1.72$ (solid) and $K_4 = 1.31$ (dotted), respectively.

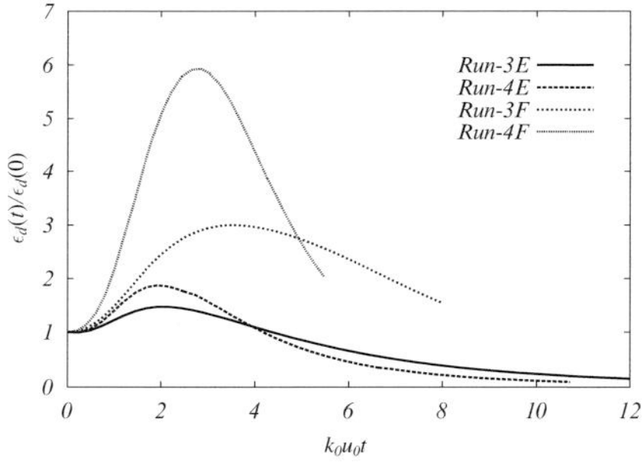

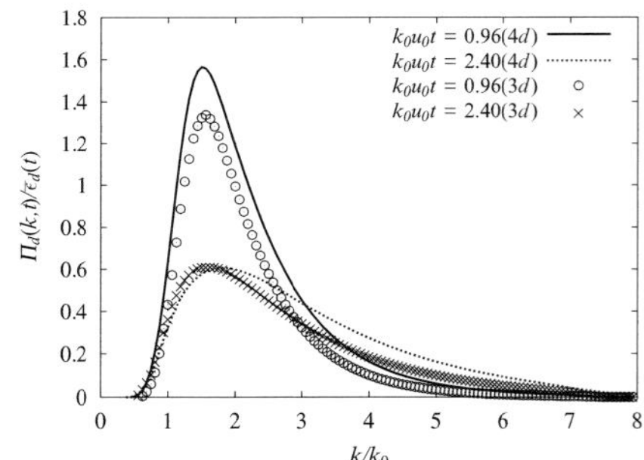

Fig. 2. Evolution of the energy dissipation rate.

Fig. 3. Evolution of the energy transfer flux function for Runs 3E and 4E.

decaying function of d and tends to a finite value of $K_\infty = 0.815$ as the spatial dimension becomes infinite.

In order to examine the efficiency of the energy transfer, it is appropriate to see the energy dissipation and the energy transfer function. Figures 2 and 3 show comparison of $\bar{\varepsilon}^{(d)}(t)/\bar{\varepsilon}^{(d)}(0)$ and the normalized energy transfer flux function $\Pi_d(k,t)/\bar{\varepsilon}^{(d)}(t)$, respectively. At $t = 0$, the ratio $\bar{\varepsilon}^{(4)}(0)/\bar{\varepsilon}^{(3)}(0)$ is about $4/3$, but at later time $\bar{\varepsilon}^{(4)}(t)/\bar{\varepsilon}^{(4)}(0)$ becomes larger than $(4/3) \times \left(\bar{\varepsilon}^{(3)}(t)/\bar{\varepsilon}^{(3)}(0)\right)$. When the Reynolds number becomes larger, the initial increase of $\bar{\varepsilon}^{(4)}(t)/\bar{\varepsilon}^{(4)}(0)$ becomes more significant than for 3d. Corresponding to this, we can see in Fig. 3 that $\Pi_d(k,t)/\bar{\varepsilon}^{(d)}(t)$ is larger in 4d than in 3d for almost all wavenumbers. We have observed the same tendency for other runs. These facts mean that the energy transfer in 4d is more efficient than in 3d, which is also consistent with the prediction of the spectral theory [15].

4 Intermittency of the Energy Dissipation

The intermittency problem is one of the most fundamental and important issues of the turbulence physics. The relevant quantities of interests here are probability density functions (PDF's) of the energy dissipation and the velocity difference. Two kinds of the energy dissipation $\varepsilon_1(\boldsymbol{x}) = \nu \sum_{i,j} \left(\frac{\partial u_j}{\partial x_i} \right)^2$ and $\varepsilon_3(\boldsymbol{x}) = \nu d(d+2) \left(\frac{\partial u_1}{\partial x_1} \right)^2$, and the velocity difference $t_1 = \sum_{i,j}(u_i(\boldsymbol{x} + r\boldsymbol{e}_j) - u_i(\boldsymbol{x}))^2$ and $t_3 = (u_1(\boldsymbol{x} + r\boldsymbol{e}_1) - u_1(\boldsymbol{x}))^2$ are examined. Figures 4 and 5 present comparison of the cumulative PDF's (cPDF's) defined by $P(> x) = \int_x^\infty P(x')\mathrm{d}x'$. It is seen that the behavior of the tails of cPDF is opposite, in that the cPDF of the total dissipation $\varepsilon_1(\boldsymbol{x})$ in 4d decays faster than that in 3d, while the cPDF of the surrogate dissipation $\varepsilon_3(\boldsymbol{x})$ in 4d decays slower than that in 3d. The same is true for the cPDF's for t_1 and t_3. These opposite behavior in the tails of cPDF's can be understood as follows. Since ε_1 and t_1 are given as the sum of components of $(\nabla \boldsymbol{u})^2$ while ε_3 and t_3 are single component, when the spatial dimension is very large, the fluctuations of the formers around their mean values would tend to the Gaussian according to the central limit theorem. This is the dimension effect. Therefore it is very likely for the cPDF tails of ε_1 and t_1 in 4d to decay faster and thus less intermittent.

The above observation leads to re-interpretation of the refined self similarity hypothesis of the Kolmogorov [16, 17], in which the longitudinal velocity difference u_r over a distance r is related to the volume averaged energy dissipation rate $\varepsilon_r(\boldsymbol{x}) = V_r^{-1} \int_{V_r} \varepsilon(\boldsymbol{x})\mathrm{d}\boldsymbol{x}$, where V_r is a volume over a sphere of radius r centered at $\boldsymbol{x}$, as $u_r \sim (r\varepsilon_r)^{1/3}$. If we take t_3 as u_r and ε_1 as ε_r, then

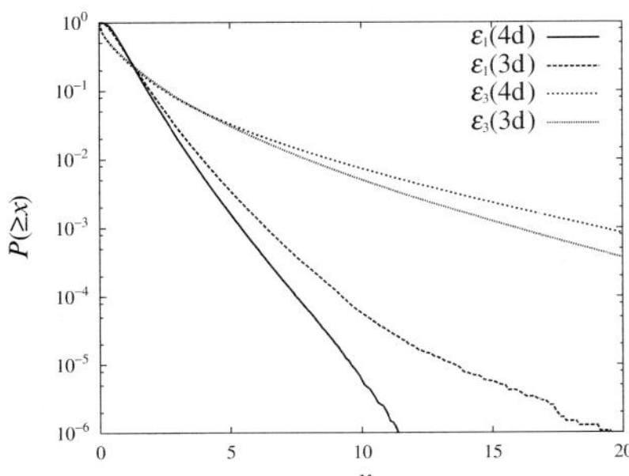

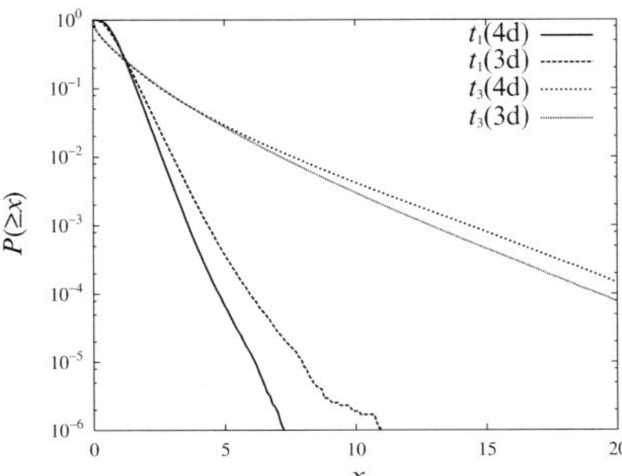

Fig. 4. CDF's of ε_1 and ε_3 for 4d (Run 4A) and for 3d (Run 3A) at $k_0 u_0 t = 2.0$. The abscissa is normalized in terms of $\bar{\varepsilon}$.

Fig. 5. CDF's of t_1 and t_3 for 4d (Run 4A) and for 3d (Run 3A) at $k_0 u_0 t = 2.0$. $r/\eta = 8.43$ in 4d and $r/\eta = 7.44$ in 3d.

the less intermittent the (total) dissipation rate is, the less so the longitudinal velocity increments are, which is in contradiction with the present observation. To reconcile the refined self-similarity hypothesis with the observations, there are two ways: (1) $t_1(r) \sim (r\varepsilon_1(r))^{1/3}$, or (2) $t_3(r) \sim (r\varepsilon_3(r))^{1/3}$. This is the new understanding obtained by increasing the spatial dimensions from 3d to 4d [18].

5 Pressure Statistics

The pressure plays an important role in the incompressible turbulence. It follows from the Lagrangian spectral theory that the relatively smaller pressure gradient leads to more persistent memory of a fluid blob along a Lagrangian trajectory, which in turn yields the effective energy transfer of the energy to smaller scales [15]. Without pressure, the shock front is formed in the velocity field and the intermittency becomes stronger as in the Burgers turbulence.

First, we look at the PDF's for the pressure (Fig. 6) and pressure gradients (Fig. 7). The normalized PDF of the pressure in 4d is slightly less skewed when compared to that in 3d. Although the right tail of the pressure PDF of 3d is shorter than that in 4d, this is partly due to the insufficient sample size and the rapid decay should be read with care. The tails of the pressure gradient PDF in 4d are just slightly narrower than those in 3d.

In order to quantify the differences in the statistics of the pressure field between 3d and 4d, we examine the moments. Variation of the moments of the pressure and the pressure gradient in time is shown in Table 2. The normalized variance, the absolute value of the skewness of the pressure in 4d are smaller than those in 3d, but it is difficult to say about the flatness because their difference between 3d and 4d is very small and the decay of the Reynolds number in 4d is faster than in 3d. The same is true for the normalized variance and the flatness of the pressure gradient.

Smallness of the normalized variance is mostly due to the dimension effects. For example, if we estimate the variance of the pressure gradients by using

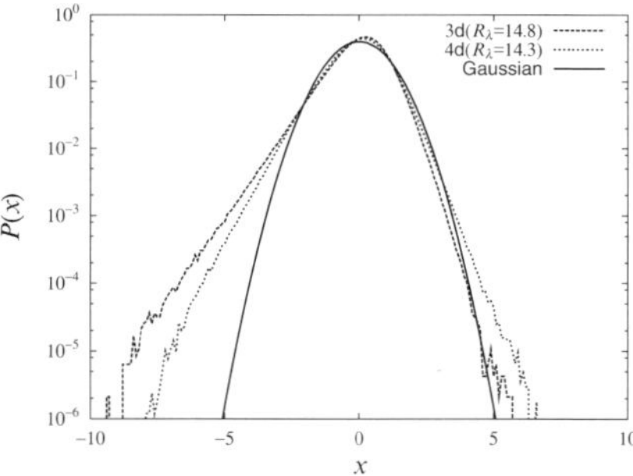

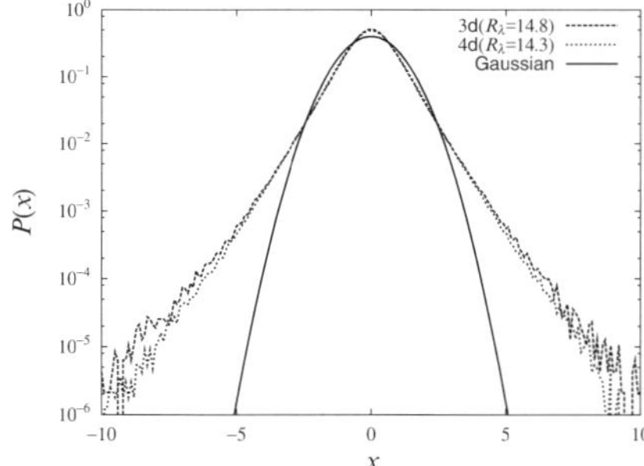

Fig. 6. PDF's of the pressure in 3d and 4d.

Fig. 7. PDF's of the pressure gradient in 3d and 4d.

Table 2. Variation of the low order moments of the pressure gradient.

Run	$k_0 u_0 t$	R_λ	$4\langle p^2\rangle/\langle \boldsymbol{u}^2\rangle^2$	$S_3(p)$	$K_4(p)$	$\dfrac{\langle(\nabla p)^2\rangle}{\bar{\varepsilon}^{3/2}\nu^{-1/2}}$	$S_3(\nabla_x p)$	$K_4(\nabla_x p)$
$3A$	2.0	15.4	0.438	-0.569	4.12	1.43	-1.31×10^{-2}	5.21
$3A$	2.4	14.6	0.429	-0.578	4.25	1.36	-5.64×10^{-2}	5.43
$3A$	2.8	13.9	0.422	-0.579	4.41	1.31	-9.77×10^{-2}	5.52
$4A$	2.0	14.3	0.214	-0.376	4.08	0.804	-3.85×10^{-2}	5.38
$4A$	2.4	13.1	0.212	-0.397	4.13	0.740	-2.64×10^{-2}	5.40
$4A$	2.8	12.2	0.210	-0.420	4.20	0.692	-1.14×10^{-3}	5.36

the quasi-normal approximation, the result normalized by the Kolmogorov variables is given by

$$\bar{a}_d^2 \equiv \bar{\varepsilon}^{-3/2}\nu^{1/2}\langle(\nabla p)^2\rangle = M_d \int_0^\infty \int_0^\infty \hat{k}\hat{p} J_d\left(\frac{\hat{p}}{\hat{k}}\right) f(\hat{p}) f(\hat{q})\,\mathrm{d}\hat{p}\mathrm{d}\hat{q}, \qquad (9)$$

$$M_d \equiv \frac{4 S_{d-1}}{(d-1)^2 S_d} B\left(\frac{d+3}{2},\frac{1}{2}\right),$$

where $B(x,y)$ is the Beta function. Since the function $J_d(x)$ satisfies $J_d(x) = J_d(1/x)$ and varies very slowly with increase of d as found in Fig. 8, the integral of the right hand side of (9) is weakly dependent on d. Then dimension effects are predominantly due to the geometric factor M_d arising from the volume integral, for example, $M_3/M_4 = 1.92$. The ratio $\bar{a}_3^2/\bar{a}_4^2$ is close to, but about 10% smaller than 1.92 as found from Table 2. The discrepancy is due to the neglect of d-dependence in J_d and the intermittency effects. Note that the difference in the variance is absorbed in the plots of the PDF's and the skewness and flatness. Unfortunately the available data of the pressure field are very limited to low Reynolds numbers. Therefore it is very difficult to draw any definite trend of the statistical law of the pressure. However, the following argument is suggestive.

The Poisson equation for the pressure is $\nabla^2 p = \frac{1}{4}\left(\Omega_{ij}^2 - S_{ij}^2\right)$ and the source term is the sum of square of the velocity gradients. Therefore as the dimension increases, it is expected from the central limit theorem that the fluctuations of the source term tend to those given by the Gaussian random velocity fields. Figure 9 shows the comparison of the PDF's of the source term in 3d and 4d. In fact, the PDF of the source term in 3d is slightly more skewed than in 4d. Then the pressure is computed as the integral of the less intermittent source term over the space wider than the one in 3d. These facts suggest that when d is large, the pressure field asymptotically approaches a statistical state generated from the random velocity field obeying the Gaussian statistics. Then the low to moderate order moments of the pressure would be described better in terms of the quasi-normal approximation. In order to answer the above problem we need the pressure data at high Reynolds numbers, and high resolution DNS of the 4d turbulence is certainly necessary and a challenge to the computational physics of turbulence.

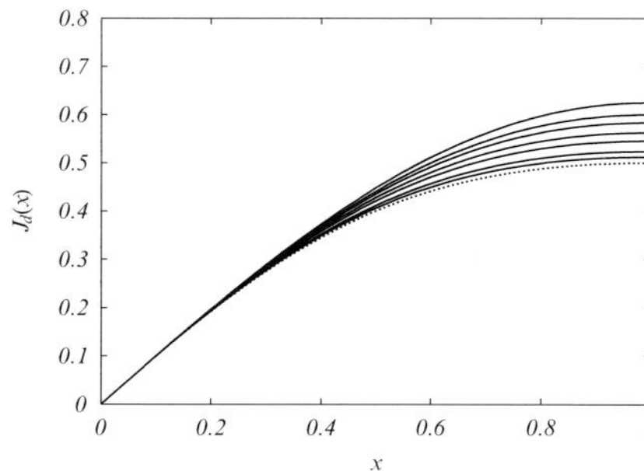

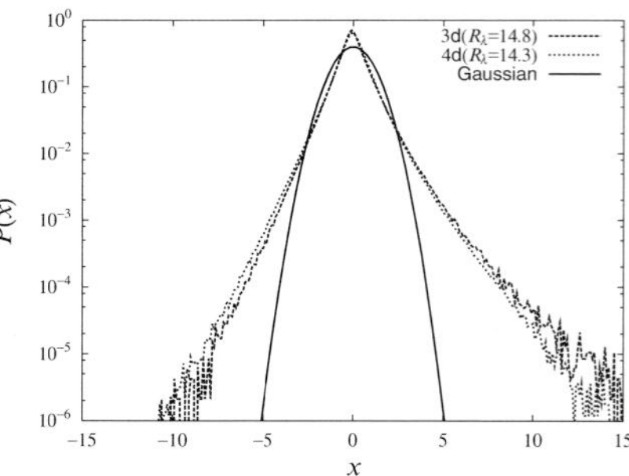

Fig. 8. $J_d(x) = J_d(1/x)$ for $d = 3, 4, 5, 7,$ 10, 20, 40 (solid) from the uppermost curve. The dotted curve is for $d = \infty$.

Fig. 9. PDF's of the source term of the Poisson equation for the pressure in 3d and 4d.

The authors express their thanks to the Theory and Computer Simulation Center, Computer and Information Network Center at National Institute for Fusion Science, and Information Technology Center of Nagoya University for their support to the numerical computation.

References

1. Kraichnan RH (1967) Phys Fluids 10: 1417–1423
2. Batchelor GK (1969) Phys Fluids Suppl 2, 12: 233–239
3. Kraichnan RH (1994) Phys Rev Lett 72: 1016–1019
4. Falkovich G, Gawedzki K, Vergassola M (2001) Rev Mod Phys 73: 913–975
5. Nelkin M (1974) Phys Rev A 9: 388, (1975) Phys Rev A 11: 1737–1743
6. Fournier JD, Frisch U (1978) Phys Rev A 17: 747–762
7. Fournier JD, Frisch U, Rose HA (1987) J Phys A 11: 187–198
8. Khesin BA, Chekanov Yu V (1989) Physica D 40: 119–131
9. Yakhot V (2001) Phys Rev E 63: 026307
10. Li Y, Meneveau C (2005) Phys Rev Lett 95: 164502
11. Li Y, Meneveau C (2006) J Fluid Mech 558: 133–142
12. Suzuki E, Nakano T, Takahashi N, Gotoh T (2005) Phys Fluids 17: 081702
13. Kaneda Y (1981) J Fluid Mech 107: 131–145
14. Kaneda Y (1986) Phys Fluids 29: 701–708
15. Gotoh T, Watanabe Y, Nakano T, Shiga Y, Suzuki E (2006) Phys Rev E 75: 016310
16. Kolmogorov AN (1962) J Fluid Mech 13: 82–85
17. Obukhov AM (1962) J Fluid Mech 13: 77–81
18. Stolovitzky G, Meneveau C, Sreenivasan KR (1998) Phys Rev Lett 80: 3883

On Intermittency in Shell Models and in Turbulent Flows

Itamar Procaccia[1], Roberto Benzi[2] and Luca Biferale[2]

[1] The Department of Chemical Physics, The Weizmann Institute of Science, Rehovot 76100, Israel, `itamar.procaccia@weizmann.ac.il`

[2] Dept. of Physics and INFN, University of Rome "Roma Tor Vergata", Via della Ricerca Scientifica 1, 00133 Rome, Italy

Abstract. We propose an approach to study the old-standing problem of the anomaly of the scaling exponents of nonlinear models of turbulence. We achieve this by constructing, for any given nonlinear model, a *linear* model of passive advection of an auxiliary field whose anomalous scaling exponents are **the same** as the scaling exponents of the nonlinear problem. The statistics of the auxiliary linear model are dominated by 'Statistically Preserved Structures' which are associated with statistical conservation laws. The latter can be used for example to determine the value of the anomalous scaling exponent of the second order structure function. The approach is equally applicable to shell models and to the Navier-Stokes equations, and it demonstrates that the scaling exponents of these nonlinear models are indeed *anomalous*. In order to adress the *universality* of these nonlinear model we study the statistical properties of a semi-infinite chain of passive vectors advecting each other and study the scaling exponents at the fixed point of this chain.

Keywords: turbulence, anomalous exponents, statistically preserved structures, shell models

1 Introduction

The calculation of the scaling exponents of structure functions of turbulent velocity fields is considered as one of the major open problems of statistical physics [1]. Dimensional considerations fail to provide the measured exponents, and present theory cannot even specify the mechanism for the so called "anomaly", i.e. the deviation of the scaling exponents from their dimensional estimates. The problem appeared sufficiently difficult to warrant the blossoming of simple models of turbulence, and in particular of shell models [2, 3] with the hope that the calculation of the scaling exponents in the latter will turn out to be an easier problem than in Navier-Stokes turbulence. Alas, so far shell models allowed accurate direct numerical calculation of their scaling exponents, including convincing evidence for their universality [4, 5, 6, 7, 8, 9], but not many inroads to the understanding of the anomaly or the evaluation of the exponents from first principles, previous attempts being mainly based

35

Y. Kaneda (ed.), *IUTAM Symposium on Computational Physics and New Perspectives in Turbulence*, 35–45.

on stochastic closures [10, 11, 12]. In this paper we review a recent theory [13] that presents a solution of this problem: we demonstrate that the scaling exponents of nonlinear models of turbulence anomalous. In addition, we show for example how to determine the anomalous scaling exponent of the second order structure function. We present the case in detail for shell models, but then show evidence that Navier-Stokes turbulence yields to the very same result. The only distinction is in the ease of numerical demonstration; for shell models we present adequate numerical confirmation of the proposed theory while for Navier-Stokes turbulence we present calculations at a resolution of 128^3. We also present some new data on the problem of a semi-infinite chain of passive vectors initally advected by a nonlinear model. This is done in order to probe the universality properties of a particular class of passive vectors under the change of the statistical features of the advecting field.

2 The Main Idea

To specify the problem more precisely, consider for example the Sabra shell model [9] which, like other shell models of turbulence, is a truncated description of the dynamics of Fourier modes, preserving some of the structure and conservation laws of the Navier-Stokes equations:

$$\left(\frac{d}{dt} + \nu k_n^2\right) u_n = i(k_{n+1} u_{n+1}^* u_{n+2} - \delta k_n u_{n-1}^* u_{n+1}$$

$$+ (1-\delta)k_{n-1} u_{n-1} u_{n-2}) + f_n. \tag{1}$$

Here u_n are the velocity modes restricted to 'wavevectors' $k_n = k_0 \mu^n$ with k_0 determined by the inverse outer scale of turbulence and $\mu = 1/2$. The model contains one additional parameter, δ, and it conserves two quadratic invariants (when the force and the dissipation term are absent) for all values of δ. The first is the total energy $\sum_n |u_n|^2$ and the second is $\sum_n (-1)^n k_n^\alpha |u_n|^2$, where $\alpha = \log_\mu(1-\delta)$. In this paper we consider values of the parameters such that $0 < \delta < 1$; in this region of parameters the second invariant contributes only with sub-leading exponents to the structure functions [9, 14, 15]. The scaling exponents characterize the structure functions,

$$S_2(k_n) \equiv \langle u_n u_n^* \rangle \sim k_n^{-\zeta_2} \quad S_3(k_n) \equiv \Im\langle u_{n-1} u_n u_{n+1}^* \rangle \sim k_n^{-\zeta_3}, \tag{2}$$

etc. for higher order $S_p(k_n) \sim k_n^{-\zeta_p}$. The values of the scaling exponents were determined accurately by direct numerical simulations. Besides ζ_3 which is exactly unity [7], all the other exponents ζ_p are anomalous, differing from $p/3$. It was established numerically that the scaling exponents are universal, i.e. they are independent of the forcing f_n as long as the latter is restricted to small n [3]. Assuming universality, our aim is to provide a theory for the anomalous exponents, and to determine the second order exponent ζ_2.

The central idea is to construct a *linear* model whose scaling exponents are the same as those of the nonlinear problem. In this linear problem the exponents are universal, and we have the mechanism for the anomaly of the scaling exponents; we use this to show that also the *nonlinear* problem must have anomalous exponents. Consider then a passive advected field which in the discrete shell space has the complex amplitudes w_n. The dynamical equations for this field are linear and constructed under the following requirements: (i) the structure of the equations is obtained by linearizing the nonlinear problem and retaining only such terms that conserve the energy; (ii) the resulting equation is identical with the sabra model when $w_n = u_n$; (iii) the energy is the only quadratic invariant for the passive field in the absence of forcing and dissipation. These requirements lead to the following linear model:

$$\frac{dw_n}{dt} = \frac{i}{3}\Phi_n(u,w) - \nu k_n^2 w_n + f_n,$$

(3)

where the advection term is defined as

$$\begin{aligned}
\Phi_n(u,w) = {}& k_{n+1}[(1+\delta)u_{n+2}w_{n+1}^* + (2-\delta)u_{n+1}^* w_{n+2}] \\
&+ k_n[(1-2\delta)u_{n-1}^* w_{n+1} - (1+\delta)u_{n+1}w_{n-1}^*] \\
&+ k_{n-1}[(2-\delta)u_{n-1}w_{n-2} + (1-2\delta)u_{n-2}w_{n-1}]
\end{aligned}$$

(4)

Observe that when $w_n = u_n$ this model reproduces the Sabra model, and also that the total energy is conserved because $\sum_n \Im[\Phi_n(u,w)w_n^*] = 0$. The second quadratic invariant is not conserved by the linear model. Finally, both models have the same 'phase symmetry' in the sense that the phase transformations $u_n \rightarrow u_n \exp(i\phi_n)$ and $w_n \rightarrow w_n \exp(i\theta_n)$ leave the equations invariant iff $\phi_{n-1} + \phi_n = \phi_{n+1}$, $\theta_{n-1} + \theta_n = \theta_{n+1}$. This identical phase relationship guarantees that the non-vanishing correlation functions of both models have precisely the same forms. Thus for example the only second and third correlation functions in both models are those written explicitly in (2).

In this linear model we know that the scaling exponents are universal, and what is the mechanism for their anomaly [16, 17, 18]. The linear model possesses "Statistically Preserved Structures" (SPS) which are evident in the decaying problem (3) with $f_n = 0$. These are *left* eigenfunctions of eigenvalue 1 of the linear propagators for each order (decaying) correlation function [16]. For example for the second order correlation function denote the propagator $P_{n,n'}^{(2)}(t|t_0)$; this operator propagates any initial condition $\langle w_n w_n^* \rangle(t_0)$ (with average over initial conditions, independent of the realizations of the advecting field u_n) to the decaying correlation function (with average over realizations of the advecting field u_n)

$$\langle w_n w_n^* \rangle(t) = P_{n,n'}^{(2)}(t|t_0)\langle w_{n'} w_{n'}^* \rangle(t_0).$$

(5)

The second order SPS, $Z_n^{(2)}$, is the left eigenfunction with eigenvalue 1,

$$Z_{n'}^{(2)} = Z_n^{(2)} P_{n,n'}^{(2)}(t|t_0).$$

(6)

Note that $Z_n^{(2)}$ is time independent even though the operator $P_{n,n'}^{(2)}(t|t_0)$ is time dependent. Each order correlation function is associated with another propagator $\boldsymbol{P}^{(p)}(t|t_0)$ and each of those has an SPS, i.e. a *left* eigenfunction $\boldsymbol{Z}^{(p)}$ of eigenvalue 1. These non-decaying eigenfunctions scale with k_n, $\boldsymbol{Z}^{(p)} \sim k_n^{-\xi_p}$, and the values of the exponents ξ_p are anomalous. Finally, it was shown that these SPS are also the leading scaling contributions to the structure functions of the *forced* problem (3) [16, 17]. Thus **the scaling exponents of the linear problem are independent of the forcing** f_n, since they are determined by the SPS of the decaying problem. Let us now consider the two coupled equations

$$\frac{du_n}{dt} = \frac{i}{3}\Phi_n(u, u) + \frac{i\lambda}{3}\Phi_n(w, u) - \nu k_n^2 u_n + f_n, \tag{7}$$

$$\frac{dw_n}{dt} = \frac{i}{3}\Phi_n(u, w) + \frac{i\lambda}{3}\Phi_n(w, w) - \nu k_n^2 w_n + \tilde{f}_n \tag{8}$$

with λ being a real number. Observe that for any $\lambda \neq 0$, (8) and (7) exchange roles under the change $\lambda w_n \leftrightarrow u_n$. The universality to forcing implies that the scaling exponents, ξ_p and ζ_p of the two fields must be the same for all $\lambda \neq 0$. For $\lambda = 0$ we recover the equations for the nonlinear and a linear models, (1) and (3). At this point we present strong evidence that the scaling exponents of either field exhibits no jump in the limit $\lambda \to 0$. Accordingly, the scaling exponents of either field can be obtained from the SPS of the linear problem.

Equations (7) and (8) were solved numerically, choosing f_n a constant complex number limited to $n = 0, 1$, and $\tilde{f}_n$ a random force with zero mean, operating on the same shells. We chose $\nu = 10^{-8}$, $\delta = 0.6$ and $\lambda = 10^{-1}, 10^{-3}, 10^{-4}, 10^{-5}, 0$. In Fig. 1 we show, for example, results for the sixth order objects $\langle |u_{n-1} u_n u_{n+1}^*|^2 \rangle$ and $\langle |w_{n-1} w_n w_{n+1}^*|^2 \rangle$. Plotted are double-logarithmic plots of these object as a function of k_n. We see that the exponents of the linear and nonlinear model at $\lambda = 0$ are the same and they do coincide with the exponents of the two coupled models (7), (8) for $\lambda > 0$ (see also inset of Fig. 2). Hence, the limit $\lambda \to 0$ is regular. In fact, in [19] it is proven rigorously that the the limit $\lambda \to 0$ is regular in the sense that if the exponents of the two fields are the same for any finite λ they must be the same also for $\lambda = 0$.

Note that the two problems do not share *exactly* the same statistics; the linear problem, being symmetric in $w_n \to -w_n$ has an even probability distribution function (pdf) and thus zero prefactors for all the odd structure functions. The statement is only about the identity of the scaling properties, neither the trajectory in phase space nor the the pdf. In Fig. 2 we demonstrate this statement: the pth order structure functions for $p \leq 10$ were computed for the linear and the nonlinear problems (with different forcing). The alleged identity of the exponents is well supported by the numeric. In the inset of Fig. 1 we also demonstrate that the linear and the nonlinear problems share

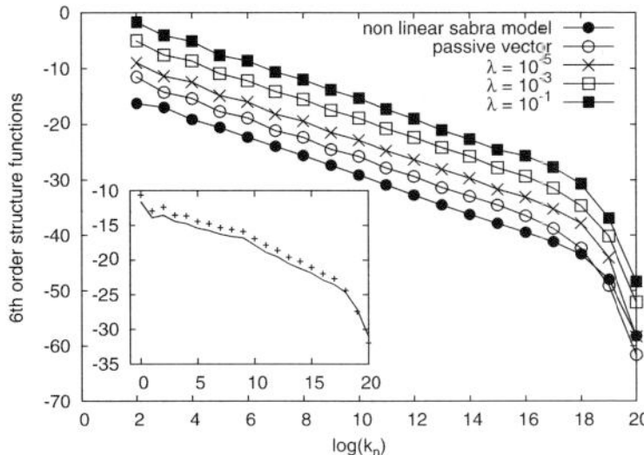

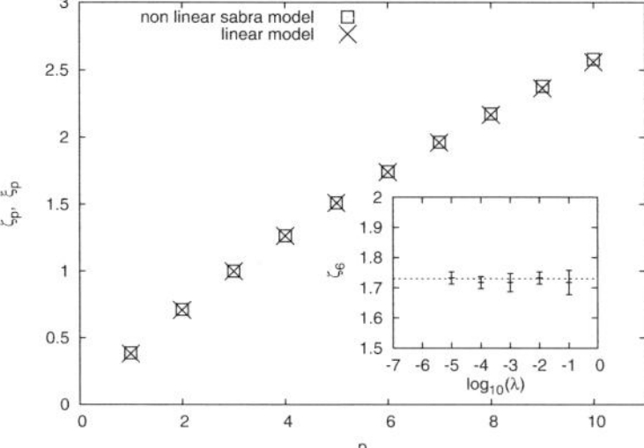

Fig. 1. The sixth order structure function of the field w_n in (8) for $\lambda = 10^{-1}, 10^{-3}$ and 10^{-5}, together with the sixth order structure function for the Sabra model (1) and for the linear model (3), respectively. The structure function of the field u_n for $\lambda > 0$ are not shown since they are indistinguishable from those of the w_n. Inset: log-log plot of the fourth-order correlation function $F_{2,2}(k_n, k_7)$ vs. k_n calculated for the linear field $(+)$ and for the nonlinear field (solid line) at $\lambda = 0$.

Fig. 2. The scaling exponents ζ_p and ξ_p of the nonlinear and linear models (the limit $\lambda \to 0$ in (7) and (8). All exponents are measured for $\langle |u_{n-1} u_n u_{n+1}^*|^{p/3} \rangle$ in the nonlinear model and $\langle |w_{n-1} w_n w_{n+1}^*|^{p/3} \rangle$ for the linear model for $p = 1, 2, \ldots, 10$. Here $\delta = 0.4$. The error bars are at worst of the symbol size. Inset: the sixth order exponent of either field in (7) and (8) for different values of λ.

the same scaling properties for correlations that depend on more than one shell. The data pertain to $F_{p,q}(k_n, k_m) \equiv \langle |u_n|^p |u_m|^q \rangle$, with $p = 2, q = 2$ for both models.

3 Connection to the Forced Nonlinear Model

The greatest asset of the present approach is that we can now forge a connection between the SPS of the linear model and the *forced* correlation function of the nonlinear problem. This underlines the anomaly of the scaling properties of the latter model, and allows us to determine ζ_2. We start with the second order quantities. We can project a generic second order **decaying** correlation function of the linear model onto the second order SPS, thus creating a statistically conserved quantity:

$$I^{(2)} \equiv \sum_n Z_n^{(2)} \langle w_n w_n^* \rangle (t) = \sum_{n,n'} Z_n^{(2)} P_{n,n'}^{(2)}(t|t_0) \langle w_{n'} w_{n'}^* \rangle (t_0)$$

$$= \sum_{n'} Z_{n'}^{(2)} \langle w_{n'} w_{n'}^* \rangle (t_0). \tag{9}$$

Where the average is over different initial conditions for the linear fields and different realization of the advecting velocity field. To show that the forced second order correlation of the nonlinear field is dominated by $Z^{(2)}$, we use this forced correlation function *instead of $Z^{(2)}$* in (9). The test is whether $I^{(2)}$ remains constant on a time window which increases with Reynolds. This is shown in Fig. 3. The success of this test demonstrates that (i) there exists a SPS for the linear problem; (ii) the SPS is well represented by the *forced* nonlinear second order correlation functions. This is a direct demonstration that the correlation function of the nonlinear model scales with the same anomalous exponent as $Z^{(2)}$. An even more stringent test can be made using SPS of orders large than 2, where also correlations between different shells are relevant for the decaying properties [17, 18]. For example $I^{(4)}$ is given by the weighted sum of three contributions:

$$I^{(4)} = \sum_{n,m} Z_{n,m}^{(a,4)} \langle |w_n|^2 |w_m|^2 \rangle (t) + \sum_n [Z_n^{(b,4)} \langle w_n w_{n+1}^2 w_{n+3}^* \rangle (t) + c.c.]$$
$$+ \sum_n [Z_n^{(c,4)} \langle w_n w_{n+1} w_{n+3} w_{n+4}^* \rangle (t) + c.c.],$$

where all the terms allowed by the phase symmetry were employed. In Fig. 3 we show results for $I^{(4)}$ where again we swapped the SPS of the linear problem for the measured *forced* correlations of the nonlinear problem: $Z_{n,m}^{(a,4)} \to \langle |u_n|^2 |u_m|^2 \rangle$ and the corresponding expressions for $Z_n^{(b,4)}$ and $Z_n^{(c,4)}$. We thus conclude that the scaling exponents of a given nonlinear shell model can be understood from the SPS of an appropriately constructed linear problem. To make this point crystal clear, we have used in fact the forced structure functions of the nonlinear model as approximants for $Z^{(2)}, Z^{(4)}$ in the calculation of $I^{(2)}$ and $I^{(4)}$ shown in Fig. 3. The constancy of both demonstrates that the forced correlation function of the nonlinear model are very well approximated by the SPS of the linear model. This demonstration can be repeated with higher order correlation functions with the same (or better) degree of success.

Finally, the existence of a conserved quantity $I^{(2)}$ can be used to calculate $\xi_2 = \zeta_2$. Starting from a given arbitrary initial condition (say a δ-function on one shell) and computing (9) with many realizations of the advecting velocity field, one finds that there exists a sharply defined ξ_2, $Z_n^{(2)} \sim k_n^{-\xi_2}$, for which $I^{(2)}$ is indeed constant. The same approach can be used to determine ζ_3 but we know that $\zeta_3 = 1$. Unfortunately, this simple approach cannot be used for higher order exponents, because the corresponding SPS depend on more than one k_n, and cannot be represented as a simple power law.

4 The Navier-Stokes Case

At this point we turn to the nonlinear Navier-Stokes equations and write the coupled model:

$$\frac{\partial \boldsymbol{u}}{\partial t} + \boldsymbol{u} \cdot \boldsymbol{\nabla} \boldsymbol{u} + \lambda \boldsymbol{w} \cdot \boldsymbol{\nabla} \boldsymbol{u} = -\boldsymbol{\nabla} p_u + \nu \nabla^2 \boldsymbol{u} + \boldsymbol{f}_u, \quad \boldsymbol{\nabla} \cdot \boldsymbol{u} = 0, \quad (10)$$

$$\frac{\partial \boldsymbol{w}}{\partial t} + \boldsymbol{u} \cdot \boldsymbol{\nabla} \boldsymbol{w} + \lambda \boldsymbol{w} \cdot \boldsymbol{\nabla} \boldsymbol{w} = -\boldsymbol{\nabla} p_w + \nu \nabla^2 \boldsymbol{w} + \boldsymbol{f}_w, \quad \boldsymbol{\nabla} \cdot \boldsymbol{w} = 0. \quad (11)$$

For $\lambda = 1$ it is easy to show that the field $\boldsymbol{u}_+ \equiv \boldsymbol{u} + \boldsymbol{w}$ satisfies the Navier-Stokes equation, while the field $\boldsymbol{u}_- \equiv \boldsymbol{u} - \boldsymbol{w}$ satisfies the equation for a passive vector advected by the field $\boldsymbol{u}_+$. Note that if the forcing terms $\boldsymbol{f}_u$ and $\boldsymbol{f}_w$ are

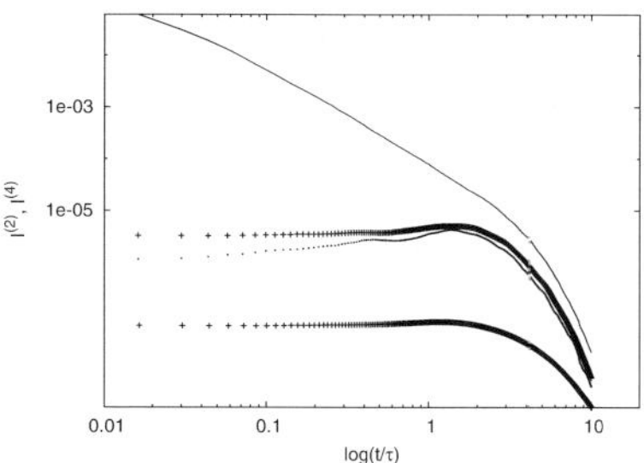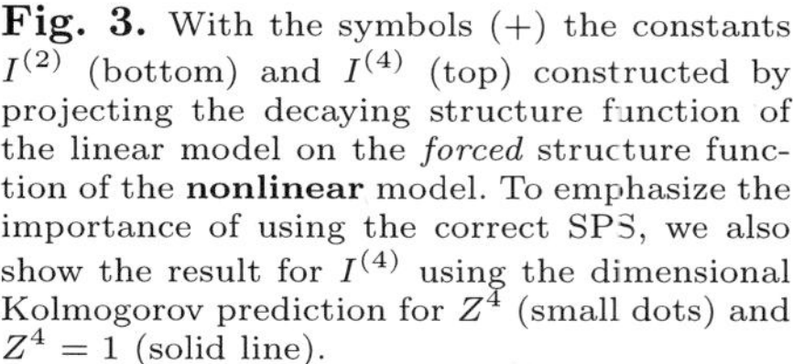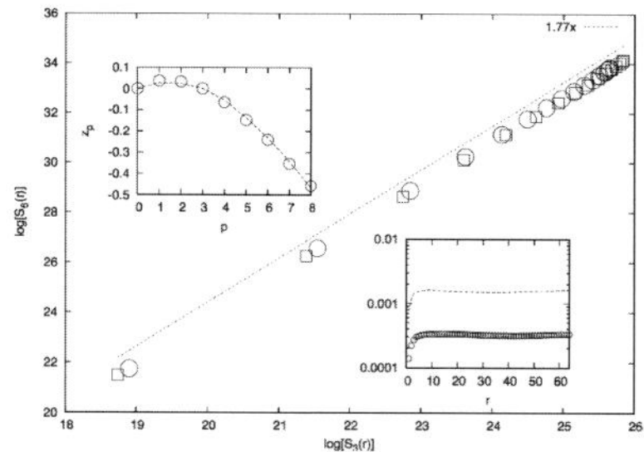

Fig. 3. With the symbols (+) the constants $I^{(2)}$ (bottom) and $I^{(4)}$ (top) constructed by projecting the decaying structure function of the linear model on the *forced* structure function of the **nonlinear** model. To emphasize the importance of using the correct SPS, we also show the result for $I^{(4)}$ using the dimensional Kolmogorov prediction for Z^4 (small dots) and $Z^4 = 1$ (solid line).

Fig. 4. Log-log plot of the sixth order structure functions of the field $\boldsymbol{u} - \boldsymbol{w}$ (circles) in (10) and (11) for $\lambda = 1$ and for the field $\boldsymbol{u} + \boldsymbol{w}$ (squares) against the third order structure functions. The dashed line corresponds to the best fit in the scaling region with slopes 1.77. Lower insert: $S_6(r)/S_3(r)^{1.77}$. Upper insert: $z_p \equiv \zeta_p/\zeta_3 - p/3$ computed for the structures functions of $\boldsymbol{u}+\boldsymbol{w}$ (line) and $\boldsymbol{u}-\boldsymbol{w}$ circles.

Gaussian random variables with the same variance, the forcing terms acting on $\boldsymbol{u}_+$ and $\boldsymbol{u}_-$ are statistically independent. We demonstrate now that the fields $\boldsymbol{u}_+$ and $\boldsymbol{u}_-$ have the same scaling exponents. To this aim we simulated (10) and (11) for $\lambda = 1$ and with two independent forcing realization generated with a Langevin process. The pseudospectral code used for the simulations has a resolution of 128^3. In Fig. 4 we show the ESS plot of the sixth order structure functions for both $\boldsymbol{u}_-$ and $\boldsymbol{u}_+$. One observes convincing scaling behavior with the same exponent for both field. In the upper insert we show the anomalous exponents $z_p \equiv \zeta_p/\zeta_3 - p/3$ for the field $\boldsymbol{u}_+$ (line) and $\boldsymbol{u}_-$ (circles) computed up to order 8: the agreement is excellent. Thus, we can confirm the conclusion that the passive vector $\boldsymbol{u}_-$ has the same scaling exponents as the Navier-Stokes field $\boldsymbol{u}_+$. Finally, we note that the equation for $\boldsymbol{u}_-$ represents a "passive vector with pressure"; such models are known to exhibit anomalous scaling [20]. Passive fields advected by Navier-Stokes turbulent velocity fields were shown to possess SPS very much in the same way as passive fields satisfying a shell model [21, 22]. We thus propose that the anomalous scaling exponents exhibited by such SPS should be the same as those characterizing the structure function of Navier-Stokes turbulence.

These results provide a clear answer to the question whether the scaling exponents of the linear model (either shell model or Navier-Stokes) are anomalous. What remains is to find out whether they are also universal, and this is what we attempt to find in the rest of this paper.

5 Fixed Point Properties

For the shell model, we consider the chain of equations

$$\frac{dw_n^{(s)}}{dt} = \frac{i}{3}\Phi_n(w^{(s-1)}, w^{(s)}) - \nu k_n^2 w_n^{(s)} + f_n^{(s)}. \tag{12}$$

Such a chain of equations has, in addition to the dependence on the initial field, also free parameters $\delta^{(s)}$. In the following we will take $\delta^{(s)} = \delta$ to be the same for all $s > 0$, but we choose $w^{(0)}$ to be the outcome of the **nonlinear** model (1), but with a value δ_0 *different* from the free parameter δ. In this way we can study the statistical properties of the vector fields for large enough s while changing the statistics of the initial field at the beginning of the semi-infinite chain. It then becomes interesting to know how the scaling properties approach a fixed point (if they do) when s increases, and whether the fixed point statitics depends on the value of δ_0. Finally, it would be of interest to find out whether the scaling exponents at the fixed point (if it exists) relate to the scaling exponents of the initial nonlinear model.

These questions were investigated by performing a direct numerical integration of (12) in the limit $s \to \infty$, which is a challenging computational task. We have performed a number of numerical simulations with different values of δ_0 and δ. In all numerical simulations, the forcings $f^{(s)}$ were chosen to be independent gaussian variables with $\langle f_n^{(s)}(t) f_{n'}^{(s')(t')} \rangle \delta_{n,n'} \delta_{s,s'} \delta(t - t')$. The scaling exponents were defined in relation to the scaling properties of the measured related structure functions and energy fluxes, respectively

$$\langle [w_n^{(s)}]^{2p} \rangle \sim k^{-\xi_{2p}(s)}, \qquad F_p^{(s)} \equiv \langle (Q_n^{(s)})^{p/3} \rangle \sim k_n^{-z_p(s)} \tag{13}$$

where $Q_n^{(s)} \equiv |w^{(s)^*}_{n+1} w_n^{(s)} w_{n-1}^{(s)}|$, is the energy flux associated with the variable $w_n^{(s)}$.

The numerical simulations show that there exists a stable fixed point to the scaling properties of the chain dynamics. In all cases, the convergence to the fixed point is achieved for s of the order of 4 or 5. Contrary to what is known [9] about the non linear problem (1), the scaling exponents do not satisfy the scaling relation $z_p = \xi_p$. That is to say, the moments of the energy flux and the structure functions exhibit different scaling exponents. We first analyze the scaling properties of the fluxes.

The convergence to the fixed point is shown in figure (5) where we plot $F_6(s)$ for $s = 0, 1, 2, 6$ with $\delta_0 = 0.8$ and $\delta = 0.4$. The figure shows a rapid convergence to the scaling behaviour $k_n^{\zeta_6}$, with a value of ζ_6 which agrees with what is measured for $\delta = 0.4$ in the non linear case.

In figure (6) we show the exponents z_6 and z_8 for two different numerical simulations, namely corresponding to $\delta_0 = 0.4$ and $\delta_0 = 0.8$, and the *same* value of $\delta = 0.4$ for (12). Figs. (5) and (6) indicate that the statistic of the passive fields far enough along the chain (for s-values large enough) are independent of the initial statistics of the nonlinear field chosen for $s = 0$.

We have also performed other numeical simulations at changing the free parameter δ along the chain (not shown). The striking feature in the latter case is that the scaling exponents z_p, at the fixed point, are also independent of δ. One possible reason for this result is that, at variance with the non linear case, the linear model (12) shows only one quadratic conserved variable in the limit of zero viscosity and forcing, independent of δ. For the non linear model

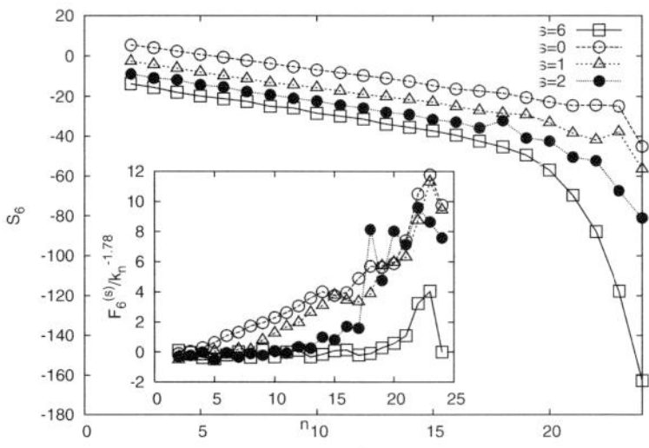

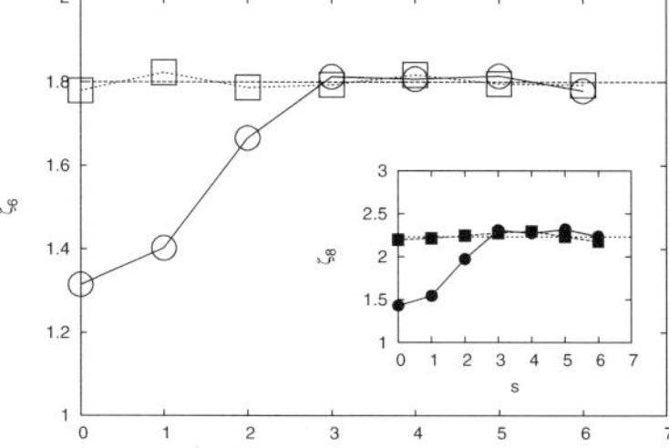

Fig. 5. Scaling properties of $F_6^{(s)}$ for $s = 0$ (initial non linear Sabra model for $\delta_0 = 0.8$) and $s = 1, 2, 6$. The scaling exponent goes from -1.5 ($s = 0$) to -1.77 ($s = 6$). In the insert, we show the compensated functions $F_6^{(s)} k_n^{1.77}$

Fig. 6. Scaling exponent $z_6(s)$ for different s ($s = 0$ corresponds to the initial Sabra model). Circles refers to the case $\delta_0 = 0.8$ and squares to the case $\delta_0 = 0.4$. In the insert, same plot for the scaling exponent $z_8(s)$. The horizontal lines refers to the scaling exponents z_6 and ζ_8 for the non linear Sabra model at $\delta = 0.4$.

there exists a second invariant which depends on δ_0 and by changing δ_0 we can change the relative importance of the second invariant with respect to the energy. This is not the case for the linear model and, therefore the fixed point solution may well be independent of δ.

Another striking feature is that the scaling exponents z_p for large s, are quite close to the numerical value of the non linear model (1) at $\delta = 0.4$. This is probably related to the fact that the scaling exponents ζ_p of (1) show a weak dependence on δ near $\delta = 0.4$. Thus, if the fixed point solution is somehow related to the non linear problem, the value of z_p should be close to the non linear model for $\delta = 0.4$, i.e. where the non linear model depends *weakly* on δ. Moreover, it is also striking to observe that the scaling exponents z_p are quite close to the scaling exponents observed for the Navier-Stokes equations.

Next we analyze the scaling exponents of the structure funcionts. As already remarked, the scaling exponens $\xi_p^{(s)}$ differ from $z_p^{(s)}$. However, a detailed investigation of the dynamical behaviour shows that:

$$\langle [w_n^{(s-1)}]^{2p} [w_n^{(s)}]^{2q} \rangle \sim \langle [w_n^{(s-1)}]^{2p} \rangle \langle [w_n^{(s)}]^{2q} \rangle \tag{14}$$

Equation (14) does not imply that $w^{(s)}$ and $w^{(s-1)}$ are uncorrelated. It simply state that the correlation *is not scale dependent*. As a consequences of (14) we can argue that:

$$F_3^{(s)} \sim \langle |w_n^{(s-1)}| |w_n^{(s)}|^2 \rangle \sim \langle |w_n^{(s-1)}| \rangle \langle |w_n^{(s)}|^2 \rangle \tag{15}$$

Thus, at the fixed point, we obtain that $z_3 = \xi_1 + \xi_2$. The above result can be generalized for all moments leading to

$$z_{3p} = \xi_p + \xi_{2p} \tag{16}$$

In Fig. 7 we show the compensated scaling of $\langle |w_n^{(s-1)}|^p \rangle \langle |w_n^{(s)}|^{2p} \rangle k_n^{z_{3p}}$ and of $F_{3p} k_n^{z_{3p}}$ for $p = 1, 2, 3$ computed at the fixed point (i.e. for $s = 6$). The figure

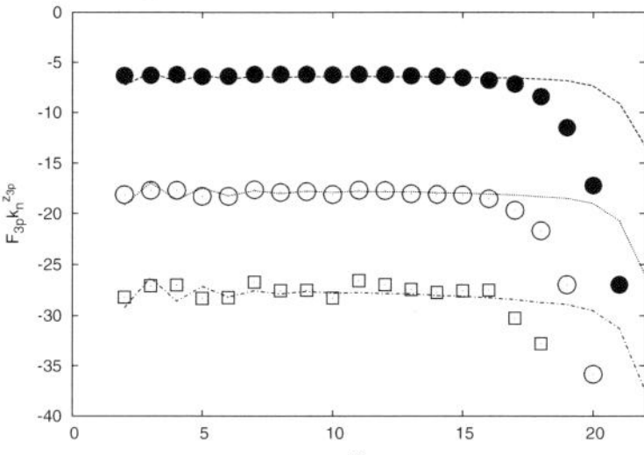

Fig. 7. A compensated plot of $\langle|w_n^{(s-1)}|^p\rangle\langle|w_n^{(s)}|^{2p}\rangle k_n^{z3p}$, symbols, and of $F_{3p}k_n^{z3p}$ for $p = 1, 2, 3$, dashed lines, computed at the fixed point in the chain model (i.e. for $s = 6$).

clearly corroborates (16) which relates the scaling of the structure functions to the moments of the energy fluxes. It remains to understand why (14) holds, and this calls for additional investigations.

6 Summary

In summary, we have shown that the passive vector (i.e. equations (10) and (11) for $\lambda = 0$) has the same multiscaling exponents as the nonlinear field. For both shell models and Navier-Stokes equations, we have provided strong evidence for stating that the forced correlation functions are dominated by the statistically preserved structures of the advecting operator. The consequences of these results are that (i) the exponents of the nonlinear problem are anomalous, and (ii) that they can be found from the homogeneity properties of the eigenfunctions of eigenvalue 1 of an associated linear problem in its decaying mode. What remains to discuss is whether the scaling exponents of the nonlinear problems are also universal. Motivated by the above results we investigated the existence of a fixed point in a chain of passive vectors. Based on numerical simulations of shell models, we found that there exists a stable fixed point in which the scaling exponents corresponds to the non linear field for a particular choice of the parameter δ. We found that the fixed point is universal (i.e. independent of δ) and we provided some qualitative arguments to explain the observed universality. At this stage, no claim can be done concerning the relation between the universal fixed point and the scaling of the non linear field. The existence of such relation would imply the universality of the scaling properties of the nonlinear shell models and the Navier-Stokes equations.

This work has been supported in part by the European Commission under a TMR grant.

References

1. Frisch U (1995) Turbulence, Cambridge University Press
2. Bohr T, Jensen MH, Paladin G, Vulpiani A (1998) Dynamical systems approach to turbulence, Cambridge University Press

3. Biferale L (2003) Ann Rev Fluid Mech 35:441–468
4. Gledzer EB (1973) Dokl Akad Nauk SSSR 20:1046–1048
5. Yamada M, Ohkitani K (1987) J Phys Soc Jpn 56:4210–4213
6. Jensen MH, Paladin G, Vulpiani A (1991) Phys Rev A 43:798–805
7. Pissarenko D, Biferale L, Courvoisier D, Frisch U, Vergassola M (1993) Phys. Fluids A 5:2533–2538
8. Procaccia I, Gat O, Zeitak R (1995) Phys Rev E 51:1148–1154
9. L'vov VS, Podivilov E, Pomyalov A, Procaccia I, Vandembroucq D (1998) Phys Rev E 58:1811–1822
10. Benzi R, Biferale L, Parisi G (1993) Physica D 65:163–171
11. Benzi R, Biferale L, Toschi F (2001) Eur Phys J B 24:125–133
12. Benzi R, Biferale L, Sbragaglia M, Toschi F (2004) Phys Rev E 68:046304
13. Angheluta L, Benzi R, Biferale L, Procaccia I, Toschi F (2006) Phys Rev Lett 97:160601
14. Ditlevsen P (1996) Phys Rev E 54:985–988
15. Biferale L, Pierotti D, Toschi F (1998) Phys Rev E 57:R2515–R2518
16. Falkovich G, Gawedzki K, Vergassola M (2001) Rev Mod Phys 73:913–975
17. Arad I, Biferale L, Celani A, Procaccia I, Vergassola M (2001) Phys Rev Lett 87:164502
18. Cohen Y, Gilbert T, Procaccia I (2002) Phys Rev E 65:026314
19. Benzi R, Levant B, Procaccia I, Titi E (2007) Nonlinearity 20:1431
20. Arad I, Procaccia I (2001) Phys Rev E 63:056302
21. Celani A, Vergassola M (2001) Phys Rev Lett 86:424–427
22. Cohen Y, Pomyalov A, Procaccia I (2003) Phys Rev E 68:036303

Enstropy Generation and Regularity
of Solutions to the 3D Navier-Stokes Equations

Charles R. Doering[1,2] and Lu Lu[1,3]

[1] Department of Mathematics, University of Michigan, Ann Arbor, MI 48109-1043
[2] Department of Physics, University of Michigan, Ann Arbor, MI 48109-1120
[3] Current address: Wachovia Investments, New York, NY

Abstract. The question of existence of smooth solutions to the 3D Navier-Stokes equations is an outstanding open problem in applied mathematics and theoretical physics. It is known that solutions remain smooth as long as the enstrophy remains finite, but it is not known whether or not the enstrophy may diverge to infinity at some finite time. In this paper we report that the state-of-the-art mathematical estimates on the growth rate of enstrophy—estimates that do not rule out the existence of a finite-time singularity—are sharp and cannot be improved.

Keywords: vorticity, enstrophy, Navier-Stokes, regularity

Consider the 3D incompressible Navier-Stokes equations

$$\dot{\mathbf{u}} + \mathbf{u} \cdot \nabla \mathbf{u} + \nabla p = \nu \Delta \mathbf{u} \quad \text{and} \quad 0 = \nabla \cdot \mathbf{u} \tag{1}$$

on a periodic box $\mathbf{x} \in [0, L]^3$. The kinetic energy $K(t)$ and enstrophy $E(t)$ are

$$K(t) = \frac{1}{2} \int |\mathbf{u}(\mathbf{x}, t)|^2 d^3 x \tag{2}$$

and

$$E(t) = \int |\nabla \mathbf{u}(\mathbf{x}, t)| d^3 x = \int |\omega(\mathbf{x}, t)| d^3 x \tag{3}$$

where $\omega = \nabla \times \mathbf{u}$ is the vorticity. Note that the L_2 norms of the velocity gradients and the vorticity are the same for these boundary conditions on this domain (we acknowledge Prof. J. G. Brasseur for insisting on this note).

The following fact is well-known [1]: given an initial divergence-free periodic velocity field $\mathbf{u}_0(\mathbf{x})$ with finite kinetic energy K_0 and enstrophy E_0, there is a time $T(K_0, E_0, \nu) > 0$ such that a unique smooth (i.e., infinitely differentiable) solution $\mathbf{u}(\mathbf{x}, t)$ exists for $t \in (0, T)$. It is also well-known that the unique smooth solution persists as long as $E(t)$ remains finite. That is, the enstrophy is an indicator or "surrogate" for the existence of unique smooth solutions to the 3D incompressible Navier-Stokes equations (1).

47

Y. Kaneda (ed.), *IUTAM Symposium on Computational Physics and New Perspectives in Turbulence*, 47–48.
© 2008 *Springer*.

What is not known is the answer to the question that comprises one of the $1M Clay Millennium Prize problems [2], namely whether or not there are smooth initial data so that $E(t)$ can possibly exhibit a finite-time singularity. Functional analysis ensures that starting from an initial flow field $\mathbf{u}_0(\mathbf{x})$ with finite K_0 and E_0, so long as $t < T(K_0, E_0, \nu)$ the energy and enstrophy obey the systems of differential inequalities

$$\frac{dK}{dt} = -\nu E \quad \text{and} \quad \frac{dE}{dt} \leq -\frac{\nu}{2}\frac{E^2}{K} + c\frac{E^3}{\nu^3} \tag{4}$$

where c is an absolute (dimensionless) constant of order 1. A simple phase-plane analysis then ensures that if the initial flow field is sufficiently weak, specifically if $K_0 E_0 \leq 3\nu^4/4c$, then the subsequent solution maintains finite energy and enstrophy for all $t < \infty$. However in case $K_0 E_0 > 3\nu^4/4c$ the system (4) only ensures finite-time smooth solutions. This analysis does not prove that there *is* a finite-time singularity; it only does not rule one out.

One natural question is whether or not the sequence of mathematical estimates (inequalities) leading to the right hand side of the dE/dt equation in (4) produces a sharp result. That analysis does not utilize incompressibility in an essential way and it is easy to imagine that such a strong constraint might have an important effect. And that analysis proceeds by comparing the highest possible enstrophy production rate due to vortex stretching with the smallest possible damping due to viscous dissipation, necessarily producing an overly conservative over-estimate of the net enstrophy generation rate.

In recent work [3] we have shown that there are smooth divergence-free velocity fields $\mathbf{u}_0(\mathbf{x})$ that satisfy

$$\left(\frac{dE}{dt}\right)_{t=0} = \tilde{c}\frac{E_0^3}{\nu^3} \tag{5}$$

where the positive constant $\tilde{c} < c$. This shows that the analysis leading to the production term in (4) *is* sharp; other than reducing prefactors that part of the estimate cannot be improved. Surprisingly, flow fields that realize the maximal enstrophy generation rate (5) are (locally) axisymmetric colliding vortex rings. These initial conditions cannot produce a finite time singularity, however, so the Clay Prize problem remains open.

This work was supported in part by US National Science Foundation awards PHY-0244859 and PHY-0555324.

References

1. Doering CR, Gibbon JD (1995) Applied Analysis of the Navier-Stokes Equations, Cambridge University Press, Cambridge
2. See the website http://www.claymath.org/millennium/
3. Lu L (2006) Bounds on the enstrophy growth rate for solutions of the 3D Navier-Stokes equations, PhD Dissertation, University of Michigan

Attempts at Computer-Aided Understanding of Turbulence

Yukio Kaneda, Takashi Ishihara, Koji Morishita and Yoshinori Mizuno

Department of Computational Science and Engineering, Graduate School
of Engineering, Nagoya University, Chikusa-ku, Nagoya 464-8603, Japan
kaneda@cse.nagoya-u.ac.jp

Abstract. A review is presented on some attempts to understand turbulence by using high-performance computing (HPC). Among them is the study of turbulence by the direct numerical simulation (DNS) of incompressible turbulence in a periodic box with the number of grid points up to 4096^3. The DNS data shed some light on the Reynolds number (Re)-dependence of the statistics of turbulence at high Re and on the intermittency of energy dissipation. HPC is shown to be useful also for solving a multidimensional singular partial differential equation that determines the anomalous scaling of a passive scalar field advected by a random velocity field with an infinitely small correlation time.

Keywords: DNS of turbulence, computational approach, intermittency of energy dissipation, anomalous scaling, passive scalar

1 Introduction

Computational approaches are useful not only for practical applications, known as CAE (computer-aided engineering) or CAD (computer-aided design), but also for the fundamental understanding of turbulence, which may be called CAU (computer-aided understanding).

Because of the rapid development of high-performance computing (HPC), the number of degrees of freedom (DOF) or of unknown variables that can be handled in computation has increased rapidly. Regarding direct numerical simulation (DNS), high-resolution DNS's of incompressible turbulence obeying the Navier-Stokes equations in a periodic box with the number N^3 of grid points up to 4096^3 were performed recently on the Earth Simulator [1, 2]. The data of such high-resolution DNS are hoped to shed some light for the exploration of the possible universality on the small-scale statistics of turbulence at high Reynolds number (Re), which remains a challenge in turbulence study.

The progress of HPC implies not only the increase of DOF in DNS, but also an increase of the flexibility in analyzing the DNS data, as shown in Sec. 2 and Sec. 3, where we discuss the Re-dependence of statistics and the intermittency of energy dissipation, respectively, in the light of the data of

Y. Kaneda (ed.), *IUTAM Symposium on Computational Physics and New Perspectives in Turbulence*, 49–54.

the high-resolution DNS. The progress of HPC also implies an increase of potentiality in solving problems that have been difficult to treat. Among them is an eigenvalue problem for a multidimensional singular partial differential equation for a passive scalar problem, as discussed in Sec. 4.

2 Reynolds Number Dependence of Statistics

The high-resolution DNS noted in Sec. 1 consists of two series; one is with $k_{\max}\eta \sim 1$ (Series 1), and the other with $k_{\max}\eta \sim 2$ (Series 2), where $k_{\max}$ is the highest wave number in each simulation, and η is the Kolmogorov length scale defined by $\eta = (\nu^3/\langle\tilde{\epsilon}\rangle)^{1/4}$, in which ν is the kinematic viscosity, and $\langle\tilde{\epsilon}\rangle$ is the mean rate of energy dissipation $\tilde{\epsilon}$ per unit mass. In the 4096^3 DNS, the Taylor scale Reynolds number R_λ is about 1130 (675) and the ratio of L/η of the integral length scale L to η is about 2130 (1040), in Series 1 (Series 2).

Visualization of the intense vorticity fields of high-resolution DNS with $N = 2048$ or 4096, in contrast to those with $N = 512$ or so, suggests that it is unlikely that the field may be described as a dynamic system with a few strong eddies that dominate the entire flow field. It shows that the field consists of clusters of small eddies, and the structure of the clusters is distinctively different from the structure of small eddies.

The analysis of the DNS data is now under way. The analysis made so far with particular attention to possible Re-dependence suggests various kinds of Re-dependence of the statistics including the following (A), (B), and (C).

(A) The DNS data suggest that the normalized energy dissipation rate $D \equiv \langle\tilde{\epsilon}\rangle L/U^3$ tends to a nonzero finite constant as $Re \rightarrow \infty$, i.e.,

$$D \rightarrow \text{nonzero const}, \quad \text{as} \quad \nu \rightarrow 0, \tag{1}$$

where $(3/2)U^3$ is the average total kinetic energy per unit mass [2]. Equation (1) implies that the limit $\nu \rightarrow 0$ of real fluid is essentially different from $\nu = 0$ (ideal fluid). In regard to flows past solid walls, such a difference is well known, as clarified by Prandtl's boundary layer theory. The confusion of the two cases $\nu = 0$ and $\nu \rightarrow 0$ would give d'Alembert's paradox. Equation (1) implies that such a difference is also the case in the absence of a solid wall. Equation (1) also implies that $\langle s_{ij}^2 \rangle$ diverges; thus the velocity field's smoothness breaks down as $Re \rightarrow \infty$, because $D \propto \nu \times \langle s_{ij}^2 \rangle$, where s_{ij} is the rate of strain tensor.

Let λ be the Taylor microscale defined by $\lambda = (15\nu U^2/\langle\tilde{\epsilon}\rangle)^{1/2}$. Then

$$\frac{L}{\eta} = \frac{\langle\tilde{\epsilon}\rangle L}{\nu} = Re^{3/4}D^{1/4}, \quad R_\lambda \equiv \frac{\lambda U}{\nu} \propto \frac{U}{(\langle\tilde{\epsilon}\rangle\,\nu)^{1/2}} = Re^{1/2}D^{-1/2},$$

where $Re \equiv UL/\nu$. Thus (1) gives

$$L/\eta \propto Re^{3/4}, \text{ and } R_\lambda \propto Re^{1/2},$$

in this limit. This scaling is in good agreement with DNS [3].

(B) According to Kolmogorov's idea (K41) [4], the energy spectrum $E(k)$ in turbulence at high Re must be a universal function of only $\langle \tilde{\epsilon} \rangle$, the wave numbers k and ν, at $k \gg 1/L$. In particular, in the inertial subrange $1/L \ll k \ll 1/\eta$, we have $E(k) \propto k^{-5/3}$. The DNS data are in fairly good agreement with K41 for $E(k)$. However a close inspection of DNS data with $N = 2048$ and 4096 shows that $E(k) \propto k^{-5/3+\mu}$ in the inertial subrange with $\mu \approx -0.1$. In the near dissipation range where $k \sim 1/\eta$, the DNS data fit well to the form

$$E(k)/(\langle \tilde{\epsilon} \rangle \, \nu^5)^{1/4} = C(k\eta)^a \exp(-bk\eta),$$

where C, a, and b are constants independent of k and approach certain values independent Re, as $Re \to \infty$, in agreement with K41. But the approach is very slow [5].

(C) The probability distribution functions (PDFs) of velocity gradients depend on Re. The DNS data suggest persistent Re-dependence of the PDF, even at very high Re. They also show that the skewness and flatness factors of the velocity gradients, which are among the simplest measures showing the Re-dependence of the PDF, have simple power law dependence on Re, but the exponents at high Re may be different from those inferred from DNS data for $R_\lambda < 400$ or so [6].

3 Intermittency of Energy Dissipation

DNS may also provide us with detailed statistics of kinetic energy dissipation rate without experimental uncertainties. Let $\epsilon(r)$ be the dissipation rate averaged over a sphere of radius r, defined as

$$\epsilon(r|\mathbf{x}, t) = \frac{1}{V(r)} \int_{|\mathbf{x}-\mathbf{x}'|<r} \tilde{\epsilon}(\mathbf{x}', t) \mathrm{d}^3\mathbf{x}',$$

where $V(r) = 4\pi r^3/3$ is the volume of the sphere of radius r, and $\tilde{\epsilon}(\mathbf{x}', t)$ is the kinetic energy dissipation rate per unit mass at position $\mathbf{x}'$ and time t. In the following, we omit the arguments $\mathbf{x}$ and t in $\epsilon(r|\mathbf{x}, t)$.

For ϵ_n defined as $\epsilon_n \equiv \epsilon(r_n)$, $n = 0, 1, 2, ..$, with $r_n = r_0 a^{-n}$, where $a > 1$ is a constant and $r_0 = L$, the well-known decomposition

$$\frac{\epsilon_n}{\epsilon_0} = \frac{\epsilon_n}{\epsilon_{n-1}} \frac{\epsilon_{n-1}}{\epsilon_{n-2}} ... \frac{\epsilon_1}{\epsilon_0} \tag{2}$$

gives

$$\log \frac{\epsilon_n}{\epsilon_0} = \log \alpha_{n-1} + \log \alpha_{n-2} + ... + \log \alpha_0, \tag{3}$$

where $\alpha_n \equiv \epsilon_{n+1}/\epsilon_n$. In the limit of $a \to 1$, we have for $\hat{\epsilon}(\tau) \equiv \epsilon(r)$,

$$\frac{\mathrm{d}\hat{\epsilon}(\tau)}{\mathrm{d}\tau} = u(\tau)\hat{\epsilon}(\tau), \tag{4}$$

where τ is the 'virtual time' defined by $\tau \equiv \log(r_0/r)$ and $u(\tau) \equiv \mathrm{d} \log \hat{\epsilon}(\tau)/\mathrm{d}\tau$.

The relations (2) and (3) suggest the importance of quantitative knowledge on the statistics of $\{\alpha_n\}$ or equivalently $\{\log \alpha_n\}$ for our understanding of the intermittency of ϵ. By assuming the certain statistical nature of $\{\alpha_n\}$ or $u(\tau)$, one may derive various intermittency models, such as the log-normal hypothesis [7, 8], fractal models, and large deviation theories (see, for example, [9, 10]). If we assume that the characteristic correlation time of $u(\tau)$ (in virtual time unit τ) is negligibly small, we obtain an approximation based on a Fokker-Planck equation [11]. In spite of the importance of the role played by $\epsilon(r)$ and $\{\alpha_n\}$ in theories of intermittency and of pioneering studies (see, e.g., [12] and references cited therein), little seems to be known about their statistics in experiments or direct numerical simulation (DNS) at high Re.

The analysis of the DNS data within the DNS with R_λ up to 732 noted in Sec. 2 suggests the following.

(i) The characteristic correlation time of $u(\tau)$ is about $0.5(\sim 3 \log a)$, for $a = 2^{1/4}$. Thus it is not negligibly small.

(ii) The DNS does not support the conjecture of the existence of a scale range where the average and the standard deviations of $\log \alpha$ are independent of r.

(iii) But, they suggest the existence of a range where the correlation between $u(\tau)$ and $u(\tau + s)$ is insensitive to τ.

An analogy of Corssin's conjecture for turbulent diffusion [13] leads one to assume that the integral $\int_0^\tau u(s) \mathrm{d}s$ is statistically independent from the instantaneous 'virtual velocity' $u(\tau)$ in (4). This assumption yields a simple approximation for $\epsilon(\tau)$ that is consistent with the above observations (i)-(iii). A preliminary analysis suggests that the approximation is in good agreement with DNS [14].

4 Eigenmode Analysis of Singular Multidimensional PDE for Anomalous Scaling

The increase of the number of variables treated in computation also implies the increase of potentiality in analyzing equations. Among the equations is a singular partial differential equation (PDE) in multidimensional space associated with the passive scalar field ψ obeying

$$\frac{\partial \psi}{\partial t} + \mathbf{u} \cdot \nabla \psi = \kappa \nabla^2 \psi.$$

If the velocity field $\mathbf{u}$ is a white noise process, we can derive exact closure equations for the moments $(1, 2, 3, ...) \equiv \langle \psi(\mathbf{x}_1, t) \psi(\mathbf{x}_2, t) \psi(\mathbf{x}_3, t) \cdots \rangle$, as shown by Kraichnan [15].

Let S_n be the structure function defined by $S_n \equiv \langle [\psi(\mathbf{x} + \mathbf{r}, t) - \psi(\mathbf{x}, t)]^n \rangle$. It can be expressed in terms of the $(1, 2, \cdot, n)$. For example, $S_2 = 2(1, 1) - 2(1, 2)$ and $S_4 = 2(1, 1, 1, 1) - 8(1, 1, 1, 2) + 6(1, 1, 2, 2)$ in homogeneous isotropic turbulence (HIT). If we assume the existence of a range where S_n scales as $\propto r^{\zeta_n}$ and call the scaling to be anomalous if $\zeta_n \neq n\zeta_2/2$, then, since $S_3 = 0$

in HIT, the lowest n for which we can detect any possible anomalous scaling in HIT is $n = 4$.

For $\Psi_4 = (1, 2, 3, 4)$ in HIT, we have

$$(\mathcal{L}_A + \mathcal{L}_\kappa)\Psi_4 = 0,$$

where

$$\mathcal{L}_A \equiv \sum_{\alpha \neq \beta, i, j} h_{ij}(\mathbf{x}_\alpha - \mathbf{x}_\beta) \frac{\partial^2}{\partial x_{\alpha,i} \partial x_{\beta,j}}, \quad \mathcal{L}_\kappa \equiv -\kappa \sum_{\alpha=1}^{4} \nabla_\alpha^2,$$

and h_{ij} is determined by the 2nd order moments of the velocity $\mathbf{u}$. Such exact closure equations are rare in turbulence study.

Let $h_{ij}(\mathbf{r}) \propto r^\xi$ in the inertial subrange. Then it is known that the scaling of S_4 in the range is determined by the scaling of the so-called zero modes Ψ_4's satisfying $\mathcal{L}_A \Psi_4 = 0$. One may assume that each of the zero modes, say $\tilde{\Phi}$, satisfies the scaling relation $\tilde{\Phi}(s\mathbf{x}_1, s\mathbf{x}_2, s\mathbf{x}_3, s\mathbf{x}_4) = s^\lambda \tilde{\Phi}(\mathbf{x}_1, \mathbf{x}_2, \mathbf{x}_3, \mathbf{x}_4)$, where λ is determined by an eigenvalue problem of the form

$$\mathcal{L}(\lambda)\Phi = 0, \tag{5}$$

in which $\mathcal{L}(\lambda)$ is a linear differential operator, and λ and Φ are the eigenvalue and the eigenfunction, respectively. The structure function S_4 for small r is dominated by the zero mode(s) with the smallest positive eigenvalue.

To see the role of the zero modes or the equation $\mathcal{L}_A \Psi_4 = 0$, it is instructive to ask, for example, 'Why does the gravitational field obey the r^{-2} law, but not $r^{-1.98}$?', or 'Why does the electric-static dipole field have particular simple angular dependence?' The answer is built in the Laplace equation in the sense that they can be explained by the scaling and angular dependence of the zero-mode Ψ satisfying the Laplace equation $\nabla^2 \Psi = 0$.

However, in contrast to the Laplace equation, for which we well know not only the scaling, but also the angular dependence of the zero modes, little is known about the zero modes for the operator $\mathcal{L}_A$. Although the scaling of various fourth-order moments like S_4 is determined by solving the eigenvalue problem (5), the analysis has been difficult because of the singularity of the coefficients and the largeness of the number of independent variables involved in $\mathcal{L}_A$.

Because of the difficulty in analytically solving the eigenvalue problem, it is worthwhile to try solving the problem by the aid of HPC. For homogeneous and isotropic turbulence, the number of independent variables of the eigenfunction can be reduced by symmetry consideration. We have solved the problem numerically for a two-dimensional case, in which the number of independent variables of the eigenfunction is four, for several given values of the exponent ξ (under collaboration with Y. Yamamoto, K. Ohi and T. Sogabe). By discretizing the PDE (5), we can reduce the problem to an eigenvalue problem of a linear matrix. The size of the matrix solved is as large

54   Y. Kaneda et al.

as $470,000 \times 470,000$. The eigenvalues thus obtained are in agreement with a perturbation analysis [16] and also with numerical simulations [17, 18].

The solution yields not only the eigenvalues, but also the eigenfunctions. The eigenfunctions give the dependence of the moment on the positions of the four points $(\mathbf{x}_1, \mathbf{x}_2, \mathbf{x}_3, \mathbf{x}_4)$. For example, they determine the correlation between two scalar differences $\{\psi(\mathbf{x}_1) - \psi(\mathbf{x}_2)\}^2$ and $\{\psi(\mathbf{x}_1) - \psi(\mathbf{x}_4)\}^2$. The numerical analysis shows that the correlation for $|\mathbf{x}_1 - \mathbf{x}_4|/|\mathbf{x}_1 - \mathbf{x}_2| \sim 1$ is weaker for larger ξ, and is larger than the value for the Gaussian fields independently of ξ.

The detail of the analysis of Sections 3 and 4 will be presented elsewhere. The computations were carried out on the Earth Simulator and on the HPC2500 system at the Information Technology Center of Nagoya University. This work was partially supported by Grant-in-Aids for the 21st COE "Frontiers of Computational Science," and (B)17340117 from the Japan Society for the Promotion of Science.

References

1. Yokokawa M, Itakura K, Uno A, Ishihara T, Kaneda Y (2002) Proc IEEE/ACM SC2002 Conf, Baltimore; http://www.sc-2002.org/paperpdfs/pap.pap273.pdf
2. Kaneda Y, Ishihara T, Yokokawa M, Itakura K, Uno A (2003) Phys Fluids 15:L21–24
3. Kaneda Y, Ishihara T (2006) J Turbul 7(20):1–17
4. Kolmogorov AN (1941) Dokl Akad Nauk SSSR 30:301–305
5. Ishihara T, Kaneda Y, Yokokawa M, Itakura K, Uno A (2005) J Phys Soc Jpn 74:1467–1471
6. Kaneda Y, Ishihara T (2007) Small-scale statistics in high-resolution direct numerical simulation of turbulence. In: Oberlack M, Khujadze G, Guenther S, Weller T, Frewer M, Peinke J, Barth S (eds) Progress in Turbulence II. Springer, Berlin Heidelberg New York
7. Obukhov AM (1962) J Fluid Mech 13:77–81
8. Kolmogorov AN (1962) J Fluid Mech 13:82–85
9. Meneveau C, Sreenivasan KR (1991) J Fluid Mech 224:429–484
10. Frisch U (1995) Turbulence: the Legacy of A. N. Kolmogorov. Cambridge University Press, Cambridge
11. Naert A, Friedrich R, Peinke J (1997) Phys Rev E 56:6719–6722
12. Jiménez J (2001) Self-similarity and coherence in the turbulent cascade. In: Kambe T, Nakano T, Miyauchi T (eds) Geometry and Statistics of Turbulence. Kluwer Academic Press
13. Corssin S (1959) Adv Geophys 6, Academic, New York:161–164
14. Kaneda Y, Morishita K (2007) J Phys Soc Jpn 76:073401-1-4
15. Kraichnan RH (1968) Phys Fluids 11:945–953
16. Gawędzki K, Kupiainen A (1995) Phys Rev Lett 75:3834–3837
17. Gat O, Procaccia I, Zeitak R (1998) Phys Rev Lett 80:5536–5539
18. Frisch U, Mazzino A, Noullez A, Vergassola M (1999) Phys Fluids 11:2178–2186

Effect of Large-Scale Structures upon Near-Wall Turbulence

Kaoru Iwamoto[1,2], Takahiro Tsukahara[1], Hideki Nakano[1], and Hiroshi Kawamura[1]

[1] Department of Mechanical Engineering, Tokyo University of Science, 2641 Yamazaki, Noda, Chiba 278-8510, Japan
[2] Present address: Department of Mechanical Systems Engineering, Tokyo University of Agriculture and Technology, 2-24-16 Nakacho, Koganei, Tokyo 184-8588, Japan
iwamotok@cc.tuat.ac.jp

Abstract. Direct numerical simulation (DNS) of a turbulent boundary layer has been carried out to examine the effect of the large-scale structures (LSSs) upon the near-wall turbulence. It is found by comparison with the DNS data of the Poiseuille and the Couette flows that the spanwise scales of LSSs are different among the canonical wall turbulences when normalized with the channel half width or the boundary layer thickness, and that the effect of LSSs on the near-wall turbulence depends upon the spanwise scales. We introduce a Reynolds number based on the spanwise scales, and suggest that the new Reynolds number dependence of the near-wall turbulence can be analyzed irrespective of the flow configurations at least at low-to-moderate Reynolds numbers.

Keywords: wall turbulence, large-scale structures, direct numerical simulation, turbulent boundary layer, Reynolds number dependence

1 Introduction

It is well known that near-wall streamwise vortices and streaky structures play a primary role in the transport mechanism on near-wall turbulence, at least, at low Reynolds number flows [1]. On the other hand, the large-scale outer-layer structures and their relationship to the near-wall structures still remain unresolved. Adrian *et al.* [2] showed that packets of large-scale hairpin vortices around the low-speed large-scale structures are often observed in high-Reynolds-number turbulent boundary layers. It is important to note that the large-scale structures are much different even among canonical wall turbulences, i.e., turbulent Poiseuille, Couette and boundary layer flows.

In the present study, direct numerical simulation (DNS) of a turbulent boundary layer (TBL) is carried out to examine the effect of the large-scale structures upon the near-wall turbulence. The influence of the flow

55

Y. Kaneda (ed.), *IUTAM Symposium on Computational Physics and New Perspectives in Turbulence*, 55–60.

configurations are also investigated as compared with the DNS data of the Poiseuille [3, 4] and the Couette flows [5].

2 Numerical Method

The numerical method used for the TBL in the present study is almost the same as that of Lund [6]; a finite difference method in which three-dimensional, time-dependent turbulent inlet-velocity data is generated by the rescaling scheme, is employed. The Reynolds number is set to $Re_{\delta,in} \equiv U_\infty \delta_{in}/\nu = 2600$, where U_∞ is the free-stream velocity, δ_{in} the 99% boundary layer thickness at the inlet and ν the kinematic viscosity. The size of the computational domain is $21.6\delta_{in} \times 3\delta_{in} \times 6.4\delta_{in}$ in the streamwise (x), wall-normal (y) and spanwise (z) directions, respectively. The number of the grid points is $432 \times 128 \times 128$. Hereafter, u, v, and w denote the velocity components in the x-, y- and z-directions, respectively. The superscript of $(^+)$ represents normalization by the local friction velocity u_τ and ν.

For the Couette flow, the flow is driven by the shear stress due to the relative movement between the top and bottom walls. The finite difference method is used for the spatial discretization. The numerical scheme with the 4th-order accuracy is adopted in the horizontal direction and the 2nd-order one is applied in the wall-normal direction [5]. The Reynolds number is $Re_\tau \equiv u_\tau \delta/\nu = 126$, where δ is the channel half width. The size of the computational domain and the number of the grid points are $192\delta \times 2\delta \times 25.6\delta$ and $2048 \times 96 \times 512$, respectively.

3 Results and Discussion

Figure 1 shows $(y - z)$ cross-stream planes of instantaneous flow fields for the Poiseuille [3], the Couette [5] and the boundary layer flows calculated in the present study. Here, δ is defined as the channel half width or the boundary layer thickness. It is found that the large-scale low- and high-speed structures extend from the near-wall region to the center of the channel or the upper part of the boundary layer irrespective of the flow configurations. However, these spanwise scales Λ are different from each other when normalized with δ, where Λ is defined as twice of the spanwise spacing of the first negative peak for the spanwise two-point correlation of u' at $y = \delta$. They are $\Lambda \approx 1.2\delta$, 5δ and 2δ for Poiseuille [4], Couette [5] and boundary layer flows [7], respectively. Note that Λ/δ is almost independent upon the Reynolds number for each of the flow configurations [4, 5, 7]. The large-scale structures affect the near-wall turbulence [8]. The present study suggests that the effect depends upon the flow configurations.

Figure 2 shows the velocity and pressure fluctuation profiles. The near-wall ($y^+ < 50$) qualitative trends of the Couette flow and the boundary layer are consistent with that of the Poiseuille flow. For the almost same Reynolds

(a)

(b)

(c)

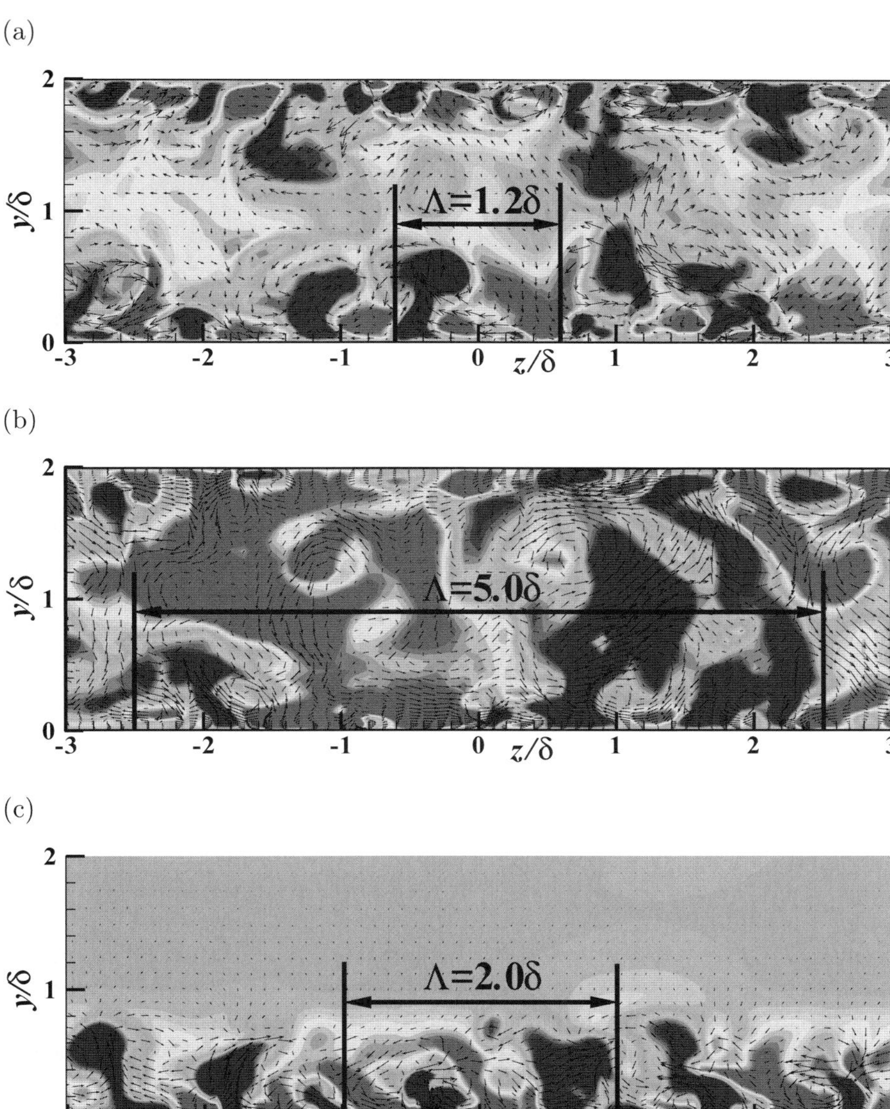

Fig. 1. Instantaneous velocity vectors in the cross-stream plane and contours of the streamwise velocity fluctuation; u'^+, dark gray to light gray, -2 to 2. (a) Poiseuille flow at $Re_\tau = 150$ [3]; (b) Couette flow at $Re_\tau = 126$ [5]; (c) boundary layer at $Re_\tau = 150$.

number of $Re_\tau = 126 - 150$, however, the profiles do not correspond quantitatively among the three flow configurations: those of the Couette flow are largest, the boundary layer middle and the Poiseuille flow smallest.

(a)

(b)

(c)

(d)

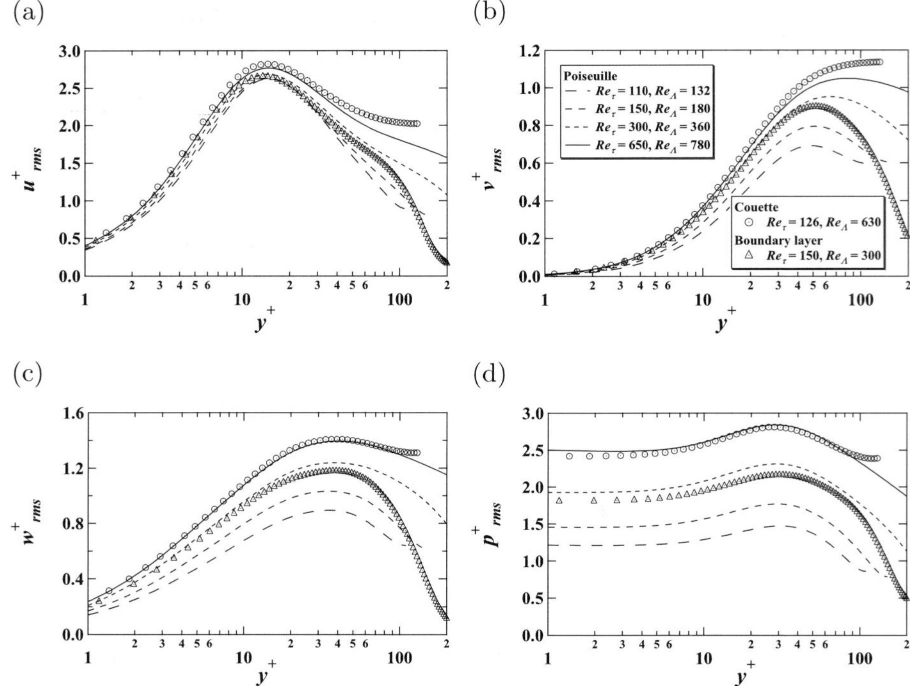

Fig. 2. Velocity and pressure fluctuation profiles: (a) u'_{rms}; (b) v'_{rms}; (c) w'_{rms}; (d) p'_{rms}. DNS data of Poiseuille [3] and Couette flows [5] are also shown.

Figure 3(a) shows the Reynolds number (Re_τ) dependence of the maximum values of $u_{i,rms}$ and p_{rms}. The wall-normal positions of these peaks except for that of v_{rms} are located at $y^+ = 15 - 40$ as shown in Fig. 2, indicating that these peaks represent the magnitude of the near-wall turbulence. The peak values are not in accordance among the three flow configurations for the same Reynolds number Re_τ. This is mainly because the effects of the large-scale structures on the near-wall turbulence are different as described in Fig. 1. For the Poiseuille flow, these values increase with the increase of the Reynolds number. Hence, the near-wall turbulence is neither scaled only with the wall unit nor with the outer scale.

In order to examine the difference of the flow configuration, a Reynolds number based on the spanwise scale of the large-scale structures is introduced as $Re_\Lambda \equiv u_\tau \Lambda / \nu$. Figure 3(b) shows the Reynolds number (Re_Λ) dependence of the peak values. Those of the Couette flow and the boundary layer are consistent with the trend of the Poiseuille flow. Figure 3(b) strongly suggests that the Reynolds number (Re_Λ) dependence is analyzed irrespective of the flow configurations at least at low-to-moderate Reynolds numbers.

(a)

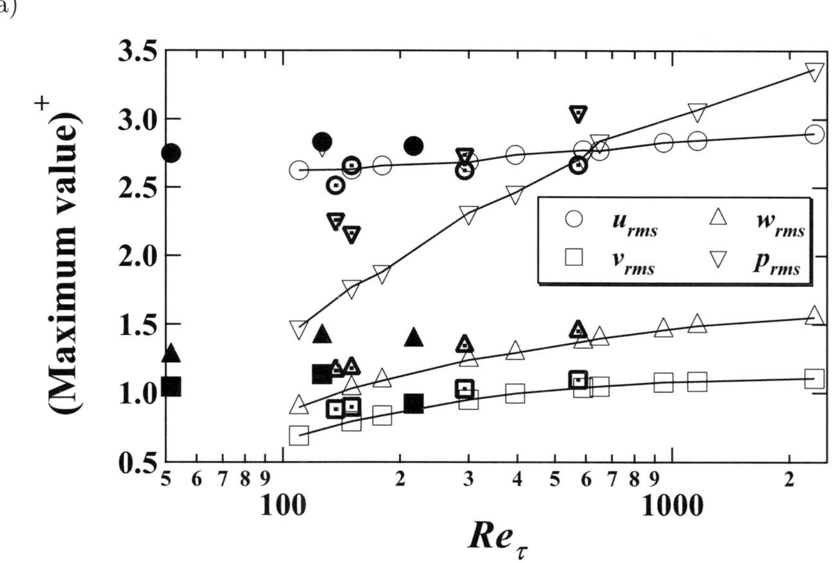

(b)

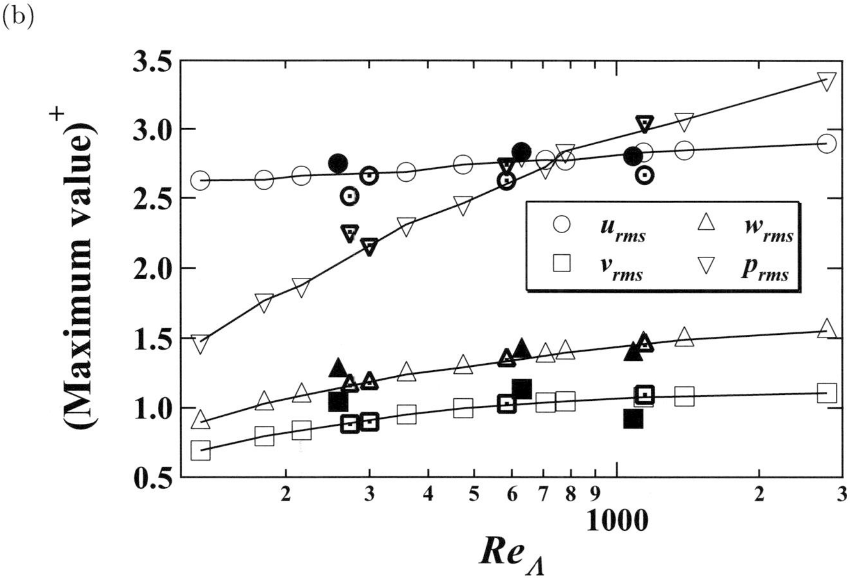

Fig. 3. Dependence of the maximum values of $u_{i,rms}$ and p_{rms} on the Reynolds numbers (a) Re_τ; (b) Re_Λ. Open symbols with a line, DNS data of Poiseuille flow at $Re_\tau = 110, 150, 300, 650$ [3], $1160, 2320$ [4], $180, 395, 590$ [9] and 950 [10]. Filled symbols, data of Couette flow at $Re_\tau = 52$ [11], 126 [5] and 217 [12]. Open thick symbols, DNS data of boundary layer at $Re_\tau = 136, 292, 572$ [13] and 150 (present DNS data).

4 Conclusions

Direct numerical simulation of a turbulent boundary layer was made in order to examine the effect of the large-scale structures upon the near-wall turbulence. The following conclusions are derived by comparison with the DNS data of the Poiseuille [3, 4] and the Couette flows [5]:

1. The spanwise scales Λ of the large-scale structures are different even among the canonical wall turbulences when normalized with the channel half width or the boundary layer thickness δ, i.e., $\Lambda = 1.2\delta$ for turbulent Poiseuille, 5δ for Couette and 2δ for boundary layer flows. The effect of the large-scale structures upon the near-wall turbulence depends upon the spanwise scales Λ.
2. A Reynolds number based on the spanwise scale of the large-scale structures is introduced as $Re_\Lambda \equiv u_\tau \Lambda / \nu$ in order to examine the Reynolds number dependence of the peak values of the velocity and the pressure fluctuations. Those of the Couette flow and the boundary layer are consistent with the trend of the Poiseuille flow at least at low-to-moderate Reynolds numbers ($130 < Re_\Lambda < 1150$).

Acknowledgement. This work was supported in part through Research Center for the Holistic Computational Science (Holcs) at Tokyo University of Science supported by the Ministry of Education, Culture, Sports, Science and Technology of Japan (MEXT), and the Grant-in-Aid for Young Researcher (B) by MEXT.

References

1. Robinson SK (1991) Annu Rev Fluid Mech 23:601–639
2. Adrian RJ, Meinhart CD, Tomkins CD (2000) J Fluid Mech 422:1–54
3. Iwamoto K, Suzuki Y, Kasagi N (2002) Int J Heat Fluid Flow 23:678–689
4. Iwamoto K, Kasagi N, Suzuki Y (2005) Proc 6th Symp Smart Control of Turbulence:327–333
5. Tsukahara T, Kawamura H, Shingai K (2006) J Turbul 7:1–16
6. Lund TS (1998) J Comput Phys 140:233–258
7. Tomkins CD, Adrian RJ (2003) J Fluid Mech 490:37–74
8. Abe H, Kawamura H, Choi H (2004) J Fluid Eng Trans ASME 126:835–843
9. Moser RD, Kim J, Mansour MN (1999) Phys Fluids 11 4:943–945
10. del Alamo JC, Jimenez J (2003) Phys Fluids 15 6:L41–L44
11. Komminaho J, Lundbladh A, Johansson AV (1996) J Fluid Mech 320:259–285
12. El Telbany MMM, Reynolds AJ (1982) J Fluid Eng Trans ASME 104:367–372
13. Spalart PR (1988) J Fluid Mech 187:61–98

Multifractal Analysis by Using High-Resolution Direct Numerical Simulation of Turbulence

Takashi Ishihara and Hirotaka Higuchi

Department of Computational Science and Engineering, Graduate School
of Engineering, Nagoya University, Chikusa-ku, Nagoya 464-8603, Japan
ishihara@cse.nagoya-u.ac.jp

Abstract. Multifractal analysis based on the energy dissipation ϵ and the enstrophy Ω in isotropic turbulence is made by using high-resolution direct numerical simulation of turbulence with the number of grid points up to 2048^3. The analysis shows that the singularity exponents α and α' that characterize the intermittencies associated with ϵ and Ω, respectively, agree well with each other in the inertial subrange. This result is consistent with $\log\epsilon_r$ and $\log\Omega_r$ correlating well with each other for the scale r in the inertial subrange, where the subscript denotes the local average over a cubic domain of size r.

Keywords: multifractal analysis, intermittency, isotropic turbulence, high-resolution DNS

1 Introduction

The intermittency of energy dissipation rate $\epsilon(\equiv 2\nu s_{ij}s_{ij})$ and that of enstrophy $\Omega(\equiv |\boldsymbol{\omega}|^2/2)$ in fully developed turbulence, and its scale-similar facets, can be well characterized by using a general formulation of multifractals [1, 2]. Here ν is the kinematic viscosity, s_{ij} the rate of strain tensor, and $\boldsymbol{\omega}$ the vorticity. The multifractal formulation has been applied extensively to both experimental and direct numerical simulation (DNS) data of turbulence since the late 1980s, and it has been found to be useful in describing intermittent distributions of ϵ and Ω (e.g., see [3] and the references cited therein). In experiments, however, one has not been able to analyze ϵ, but its one dimensional surrogate, while in DNS the resolution as well as the Reynolds number (Re) has so far been severely limited by available computer resources, so that only an insufficient scale range has been simulated in most DNS's.

Recently a series of high-resolution DNS's of incompressible turbulence in a periodic box was performed using an alias-free spectral method on the Earth Simulator (ES) [4, 5]. The number of grid points N^3 and the Taylor microscale Reynolds number R_λ of the DNS are up to 4096^3 and approximately 1130, respectively. In this paper, we perform multifractal analysis by

61

Y. Kaneda (ed.), *IUTAM Symposium on Computational Physics and New Perspectives in Turbulence*, 61–66.

using the high-resolution DNS data of turbulence with special emphasis on the similarities and differences between ϵ and Ω. It is expected that the analysis may shed some light on the intermittency of turbulence because (i) the DNS provides us with detailed turbulence data that are free from experimental ambiguities, such as the effects of using Taylor's hypothesis and one-dimensional surrogates, and (ii) the resolution of our DNS is 64 times larger than previous DNS's, with 1024^3 grid points; thus the turbulence field of our DNS has a wider inertial subrange.

2 Joint PDF of $\log \epsilon_r$ and $\log \Omega_r$

Figures 1(a) and 1(b) show the joint probability distribution function (PDF) of $\log \epsilon$ and $\log \Omega$, and their PDFs, respectively, computed by the DNS data of turbulence with 2048^3 grid points and $R_\lambda = 732$. It is observed in Fig. 1(a) that ϵ and Ω are almost uncorrelated. The difference between the two PDFs in Fig. 1(b) suggests that ϵ and Ω have a different small-scale structure from each other. Figures 1(c) and 1(d) are the same as Figs. 1(a) and 1(b), respectively, but ϵ_r and Ω_r are used instead of ϵ and Ω, where A_r is the average of A over

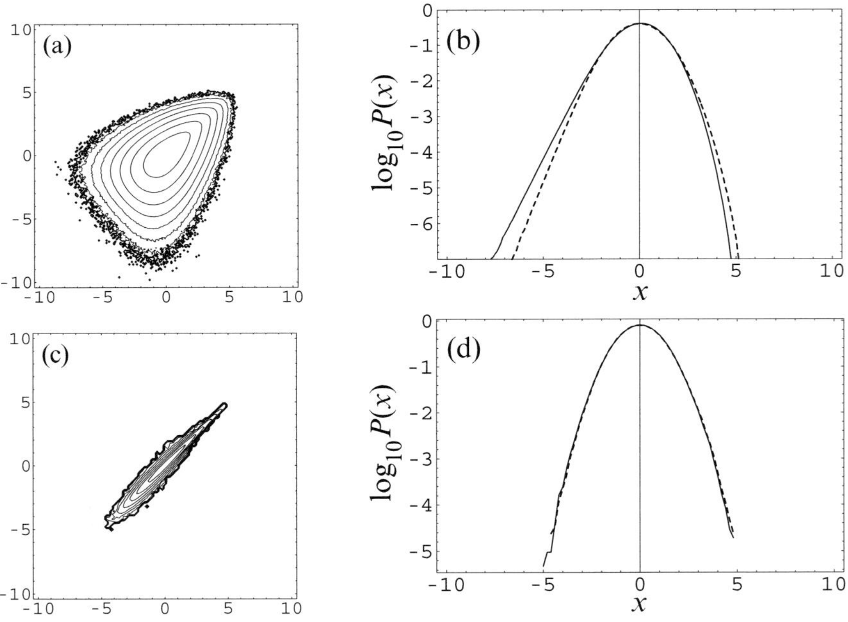

Fig. 1. (a) Joint PDF of $(a - \langle a \rangle)/\sigma_a$ (horizontal) and $(b - \langle b \rangle)/\sigma_b$ (vertical), where $a = \log \epsilon$, $b = \log \Omega$, and $\langle a \rangle$ and σ_a are the mean and standard deviation of a, respectively. The data are obtained by the 2048^3 DNS with $R_\lambda = 732$. (b) The PDF $P(x)$ of $x = (a - \langle a \rangle)/\sigma_a$ (broken line) and that of $x = (b - \langle b \rangle)/\sigma_b$ (solid line). (c) and (d) are the same as (a) and (b), respectively, but ϵ_r and Ω_r are used instead of ϵ and Ω, where $r \approx 0.88\lambda \approx 47\eta$.

a cubic domain of size r with $r \approx 0.88\lambda \approx 47\eta$, where λ and η are the Taylor microscale and the Kolmogorov length scale, respectively. Figure 1(c) shows that ϵ_r and Ω_r are strongly correlated with each other, and Fig. 1(d) suggests that ϵ_r and Ω_r have almost the same spatial distribution for the scale r in the inertial subrange as would be expected from Fig. 1(c).

In [6], the spectra of $f = \epsilon$ and Ω defined by $E_f(k) = \sum_k \hat{f}(\mathbf{k})\hat{f}(-\mathbf{k})$ were studied by using the DNS data of isotropic turbulence, where $\hat{f}$ is the Fourier transform of f, and it was shown that (i) $E_{\epsilon/(2\nu)}(k)$ and $E_\Omega(k)$ are similar to each other in the inertial subrange, but (ii) they are significantly different from each other in the dissipation range. These results are consistent with Fig. 1.

3 Generalized Dimension and $f(\alpha)$ Spectrum

In multifractal formulation, the so-called generalized dimension D_q is defined as an r-independent exponent according to

$$\sum_i (E_r(\mathbf{x}_i)/E_L)^q \sim (r/L)^{(q-1)D_q}, \tag{1}$$

where E_r is the integral of $E(=\epsilon, \Omega)$ over the cubic domain of size r around location $\mathbf{x}_i$ and E_L is the sum of E over the cubic domain of size L. Figure 2(a)

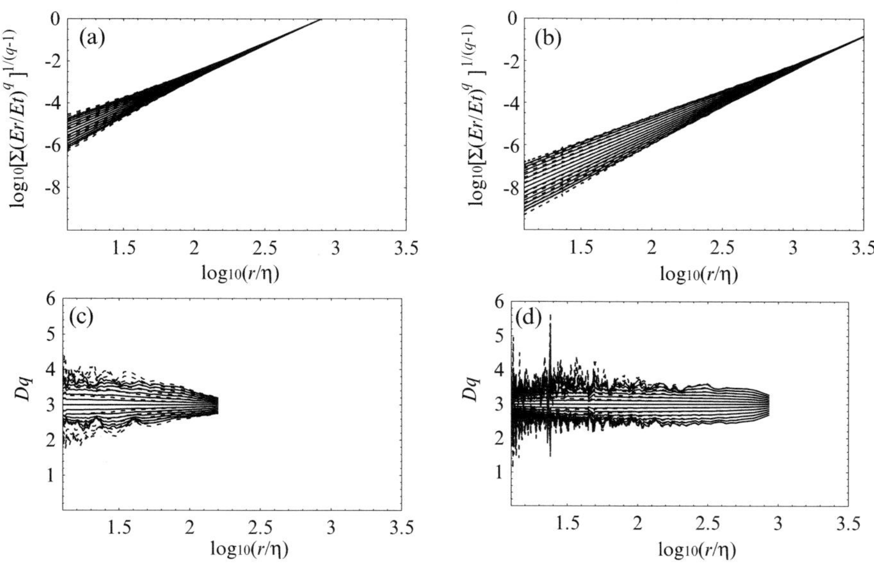

Fig. 2. $\log_{10}\left[\sum_i (E_r(\mathbf{x}_i)/E_t)^q\right]^{1/(q-1)}$ vs. $\log_{10}(r/\eta)$ by the DNS's with (a) $N^3 = 256^3$, $R_\lambda = 167$, and (b) $N^3 = 2048^3$, $R_\lambda = 732$; solid and broken lines are for $E = \epsilon$ and Ω, respectively, and $q = -6, -4, \cdots, 6$ from bottom to top. (c) D_q [the local slopes of the lines in (a)] vs. $\log_{10}(r/\eta)$. (d) The same as (c) but for the lines in (b).

suggests that the left-hand side of (1) obtained from the 256^3 DNS ($R_\lambda = 167$) has a simple power-law dependence on r, but Fig. 2(c) reveals that the local slopes of the lines in Fig. 2(a) depend on r. On the other hand, Fig. 2(d) shows that there exist an r region for which r dependence of D_q is weak and suggests that the scaling (1) holds in the inertial subrange of high Re turbulence. Figure 2 shows that the values plotted for ϵ approximately coincide with those for Ω for $r/\eta > 10^{1.8} \approx 63$ and also that the (r-independent) value of D_q for ϵ agrees well with that for Ω in the inertial subrange.

In the multifractal model based on $E(=\epsilon, \Omega)$, the intermittency is characterized by local singularity exponents α and the fractal dimension $f(\alpha)$, i.e., there is a set ϑ_α of dimension $f(\alpha)$, such that as $r \to 0$, $E_r(\mathbf{x})/E_L \sim (r/L)^\alpha, \mathbf{x} \in \vartheta_\alpha$. The dimension $f(\alpha)$ is obtained from D_q using relations $f(\alpha) = q\alpha - (q-1)D_q$ and $\alpha = (\mathrm{d}/\mathrm{d}q)\left[(q-1)D_q\right]$. As expected, our DNS data show that $f(\alpha)$ of ϵ almost coincides with that of Ω in the inertial subrange. The $f(\alpha)$ agrees reasonably well with that of [1] (figures omitted).

4 Joint Multifractal Analysis

To characterize the mutual relation between the singularity exponents associated with the measures ϵ and Ω, we perform the joint multifractal analysis in this section following [2]. In the analysis (i) the whole domain is divided into boxes of size r, and (ii) $\epsilon_{r,i}$ and $\Omega_{r,i}$ are defined as the integrated measures in the ith box of size r. The local singularity exponents α and α' are defined according to $\epsilon_{r,i} \sim (r/L)^\alpha$ and $\Omega_{r,i} \sim (r/L)^{\alpha'}$. Let us denote by $N_r(\alpha, \alpha')\mathrm{d}\alpha\mathrm{d}\alpha'$ the number of boxes of size r where the variable α has values in a band $\mathrm{d}\alpha$ around α, and α' in a band $\mathrm{d}\alpha'$ around α'. Then $f(\alpha, \alpha')$ is defined as the scaling exponent of $N_r(\alpha, \alpha')$ according to $N_r(\alpha, \alpha')\mathrm{d}\alpha\mathrm{d}\alpha' \sim (r/L)^{-f(\alpha,\alpha')}\mathrm{d}\alpha\mathrm{d}\alpha'$. From this, $f(\alpha, \alpha')$ can be regarded as the fractal dimension of a set of boxes with size r where α has a certain value and α' has another given value. Therefore the curve $f(\alpha, \alpha')$ characterizes the scaling properties of the joint distribution of the local exponents α and α'.

4.1 Data Processing

To obtain $f(\alpha, \alpha')$, we use the method of μ-weighted averaging as in [2], where a new normalized measure μ, which depends on the box size r and on parameters q and p, is defined according to

$$\mu_{r,i}(q,p) = [\epsilon_{r,i}]^q[\Omega_{r,i}]^p / \sum_i [\epsilon_{r,i}]^q[\Omega_{r,i}]^p. \tag{2}$$

The control parameters q and p have the following meaning. For high-positive values of q and p, $\mu_{r,i}(q,p)$ emphasizes the regions where both measures are very intense. Large negative values of q and p emphasizes the regions where both measures are very sparse. The average values of $\alpha = \ln[\epsilon_{r,i}]/\ln(r/L)$ and $\alpha' = \ln[\Omega_{r,i}]/\ln(r/L)$ with respect to $\mu_{r,i}(q,p)$ are given by

$$\left\{ \begin{array}{c} \alpha(q,p) \\ \alpha'(q,p) \end{array} \right\} = \sum_i [\mu_{r,i}(q,p) \ln \left\{ \begin{array}{c} \epsilon_{r,i} \\ \Omega_{r,i} \end{array} \right\}] / \ln(r/L). \tag{3}$$

Similarly, $f(\alpha, \alpha')$ is related to the average value of $\ln[\mu_{r,i}(q,p)]/\ln(r/L)$ with respect to $\mu_{r,i}(q,p)$ according to

$$f(\alpha, \alpha') = \lim_{r \to 0} \sum_i \{\mu_{r,i}(q,p) \ln[\mu_{r,i}(q,p)]\}/\ln(r/L). \tag{4}$$

A detailed derivation of the relations is written in [2]. The exponents α and α' are measured from the plots of $\sum_i \{\mu_{r,i}(q,p) \ln[\epsilon_{r,i}]\}$ and $\sum_i \{\mu_{r,i}(q,p) \ln[\Omega_{r,i}]\}$ vs. $\ln(r/L)$, and $f(\alpha, \alpha')$ from the plots of $\sum_i \{\mu_{r,i}(q,p) \ln[\mu_{r,i}(q,p)]\}$ vs. $\ln(r/L)$. Different pairs of (α, α') are obtained by changing q and p.

The calculation of (2)–(4) was performed repeatedly for a variety of box sizes r and values of q and p. The values of q and p were altered from -4 to 4 with intervals of 0.5. A total of 17^2 pairs of (q,p) was calculated for each box-size r. About 100 different box sizes between $r/\eta = 10$ and $r/\eta = 500$ were used.

4.2 Results

The values of $\alpha(q,p)$, $\alpha'(q,p)$ and $f(\alpha, \alpha')$ were obtained by measuring the slopes of the plots described in Sect. 4.1. The resulting contour plot is shown in Fig. 3. It was obtained from 2048^3 DNS data with different scaling ranges (a) $r/\eta = 12$ to 84 (from the dissipation range to the inertial range) and (b) $r/\eta = 86$ to 500 (the inertial subrange). If ϵ and Ω have completely dependent distributions, there is a one-to-one relation between α and α', i.e., $\alpha' = \alpha'(\alpha)$, and we have also $f(\alpha, \alpha') = f(\alpha) = f'(\alpha')$, (e.g., see [2]). Figure 3(b) shows that $\alpha' \sim \alpha$ and $f(\alpha, \alpha') \sim f(\alpha) \sim f'(\alpha)$ in the inertial subrange as expected from the results in Sect. 2. This result implies that ϵ and Ω have almost the same distribution for the scale r in the inertial subrange. Note that as shown in Fig. 3(a), they have different distributions for the scale r in the near dissipation range.

In [2], the joint-multifractal analysis of ϵ' and ω_x^2 was performed using atmospheric data at a high Reynolds number, where $\epsilon' \sim (\partial u_1/\partial t)^2$ and ω_x is the streamwise vorticity, and it was shown that the correlation coefficient between α (for ϵ') and α' (for ω_x^2) defined as

$$\rho = \frac{\langle (\alpha - \alpha_0)(\alpha' - \alpha_0') \rangle}{\langle (\alpha - \alpha_0) \rangle^{1/2} \langle (\alpha' - \alpha_0') \rangle^{1/2}}$$

is approximately 0.3, where $\alpha_0 = \langle \alpha \rangle$ and $\alpha_0' = \langle \alpha' \rangle$. Our analysis based on the 2048^3 DNS data have shown that $\alpha \sim \alpha'$ and $\rho \sim 1$ for ϵ and Ω in the inertial subrange. The difference between $\rho \approx 0.3$ for ϵ' and ω_x^2 and $\rho \sim 1.0$ for ϵ and Ω presents an example that the statistics obtained experimentally, using the one-dimensional surrogates, are different from the original ones. Since $-\nabla^2 p = \epsilon/(2\nu) - \Omega$, the statistics of the pressure p remains to be studied to explain the results in this paper.

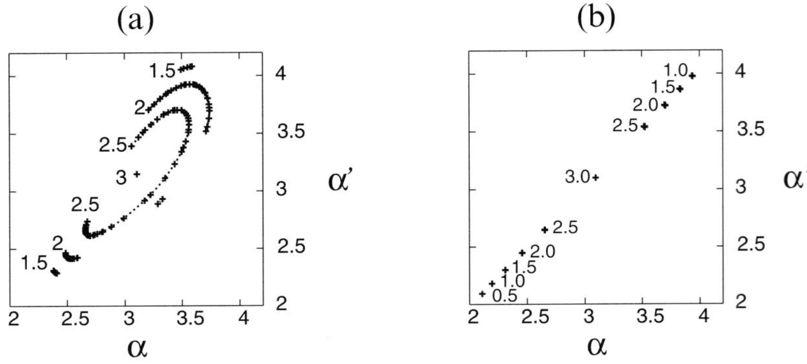

Fig. 3. The contour plot of $f(\alpha, \alpha')$ obtained from the 2048^3 DNS ($R_\lambda = 732$), where α and α' are the local singularity exponents of energy dissipation and enstrophy, respectively. Scaling ranges are (a) $r/\eta = 12$ to 84 (from the dissipation range to the inertial range) and (b) $r/\eta = 86$ to 500 (the inertial subrange).

Acknowledgments

The computations were carried out on the ES and on the HPC2500 system at the Information Technology Center of Nagoya University. The authors would like to express their thanks to Prof. Y. Kaneda for fruitful discussions on turbulence. This work was supported by Grant-in-Aids for Scientific Research (C)17560051 from the Japan Society for the Promotion of Science, and also by a Grant-in-Aid for the 21st Century COE "Frontiers of Computational Science."

References

1. Meneveau C, Sreenivasan KR (1991) J Fluid Mech 224:429–484
2. Meneveau C, Sreenivasan KR, Kailasnath R, Fan MS (1990) Phys Rev A 41:894–913
3. Sreenivasan KR (1991) Annu Rev Fluid Mech 23:539–600
4. Yokokawa M, Itakura K, Uno A, Ishihara T, Kaneda Y (2002) Proc IEEE/ACM SC2002 Conf, Baltimore; http://www.sc-2002.org/paperpdfs/pap.pap273.pdf
5. Kaneda Y, Ishihara T, Yokokawa M, Itakura K, Uno A (2003) Phys Fluids 15:L21–L24
6. Ishihara T, Kaneda Y, Yokokawa M, Itakura K, Uno A (2003) J Phys Soc Jpn 72:983–986

Fine Scale Eddy Cluster and Energy Cascade in Homogeneous Isotropic Turbulence

Mamoru Tanahashi, Kuniharu Fujibayashi, and Toshio Miyauchi

Department of Mechanical and Aerospace Engineering,
Tokyo Institute of Technology, Meguro-ku, Tokyo 152-8550, Japan
`mtanahas@mes.titech.ac.jp`

Abstract. To investigate statistics of the cluster of coherent fine scale eddies and its relation to energy cascade, direct numerical simulation data of homogeneous isotropic turbulence up to $Re_\lambda = 287.6$ have been analyzed. The cluster of the strong coherent fine scale eddies, which is of the order of the integral length, tends to exist in the region with high strain rate of large scales. Energy transfer from large scales to small scales is also high in the cluster regions. The clusters are composed of the small eddies ($<20\eta$), while the large scale eddies ($>80\eta$) are distributed outside of the cluster. The high strain rate regions due to different scales in the inertial subrange do not coincide except for the cluster regions. The importance of these structures in SGS stress models of large eddy simulation is discussed.

Keywords: turbulence, fine scale structure, energy cascade, direct numerical simulation, SGS model

1 Introduction

Recent huge DNS of turbulence have shown that turbulence is composed of universal fine scale eddies which are verified in homogeneous isotropic turbulence [1, 2], turbulent mixing layer [3] and turbulent channel flow [4, 5]. The characteristics of these eddies can be scaled by Kolmogorov length (η) and Kolmogorov velocity (u_k), and the most expected diameter and maximum azimuthal velocity are about 8η and $1.2u_k$ except for the near-wall eddy [4]. The intense fine scale eddy, which is closely related with the intermittency of turbulent energy dissipation rate, can be scaled by r. m. s. of velocity fluctuation (u_{rms}). From the existence of the inertial subrange in energy spectrum, large and medium eddies have been supposed to explain the energy cascade. However, there are no detailed information about large scale (or medium scale) structure in the physical space. In high Reynolds number turbulence, large scale clusters of fine scale eddy have been observed in various turbulent flows [3, 4, 5]. As for homogeneous isotropic turbulence, clusters of the order of integral length scale (l_E) appear. In this study, relation between fine scale eddy cluster and energy cascade is investigated using DNS data of homogeneous isotropic turbulence.

67

Y. Kaneda (ed.), *IUTAM Symposium on Computational Physics and New Perspectives in Turbulence*, 67–72.

2 DNS Data and Eddy Identification

DNS data of decaying homogeneous isotropic turbulence up to $Re_\lambda = 287.6$ are analyzed. Numerical conditions are shown in Table 1. All DNS have been conducted by a spectral method with fully de-aliasing. Time advancements are performed using a third-order Range-Kutta scheme. The spatial resolution have kept to $\eta k_{max} > 1$ for all Reynolds number and the size of computational box have been selected to be larger than $6l_E$ which is enough to resolve large scale motions of turbulence.

In this study, coherent fine scale eddies are educed by using the method similar to our previous studies [2, 4]. Figure 1 shows distribution of axis of fine scale eddy in a box with $3.24l_E \times 3.24l_E \times 3.24l_E$. The thickness of the axis is drawn to be proportional to square root of second invariant of the velocity gradient tensor on the axis. Therefore, stronger eddies are drawn to be thicker. The intense eddies which have $D \approx 8$ and $u_{\theta,m} > u_{rms}$ tend

Table 1. DNS data of decaying homogeneous isotropic turbulence. Re_λ: Reynolds number based on λ and u_{rms}, Re_{l_E}: Reynolds number based on l_E and u_{rms}, N^3: total gird number, L/l_E: ratio of computational domain to l_E, $S_{u'}$ and $F_{u'}$: skewness and flatness factors of longitudinal velocity derivative.

ID	Re_λ	Re_{l_E}	N^3	L/l_E	l_E/η	$S_{u'}$	$F_{u'}$
HIT7	222.7	2087.6	640^3	6.63	276	-0.534	6.34
HIT8	256.1	2899.1	800^3	6.36	361	-0.538	6.47
HIT9	287.6	3694.1	960^3	6.47	428	-0.564	6.97

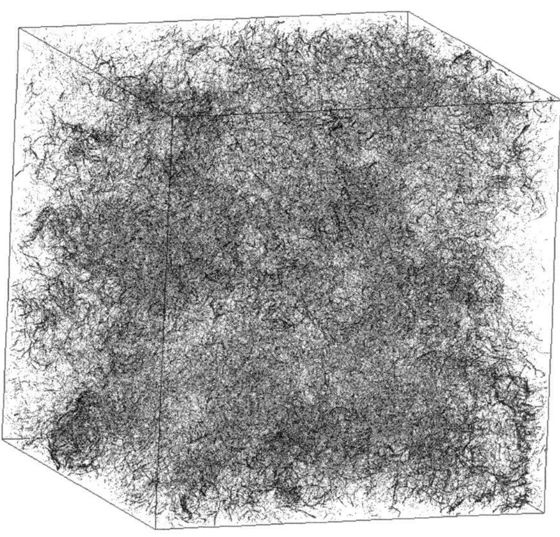

Fig. 1. Distribution of axis of coherent fine scale eddy for $Re_\lambda = 287.6$.

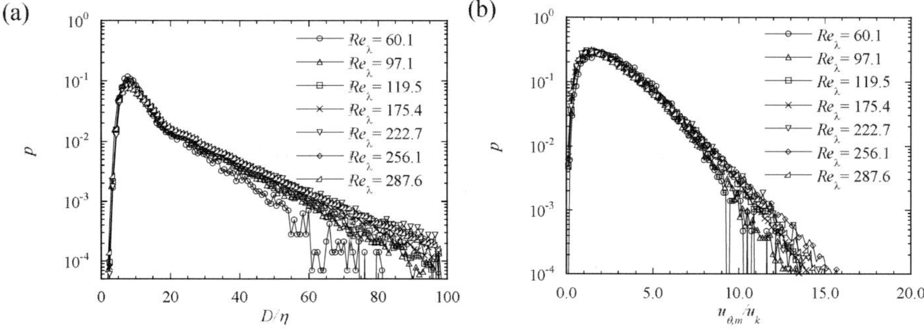

Fig. 2. Pdfs of diameter (a) and maximum azimuthal velocity (b) of the coherent fine scale eddy.

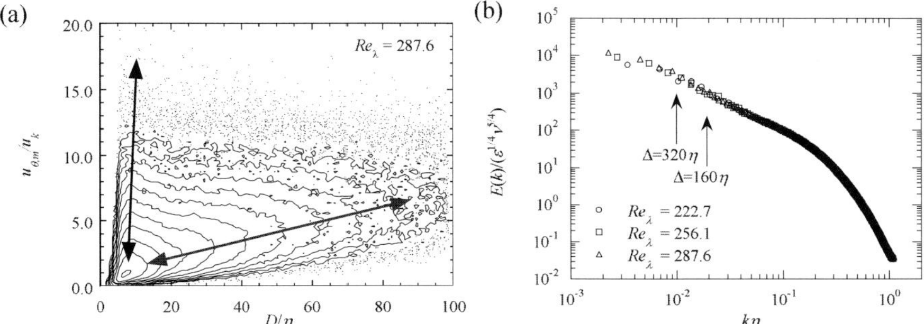

Fig. 3. Joint pdf of diameter and maximum azimuthal velocity (a) and energy spectrums with cutoff filters (b).

to create cluster of $O(l_E)$. Here, D represents eddy diameter defined from location with maximum value of mean azimuthal velocity and $u_{\theta,m}$ denotes the maximum of mean azimuthal velocity.

In Fig. 2, probability density functions (pdfs) of D and $u_{\theta,m}$ are shown with previous our results for lower Re_λ cases [6]. The diameter and the maximum azimuthal velocity are normalized by η and u_k. The most expected diameter of the coherent fine scale eddies is about 8η and is independent on Re_λ. Figure 3 (a) shows joint pdf of the diameter and the maximum azimuthal velocity for $Re_\lambda = 287.6$. Since the intervals of contour lines are selected to be $\log_2 p$ in Fig. 3(a), probability densities on neighboring two contour lines are different 2 times. It is clear that diameter and maximum azimuthal velocity of fine scale eddies shows a strong correlation. The diameter and maximum azimuthal velocity of the most expected fine scale eddies in homogeneous isotropic turbulence is $D \approx 8\eta$ and $u_{\theta,m} \approx u_k$, which is also independent on Re_λ (not shown here).

The interesting behaviors can be observed for $D/\eta \approx 10$. The coherent fine scale eddies with $D \approx 10\eta$ can posses a very large maximum azimuthal velocity. For the highest Re_λ case, the maximum azimuthal velocity is about $2.5u_{rms}$. There are very intermittent structures which produced large velocity difference of $5u_{rms}$ within the distance of 10 times Kolmogorov length. For larger eddies, $u_{\theta,m}$ tend to approach $1.0 \sim 1.5u_{rms}$, which coincides with a conventional idea for so-called 'large scale eddies' in turbulence because these eddies have large velocity and length scales. However, real feature of these large eddies have not been clarified. By applying a sharp cutoff filter or a bandpass filter to DNS data, Tanahashi et al. [6] have investigated vorticity and strain fields in the inertial subrange and showed that locations of high vorticity region is different for each scales and those of high strain rate coincide at the fine scale eddy cluster.

3 Fine Scale Eddy Cluster and Energy Transfer

To investigate energy transfer, velocity field is split into two components by applying a sharp cutoff filter with filter width Δ. Energy transfer between large and small scales divided by a cutoff filter can be written as follow;

$$E_\tau = -\tau_{ij}\overline{S}_{ij} = -L_{ij}\overline{S}_{ij} - C_{ij}\overline{S}_{ij} - R_{ij}\overline{S}_{ij}, \tag{1}$$

where an overbar represents a filtered variable which includes scales larger than Δ and $\tau_{ij} = L_{ij} + C_{ij} + R_{ij}$ are SGS stresses. L_{ij}, C_{ij} and R_{ij} are called as Leonard term, cross term and Reynolds term respectively and defined by;

$$L_{ij} = \overline{\overline{u}_i\overline{u}_j} - \overline{u}_i\overline{u}_j, \quad C_{ij} = \overline{u_i'\overline{u}_j - \overline{u}_i u_j'}, \quad R_{ij} = \overline{u_i' u_j'}. \tag{2}$$

E_τ is GS-SGS energy transfer in large eddy simulation (LES). The volume averaged energy transfer due to the Leonard term is zero for all filter widths and have no contribution to the net energy transfer. Leonard and Reynolds terms show negative values on the average and GS (large scale) energy is transferred to SGS (small scale) energy due to these terms. The magnitudes of these terms depend on the filter width and have investigated by Tanahashi et al. [6]. The cross term dominates the GS-SGS energy transfer in the case of small filter width ($\Delta \approx 20\eta$), whereas energy transfer due to Reynolds term becomes significant for larger filter width. These two terms crossover at about $\Delta \approx 30\eta$. In this study, filter width of the cutoff filter is selected to be in the inertial subrange as shown in Fig. 3 (b).

Figure 4 (a) shows distribution of axis of fine scale eddies smaller than 20η in a box with $3l_E \times 3l_E \times 0.25l_E$ with GS-SGS energy transfer on a cross section near fine scale eddy clusters for $Re_\lambda = 287.6$. In this figure, dark and light coloring of axis represents second invariant on the axis. Light color denotes strong eddy and dark color does weak one. Here, $\Delta = 320\eta$ is $O(l_E)$ and darker regions represent stronger forward scatter. The cluster of the intense coherent fine scale eddies exist in the region with large forward scatter, which may suggest that energy in large scales is transferred directly to small scales due to the large scale strain field [6]. In Fig. 4(b), medium scale eddies larger than

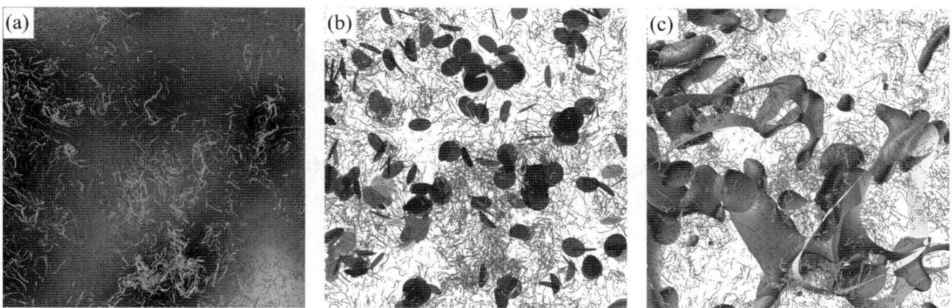

Fig. 4. Distribution of axis of coherent fine scale eddy with GS-SGS energy transfer (a), with medium scale eddies (b) and with strain rate of different scales (c).

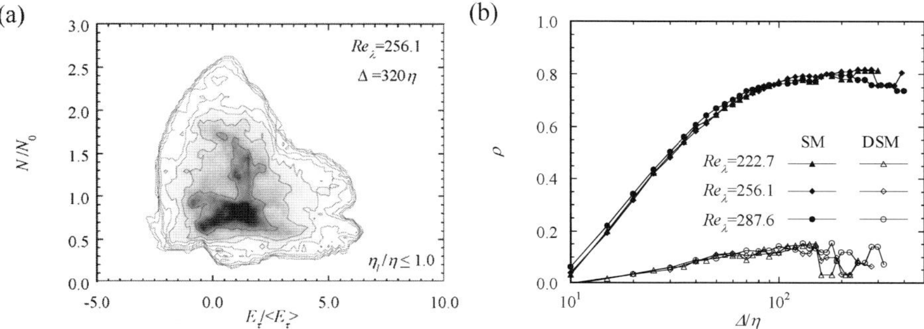

Fig. 5. Joint pdf of number density of medium scale eddy and GS-SGS energy transfer (a) and correlation coefficient between DNS and SGS models (b).

80η are shown by colored disks with axis of fine scale eddies. Disk diameter and color represents eddy size and eddy strength, respectively. The medium scale eddy tends to exist around the cluster of the fine scale eddy and the energy transfer is relatively low near the medium scale eddy.

In Fig. 5 (a), joint pdf of number density of the medium scale eddy and energy transfer which are averaged in each box with l_E^3. The number density of the medium scale eddy is normalized by an average in the whole flow field (N_0). The joint pdf is conditioned with the local Kolmogorov length (η_l) which is defined from mean energy dissipation rate (ε_l) in a l_E^3 box and is shown only for $\eta_l < \eta$. Note that the l_E^3 box with smaller η_l (or larger ε_l) coincides with the region with the fine scale eddy cluster. The energy transfer shows large values in the regions with low number density of the medium scale eddy, whereas that in the regions with lots of the medium eddy is fluctuating around the mean energy transfer ($< E_\tau >$). These results imply that the medium scale eddy may not dominate the energy transfer between large and small scales. Figure 4 (c) shows contour surfaces of strain rate($=\overline{S}_{ij}\overline{S}_{ij}$) of different scales ($\Delta/\eta = 80$(dark gray), 160(gray) and 320(light gray)). Spatial distribution of the strain rate strongly depends on the scale and do not coincide each other except for the regions with the fine scale eddy cluster.

These structures in the inertial subrange are very important for SGS modeling in LES. In Fig. 5 (b), the results of *a priori* tests of Smagorinsky model and dynamic model [7, 8] are shown for different filter widths. Smagorinsky model well correlate with the exact energy transfer for $\Delta > 80\eta$, whereas dynamic model shows very low correlation even in the inertial subrange. These results show that an assumption of M_{ij} in the dynamic model is not always correct.

4 Summary

In this study, relations between fine scale eddy cluster and energy cascade were investigated by using DNS data. The cluster of the intense coherent fine scale eddy exists in the region with large forward scatter, whereas the medium scale eddy tends to exist in the region with relatively low energy transfer around the fine scale eddy cluster. Spatial distribution of the strain rate strongly depends on the scale and is related with fine scale eddy cluster.

The GS-SGS energy transfer by Smagorinsky model shows strong correlation with exact DNS results because strain rate in large scales shows high values at the cluster where the energy transfer from GS to SGS is high. However, energy transfer described by dynamic model does not show good correlation with DNS data even in the inertial subrange.

References

1. Jiménez J, Wray AA, Saffman PG, Rogallo RS (1993) J Fluid Mech 255:65–90
2. Tanahashi M, Miyauchi T, Ikeda J (1999) Identification of Coherent Fine Scale Structure in Turbulence. In: Simulation and Identification of Organized Structures in Flows. Kluwer Academic Publishers 131–140
3. Tanahashi M, Iwase S, Miyauchi T (2001) J Turbul 2:6
4. Tanahashi M, Kang SJ, Miyamoto T, Shiokawa S, Miyauchi T (2004) Int J Heat and Fluid Flow 25:331–340
5. Del Álamo J C, Jimenez J, Zandonade P, Moser R D (2006) J Fluid Mech 561:329–358
6. Tanahashi M, Tsukamoto S, Iwase S, Miyauchi T (2002) Coherent Fine Scale Eddies and Energy Cascade in Homogeneous Isotropic Turbulence, In: Proc. Int. Symp. Dynamics and Statistics of Coherent Structures in Turbulence: Roles in Elementary Vortices, ed. Kida S, 259–268
7. Germano M, Piomelli U, Moin P, Cabot WH (1991) Phys Fluids A3:1760–1765
8. Lilly DK (1992) Phys Fluids A4: 633–635

Acceleration Statistics of Inertial Particles from High Resolution DNS Turbulence

Federico Toschi[1], Jeremie Bec[2], Luca Biferale[3], Guido Boffetta[4], Antonio Celani[5], Massimo Cencini[6], Alessandra S. Lanotte[7], and Stefano Musacchio[8]

[1] CNR-IAC, Viale del Policlinico 137, I-00161 Roma, Italy and INFN, Sezione di Ferrara, via G. Saragat 1, I-44100, Ferrara, Italy. `toschi@iac.cnr.it`
[2] CNRS Observatoire de la Côte d'Azur, B.P. 4229, 06304 Nice Cedex 4, France
[3] Dept. of Physics and INFN, University of Rome "Tor Vergata", Via della Ricerca Scientifica 1, 00133 Roma, Italy
[4] Dept. of Physics and INFN, University of Torino, Via Pietro Giuria 1, 10125, Torino, Italy
[5] CNRS, INLN, 1361 Route des Lucioles, F-06560 Valbonne, France
[6] SMC-INFM c/o Dept. of Physics University of Rome "La Sapienza", Piazz.le A. Moro, 2, I-00185 Roma, Italy, and CNR-ISC via dei Taurini, 19 I-00185 Roma, Italy
[7] CNR-ISAC and INFN, Sezione di Lecce, Str. Prov. Lecce-Monteroni km.1200, I-73100 Lecce, Italy
[8] Dept. of Physics and INFN, University of Torino, Via Pietro Giuria 1, 10125, Torino, Italy

Abstract. We present results from recent direct numerical simulations of heavy particle transport in homogeneous, isotropic, fully developed turbulence, with grid resolution up to 512^3 and $R_\lambda \approx 185$. By following the trajectories of millions of particles with different Stokes numbers, $St \in [0.16 : 3.5]$, we are able to characterize in full detail the statistics of particle acceleration. We focus on the probability density function of the normalised acceleration $a/a_{\rm rms}$ and on the behaviour of their root-mean-squared acceleration $a_{\rm rms}$ as a function of both St and R_λ. We explain our findings in terms of two concurrent mechanisms: particle clustering, very effective for small St, and filtering induced by finite particle response time, taking over at larger St.

Keywords: Lagrangian turbulence, heavy particles, Stokes particles, acceleration statistics

1 Introduction

Small impurities, as for example dust, droplets or bubbles, suspended in an incompressible flow are finite-size particles whose density, in general, differs from that of the underlying advecting fluid. Hence such particles cannot be modeled as point-like tracers. The description of their motion must account for

73

Y. Kaneda (ed.), *IUTAM Symposium on Computational Physics and New Perspectives in Turbulence*, 73–78.

inertia whence the name inertial particles. Ultimately such inertial particles concentrate onto nonuniform sets that evolve together with the fluid motion and display strong spatial inhomogeneity, often called preferential concentration (see [1] in this same volume). The role of inertia is also reflected into a slow response of these particles to changes in the fluid velocity. The combined effects of non-uniform sampling of the physical space (preferential concentration) and of the finite response time of the particles explain the observed behaviour of particle acceleration.

2 Heavy Particle Dynamics and Numerical Simulations

The equations of motion of a small, rigid, spherical particle immersed in an incompressible flow have been derived from first principles in [2]. For particles much heavier than the surrounding fluid, these equations take the form

$$\frac{d\boldsymbol{X}}{dt} = \boldsymbol{V}(t)\,, \qquad \frac{d\boldsymbol{V}}{dt} = -\frac{\boldsymbol{V}(t) - \boldsymbol{u}(\boldsymbol{X}(t), t)}{\tau_s}\,. \tag{1}$$

Here, $\boldsymbol{X}(t)$ denotes the particle trajectory, $\boldsymbol{V}(t)$ its velocity, $\boldsymbol{u}(\boldsymbol{x}, t)$ is the fluid velocity. The Stokes response time is $\tau_s = 2\rho_p a^2/(9\rho_f \nu)$ where a is the particle radius ρ_p and ρ_f are the particle and fluid density, respectively, and ν is the fluid kinematics viscosity. The Stokes number is defined as $St = \tau_s/\tau_\eta$ ($\tau_\eta = (\nu/\epsilon)^{1/2}$ being the Kolmogorov timescale and ϵ the average rate of energy injection). Equation (1) holds for very dilute suspensions, where particle-particle interactions (collisions) and hydrodynamic coupling can be neglected. The fluid evolves according to the incompressible Navier-Stokes equations:

$$D_t\boldsymbol{u} = (\partial_t\boldsymbol{u} + \boldsymbol{u}\cdot\boldsymbol{\nabla}\boldsymbol{u}) = -\frac{1}{\rho_f}\boldsymbol{\nabla}p + \nu\Delta\boldsymbol{u} + \boldsymbol{f}\,, \quad \boldsymbol{\nabla}\cdot\boldsymbol{u} = 0\,, \tag{2}$$

where p is the pressure field and $\boldsymbol{f}$ is the external energy source, $\langle \boldsymbol{f}\cdot\boldsymbol{u}\rangle = \epsilon$.

The Navier-Stokes equations are solved on a (three-periodic) cubic grid of size N^3. Forcing is realized by keeping constant the spectral content of the two smallest wavenumber shells [3]. Viscosity is chosen by requiring a Kolmogorov lengthscale $\eta \approx \Delta x$ (Δx being the grid spacing): this choice ensures that the small-scale velocity dynamics is well resolved. We used a fully dealiased pseudospectral algorithm with $2nd$ order Adam-Bashforth time-stepping.

Point particles are seeded homogeneously and with velocities equal to the local fluid velocity of a thermalized configuration. After a transient of about half large scale eddy turn over time the Lagrangian dynamics becomes stationary, and measurements are performed. We follow 15 sets of inertial particles with Stokes numbers in the range [0.16 : 3.5] and for each set, we store at high-frequency the position, the velocity of the particles, the velocity of the carrier fluid. Fluid tracers ($St = 0$), evolving as $d\boldsymbol{x}(t)/dt = \boldsymbol{u}(\boldsymbol{x}(t), t)\,$, are also followed for comparison. A summary of relevant physical parameters is given in Table 1.

Table 1. Parameters of our DNS. Microscale Reynolds number R_λ, root-mean-square velocity u_{rms}, energy dissipation ε, viscosity ν, Kolmogorov lengthscale $\eta = (\nu^3/\varepsilon)^{1/4}$, integral scale L, large-eddy Eulerian turnover time $T_E = L/u_{\mathrm{rms}}$, Kolmogorov timescale τ_η, total integration time T_{tot}, duration of the transient regime T_{tr}, grid spacing Δx, resolution N^3, number of trajectories of inertial particles for each Stokes N_t dumped at frequency $\tau_\eta/10$, number of particles N_p per Stokes dumped at frequency $10\tau_\eta$, total number of advected particles N_{tot}. Typical errors on all statistically fluctuating quantities are of the order of 10%.

R_λ	u_{rms}	ε	ν	η	L	T_E	τ_η	T_{tot}	T_{tr}	Δx	N^3	N_t	N_p	N_{tot}
185	1.4	0.94	0.00205	0.010	π	2.2	0.047	14	4	0.012	512^3	$5 \cdot 10^5$	$7.5 \cdot 10^6$	$12 \cdot 10^7$
105	1.4	0.93	0.00502	0.020	π	2.2	0.073	20	4	0.024	256^3	$2.5 \cdot 10^5$	$2 \cdot 10^6$	$32 \cdot 10^6$
65	1.4	0.85	0.01	0.034	π	2.2	0.110	29	6	0.048	128^3	$3.1 \cdot 10^4$	$2.5 \cdot 10^5$	$4 \cdot 10^6$

3 Results

Here we briefly recall recent results on the statistical properties of acceleration [7]. The main signature of inertia on acceleration can be appreciated from Fig. 1 (left), where we show the normalized root mean squared acceleration, $a_{rms} = \sqrt{\langle|\boldsymbol{a}|^2\rangle/3}$. As the Stokes number increases a sharp decrease of a_{rms} can be observed (particularly for small St values). At changing the Reynolds number, we find an overall dependence very similar to that observed for tracers [4]. Let us now understand how this fall off is generated. In Fig. 1 (right) we plot the root mean square fluid acceleration conditioned on the particle positions, $D_t\boldsymbol{u}(\boldsymbol{X}(t), t)$. For small Stokes numbers such quantities collapse onto the curve of the particle acceleration. This tells us that the fast fall off of acceleration for small St can be fully explained in terms of the "preferential concentration" of particles: particles do not sample uniformly the space. Indeed they tend to be ejected from vortex filaments, which are characterized by high acceleration [5].

However, at larger Stokes number the conditional acceleration tends to recover the value it has in the case of tracers, meaning that as St increases - in the range here explored-, an homogeneous sampling of the small scale flow structures is recovered (see also [1]). Therefore preferential concentration is not the only mechanisms influencing particle acceleration statistics. For this we note that the particle equation of motion (1) can be formally solved as $\boldsymbol{V}(t) = \int_{-\infty}^{t} e^{-(t-s)/\tau_s}\boldsymbol{u}(\boldsymbol{X}(s), s)\,ds$ which tells us that inertia plays a role similar to that of an exponential low-pass filter for the fluid velocity. Of course this is not exactly true because particle trajectories are different from those of tracers. We can define a filtered fluid velocity along tracers trajectories, i.e. $\boldsymbol{u}^F(t) = \int_{-\infty}^{t} e^{-(t-s)/\tau_s}\boldsymbol{u}(\boldsymbol{x}(s), s)\,ds$ and compute the acceleration $\boldsymbol{a}^F = \frac{d}{dt}\boldsymbol{u}^F$. By averaging $|\boldsymbol{a}^F|^2$ along the tracer trajectories we eliminate any effect of preferential concentration and focus on the effect of filtering. As one can see from 1 (right), for large St, one has a fairly good agreement between the a_{rms} of the inertial particles and the root mean squared acceleration computed on

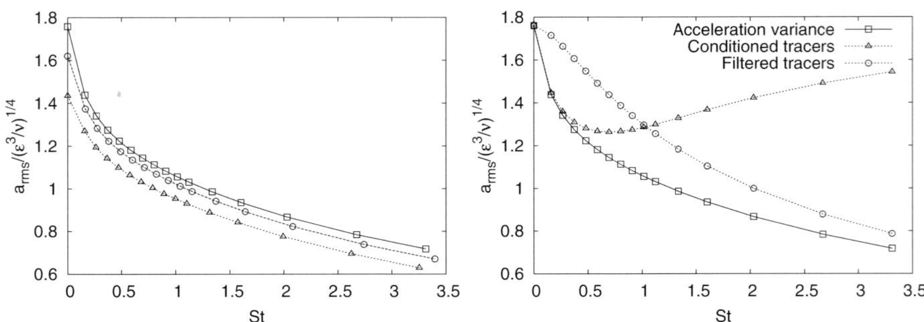

Fig. 1. (left) Normalised acceleration variance $a_{\mathrm{rms}}/(\epsilon^3/\nu)^{1/4}$ vs. St at varying the Reynolds number: $R_\lambda = 185$ (□); $R_\lambda = 105$ (○); $R_\lambda = 65$ (△). (right) Acceleration variance, a_{rms} (□), vs. St, compared with the fluid acceleration at particle position, $\langle (D\boldsymbol{u}/Dt(\boldsymbol{X}(t),t)^2\rangle^{1/2}$ (△) and with the root mean square acceleration of filtered tracer trajectories, a_{rms}^F (○). Data in this figure refer to $R_\lambda = 185$.

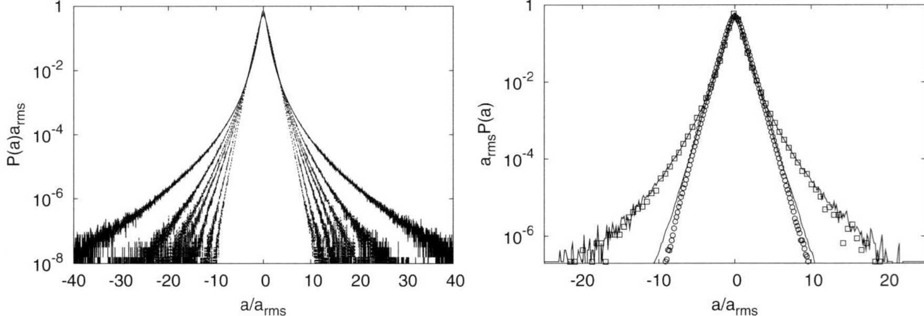

Fig. 2. (left panel) Acceleration probability distribution function for a subset of Stokes values ($St = 0, 0.16, 0.37, 0.58, 1.01, 2.03, 3.31$ from top to bottom, respectively) and for the case $R_\lambda = 185$. (right panel) The two more external curves correspond to the acceleration pdf for $St = 0.16$ (□) and the pdf of the fluid tracers acceleration measured at the same position of the inertial particles, $\frac{D\boldsymbol{u}}{Dt}$ (solid line). The two inner curves are the acceleration pdf for the highest Stokes number, $St = 3.31$, (○) and the pdf of the filtered fluid acceleration (solid line). All curves are normalised to unit variance.

filtered (over a time window τ) tracer trajectory. This shows that, for those values of St, the main role of inertia is to act as a filter for the high-frequency (intense) fluctuations of the fluid acceleration.

In Fig. 2 we show the acceleration probability density functions for some particular St values. It is evident that the effect of increasing St does produce a tendency for the acceleration pdfs to have lower and lower tail. Thanks to the previous observations, one is tempted to interpret this steepening of the pdf's tails as due to the combined effect of preferential concentration and filtering. Figure 2 demonstrates that this is indeed correct. Here we compare the normalised pdf of the particle acceleration with that of the fluid tracers conditioned to be at particle positions (for a small St value) and the filtered tracer acceleration for a large St value. As one can see the agreement between the entire functional form of the acceleration pdf is striking and confirms the relevance of preferential concentration or filtering in the two opposite limiting case of, respectively, small or large St value. Besides the phenomenological picture for the behaviour of the acceleration, one may be interested in assessing how much our simplified model, represented by (1), do indeed describe physical reality. For the sake of this comparison we refer to recent experiments reported in [6]. A first observation is that in experiments it is impossible to deal with an exactly monodispersed phase, i.e. the radii of the particles (droplets) is distributed around some mean value. As a consequence this implies that the particles will not have the same Stokes number but a distribution peaked around a mean value. However experiments [6] were so accurate that the variance of the particle size distribution was precisely controlled and the particles could almost be regarded as monodisperse. Indeed we do find an excellent agreement between experimental and numerical pdf for the acceleration (see Fig. 3) at comparable Stokes numbers.

As a side note, the huge database that we have collected, in principle, allows us to make predictions also for polydispersed solution. If the solution is polydispersed one would measured, e.g. for the acceleration, the convolution of the acceleration measured for a give Stokes with the relative probability to find such a Stokes particle. Analysis in this direction has beet attempted in [8].

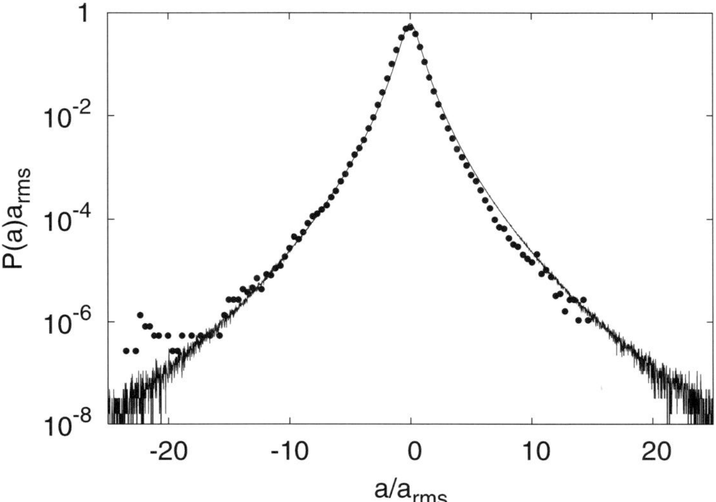

Fig. 3. Comparison between experimental acceleration pdf (•) at $\langle St \rangle = 0.09 \pm 0.03$ from [6] and numerical data (solid line) at $St = 0.16$ from [7, 8].

4 Conclusions

We presented results on the statistical properties of acceleration for inertial particles in turbulent flows. We have shown that for small St values the predominant effect can be associated to a tiny change in the space distribution of particles (not anymore uniform) which, however, does produce a remarkable effect on both the variance and the tails of the acceleration pdfs. At large St values filtering effect are instead capable to explain the asymptotic behaviour of the acceleration. Present results do compare nicely (see Fig. 3) with recent state-of-the-art experimental measurements [6]. In the future it will be interesting to assess the relevance of finite density effect or of the back-reaction of the particle on the fluid itself.

Acknowledgements

We wish to thank Zellman Warhaft for sharing the experimental data. This work has been partially supported by the EU under contract HPRN-CT-2002-00300 and the Galileo program on particle suspensions. DNS were performed at CINECA (Italy) and IDRIS (France) under the HPC-Europa programme (RII3-CT-2003-506079). The unprocessed data of this study are freely available from http://cfd.cineca.it.

References

1. Bec J, Biferale L, Cencini M, Lanotte AS, Musacchio S, Toschi F (2006) Clusters and voids of heavy particles in turbulent flows. In this same volume.
2. Maxey MR, Riley J (1983) Phys Fluids 26:883–889
3. Chen S, Doolen GD, Kraichnan RH, She ZS (1993) Phys Fluids A 5:458–463
4. Sawford BL, Yeung PK, Borgas MS, Vedula P, La Porta A, Crawford AM, Bodenschatz E (2003) Phys Fluids 15: 3478–3489
5. La Porta A, Voth GA, Crawford AM, Alexander J, Bodenschatz E (2001) Nature 409:1017
6. Ayyalasomayajula S, Gylfason A, Collins LR, Bodenschatz E, Warhaft Z (2006) Phys Rev Lett 97:144507
7. Bec J, Biferale L, Boffetta G, Celani A, Cencini M, Lanotte A, Musacchio S, Toschi F (2006) J Fluid Mech 550:349–358
8. Cencini M, Bec J, Biferale L, Boffetta G, Celani A, Lanotte A, Musacchio S, Toschi F (2006) J Turbul 7:1–16

Heavy Particle Clustering in Turbulent Flows

Jérémie Bec[1], Luca Biferale[2], Massimo Cencini[3], Alessandra Lanotte[4], Stefano Musacchio[5], and Federico Toschi[6]

[1] CNRS UMR6202, Observatoire de la Côte d'Azur, BP4229, 06304 Nice Cedex 4, France. `jeremie.bec@obs-nice.fr`
[2] Dept. of Physics and INFN, University of Rome "Tor Vergata", Via della Ricerca Scientifica 1, 00133 Roma, Italy.
[3] INFM-CNR, SMC Dept. of Physics, University of Rome "La Sapienza", Piazzale A. Moro 2, 00185 Roma, Italy.
[4] CNR-ISAC and INFN, Sezione di Lecce, Str. Prov. Lecce-Monteroni, 73100 Lecce, Italy.
[5] INLN, 1361 Route des Lucioles, 06560 Valbonne, France.
[6] CNR-IAC, Viale del Policlinico 137, 00161 Roma, Italy, and INFN, Sezione di Ferrara, Via G. Saragat 1, 44100 Ferrara, Italy.

Abstract. Distributions of heavy particles suspended in incompressible turbulent flows are investigated by means of high-resolution direct numerical simulations. It is shown that particles form fractal clusters in the dissipative range, with properties independent of the Reynolds number. Conversely, in the inertial range, the particle distribution is not scale-invariant. It is however shown that deviations from uniformity depends only on a rescaled contraction rate, and not on the local Stokes number given by dimensional analysis. Particle distribution is characterized by voids spanning all scales of the turbulent flow; their signature on the coarse-grained mass probability distribution is an algebraic behavior at small densities.

Keywords: particle-laden flows, preferential concentration, inertial particles, turbulent transport

1 Introduction

Spatial disributions of finite-size heavy impurities suspended in incompressible flows is a crucial issue in engineering [1], planetology [2] and cloud physics [3]. Such particles possess inertia, and generally distribute in a strongly inhomogeneous manner. The common understanding of such *preferential concentrations* relies on the idea that, in a turbulent flow, vortices act as centrifuges ejecting such heavy particles [4]. This picture was successfully used to describe the small-scale particle distribution and, in particular, to show that it depends only on the Stokes number $\mathcal{S}_\eta = \tau/\tau_\eta$ which is obtained by non-dimensionalizing the particle response time τ with the characteristic time τ_η of the small turbulent eddies.

79

Y. Kaneda (ed.), *IUTAM Symposium on Computational Physics and New Perspectives in Turbulence*, 79–84.
© 2008 *Springer*.

We confirm here that such a description is relevant at length scales which are smaller than the dissipative scale η of the fluid turbulent flow. In particular, maximal clustering is found for Stokes numbers of the order of unity. However, we show that particle concentration experiences also very strong fluctuations at scales within the inertial range of turbulence. In analogy with small-scale clustering, it is expected that for $r \gg \eta$ the relevant parameter is the local Stokes number $\mathcal{S}_r = \tau/\tau_r$, where τ_r is the characteristic time of eddies of size r [5]. Surprisingly, we present evidences which are reported with more details in [6], that such a dimensional argument does not apply to describe the organization of particles in the inertial range of turbulence.

2 Model and DNS

In very dilute suspensions, small particles which are much heavier than the fluid evolve according to the Newton equation [7]

$$\tau\ddot{\boldsymbol{X}} = \boldsymbol{u}(\boldsymbol{X}, t) - \dot{\boldsymbol{X}}\,, \tag{1}$$

where buoyancy is neglected. The response time τ is proportional to the square of the particles size and to their density contrast with the fluid. The fluid velocity $\boldsymbol{u}$ is a given solution to the incompressible Navier-Stokes equation obtained numerically by direct pseudo-spectral simulations on cubic grids of size $\mathcal{L} = 2\pi$ with 128^3, 256^3 and 512^3 collocation points corresponding to Taylor micro-scale Reynolds numbers $Re_\lambda \approx 65$, 105 and 185, respectively. The fluid flow is forced by keeping constant the energy content in the two first shells in Fourier space. Particles ($N = 7.5$ millions for each set of 15 different Stokes number in the range $\mathcal{S}_\eta \in [0.16 : 3.5]$), initially seeded homogeneously in space with velocities equal to the local fluid velocity, are evolved with (1) for about two large-scale eddy turnover times. After this time, a statistical steady state is reached and measurements are performed. Details on the numerics are reported in [8].

3 Small-Scale Clustering

Below the Kolmogorov scale η, the velocity field is differentiable and the motion of particles is governed by the fluid strain. Their dissipative dynamics leads their trajectories to converge to a dynamically evolving attractor, so that their mass distribution is singular and generically scale-invariant with fractal properties at small scales. In order to characterize such particle clusters we measured the correlation dimension $\mathcal{D}_2$, which is estimated through the small-scale behavior of the probability to find two particles at a distance less than a given r: $P_2(r) \sim r^{\mathcal{D}_2}$. The dependence of $\mathcal{D}_2$ on $\mathcal{S}_\eta$ and Re_λ is shown in Fig. 1 (Left). We first notice that $\mathcal{D}_2$ depends very weakly on Re_λ, at least in the range of Reynolds numbers explored here. This observation agrees with a recent numerical study [9], where particle clustering was equivalently characterized in terms of the radial distribution function. This confirms

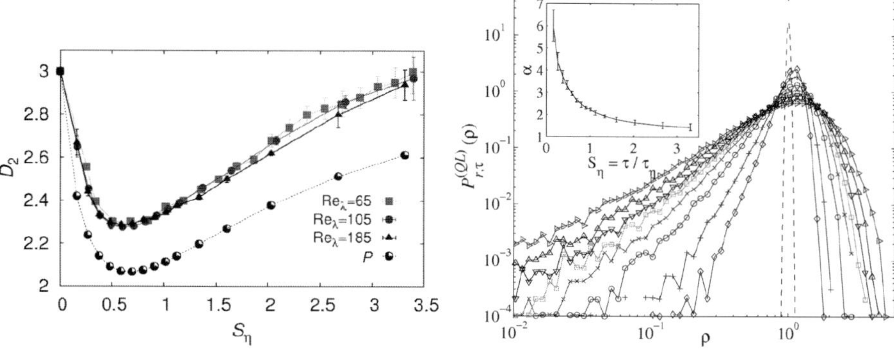

Fig. 1. Left: Correlation dimension $\mathcal{D}_2$ vs $\mathcal{S}_\eta$ for three different Re_λ, and probability $\mathcal{P}$ to find particles in non-hyperbolic (rotating) regions of the flow, shown for $Re_\lambda \approx 185$ (multiplied by an arbitrary factor for plotting purposes). Right: Quasi-Lagrangian PDF of the coarse-grained mass density ρ_r in log-log scale for $\mathcal{S}_\eta = 0.27$, 0.37, 0.58, 0.80, 1.0, 1.33, 2.03, 3.31 (from bottom to top) at scale $r = 32$ grid points and for $Re_\lambda \approx 185$. The solid line represent the Poisson distribution. Inset: exponent α of the left algebraic tail vs $\mathcal{S}_\eta$.

that τ_η, which varies by more than a factor 2 between the smallest and the largest Reynolds number considered here, is the only relevant time scale to characterize clustering below η and in particular that the effects of the flow intermittency cannot be detected. For all values of Re_λ, a maximum of clustering (corresponding to a minimum of $\mathcal{D}_2$) is observed for $\mathcal{S}_\eta \approx 0.6$. Particle positions strongly correlate with the local structure of the fluid velocity field. This is evidenced in Fig. 1 where is plotted the probability $\mathcal{P}$ to find particles in non-hyperbolic regions of the flow, i.e. at those points where the strain matrix has two complex conjugate eigenvalues. This is consistent with the traditional view relating particle clustering to vortex ejection.

4 Inertial Range Clustering

Fluctuations in the particle spatial distribution are qualitatively observed to extend far inside the inertial range. To quantify this effect, we consider the probability density function (PDF) $P_{r,\tau}^{(QL)}(\rho)$ of the quasi-Lagrangian particle density coarse-grained on a scale r inside the inertial range. This quantity is obtained by looking at the mass distribution in balls centered on a given particle trajectory. This amounts to weighting each cell with the mass it contains; for statistically homogeneous distributions, the quasi-Lagrangian density is related to the Eulerian density by $\langle \rho_r^p \rangle_{QL} = \langle \rho_r^{p+1} \rangle_{Eul}$ (see [10] for more details and a precise definition of quasi-Lagrangian averages). For tracers, which are uniformly distributed, this PDF tends for infinite number of particles, $N \to \infty$, to a delta function centered at $\rho = 1$. For a large but finite N, it is given by the asymptotic behavior of the binomial distribution.

As seen from Fig. 1 (Right), strong deviations from a uniform distribution can clearly be measured at both moderately small response times and for length scales inside the inertial range. This indicates that concentration fluctuations are important not only at dissipative scales but also in the inertial range. A noticeable observation is that the low-density tail of the PDF (related to voids) displays an algebraic behavior $P_{r,\tau}^{(QL)}(\rho) \sim \rho^{\alpha(r,\tau)}$. The dependence of such an exponent for fixed r and varying the Stokes number $\mathcal{S}_\eta$ is shown in the inset of Fig. 1 (Right). For low inertia ($\mathcal{S}_\eta \to 0$) the exponent tends to infinity in order to recover the non-algebraic behavior of tracers. At larger Stokes numbers the exponent approaches $\alpha = 1$, indicating a non-zero probability for totally empty areas.

Fixing the response time τ and increasing the observation scale r reproduces the same qualitative picture as fixing r and decreasing τ. A uniform distribution is recovered in both limits $r \to \infty$ or $\tau \to 0$. These two limits are actually equivalent. At length-scales $r \gg \eta$ within the inertial range, the fluid velocity field is not smooth: according to Kolmogorov (K41) theory, velocity increments behave as $\delta_r u \sim (\varepsilon r)^{1/3}$. Standard turbulence theory suggests that the physics at scale r is associated to time scales of the order of the turnover time $\tau_r = r/\delta_r u \sim \varepsilon^{-1/3} r^{2/3}$. It is then clear that, for any finite particle response time τ, the local inertia measured by $\mathcal{S}_r = \tau/\tau_r$ becomes so small at sufficiently large scales that particles should behave as tracers and so distribute uniformly in space [5]. Deviations from uniformity for finite $\mathcal{S}_r$ are expected not to be scale-invariant [11]. In particular, phenomenology suggests that the particle distribution should depend only on the local Stokes number $\mathcal{S}_r$ as observed in random δ-correlated in time flows [12]. However this argument does not take into consideration some important aspects of realistic flows.

We first notice that (1) can be rewritten as $V = u - \tau A$, where $V = \dot{X}$ denotes the velocity of the particle and $A = \ddot{X}$ its acceleration. Far enough in the inertial range, we have $\mathcal{S}_r \ll 1$ and the particle acceleration can be approximated with the fluid acceleration: $A \approx a = \partial_t u + u \cdot \nabla u$. As a consequence, the particles evolve as if transported by a synthetic compressible flow [7, 11] whose divergence is $\nabla \cdot V = -\tau \nabla \cdot (u \cdot \nabla u) \approx \tau \nabla^2 p$. In this effective velocity field, the only time scale $\mathcal{T}_{r,\tau}$ relevant to the distribution of particles at scale r is given by the inverse of the contraction rate of a volume of size r, i.e.

$$1/\mathcal{T}_{r,\tau} = (1/r^3) \int_{[0,r]^3} \nabla \cdot V \, d^3x \approx (1/r^3) \int_{[0,r]^3} \tau \nabla^2 p \, d^3x \,. \qquad (2)$$

Figure 2 (Left) displays the power spectrum of the pressure $E_p(k)$, pressure gradients $E_{\nabla p}(k)$ and of the velocity field $E_u(k)$. As one can see in the range of wavenumbers where the K41 scaling is observed, i.e. where $E_u(k) \sim k^{-5/3}$, we find that $E_p(k) \sim k^{-5/3}$ and $E_{\nabla p}(k) \sim k^{1/3}$, suggesting that $\delta_r p \sim r^{1/3}$ and $\delta_r \nabla p \sim r^{-2/3}$, so that the scaling of pressure is there dominated by sweeping, i.e. $\delta_r \nabla p = \nabla \delta_r u^2 \sim U(\varepsilon r)^{1/3}/r$, where U is the root mean square velocity.

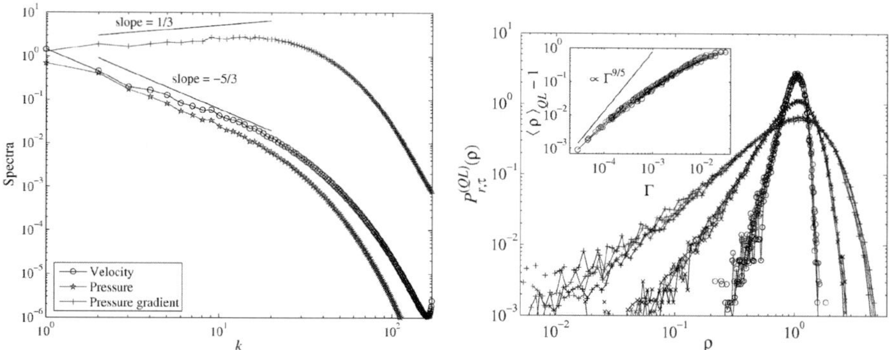

Fig. 2. Left: Pressure, pressure gradient and energy spectra vs the wavenumber k. The straight lines indicate the scaling $k^{-5/3}$ and $k^{1/3}$. Right: PDF of the coarse-grained mass in the inertial range for three values of the non-dimensional contraction rate Γ (different collapsing curves refer to different Stokes numbers and different scales). From bottom to top: $\Gamma = 4.8\,10^{-4}$, $\Gamma = 2.1\,10^{-3}$, and $\Gamma = 7.9\,10^{-3}$ (for $\mathcal{S}_\eta = 1.60,\ 2.03,\ 2.67,\ 3.31$). Inset: deviation from unity $\langle\rho\rangle_{QL} - 1$ of the first-order QL moment for scales r within the inertial range. For comparison, the behavior $\propto \Gamma^{9/5}$ obtained when assuming point clusters of particles is shown as a solid line. Both figures refer to $Re_\lambda \approx 185$.

Note that the scaling we observe is likely to be due to finite Reynolds number effects [13]. After one integration by part of (2), and plugging the above scaling, the integral can be dimensionally estimated as $1/\mathcal{T}_{r,\tau} \sim \tau U \varepsilon^{1/3} r^{-5/3}$. We thus expect that the PDF of the coarse-grained mass density is not a function of r and τ separately, but of the dimensionless contraction rate $\Gamma = \tau \lambda \varepsilon^{1/3} r^{-5/3}$, where we have adimensionalized $\mathcal{T}_{r,\tau}$ with the reference time λ/U.

Figure 2 (Right) shows $P_{r,\tau}^{(QL)}(\rho)$ for three choices of Γ obtained from different sets of values of r and τ. The different curves collapse, giving strong evidence in favor of the above argument. In particular, as represented in the inset of Fig. 2 (Right), the deviations from unity of the first moment of density collapse for all $\mathcal{S}_\eta$ investigated and for all scales inside the inertial range of our simulation. This quantity is the same as the Eulerian 2nd-order moment and gives the probability $P_2(r)$ to have two particles within a distance r (here inside the inertial range). The particle distribution recovers uniformity at large scales very slowly: much slower than if they were distributed as Poisson point-like clusters, for which $\langle\rho^2\rangle - 1 \propto r^{-3} \propto \Gamma^{9/5}$ (shown in the inset for comparison).

5 Summary and Conclusions

In conclusions we have shown that fluctuations of the concentration of heavy particles are important, not only in the dissipative range of turbulence where they can be described by using tools borrowed by dynamical systems theory, but also in the inertial range. Moreover, the PDF of the coarse-grained mass density displays a rescaling property in the inertial range and only depends on $\Gamma = \tau\lambda\varepsilon^{1/3}r^{-5/3}$, in relation with the scaling properties of the pressure field. The presence of intermittency corrections cannot be excluded for high-order statistics but they are not detectable by the present investigation. These findings may be important for developing models for water droplet growth and the scavenging of aerosol particles. Worth of further investigation is a characterization of the large fluctuations of the particle density field by means of statistical cluster analysis tools, such as minimum spanning trees or Minkovski functionals.

We benefited from useful discussions with G. Boffetta, A. Celani and A. Fouxon. This work has been partially supported by the EU under contract HPRN-CT-2002-00300 and the Galileo program on particle suspensions. DNS were performed at CINECA (Italy) and IDRIS (France) under the HPC-Europa programme (RII3-CT-2003-506079). The unprocessed data of this study are freely available from http://cfd.cineca.it.

References

1. Crowe C, Sommerfeld M, Tsuji Y (1998) Multiphase Flows with Particles and Droplets. CRC Press, New York
2. de Pater I, Lissauer J (2001) Planetary Science. Cambridge University Press, Cambridge
3. Pruppacher H, Klett J (1996) Microphysics of Clouds and Precipitation, Kluwer Academic Publishers, Dordrecht
4. Fessler J, Kulick J, Eaton J (1994) Phys Fluids 6:3742–3749
5. Falkovich G, Fouxon A, Stepanov M (2003) Statistics of turbulence-induced fluctuations of particle concentration. In Gyr A, Kinzelbach W (eds) Sedimentation and Sediment Transport. Kluwer Academic Publishers, Dordrecht
6. Bec J, Biferale L, Cencini M, Lanotte A, Musacchio S, Toschi F (2006) Preprint nlin.CD/0608045
7. Maxey M (1987) J Fluid Mech 174:441–465
8. Bec J, Biferale L, Boffetta G, Celani A, Cencini M, Lanotte A, Musacchio S, Toschi F (2006) J Fluid Mech 550:349–358
9. Collins L, Keswani A (2004) New J Phys 6:119
10. Bec J, Gawędzki K, Horvai P (2004) Phys Rev Lett 92:224501
11. Balkovsky E, Falkovich G, Fouxon A (2001) Phys Rev Lett 86:2790–2793
12. Bec J, Cencini M, Hillerbrand R (2006) Phys Rev E in press nlin.CD/0606038
13. Gotoh T, Fukayama D (2001) Phys Rev Lett 86:3775–3779

Reynolds Number Effects on the Turbulent Mixing of Passive Scalars

Jörg Schumacher

Department of Mechanical Engineering, Technische Universität Ilmenau, D-98684 Ilmenau, Germany
joerg.schumacher@tu-ilmenau.de

Abstract. The effects of the Reynolds number Re of the advecting turbulent flow on the small-scale properties of passive scalars fields are studied numerically for the case of Schmidt numbers Sc larger than unity. The deviations from local isotropy as quantified by odd-order derivative moments are found to decrease. The intermittency of the scalar gradient and dissipation rate fields is measured by the fourth order derivative moment and the multifractal formalism, respectively. Both analysis methods indicate an increasing small-scale intermittency of the mixing for increasing Re at fixed Sc.

Keywords: scalar mixing, high Schmidt numbers, direct numerical simulation, local isotropy, small-scale intermittency

1 Introduction

The mixing of passive scalars in turbulent flows is a process with numerous applications reaching from spreading of pollutants in the stratosphere and the transport of salinity in the ocean to non-premixed combustion of fuel. Frequently the Schmidt number, Sc, which is formed as the ratio of the kinematic viscosity ν of the flow to the diffusivity κ of the passive scalar, exceeds the unit value significantly. In this case, the viscous-convective Batchelor range is established below the Kolmogorov scale η of the flow (see e.g. [1]). In addition, mixing processes are often accompanied by an outer gradient, such as a mean temperature or concentration profile. Such a gradient is a large-scale source of anisotropy and it is therefore of interest to know how strongly deviations from local isotropy at the smallest scales of scalar turbulence depend on it. When the Schmidt number is increased while keeping the Reynolds number fixed the isotropy properties of mixing were found to be ever less affected by such mean gradient [2]. This was quantified by the magnitude of odd order scalar derivative moments [3].

But how is the statistics of scalar gradients at small scales varying when $Sc > 1$ remains unchanged and Re is increased? The present work aims at finding some answers to exactly this question. In contrast to the pure flow

85

Y. Kaneda (ed.), *IUTAM Symposium on Computational Physics and New Perspectives in Turbulence*, 85–90.

case where a variation of the large scale shear rate S will change the Reynolds number and consequently the degree of local isotropy, variations of the mean scalar gradient magnitude leave the isotropy deviations for the passive scalar unaffected. This is due to the linearity of the passive scalar dynamics. We might then conclude that if the physics of velocity field on sub-Kolmogorov scales is independent of the Reynolds number, the degree of isotropy of scalar mixing is depending on the Schmidt number only. As it will turn out this is not the case.

2 Numerical Simulations

Our study is based on direct pseudospectral simulations in a periodic box of sidelength $L = 2\pi$ at large resolutions of $N^3 = 512^3$ and 1024^3. We integrate the Navier-Stokes equations for an incompressible fluid $\mathbf{u}(\mathbf{x}, t)$ in combination with an advection-diffusion equation for the passive scalar field $\theta(\mathbf{x}, t)$. The flow is in a state of homogeneous isotropic turbulence generated by large-scale random forcing. The turbulent fluctuations of the passive scalar are sustained by a constant mean scalar gradient in y direction. The Schmidt numbers take values of $Sc = 2, 8, 32$. Much effort has been put into the sufficient spectral resolution here, which significantly exceeds the usually adopted ones. Recent works demonstrated that this is indeed necessary when extreme fluctuations of scalar derivatives are in the focus of the study [4, 5]. This limits the range of accessible Taylor microscale Reynolds numbers R_λ to values reaching from 10 to 64.

3 Statistics of Passive Scalar Gradients

Two statistical quantities of scalar gradient fields are discussed. We investigate the normalized derivative moments of orders 3, 4 and 5 first. The derivative is taken in the direction of the outer mean scalar gradient and the normalized moments are defined as

$$M_n(\partial\theta/\partial y) = \frac{\langle(\partial\theta/\partial y)^n\rangle}{\langle(\partial\theta/\partial y)^2\rangle^{n/2}} \tag{1}$$

The results for M_3, M_4 and M_5 are listed in Table 1. The odd order moments have a positive sign which reflects preferential spatial variations parallel to the outer mean gradient. For the third order moment, we can see a clear decrease of the moment magnitude indicating a more isotropic small-scale mixing for both cases, $Sc = 8$ and 32, respectively. As the Reynolds number of the advecting flow increases, the asymmetry of the probability density function (PDF) of $\partial\theta/\partial y$ decreases which indicates that large negative and positive gradients become more and more equiprobable. The trends for the fifth order moment are however less clear. Higher Schmidt number values would have been necessary in order to get a clear behaviour. Nevertheless, the data for M_n imply that high-Sc mixing becomes more isotropic with increasing Reynolds

Table 1. Values of the normalized scalar derivative moments. Listed are the orders 3, 4 and 5 for different Schmidt and Reynolds numbers.

R_λ	10	24	42	64
M_3 $(Sc = 8)$	1.57	1.27	-	0.77
M_3 $(Sc = 32)$	1.33	0.84	0.76	-
M_4 $(Sc = 8)$	10.1	11.0	-	13.8
M_4 $(Sc = 32)$	12.4	11.5	14.1	-
M_5 $(Sc = 8)$	44.7	49.0	-	43.7
M_5 $(Sc = 32)$	59.2	40.8	44.9	-

number. This is the same trend that has been found for increasing Schmidt number at fixed Reynolds number [2].

The fourth order moment measures the deviations of the gradient statistics from the Gaussian case. Values larger than 3 indicate intermittency. We can conclude that the small-scale intermittency of the mixing increases with growing Reynolds number as quantified by the growing magnitude of M_4. The latter aspect can be also supported by a statistical analysis of the scalar dissipation rate field which brings us to the second quantity of interest. The field is defined as

$$\epsilon_\theta(\mathbf{x}, t) = \kappa \sum_{i=1}^{3} \left(\frac{\partial \theta}{\partial x_i} \right)^2 , \tag{2}$$

with $x_1, x_2, x_3 = x, y, z$. In Ref. [5], we reported significant deviations of the probability density function $p(\epsilon_\theta)$ from log-normality for both tails, the very small and the very large amplitudes of the scalar dissipation rate, respectively. Figure 1 illustrates the spatial arrangement of the largest dissipation field amplitudes. Both snapshots display the isovolumes at the level of $7\langle\epsilon_\theta\rangle$. We observe that the maxima are found to be less connected and to fill the three-dimensional space more densely with increasing Reynolds number.

As can be seen in the figure, the maxima of the scalar dissipation rate appear in form of sheets and correspond to the potentially most intensive scalar mixing events. Their cross-section extension determines a locally varying diffusion scale of the mixing process and extends the classical Batchelor picture of one mean diffusion scale [6]. In [7], the distribution of the local diffusion scales is analysed for different Reynolds and Schmidt numbers with a fast multiscale image segmentation technique applied to two-dimensional cuts of the three-dimensional very high-resolution simulation data. The local dissipation scales take always values across the whole Batchelor range, i.e., between the Kolmogorov scale η and the Batchelor scale $\eta_B = \eta/\sqrt{Sc}$ and even beyond. The resulting distributions are shown in Fig. 2. These distributions can be traced back to the distribution of the contractive finite-time Lyapunov exponent of the flow.

Fig. 1. Scalar dissipation rate fields ϵ_θ for a Schmidt number $Sc = 8$. Both isovolume plots are at a level of $7\langle\epsilon_\theta\rangle$. Top: $R_\lambda = 24$. Bottom: $R_\lambda = 64$.

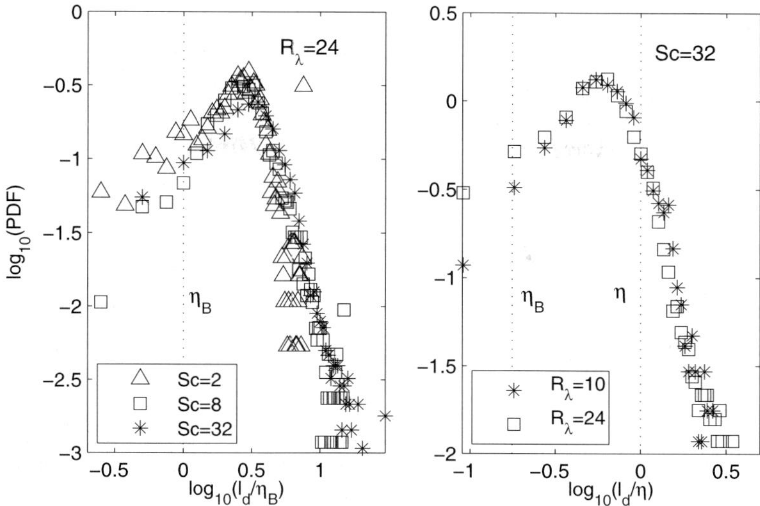

Fig. 2. Distribution of the local cross-section thickness l_d of the scalar dissipation rate filaments for level sets $\epsilon_\theta \geq 4\langle\epsilon_\theta\rangle$. Left panel: Probability density function (PDF) $p(l_d/\eta_B)$ for three different Schmidt numbers at $R_\lambda = 24$. Right panel: PDF $p(l_d/\eta)$ for two different Reynolds numbers at $Sc = 32$.

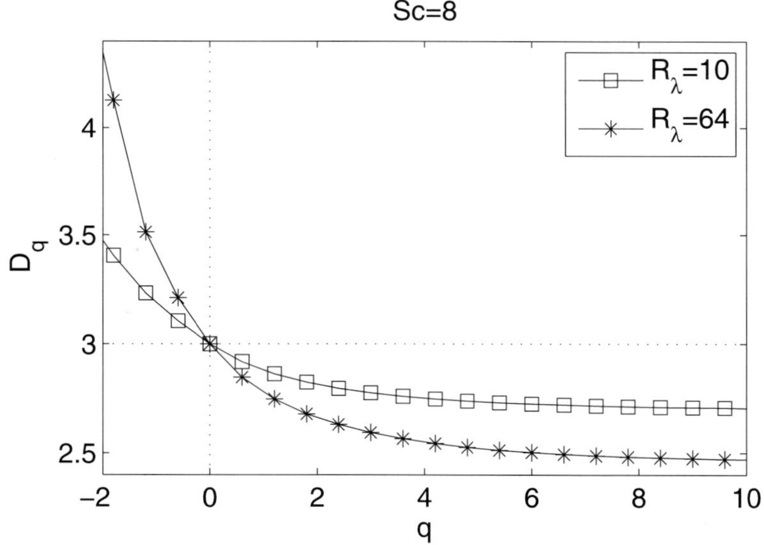

Fig. 3. Spectrum of generalized dimensions D_q as obtained from a multifractal analysis of the scalar dissipation rate field. Data for two different Reynolds numbers at $Sc = 8$ are shown. The box counting dimension D_0 has to be 3 for our analysis.

In order to further quantify this geometric feature we have conducted a multifractal analysis for the scalar dissipation field on scales between η_B and η. The scalar dissipation field is coarse grained over subvolumes of size r. Moments of order q of the resulting measure are taken where $-\infty \leq q \leq \infty$. The scaling of the moments with respect to coarse graining scale is found to be algebraic with an exponent D_q, the generalized dimension. In Fig. 3, we plot the spectrum of generalized dimensions for two Taylor microscale Reynolds numbers. Smaller D_q values for $q > 0$ indicate spatially rougher fields. This confirms the statistical findings for the scalar derivatives. The small-scale mixing becomes more intermittent with increasing R_λ (see also Fig. 1).

4 Summary

We have discussed some aspects of the Reynolds number dependence of the mixing of passive scalars in the viscous-convective range which is established for Schmidt number larger than unity. In summary, our numerical experiments indicate that the mixing becomes more isotropic and intermittent at the smallest turbulent scales when the Reynolds number is increased at fixed Schmidt number $Sc > 1$. Finally, we want to mention that the concept of local dissipation scales can be applied to the smallest scales of the advecting flow as well. Investigations on this subject and their impact on scalar mixing is part of future work.

The author wishes to thank A. Brandt, D. Kushnir, K.R. Sreenivasan, V. Yakhot and P.K. Yeung for the fruitful collaboration on this subject and topics which are related to it. Furthermore, discussions with J. Davoudi, C.R. Doering, B. Eckhardt and E. Villermaux are acknowledged. The numerical computations were done on the JUMP cluster of the John von Neumann Institute for Computing of the Forschungszentrum Jülich. The work was supported in parts by the Deutsche Forschungsgemeinschaft and the US National Science Foundation.

References

1. Davidson PA (2004) Turbulence: An Introduction for Scientists and Engineers. Oxford University Press, Oxford
2. Schumacher J, Sreenivasan KR (2003) Phys Rev Lett 91:174501
3. Warhaft Z (1999) Proc Nat Acad Sci 91:2481–2486
4. Schumacher J, Sreenivasan KR, Yeung PK (2005) J Fluid Mech 531:113–122
5. Schumacher J, Sreenivasan KR (2005) Phys Fluids 17:125107
6. Batchelor GK (1959) J Fluid Mech 5:113–133
7. Kushnir D, Schumacher J, Brandt A (2006) Phys Rev Lett 97:124502

Statistical Theory of Turbulence Based on Cross-Independence Closure Hypothesis

Tomomasa Tatsumi[1] and Takahiro Yoshimura[2]

[1] International Institute for Advanced Studies, 9-3 Kizugawadai, Kizu-cho, Kyoto 619-0225 Japan
`tatsumi@skyblue.ocn.ne.jp`
[2] 2744-7 Izumi-cho, Izumi-ku, Yokohama 245-0016, Japan

Abstract. The statistical theory of homogeneous isotropic turbulence based on the cross-independence closure hypothesis is outlined and discussed. The one-point velocity distribution is obtained as an inertial normal distribution N1 associated with the energy dissipation rate ϵ as the variance and no viscosity ν. The two-point velocity distributions, or the velocity-sum and velocity-difference distributions, are obtained for large and small distances r between the two points. For large r compared with Kolmogorov's length $\eta = (\nu^3/\epsilon)^{1/4}$, the both distributions are obtained as another inertial normal distribution N2 with half a variance of N1. In the local range comparable with the length η, the velocity-sum distribution and the lateral velocity-difference distribution are obtained as the normal distributions N3 and N4 in the local coordinate $r^* = r/\eta$ associated with the self energy-dissipation rates $\epsilon_+(r^*)$ and $\epsilon_-(r^*)$ as the variances respectively. The longitudinal velocity-difference distribution is obtained in three different forms in the local range: the local normal distribution N5, the inertial algebraic distribution A1 and the viscous algebraic distribution A2, in the order of decreasing distance r^*.

Keywords: cross-independence closure hypothesis, homogeneous isotropic turbulence, inertial normality, local similarity, self energy-dissipation

1 Introduction

In spite of great developments in turbulence research during the last half a century since Kolmogorov's epoch-making work [1], the theory of turbulence seems still lacking in the total consistency, statistical theories being mainly concerned with small components of turbulence, while large components mostly left to numerical analysis. The present approach intends to deal with large and small components of turbulence on equal footing by making use of the Lundgren-Monin equations for the velocity distributions [2, 3] together with the cross-independence closure hypothesis proposed by Tatsumi [4].

Y. Kaneda (ed.), *IUTAM Symposium on Computational Physics and New Perspectives in Turbulence*, 91–97.

2 Velocity Distributions

We consider turbulence in an incompressible viscous fluid which is governed by the Navier-Stokes equation of motion and the non-divergence condition,

$$\partial\mathbf{u}/\partial t + (\mathbf{u}\cdot\partial/\partial\mathbf{x})\mathbf{u} - \nu|\partial/\partial\mathbf{x}|^2\mathbf{u} = -(1/\rho)\partial p/\partial\mathbf{x}, \quad (\partial/\partial\mathbf{x})\cdot\mathbf{u} = 0, \quad (1)$$

for the velocity $\mathbf{u}(\mathbf{x},t)$ and the pressure $p(\mathbf{x},t)$, where ρ and ν denote the density and the kinetic viscosity of the fluid respectively.

If we take the velocities at two points $\mathbf{u}_1 = \mathbf{u}(\mathbf{x}_1,t)$ and $\mathbf{u}_2 = \mathbf{u}(\mathbf{x}_2,t)$, the one- and two-point velocity distributions are defined as $f(\mathbf{v}_1,\mathbf{x}_1,t) = \langle\delta(\mathbf{u}_1-\mathbf{v}_1)\rangle$ and $f^{(2)}(\mathbf{v}_1,\mathbf{v}_2;\mathbf{x}_1,\mathbf{x}_2;t) = \langle\delta(\mathbf{u}_1-\mathbf{v}_1)\delta(\mathbf{u}_2-\mathbf{v}_2)\rangle$, where $\mathbf{v}_1$ and $\mathbf{v}_2$ are the probability variables and the brackets $\langle\,\rangle$ denote the probability mean with respect to an initial distribution and δ the delta function. The higher distributions $f^{(n)}$, $n \geq 3$ can be defined accordingly.

The Lundgren-Monin equations for the velocity distributions are not closed since the equation for $f^{(n)}$ includes the higher distribution $f^{(n+1)}$, and in order to make the equations solvable we have to introduce a *closure hypothesis* for expressing $f^{(n+1)}$ in terms of $f^{(n)}$ and those of lower orders.

A commonly used hypothesis is the independent decomposition such that

$$f^{(2)}(\mathbf{v}_1,\mathbf{v}_2;\mathbf{x}_1,\mathbf{x}_2;t) = f(\mathbf{v}_1,\mathbf{x}_1,t)f(\mathbf{v}_2,\mathbf{x}_2,t). \quad (2)$$

Equation (2) is known to be valid at large distance $r = |\mathbf{x}_2 - \mathbf{x}_1|$ but not for small distance r.

On the other hand, if we take the sum and difference of the velocities $\mathbf{u}_1$ and $\mathbf{u}_2$ as $\mathbf{u}_+ = (1/2)(\mathbf{u}_1 + \mathbf{u}_2), \mathbf{u}_- = (1/2)(\mathbf{u}_2 - \mathbf{u}_1)$, and consider the one- and two-velocity distributions of the cross-velocities $(\mathbf{u}_+,\mathbf{u}_-)$ as

$$g_\pm(\mathbf{v}_\pm;\mathbf{x}_1,\mathbf{x}_2;t) = \langle\delta(\mathbf{u}_\pm - \mathbf{v}_\pm)\rangle,$$
$$g^{(2)}(\mathbf{v}_+,\mathbf{v}_-;\mathbf{x}_1,\mathbf{x}_2;t) = \langle\delta(\mathbf{u}_+ - \mathbf{v}_+)\delta(\mathbf{u}_- - \mathbf{v}_-)\rangle, \quad (3)$$
$$\mathbf{v}_+ = (1/2)(\mathbf{v}_1 + \mathbf{v}_2), \quad \mathbf{v}_- = (1/2)(\mathbf{v}_2 - \mathbf{v}_1), \quad (4)$$

we can define the *cross-independence decomposition* of the distribution $f^{(2)}$ by

$$f^{(2)}(\mathbf{v}_1,\mathbf{v}_2;\mathbf{x}_1,\mathbf{x}_2;t) = \{\partial(d\mathbf{v}_+,d\mathbf{v}_-)/\partial(d\mathbf{v}_1,d\mathbf{v}_2)\}g^{(2)}(\mathbf{v}_+,\mathbf{v}_-;\mathbf{x}_1,\mathbf{x}_2;t)$$
$$= 2^{-3}g^{(2)}(\mathbf{v}_+,\mathbf{v}_-;\mathbf{x}_1,\mathbf{x}_2;t), \quad (5)$$

$$g^{(2)}(\mathbf{v}_+,\mathbf{v}_-;\mathbf{x}_1,\mathbf{x}_2;t) = g_+(\mathbf{v}_+;\mathbf{x}_1,\mathbf{x}_2;t)g_-(\mathbf{v}_-;\mathbf{x}_1,\mathbf{x}_2;t). \quad (6)$$

The decomposition of (5) and (6) is shown to be valid at both large and small r.

A close similarity may be noted between this hypothesis and that of Kolmogorov's theory [1] which assumes independence between small eddies represented by $\Delta\mathbf{u} = \mathbf{u}_2 - \mathbf{u}_1 (= 2\mathbf{u}_-)$ and large eddies represented by $\mathbf{u}_1$ and $\mathbf{u}_2$. Apart from this similarity, turbulence is assumed to be steady and nearly homogeneous isotropic in Kolmogorov's theory while it is assumed here to be non-steady and strictly homogeneous isotropic.

3 Inertial Similarity of Velocity Distributions

In *homogeneous isotropic turbulence* which is governed by (1) with no external forcing, the *mean energy* of turbulence $E(t)$ and the *mean energy-dissipation rate* $\epsilon(t)$ are related with each other by the equation, $\mathrm{d}E(t)/\mathrm{d}t = -\epsilon(t)$.

If we assume the local equilibrium for homogeneous isotropic turbulence after Kolmogorov [1], the turbulent velocity field is statistically determined by the *energy-dissipation rate* ϵ and the *viscosity* ν. Then, the fundamental variables of the field are expressed in terms of ϵ and ν as Kolmogorov's length: $\eta = (\nu^3/\epsilon)^{1/4}$, time: $\tau = (\nu/\epsilon)^{1/2}$.

Then, the energy-dissipation rate $\epsilon(t)$ is expressed dimensionally as

$$[\epsilon(t)] = [\nu] \times [\{(\nu^3/\epsilon)^{1/4}\tau^{-1}/(\nu^3/\epsilon)^{1/4}\}^2] = [\nu] \times [\tau^{-2}], \tag{7}$$

where [ ] denote the dimension. Then it follows that

$$\epsilon(t) = \epsilon_0 t^{-2}, \quad E(t) = E_0 t^{-1}, \tag{8}$$

where $\epsilon_0 = E_0$ are constants of the dimension of ν. Thus, we obtain the *inverse-linear energy-decay law* of decaying turbulence from the local equilibrium hypothesis.

On substitution from (5) and (6) into the Lundgren-Monin equation for the *one-point velocity distribution* f, we obtain the following closed equation for f:

$$[\partial/\partial t + \alpha(t)|\partial/\partial\mathbf{v}|^2]f(\mathbf{v}, t) = 0, \tag{9}$$

$$\alpha(t) = (2/3)\nu \lim_{r \to 0} |\partial/\partial\mathbf{v}|^2 \int |\mathbf{v}_-|^2 g_-(\mathbf{v}_-, \mathbf{r}, t)\mathrm{d}\mathbf{v}_-, \tag{10}$$

with $\alpha(t) = \epsilon(t)/3$. Equation (9) is *inertial* in the sense that it is independent of the viscosity ν.

The self-similar solution of (9) in accordance with (8) is obtained as

$$f(\mathbf{v}, t) = f_0(\mathbf{v}, t) \equiv (t/4\pi\alpha_0)^{3/2} \exp[-|\mathbf{v}|^2 t/4\alpha_0], \quad \alpha(t) = \alpha_0 t^{-2}. \tag{11}$$

with $\alpha_0 = \epsilon_0/3$. Equation (11) gives an *inertial normal distribution*, say N1 for the distribution f.

The distribution N1 or (11) changes in time t, starting from a uniform distribution with zero probability density at $t = 0$, growing up as a normal distribution for $t > 0$, and tending to the delta distribution corresponding to the dead still state for $t \to \infty$. During this process, the energy $E(t)$ decreases as t^{-1} due to the energy-dissipation $\alpha(t) = \epsilon(t)/3$ which satisfies the "fluctuation-dissipation theorem" (10). (see [4] and [5])

The closed equation for the two-point velocity distribution $f^{(2)}$ is derived by substituting the relations similar to (5) and (6) into the Lundgren-Monin equation for $f^{(2)}$ as

$$[\partial/\partial t + (\mathbf{v}_2 - \mathbf{v}_1) \cdot \partial/\partial \mathbf{r} + \alpha(t)(|\partial/\partial \mathbf{v}_1|^2 + |\partial/\partial \mathbf{v}_2|^2)$$
$$- \{\partial/\partial \mathbf{v}_1 \cdot \partial/\partial \mathbf{x}_1 \beta_1(\mathbf{v}_1, t) + \partial/\partial \mathbf{v}_2 \cdot \partial/\partial \mathbf{x}_2 \beta_2(\mathbf{v}_2, t)\}] f^{(2)}(\mathbf{v}_1, \mathbf{v}_2; \mathbf{r}, t) = 0,$$
$$(12)$$

where the parameters $\beta_1(\mathbf{v}_1, t)$ and $\beta_2(\mathbf{v}_1, t)$ are defined by

$$\beta_1(\mathbf{v}_1, t) = \frac{1}{4\pi} \iint |\mathbf{r}'|^{-1}((\mathbf{v}_1 + 2\mathbf{v}'_-) \cdot \partial/\partial \mathbf{r}')^2 g_-(\mathbf{v}'_-, \mathbf{r}', t) \mathrm{d}\mathbf{v}'_- \mathrm{d}\mathbf{r}', \qquad (13)$$

$$\beta_2(\mathbf{v}_2, t) = \frac{1}{4\pi} \iint |\mathbf{r}''|^{-1}((\mathbf{v}_2 + 2\mathbf{v}''_-) \cdot \partial/\partial \mathbf{r}'')^2 g_-(\mathbf{v}''_-, \mathbf{r}'', t) \mathrm{d}\mathbf{v}''_- \mathrm{d}\mathbf{r}''. \qquad (14)$$

The distribution $f^{(2)}$ is conveniently expressed in terms of the distributions g_+ and g_- under the present closure. The closed equations for g_+ and g_- are obtained by integrating (12) substituted from (5) and (6) with respect to $\mathbf{v}_-$ and $\mathbf{v}_+$ respectively as follows:

$$[\partial/\partial t + (1/2)\alpha(t)|\partial/\partial \mathbf{v}_+|^2] g_+(\mathbf{v}_+, \mathbf{r}, t) = 0, \qquad (15)$$

$$[\partial/\partial t + (1/2)\alpha(t)|\partial/\partial \mathbf{v}_-|^2 + 2\mathbf{v}_- \cdot \partial/\partial \mathbf{r} + (1/2)\partial/\partial \mathbf{v}_- \cdot \{\partial/\partial \mathbf{x}_1 \beta_1(\mathbf{v}_-, t)$$
$$- \partial/\partial \mathbf{v}_2 \cdot \partial/\partial \mathbf{x}_2 \beta_2(\mathbf{v}_-, t)\}] g_-(\mathbf{v}_-, \mathbf{r}, t) = 0. \qquad (16)$$

Velocity-Sum Distribution

The *velocity-sum distribution* g_+ is obtained as the self-similar solution of (15),

$$g_+(\mathbf{v}_+, \mathbf{r}, t) = g_0(\mathbf{v}_+, t) \equiv (t/2\pi\alpha_0)^{3/2} \exp[-|\mathbf{v}_+|^2 t/2\alpha_0], \qquad (17)$$

which gives another *inertial normal distribution*, say N2 having half a variance of N1.

Velocity-Difference Distribution

Equation (16) for the distribution g_- is r-dependent, but since such dependence takes place in the local range, g_- is obtained as the solution of the r-independent part of (16) as

$$g_-(\mathbf{v}_-, \mathbf{r}, t) = g_0(\mathbf{v}_-, t) \equiv (t/2\pi\alpha_0)^{3/2} \exp[-|\mathbf{v}_-|^2 t/2\alpha_0], \qquad (18)$$

for the region $r > 0$. Equation (18) gives again the *inertial normal distribution* N2 for the *velocity-difference distribution* g_-.

Inertial Similarity and Local Similarity

Although the distributions g_+ and g_- are obtained as the r-independent functions (17) and (18) for $r > 0$, they must satisfy the boundary conditions for $r \to 0$,

$$g_+(\mathbf{v}_+, 0, t) = f_0(\mathbf{v}_1, t), \quad g_-(\mathbf{v}_-, 0, t) = \delta(\mathbf{v}_-), \tag{19}$$

which is required by the conditions $\mathbf{v}_+ \to \mathbf{v}_1$ and $\mathbf{v}_- \to 0$ in this limit. Such discontinuous changes of $g_\pm$ for $r \to 0$ are due to the *inertial similarity*, but actually the changes should occur continuously in the finite *local range* with the finite viscosity $\nu > 0$. (see [5])

4 Local Similarity of Velocity Distributions

The distributions g_+ and g_- in the local range are described in terms of the local variables based on Kolmogorov's length and time which are indicated by the suffix $*$.

The *velocity-sum distribution* g_+ is obtained as the *local normal distribution* N3,

$$g_+(\mathbf{v}_+^*, \mathbf{r}^*, t^*) = (t^*/4\pi\alpha_{+0}^*(r^*))^{3/2} \exp[-|\mathbf{v}_+^*|^2 t^*/4\alpha_{+0}^*(r^*)],$$
$$\alpha_+^*(r^*, t^*) = \alpha_{+0}^*(r^*)t^{*-2}, \tag{20}$$

where $3\alpha_+^*(r^*, t^*)$ gives the *dissipation* of the *self-energy* $E_+^*(r^*, t^*) = \langle|\mathbf{u}_+^*(\mathbf{x}^*, \mathbf{r}^*, t^*)|^2\rangle/2$.

The velocity-difference distribution g_- is axi-symmetric in the local range. The *lateral velocity-difference distribution* $g_\perp$ is found to be identical with the one-dimensional component of N3, that is,

$$g_\perp(v_\perp^*, r^*, t^*) = (t^*/4\pi\alpha_{-0}^*(r^*))^{1/2} \exp[-v_\perp^{*2} t^*/4\alpha_{-0}^*(r^*)],$$
$$\alpha_-^*(r^*, t^*) = \alpha_{-0}^*(r^*)t^{*-2}, \tag{21}$$

where $3\alpha_-^*(r^*, t^*)$ gives the *dissipation* of the *self-energy* $E_-^*(r^*, t^*) = \langle|\mathbf{u}_-^*(\mathbf{x}^*, \mathbf{r}^*, t^*)|^2\rangle/2$.

The equation for the *longitudinal velocity-difference distribution* $g_\parallel$ is simplified by localizing the non-local parameters β_1 and β_2 defined by (14) and the distribution $g_\parallel$ is expressed in the self-similar forms at the three subranges:

$$g_\parallel(v_\parallel^*, r^*, t^*) = t^{*1/2}(r^* t^{*-1/2})^{-\theta} H(\xi),$$
$$\xi = t^{*1/2}(r^* t^{*-1/2})^{-\theta} v_\parallel^*, \alpha_{-0}^*(r^*) = a_\theta(r^* t^{*-1/2})^{2\theta}, \tag{22}$$

with a value of θ for a subrange. Then, the equation of $H(\xi)$ is solved at each subranges.

96   T. Tatsumi and T. Yoshimura

Intermediate Subrange

The distribution is obtained as the *inertial normal distribution* N5

$$H(\xi) = H_0(\xi) \equiv (4\pi a_\theta/(1+\theta))^{-1/2} \exp[-((1+\theta)/4a_\theta)\xi^2], \qquad (23)$$

with $0 \leq \theta \leq 1/3$, 0 corresponding to the outer boundary and $1/3$ to the inner boundary.

Inertial Subrange

The distribution is obtained as the *inertial algebraic distribution* A1,

$$H(\xi) = H_c(\xi) \equiv 12\pi a_{1/3}^2(\xi^2 + 4a_{1/3})^{-5/2}, \qquad (24)$$

with $\theta = 1/3$. The distribution A1 has the algebraic tails $|\xi|^{-5}$ for $|\xi| \to \infty$ and hence the divergent moments for $n \geq 4$. The symmetry of A1 seems to raise a question on the accuracy of the solution, which will be dealt with in a separate paper (see [6]).

Viscous Subrange

The distribution A2 is obtained by solving numerically the equation,

$$\{1 - \frac{10}{3}\xi\}H + \xi\{1 - \frac{14}{3}\xi\}H' + \frac{9}{128}\{1 - \frac{8}{3}\xi - \frac{256}{27}\xi^3\}H'' = 0, \qquad (25)$$

with $\theta = 1, a_1 = 9/128$. The distribution A2 is similar to A1 but asymmetric with respect to ξ and also has the tails $|\xi|^{-5}$ for $|\xi| \to \infty$ and the divergent moments for $n \geq 4$. (see [6])

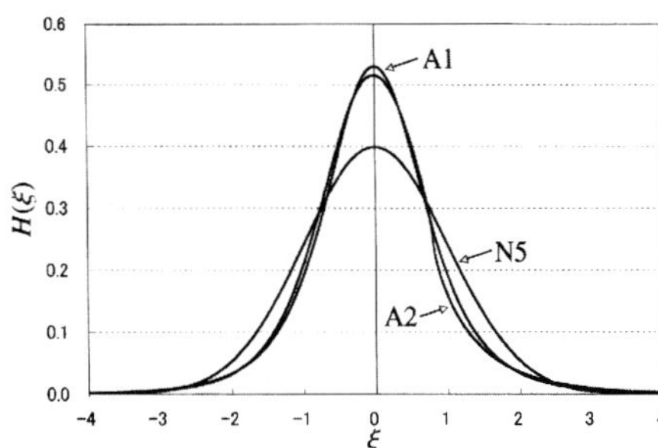

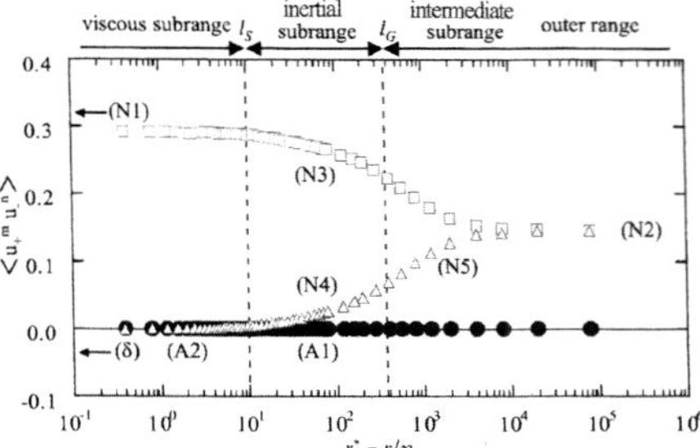

Fig. 1. Longitudinal velocity-difference distributions in the local range. N5: Inertial normal distribution (23). A1: Inertial algebraic distribution (24). A2: Viscous algebraic distribution obtained from (25).

Fig. 2. Overview of the two-point velocity distributions on the figure of the longitudinal cross-velocity correlations measured by Makita et al. [7] at $R_\lambda = 350$. $\square$: $\langle u_+^{*2}\rangle$, $\triangle$: $\langle u_-^{*2}\rangle$, $\bullet$: $\langle u_+^* u_-^*\rangle$.

5 Summary and Discussion

The longitudinal velocity-difference distributions in the local range obtained in the previous section are shown graphically in Fig. 1. The non-normality of the distributions in the inertial and viscous subranges may clearly be observed.

An overview of the various velocity distributions of *homogeneous isotropic turbulence* obtained here may be given by Fig. 2, or the diagram of the longitudinal cross-velocity variance $\langle u_+^{*\,2} \rangle$ ($\square$) and $\langle u_-^{*\,2} \rangle$ ($\triangle$) versus the distance r^* measured by Makita et al. [7]. The prevailing inertial normality of the distributions may be noted. The cerebrated *2/3 power variance* or the $-5/3$ *power energy spectrum* of Kolmogorov [1] is realized as the longitudinal velocity-difference distribution A1 in the inertial subrange. In conclusion, the senior author wishes to record his hearty thanks to the authors of the experimental and numerical results cited in [4], [5], [6] and the present paper which greatly helped him by indicating real situations and plausible principles of this still unknown subject. It is his earnest hope to able to construct a firm and concrete system of theoretical results in turn.

References

1. Kolmogorov AN (1941) Dokl Akad Nauk SSSR 30: 301–305
2. Lundgren TS (1967) Phys Fluids 10: 969-975
3. Monin AS (1967) PMM J Appl Math Mech 31: 1057–1068
4. Tatsumi T (2001) Mathematical physics of turbulence. In: Kambe T et al. (eds) Geometry and Statistics of Turbulence Kluwer Acad Pub Dordrecht
5. Tatsumi T, Yoshimura T (2004) Fluid Dyn Res 35:123–158
6. Tatsumi T, Yoshimura T (2007) Fluid Dyn Res 39:221–266
7. Makita H, Takasa S, Sekishita N (2005) JAXA Special Publication 04-002:71–74

Multi-Scale Analysis of Turbulence

Malte Siefert[1], Stephan Lück, Andreas P. Nawroth[2], and Joachim Peinke[2]

[1] Institute of Physics, University of Potsdam, Potsdam, Germany
`malte.siefert@gmail.com`
[2] Institute of Physics, University of Oldenburg, Oldenburg, Germany

Abstract. The basic idea of the processes taking place in fully developed turbulence is still that of a cascade in which vortices break up successively into smaller ones. A result of this directed process is the emergence of intermittency, i.e. extraordinary strong fluctuations on small scales. An objective is to describe and to understand the statistics of the velocity fluctuations on differerent scales. Based on the pioneer work of Kolmogorov the cascade resembles a process being self-similar on different scales but with deviations due to intermittency. But this description, which is based on scaling property of structure functions, is too idealized for moderate Reynolds numbers and is not complete as discussed below.

It has been shown by Friedrich and Peinke [1] that the measured velocity-fluctuations on different scales can be well described by a diffusion process. Experimentally almost no deviation from this description can be found for a variety of flows and for a large range of scales [2]. Moreover this description is complete in the sense that it grasps also all statistical relations *between* the different scales. The result is a Fokker-Planck equation, which can be estimated in a parameter-free way directly from data and which can be used for subsequent analytical considerations.

In this paper we give an overview of some central results using this analysis. We deal in essential with four different points:

- At scales below the Taylor-scale, the here proposed statistical description of the cascade is not possible anymore. We discuss this point and relate the limiting scale to a new Einstein-Markov-coherence length.
- We show how this diffusion process in scale can be used to generate time series with identical scale dependent statistics.
- We discuss the interactions between different velocity components in terms of a multi-variate Fokker-Planck equation and explain their debated differences.
- We discuss the Reynolds number dependence of a turbulent field and discuss the asymptotic state for infinite high Reynolds numbers.

Keywords: multi-scale statistics, Taylor length, transversal increments, Reynolds number dependence, time series reconstruction

99

Y. Kaneda (ed.), *IUTAM Symposium on Computational Physics and New Perspectives in Turbulence*, 99–104.
© 2008 *Springer*.

1 Multi-Scale Statistics

Turbulence is often characterized by velocity increments $[\mathbf{U}(\mathbf{x}+\mathbf{r}) - \mathbf{U}(\mathbf{x})]\,\mathbf{e}$, where the velocity differences over a certain scale $\mathbf{r}$ are projected onto a certain direction $\mathbf{e}$. Here we denote longitudinal increments with u, for which $\mathbf{r}$ and $\mathbf{e}$ are parallel and transverse increments with v for which $\mathbf{r}$ is perpendicular to $\mathbf{e}$.

It has been shown that it is possible to get access to the joint probability distribution $p(u(r_1), u(r_2), \ldots, u(r_n))$ of these increments via a Fokker-Planck equation, which can be estimated directly from measured data. This method is more general than the analysis by structure functions $\langle u(r)^n \rangle$, which is almost equivalent to the one-scale statistics $p(u(r))$ or $p(v(r))$. The Fokker-Planck equation (here written for vector quantities) reads as

$$-r\frac{\partial}{\partial r}p(\mathbf{u}, r|\mathbf{u_0}, r_0) \tag{1}$$

$$= \left(-\sum_{i=1}^{n} \frac{\partial}{\partial u_i} D_i^{(1)}(\mathbf{u}, r) + \sum_{i,j=1}^{n} \frac{\partial^2}{\partial u_i \partial u_j} D_{ij}^{(2)}(\mathbf{u}, r) \right) p(\mathbf{u}, r|\mathbf{u_0}, r_0).$$

(i labels the components of $\mathbf{u}$, we fix $i = 1$ for the longitudinal and $i = 2$ for the transverse increments.) This representation of a stochastic process is different from the usual one: instead of the time t, the independent variable is the scale variable r. The minus sign appears from the evolution of the probability distribution from large to small scales. In this sense, this Fokker-Planck equation may be considered as an equation for the "dynamics" of the cascade, which describes how the increments evolve from large to small scales under the influence of deterministic ($D^{(1)}$) and noisy ($D^{(2)}$) forces. The whole equation is multiplied without loss of generality by r to get power laws for the moments in a more simple way. Both coefficients, the so-called drift term $D_i^{(1)}(\mathbf{u}, r)$ and diffusion term $D_{ij}^{(2)}(\mathbf{u}, r)$, can be estimated directly from the measured data using its mathematical definition. With the notation $\Delta u_i(r, \Delta r) := u_i(r - \Delta r) - u_i(r)$ the coefficients are given by

$$D_i^{(1)}(\mathbf{u}, r) = \lim_{\Delta r \to 0} \frac{r}{\Delta r} \langle \Delta u_i(r, \Delta r) \rangle|_{\mathbf{u}(r)}, \tag{2}$$

$$D_{ij}^{(2)}(\mathbf{u}, r) = \lim_{\Delta r \to 0} \frac{r}{2\Delta r} \langle \Delta u_i(r, \Delta r) \Delta u_j(r, \Delta r) \rangle|_{\mathbf{u}(r)}. \tag{3}$$

It is important to note that only two assumptions are needed to get the Fokker-Planck equation for the velocity increments; both can be verified by data analysis. Firstly, we have tested the Markovian properties for different scales and for different flows, they are well fulfilled. Secondly, the higher order Kramers-Moyal coefficients are small so that the general expansion reduces to the Fokker-Planck equation. As a final test of self-consistency we have shown that the increments' distribution, as well as the joint distribution of the measured increments, is well described by the numerical solution of the Fokker-Planck equation.

2 Small-Scale Statistics

To verify the validity of a Markov process from experimental data we use the Wilcoxon test. It reveals a lower scale for the validity of the Markovian properties, which gives additionally an estimation of the smallest scale, called Einstein-Markovian (EM) coherence length l_{em}.

An investigation of the Reynolds number dependence of l_{em} is shown in Figure 1a, for further details see [2]. It can be seen that within the estimation errors the EM coherence length is proportional to the Taylor length λ with a factor of about 0.8. This has been found in different turbulent flows types like the wakes of a grid or a cylinder, as well as the free jet.

The ratio of the EM coherence length with respect to λ and the Kolmogorov length η in dependence of Re is shown in Figure 1b). This dependence is within the erros $Re^{1/4}$, which is the expected value for the ratio λ/η. This shows again that the Einstein-Markov coherence length is closly related to λ and not to η.

These results give strong evidence that the EM coherence length l_{em} can be introduced as a new length scale in turbulence and it is closely linked to the Taylor length which gives a novel physical interpretation to this well established but empirical quantity. The violation of the Markovian properties on step sizes smaller than l_{em} indicates that the disorder of two increments are linked via memory effects, if their scales r and r' are too close. Such memory effects may be connected with the presence of small scale coherent structures. In turbulence, vorticity has the tendency to accumulate in thin sets like vortex filaments and vortex sheets. The thickness of these structures all fall in the range of the Taylor length and therefore we propose that they are responsible for the existence of the finite Markov length l_{em}.

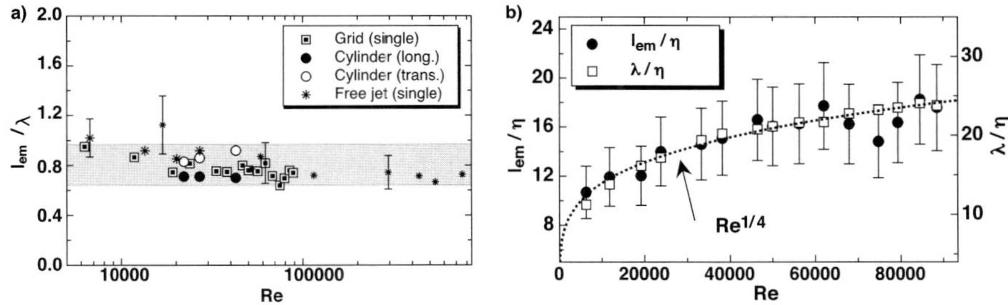

Fig. 1. a) Ratio l_{em}/λ versus Reynolds number Re for different flows (see legend). Some error bars are shown exemplarily. b) Ratios l_{em}/η and λ/η versus Reynolds number Re for grid turbulence. Dotted curve - theoretical behavior of λ/η (after [2]).

3 Multi-Scale Reconstruction of Time Series

In this section we restrict the discussion to Markov processes in scale and to right-justified increments. To determine a new element of an existing series $U(t)$ at time t^*, $p(U(t^*)|U(t^* - \tau_n), \ldots, U(t^* - \tau_1))$ is needed. $\{U(t^* - \tau_1), \ldots, U(t^* - \tau_n)\}$ are known, as $p(u(\tau_1), \ldots, u(\tau_n))$ is provided by the data itself or by the solution of Eq. (1). Because $U(t^*)$ is not known, a certain realisation is assumed, where $U(t^*)$ takes the value $\tilde{U}(t^*)$. $p(u(\tau_1), \ldots, u(\tau_n))$ is then used to determine the probability of such a realisation. Repeating this for many realisations $\tilde{U}(t^*)$, an estimate for $p(U(t^*)|U(t^* - \tau_n), \ldots, U(t^* - \tau_1))$ is obtained. Choosing now a random value from this distribution, the time series will be extended correctly by another point. For further details see [3]. As an example we take data from a free jet. The results of a time series reconstruction are shown in Fig. 2a. In order to verify our results, we compare the distributions of the experimental data with the distributions of the generated time series. The corresponding pdfs for eight different scales are shown in Fig. 2b. The agreement between the original and reconstructed pdfs is quite good and even improves with the size of the original data.

In a next step we compare this method with a one-scale method. A random walk is constructed, such that the distribution of the increments on the smallest scale is identical to that of the empirical time series, results are shown in Fig. 2c. The adapted random walk provides for small scales a good approximation, but for the larger scales it becomes quite poor. This demonstrates clearly the achieved improvement of a multi-scale reconstruction.

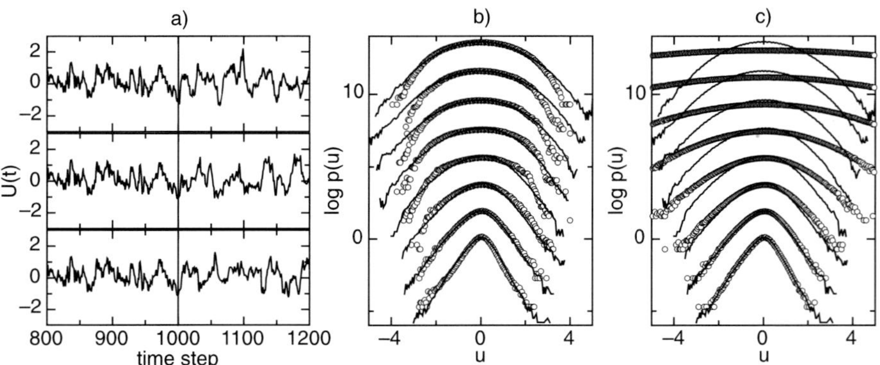

Fig. 2. a) Three runs of reconstruction. The data points used as initial condition are shown left to the vertical line, while the reconstructed one are to the right. b) The pdfs from the original (solid black line) and the reconstructed (circles) data set are shown. The considered scales are $2^n \cdot l_{mar}$ with $n = 7, 6, .., 0$ from top to bottom. The pdfs for different scales are shifted for clarity of presentation. c) Pdfs in analogy to b) are shown for the reconstructed data from a simple one-scale process was used (after [3]).

4 Differences of Longitudinal and Transverse Increments

There was a long debate about the statistical differences between longitudinal and transverse increments. The final interpretation has been that the transverse increments are more intermittent than the longitudinal ones. Next, we discuss this differences in term of a multivariate diffusion process.

The drift and diffusion coefficients are estimated from the measured data according to Eqs. (2) and (3). They can be well approximated by the following low dimensional polynoms, (further details see [4])

$$D_1^{(1)}(u,v,r) = d_1^u(r)u, \quad D_2^{(1)}(u,v,r) = d_2^v(r)v$$
$$D_{11}^{(2)}(u,v,r) = d_{11}(r) + d_{11}^u(r)u + d_{11}^{uu}(r)u^2 + d_{11}^{vv}(r)v^2 \qquad (4)$$
$$D_{22}^{(2)}(u,v,r) = d_{22}(r) + d_{22}^u(r)u + d_{22}^{uu}(r)u^2 + d_{22}^{vv}(r)v^2.$$

The coefficients $d_{11}^{uu}(r)$ and $d_{22}^{vv}(r)$ are responsible for intermittency, the coefficients $d_{11}^u(r)$ and $d_{22}^u(r)$ are essential for the skewness of the probability distribution (Kolmogorov's 4/5-law).

The analysis reveals a remarkable symmetry between the d-coefficients: the r-dependence of related longitudinal and transverse d-coefficients coincides if $d_{\text{long}}(r) \approx d_{\text{transv}}(2/3r)$, e.g. $d_1^u(r) = d_2^v(2/3r)$ etc.; only the coefficients, which are responsible for the skewness and for the intermittency, deviate slightly from this symmetry. The main differences between longitudinal and transverse increments can be found in a 3/2-times faster transverse cascade.

If this 3/2-rescaling is taken into account in the phenomenological estimation of the structure-functions' scaling-exponents, then the differences vanishes almost within the error-bars, see Fig. 3. Thus the intermittency of longitudinal and transverse increments are almost the same. The fact that the scaling-exponents change with rescaling indicates that no proper scaling feature can be present.

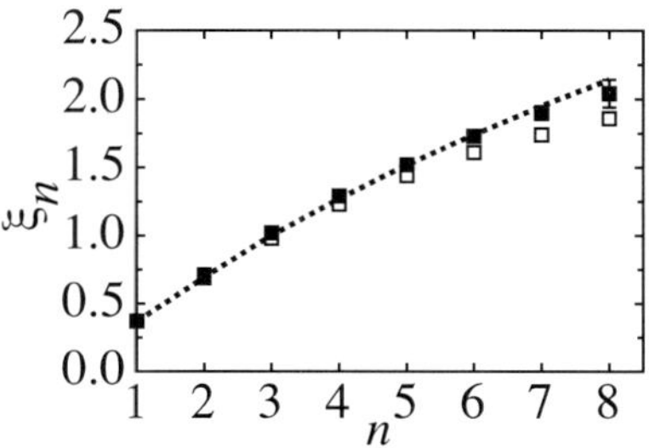

Fig. 3. The scaling exponents of the longitudinal exponents (line) in comparison to the transverse exponents (symbols). If the 3/2-rescaling is taken into account in ESS, the transverse exponents come close to the longitudinal ones (white squares: without, black squares: with 3/2-rescaling; after [4]).

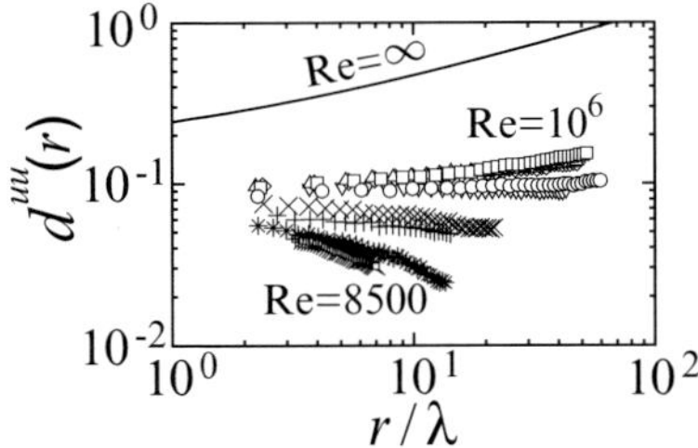

Fig. 4. Intermittency-coefficients $d_2^{uu}(r, Re)$ exhibit a strong dependence on the Re-number with a tendency towards the limiting value $d_2^{uu}(r, Re \to \infty)$ (after [5]).

5 Universality of Turbulence

Turbulence is usually assumed to be universal, and thus independent of large scale boundary conditions, the mechanism of energy dissipation and the Reynolds number. We examine the universality of the Re number dependence of a turbulent flow measured in a cryogenic free jet in a range of Re numbers from 8500 up to 10^6 [5]. Especially we study the Re number-dependence of the drift and diffusion coefficients:

$$D^{(1)}(u, r, Re) = d_1^u(r, Re)u \tag{5}$$

$$D^{(2)}(u, r, Re) = d_2(r, Re) + d_2^u(r, Re)u + d_2^{uu}(r, Re)u^2. \tag{6}$$

It has been found that d_1^u is independent of the Re number, whereas the both coefficients d_2 and d_2^u vanish proportional to $Re^{-3/8}$. Consequently, the intermittency term d_2^{uu} dominates the statistics for high Re numbers. With Kolmogorov's 4/5-law, an asymptotic value for d_2^{uu} can be given. In Fig. 4 it can bee seen that the measured values of d_2^{uu} reach the asymptotic value very slowly even for high $Re = 10^6$. It is therefore important to develop a better understanding of the limiting case $Re \to \infty$. It seems questionable whether models on turbulence established under the assumption of infinite Reynolds numbers can be tested in real-life experimental situations at all.

References

1. Friedrich R, Peinke J (1997) Physica D 102:147–155
2. Lück S, Renner C, Peinke J, Friedrich R (2006) Phys Lett A 359:335–338
3. Nawroth AP, Peinke J (2006) Physics Letters A 360:234–237
4. Siefert M, Peinke J (2006) J Turbul 7:1–35
5. Renner C, Peinke J, Friedrich R, Chanal O, Chabaud B (2002) Phys Rev Lett 89:art no 124502

Helicity within the Kolmogorov Phenomenology of Turbulence

Susan Kurien

Mathematical Modeling and Analysis Group
Theoretical Division
Los Alamos National Laboratory
Los Alamos, NM 87545, USA
skurien@lanl.gov

Abstract. In a phenomenology in which both energy and helicity exhibit net flux to the small scales it is natural to investigate how they might influence each other. Motivated by Kraichnan's 1971 derivation of spectral scaling laws using the timescale for energy transfer in wavenumber, we proceed by considering the timescale for helicity transfer and its potential impact on energy distribution in wavenumber. We demonstrate using resolved direct numerical simulations that the predicted effects of a second timescale related to helicity transfer are consistent with observed statistics. Both the energy and helicity spectra show to close to a $k^{-4/3}$ scaling in the higher wavenumber 'bottleneck' regime of the inertial range. The latter scaling is in agreement with our prediction for the scaling exponent based on a helicity-dependent timescale for twisting rather than shearing motions.

Keywords: isotropic turbulence, homogeneous turbulence, helical turbulence, turbulence phenomenology

1 Introduction

There are two quadratic invariants of the inviscid Navier-Stokes equations for 3D incompressible flows; the total energy, and the total helicity, $H = \int \mathbf{u}(\mathbf{x}) \cdot \boldsymbol{\omega}(\mathbf{x}) d\mathbf{x}$, where $\mathbf{u}(\mathbf{x})$ is the velocity field at point $\mathbf{x}$, and $\boldsymbol{\omega} = \nabla \times \mathbf{u}$ is the vorticity. Unlike energy, helicity is a sign-indefinite *pseudo*-scalar, and is parity breaking (or reflection-antisymmetric), that is, it changes sign if $\mathbf{x} \to -\mathbf{x}$ and $\mathbf{u} \to -\mathbf{u}$. A useful picture to keep in mind to visualize helicity is that of a fluid parcel moving along a helix; the 'handedness' of this configuration is immediately apparent. Fluid helicity appears in many phenomena at scales ranging from atmospheric to viscous dissipation scales. In fact it is difficult to conceive of a flow with identically zero helicity in all wavenumbers.

A statistical relation for velocity correlations related to helicity flux has been sought in various publications beginning in 1961 [1] and such a

105

Y. Kaneda (ed.), *IUTAM Symposium on Computational Physics and New Perspectives in Turbulence*, 105–110.
© 2008 *Springer*.

relationship now exists for antisymmetric velocity correlations which describe the helicity flux (see references [2]):

$$\langle (u_l(\mathbf{x}+\mathbf{r}) - u_l(\mathbf{x}))(\mathbf{u}_t(\mathbf{x}+\mathbf{r}) \times \mathbf{u}_t(\mathbf{x})) \rangle = \frac{2}{15} h r^2 \tag{1}$$

where u_l is the component of velocity $\mathbf{u}$ along the separation vector $\mathbf{r}$, $\mathbf{u}_t = \mathbf{u} - u_l \hat{\mathbf{r}}$ is the transverse component of the velocity, h is the mean helicity dissipation rate. This law is derived under the assumption of local statistical isotropy and homogeneity. It is analogous to the Kolmogorov 1941 4/5-law for the third-order longitudinal structure functions [3]. While the latter was derived assuming reflection-symmetry of the flow, (1) requires reflection symmetry breaking due to the presence of helicity. The 2/15-law was verified for the first time in resolved numerical simulations in [4].

In a phenomenology in which both energy and helicity exhibit net flux to the small scales (forward cascade) it is natural to investigate how they might influence each other. Motivated by Kraichnan's [5] derivation of spectral scaling laws using the timescale for energy transfer in wavenumber, we proceed to consider a timescale for helicity transfer and its potential impact on energy distribution in wavenumber. It seems reasonable that the twisting character of helical motions will give them a different characteristic timescale than, say, energy transfer by shearing motions between two parcels of fluid. We follow the arguments in [6] to construct a phenomenology with *two* timescales, τ_E for pure energy transfer and the τ_H for pure helicity transfer rates. The latter timescale implies a $k^{-4/3}$ scaling distribution of both energy and helicity. In [4] we demonstrated using DNS that a second timescale might indeed be important in the dynamics. In that work, resolved Navier-Stokes data were acquired in a periodic box at resolutions of 512^3 with controlled helicity input, and 1024^3 with random helicity input. Both the energy and helicity spectra show to close to a $k^{-4/3}$ scaling in the higher end of the inertial range. The latter scaling is consistent with our prediction for the scaling exponent based on a helicity-dependent timescale for twisting rather than shearing motions. We postulate self-consistent arguments for how such a phenomenology might arrange to affect the energy spectrum as observed. The main results summarized in this paper are discussed in detail in references [4, 7].

2 The Phenomenology of Helicity and Related Statistics

2.1 A Transfer Timescale of Helicity and Its Effect on Spectral Scaling

We recall from [5] the definition of the distortion time of an eddy of size $1/k$ where k is the wavenumber:

$$\tau_E^2 \sim \left(\int_0^k E(p) p^2 \mathrm{d}p \right)^{-1} \sim [E(k) k^3]^{-1} \tag{2}$$

This estimate was shown to lead to the well-known scaling of the energy spectrum $E(k) \sim k^{-5/3}$. It was shown in [6] that, assuming the helicity cascade was governed by the energy timescale τ_E, the helicity spectrum is expected to scale the same way, as $H(k) \sim k^{-5/3}$.

We can define an analogous timescale for distortions of an eddy due to helicity transfer [7]:

$$\tau_H^2 \sim [|H(k)|k^2]^{-1} \qquad (3)$$

This timescale may be interpreted as the time taken for helicity transfer from eddies of size $1/k$ to eddies of (say) smaller size. On a somewhat crude level, one may think of τ_H as the the time taken for twisting motions of scale $1/k$ to generate twisting motions in scales one generation smaller in the cascade process. The main point here is that the distortion or shear corresponding to τ_H is different from the distortion corresponding to τ_E of [5]. If τ_H is allowed to govern the energy and helicity timescales then we can estimate that $E(k) \sim H(k) \sim k^{-4/3}$. The ratio of the two timescales,

$$\frac{\tau_E}{\tau_H} \sim \left(\frac{|H(k)|}{2kE(k)} \right)^{-1/2}.$$

In [7] we argue for the possibility that τ_H may be comparable (though slower) than τ_E thus allowing for $k^{-4/3}$ scaling of the spectra in the high wavenumbers. In the next section we present the numerical results which support our proposed phenomenology.

3 Simulations and Results

We will discuss two simulations in this section. The parameters of the simulations are listed in Table 1. Data I at resolution of 512^3 was forced with large (maximum) helicity input of the same sign at every timestep which resulted in a large mean helicity in statistically steady turbulence. Data II at resolution of 1024^3 was forced with random (uncontrolled) helicity at each timestep resulting in nearly zero mean helicity. Other details of the simulations, including the numerical schemes are given the primary references.

In Figure 1 we demonstrate using data I that both helicity and energy display constant fluxes (balanced by their respective dissipation rates) in about

Table 1. Parameters of the numerical simulations I and II. ν - viscosity; R_λ - Taylor Reynolds number; mean total energy $E = \frac{1}{2}\sum_k |\tilde{\mathbf{u}}(\mathbf{k})|^2$; ε - mean energy dissipation rate; mean total helicity $H = \sum_k \tilde{\mathbf{u}}(\mathbf{k}) \cdot \tilde{\omega}(-\mathbf{k})$; h - mean helicity dissipation rate; $\eta_\varepsilon = (\nu^3/\varepsilon)^{1/4}$; $\eta_h = (\nu^3/h)^{1/5}$.

	N	$\nu \times 10^4$	R_λ	E	ε	H	h	$\eta_\varepsilon \times 10^3$	$\eta_h \times 10^4$
I	512	1	270	1.72	1.51	26.8	62.2	9	1.7
II	1024	0.35	430	1.87	1.75	-0.12	13.2	4	1.3

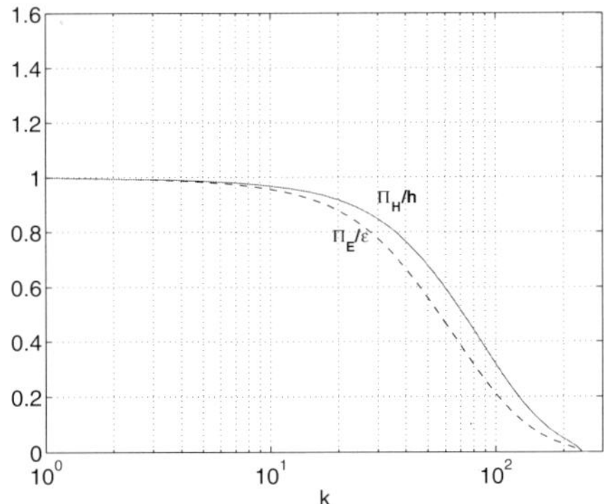

Fig. 1. Fluxes of energy and helicity for data I. Dotted line: Flux of energy Π_E normalized by mean dissipation rate of energy ε. Solid line: Flux of helicity Π_H normalized by mean dissipation rate of helicity h.

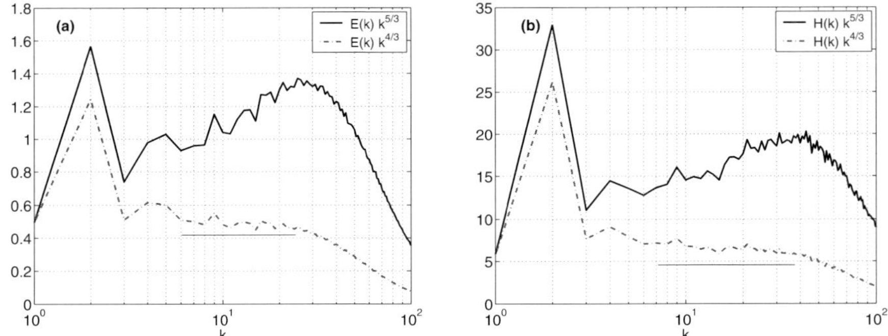

Fig. 2. Solid black lines – spectra compensated by $k^{5/3}$, dotted blue lines – same spectra compensated by $k^{4/3}$; horizontal lines indicate of the extent of scaling ranges. (a) Energy spectrum for the 512^3 simulation, which has a large mean helicity, there appears to be no scaling as $k^{-5/3}$, and very close to $k^{-4/3}$ scaling in the range $6 < k < 25$. (b) Helicity spectrum for the same simulation.

the same range of scales. This is a demonstration of the simultaneous transfer of both energy and helicity to the small scales.

Figure 2 shows the spectra of energy and helicity for data I. In order to clearly distinguish between the slightly different scalings of $k^{-5/3}$ and $k^{-4/3}$, we plot on a linear-log scale and compensate the spectra by multiplying them by both $k^{5/3}$ (black lines) and $k^{-4/3}$ (blue dotted lines). Both energy and helicity spectra plotted in this way display no $k^{-5/3}$ scaling as predicted by the K41 theory. However, both spectra display agreement with $k^{-4/3}$ scaling in wavenumbers which are well-correlated with the constant flux regime in Fig. 1.

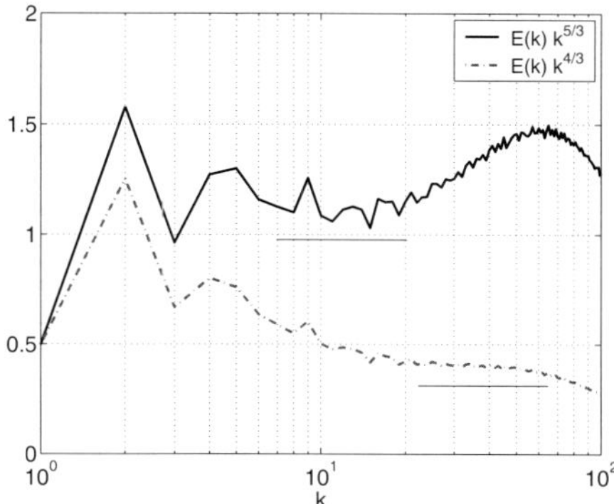

Fig. 3. Energy spectrum of data II represented in the same manner as figure 2a. Note the low-wavenumber scaling of $k^{-5/3}$ for a short range $8 < k < 20$ followed by a high wavenumber scaling of $k^{-4/3}$ for the range $20 < k < 70$. The two scaling ranges occur over comparable number of lengthscales and the latter scaling corresponds closely to the so-called 'bottleneck' feature.

4 Conclusion

The $k^{-5/3}$ scaling law and the associated K41 phenemenology have underlying assumptions of statistically isotropic and reflection symmetric, helicity-free flows. Early efforts at integrating helicity into the picture, once it was discovered that it too is a global invariant, have usually relied on the (reasonable) guess that energy governs the overall dynamics with helicity being carried along more or less passively. Until now, the only accepted effect of helicity has been that it slows down the energy cascade in the following sense: if a flow is started with initially high helicity, the time it takes for the energy to reach the small scales is longer than if the flow were initialized with low or no helicity. The slowing down of energy transfer is thought to be benign in that the final statistical behavior of the flow, the spectral scalings and so forth, remain the same whether or not the flow is helical.

The phenomenology we propose here goes a bit further. First, we propose that it is not the net helicity but the relative (or local) helicity which is important in the dynamics. Second, the effect of relative helicity is locked into the dynamics in such a way as to cause energy to pile-up in the high-wavenumbers because of the slower timescale for helicity transfer. This results in the shallower $k^{-4/3}$ spectral scaling. The latter effect may also arise not from a cascade process, but because of generation of local helicity in the small scales by the local viscous production-dissipation of helicity. The cross-over between the $k^{-5/3}$ and the $k^{-4/3}$ scaling regimes appears to be depend on the Reynolds number although we have not performed sufficiently many simulations to determine this with certainty. Higher Reynolds numbers simulations could perhaps determine whether this is indeed a finite Reynolds number effect. The main conclusion of this work is that the bottleneck of energy in the high wavenumbers may be shown to be consistent with a phenomenology in which the presence of helicity slows down the energy cascade enough to cause a pile-up and hence a shallower spectral scaling.

References

1. Betchov R (1961) Phys Fluids 4:925–926
2. Chkhetiani O (1996) JETP Lett 63:768–772; L'vov V, Podivilov E, Procaccia I (1997) http://arxiv.org/abs/chao-dyn/9707015; Gomez T, Politano H, Pouquet A (2000) Phys Rev E 61:5321–5325; Kurien S (2003) Physica D 175:167–176
3. Kolmogorov A (1941) Dokl Akad Nauk SSSR 32:16–18
4. Kurien S, Taylor M, Matsumoto T (2004) Phys Rev E 69:066313
5. Kraichnan R (1971) J Fluid Mech 47:525–535
6. Brissaud A, Frisch U, Leorat J, Lesieur M, Mazure A (1973) Phys Fluids 16:1366–1367
7. Kurien S, Taylor M, Matsumoto T (2004) J Fluid Mech 515:87–97

Anomalous Scaling Laws of Passive Scalar Intermittency in 3-Dimensional Turbulence

Takeshi Watanabe[1,2]

[1] Department of Engineering Physics, Graduate School of Engineering,
Nagoya Institute of Technology, Gokiso, Showa-ku, Nagoya, 466-8555, Japan
[2] CREST, Japan Science and Technology Agency, 4-1-8 Honcho,
Kawaguchi, Saitama 332-0012, Japan
watanabe@nitech.ac.jp

Abstract. Reynolds number and anisotropy effects on the anomalous scaling laws of passive scalar turbulence under a uniform mean scalar gradient are investigated by performing the high resolution direct numerical simulations. The local scaling exponents of high order scalar structure function are computed in the case for $R_\lambda \simeq 600$ simulation and compared to the preceding lower R_λ results. Analysis of the SO(3) decomposition of scalar structure function reveals that the anisotropy effects on the scaling behavior are not significant in the inertial convective range even at the high order statistics.

Keywords: turbulence, intermittency, passive scalar, direct numerical simulation, anomalous scaling

1 Introduction

One of the central topics of the recent numerical, experimental and theoretical studies for the passive scalar turbulence is intermittency [1, 2]. A challenging problem is to clarify the degree of universality of scaling exponent of the scalar structure function in the inertial convective range (ICR). In order to discuss this problem more carefully, it is effective to change the large scale conditions of the scalar alone while keeping the turbulent velocity field the same and to see the difference in the scalar statistics. It was clarified by the direct numerical simulations (DNSs) of passive scalar turbulence in 2D that the scaling exponents are the same to the two kinds of scalar sources and saturated at large order [3]. This is consistent with the predictions of the Kraichnan model which suggests the significance of zero-mode governing the scaling law of scalar structure function irrespective of the scalar injection mechanism [4]. These facts bear an expectation that the anomalous scaling laws of passive scalar convected by the generic turbulent flow are universal [2, 3, 4]. On the other hand, the recent DNS study in this direction in 3D shows that the anomalous scaling exponents at high order are dependent on

Y. Kaneda (ed.), *IUTAM Symposium on Computational Physics and New Perspectives in Turbulence*, 111–116.

the scalar injection mechanism up to $R_\lambda \simeq 470$, suggesting the sensitivity of scaling behavior on the changes of macro-scale condition [5]. However it seems that we need the further DNS studies to explain the discrepancy between 2D and 3D cases because we thoughtfully discuss the Reynolds number and the anisotropy effects on the scaling behavior.

The purpose of the present study is to get more insight into this problem. We investigate the anomalous scaling laws of passive scalar in 3D turbulence by performing the high resolution DNS up to 2048^3 grid points. We explore the Reynolds number effects on the anomalous scaling laws. Moreover the anisotropy effects on the scaling behavior, which cannot be considered in the preceding studies [5, 6], are discussed by carrying out the further analysis of the SO(3) decomposition of scalar structure function [7].

2 Direct Numerical Simulations

The passive scalar field $\theta(\mathbf{x}, t)$ under a uniform mean scalar gradient in the x_3 direction is governed by the advection diffusion equation $(\partial_t + u_j\partial_j)\theta = \kappa\partial_j^2\theta - Gu_3$, where κ denotes the molecular diffusivity and G being a mean scalar gradient. Incompressible velocity field $u_i(\mathbf{x}, t)$ $(i = 1, 2, 3)$ obeys the Navier-Stokes equation with the solenoidal Gaussian random force having the spectral support in $1 \leq |\mathbf{k}| \leq 2$. We consider the only case of Schmidt number $Sc = 1$ and $G = 1$. The DNS method is the same as our preceding studies [5, 6]. We performed the three cases of DNS with the different R_λ as $R_\lambda = 263$ ($N^3 = 512^3$, Run G2), 468 (1024^3, Run G3) and 589 (2048^3, Run G4), where the resolution condition based on the mean Kolmogorov scale $\overline{\eta}$ is chosen by $K_{max}\overline{\eta} = 1.0 \sim 1.4$. Durations of the temporal average are $T_{av} = 6.0T_{ed}$ (G2), $4.0T_{ed}$ (G3) and $0.35T_{ed}$ (G4), where T_{ed} denotes the large eddy turnover time. Detailed results by Run G3 are also found in [5]. Here one needs care for the large scale statistics obtained by Run G4 because T_{av} is extremely shorter. Although it is desirable to take the longer time averages to get the more reliable statistical data, our experiences tell us that the low order statistics at small scale for $r/\overline{\eta} < 300$ converges well even when the averaging time is insufficient.

In order to confirm the width of ICR, the compensated kinetic energy and scalar variance spectra, $\hat{E}(k\overline{\eta}) = \overline{\epsilon}^{-2/3}k^{5/3}E(k)$ and $\hat{E}_\theta(k\overline{\eta}_B, Sc) = \overline{\chi}^{-1}\overline{\epsilon}^{1/3}k^{5/3}E_\theta(k)$, are shown in Figure 1, where $\overline{\epsilon}$ and $\overline{\chi}$ are the mean dissipation rates for the kinetic energy and scalar variance, respectively. We expect $\hat{E}(x) = K$ and $\hat{E}_\theta(x, 1) = C_{OC}$ for $x \ll 1$, where K and C_{OC} are called the Kolmogorov and Obukhov-Corrsin constants. It is seen that the plateaus are observed in the range $0.003 < k\overline{\eta} < 0.03$, where the constants are evaluated as $K \simeq 1.61$ and $C_{OC} \simeq 0.64$. Moreover the one-dimensional scalar variance spectrum $E_\theta^{\|}(k_3)$ is also plotted in the same figure, in which the clear plateau is observed in the same range of $E_\theta(k)$. It should be noted that the relation $E_\theta(k) = -kdE_\theta^{\|}(k)/dk$ for the isotropic scalar field is satisfied very well in

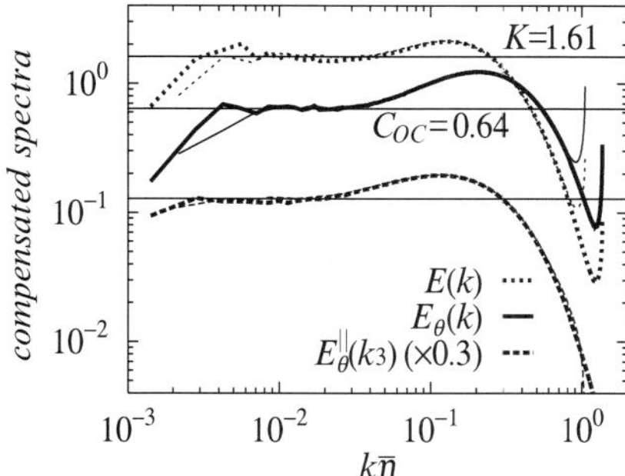

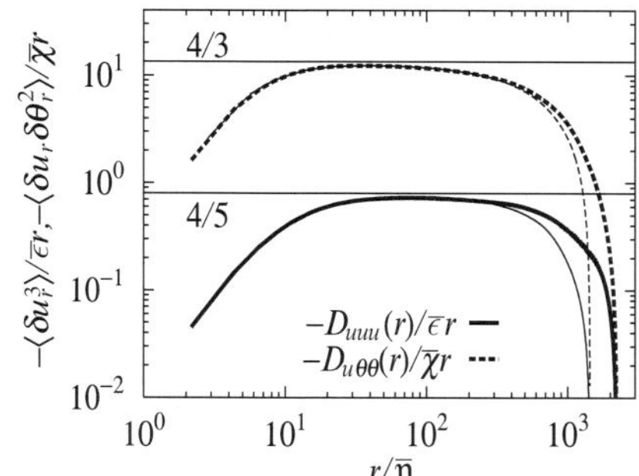

Fig. 1. Compensated plots of $E(k)$ and $E_\theta(k)$ obtained by Run G3 (thin lines) and G4 (thick lines). Curves for $E_\theta^\parallel(k_3)$ are multiplied by the factor 0.3 for clarity.

Fig. 2. Approaches to the 4/5 and 4/3 laws obtained by Run G3 (thin lines) and G4 (thick lines). Curves for $-D_{u\theta\theta}(r)/\overline{\chi}r$ are multiplied by factor 10 for clarity.

$k\overline{\eta} > 0.003$. This suggests that the anisotropy effects are negligible at the second order moment, as will be seen in the next section.

To examine the existence of ICR more precisely, we investigate the validity of asymptotically exact statistical laws derived from the fundamental equation of motions in the range $\overline{\eta} \ll r \ll L$, so-called 4/5 and 4/3 laws as $-\langle \delta u_r^3 \rangle / \overline{\epsilon}r = 4/5$ and $-\langle \delta u_r \delta \theta_r^2 \rangle / \overline{\chi}r = 4/3$. Results are shown in Figure 2, indicating that the both curves approach to the expected values of 4/5 and 4/3 in the range $30 < r/\overline{\eta} < 300$. However the extension of the valid range of 4/3 law is shorter than that of 4/5 law when R_λ is increased. This is due to the insufficient convergence of the large scale statistics in Run G4, as mentioned in the first paragraph in this section. In contrast, we can see that the curves at small scale are indistinguishable for both cases of R_λ. This encourages us to study the scaling laws of structure function at small scale evaluated by Run G4.

3 Passive Scalar Intermittency and Anisotropy Effects

We investigate the scaling laws of passive scalar structure function defined by

$$S_q^G(r) = \frac{1}{3} \sum_{i=1}^{3} \overline{\langle (\theta(\mathbf{x} + r\mathbf{e}_i) - \theta(\mathbf{x}))^q \rangle_L}, \tag{1}$$

where $\overline{f}$ and $\langle f \rangle_L$ respectively denote the temporal and the spatial averages, and $\mathbf{e}_i$ is a unit vector in the x_i direction. Figure 3 shows the variation of local slopes of $S_q^G(r)$ against R_λ for $q = 2p$ ($p = 1, \cdots, 4$). We can see that $\zeta_q^G(r)$ with $q \leq 4$ has the plateau in the range $60 < r/\overline{\eta} < 200$ irrespective of R_λ, where their values are close to each other and the flat ranges mildly extend toward the larger scales as R_λ increases. In contrast, the curves by Run G4 at high order gradually deviate from the others with increase of r and there is no clear flat behavior at $q = 8$. This behavior reminds us of the

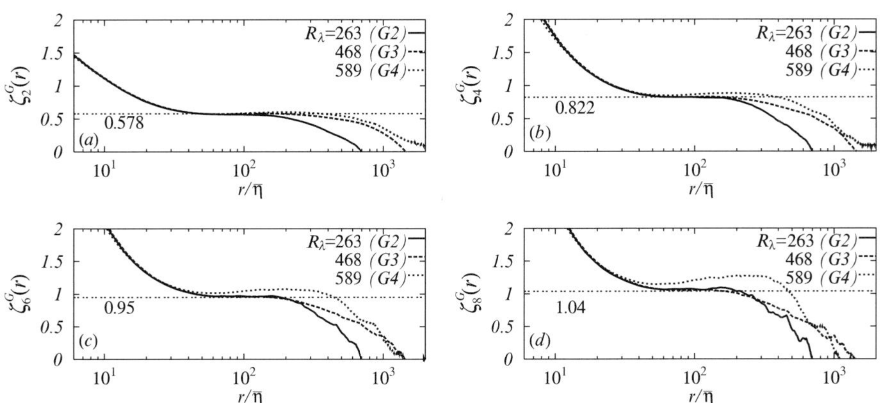

Fig. 3. Variations of the local slopes $\zeta_q^G(r)$ against R_λ for (a) $q = 2$, (b) $q = 4$, (c) $q = 6$ and (d) $q = 8$.

case of random scalar source, in which the local slopes have no clear single plateau and largely vary with R_λ [5, 6]. Although we should not draw the definite statement from Figure 3 because the time average is insufficient for getting the adequate convergence of high order statisitcs, Figure 3 suggests the possibility of the existence of scaling behavior similar to the case for the random scalar source [5, 6]. More precise results will be reported in the future paper.

As seen the formation of ramp-cliff structures [1], the passive scalar field under a uniform mean scalar gradient reveals the anisotropic natures. This implies that the degree of anisotropy is the important factor for determining the scaling laws of $S_q^G(r)$. The SO(3) decomposition of scalar structure functions is a useful tool for examining the anisotropy effects on the scaling behavior [7], which is represented by

$$S_q(\mathbf{r}) = \overline{\langle (\theta(\mathbf{x} + \mathbf{r}) - \theta(\mathbf{x}))^q \rangle_L} = \sum_{l=0}^{\infty} \sum_{m=-l}^{l} S_q^{lm}(r) Y_{lm}(\mathbf{r}/r), \qquad (2)$$

where $Y_{lm}(\mathbf{r}/r)$ is the spherical harmonic function. If the scalar field is isotropic in the x_1-x_2 plane, the form (2) is reduced to

$$S_q(r, \vartheta) = \sum_{l=0}^{\infty} S_q^{l0}(r) \sqrt{\frac{2l+1}{4\pi}} P_l(\cos \vartheta), \qquad (3)$$

where $P_l(z)$ represents the Legendre polynomial and ϑ is the angle between $\mathbf{r}$ and x_3-axis. Structure function is represented by the superposition of the different scaling forms if the coefficient $S_q^{lm}(r)$ has the scaling law in the ICR as $S_q^{lm}(r) \sim C_q^{lm} r^{\zeta_q^{lm}}$. If C_q^{lm} is the same order for different l and m values, the scaling exponent in $r \ll L$ is determined by the minimum of ζ_q^{lm}. Another case

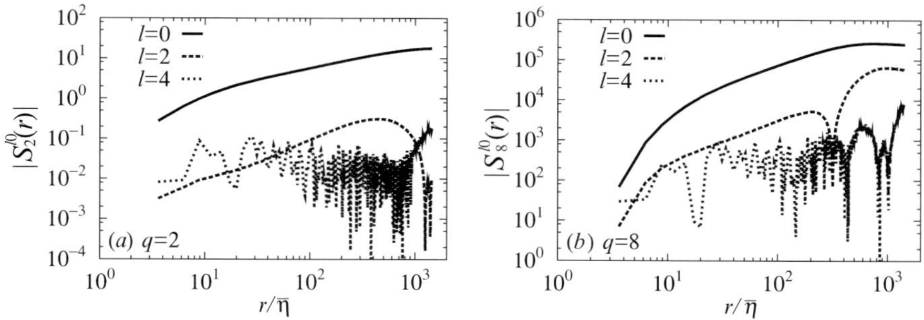

Fig. 4. Variations of $l = 0, 2$ and 4 sectors by SO(3) decomposition of $S_q^G(\mathbf{r})$ at (a) $q = 2$ and (b) $q = 8$ obtained by Run G3.

is that the scaling exponent is governed by the sector having the much larger value of prefactor than the others when the ICR is not so extended. In order to investigate the degree of contribution of each sectors, we compute $S_q(\mathbf{r})$ from the DNS data and evaluate $S_q^{l0}(r)$ via the inverse transformation of (2), where $S_q(\mathbf{r})$ is evaluated by $S_q(\mathbf{r}) = \sum_{m=0}^{q}(-1)^{q-m}\binom{q}{m}\overline{\langle \theta(\mathbf{x}+\mathbf{r})^m\theta(\mathbf{x})^{q-m}\rangle}_L$ with applying the fast Fourier transform (FFT) to the computation of pair correlations. This enables us to reduce the computational cost extremely when compared to the conventional method with $O(N^6)$ operations. Details on this method will be reported elsewhere. Figure 4 shows the variation of sectors with $m = 0$ and $l = 0, 2$ and 4 for the cases with $q = 2$ and $q = 8$ computed by the five field data of Run G3 recorded at $0.3T_{ed}$ intervals. We can confirm that the absolute values of $l = 2$ and 4 sectors at $q = 2$ are much smaller than that of the isotropic sector over the ICR scale. At high order of $q = 8$, the contribution of anisotropic sectors relatively grows and is more pronounced in $r > L$ because the local isotropy in the scalar field is not fully established due to the persistent mean scalar gradient. However the anisotropic sectors are again smaller than $S_8^{00}(r)$ in the ICR. These results suggest that the anisotropy effects are not so important even at the high order statistics as far as the ICR scale is concerned.

Finally we discuss the anisotropy effects on the structure function computed by (1) which is commonly used in the DNS analysis. Structure function evaluated by the three directional average is expressed in terms of the SO(3) decomposition as $S_q^G(r) = [S_q(r, 0) + 2S_q(r, \pi/2)]/3 = S_q^{00}(r)/\sqrt{4\pi} + (7/4)S_q^{40}(r)/\sqrt{4\pi} + \cdots$, where one should notice that the sub-leading sector $(l = 2)$ is eliminated. This suggests that $S_q^G(r)$ is a good approximation of the isotropic sector of $S_q(\mathbf{r})$ if the anisotropic ones with $l \geq 4$ are much smaller than $S_q^{00}(r)$. To test the above idea, we directly compare the curves for $S_q^G(r)$ and $S_q^{00}(r)$ in Figure 5. We see that $S_q^G(r) \simeq S_q^{00}(r)/\sqrt{4\pi}$ over the two decade of the range $6 < r/\bar{\eta} < 600$ up to $q = 10$. In order to make the more severe examination, their local slopes are also compared in the same figure. Although

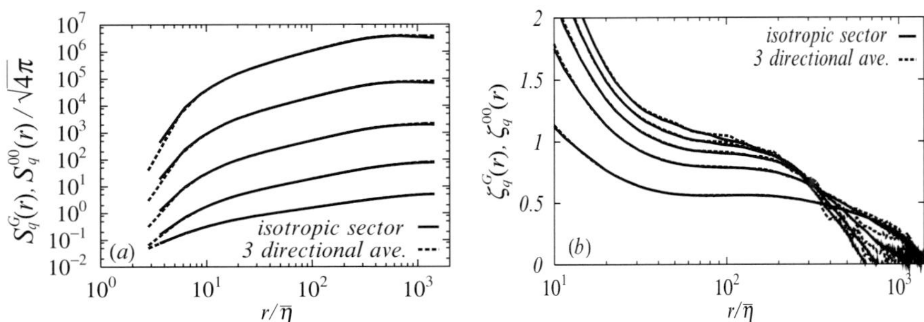

Fig. 5. (a) Comparison of the behavior of scalar structure functions for $S_q^G(r)$ and $S_q^{00}(r)/\sqrt{4\pi}$ obtained by Run G3 and (b) of their local slopes.

the curves of $\zeta_q^G(r)$ deviate from $\zeta_q^{00}(r)$ in the range $r/\overline{\eta} > 400$, they are satisfactorily on the same curves at small scale even at $q = 10$. Thus we can conclude that $\zeta_q^G(r)$ contains the negligible anisotropy effects and gives the precise estimation of anomalous scaling exponents by $S_q^{00}(r)$.

4 Conclusion

We investigated the anomalous scaling laws of passive scalar turbulence under a uniform scalar gradient by performing the high resolution DNS up to 2048^3 grid points. It was observed that the scaling behavior of scalar structure function for $R_\lambda \simeq 600$ is anomalous and similar to those for the lower R_λ results. Moreover the anisotropy effects on the scaling behavior were examined in terms of the SO(3) decomposition. Although the scalar field indicates the anisotropic natures due to the existence of a mean scalar gradient, it was clarified that the anisotropy effects on the even order scalar structure function evaluated by (1) is not so significant in the ICR even at the high order ones.

The author thanks the Earth Simulator Center, the Theory and Computer Simulation Center of the National Institute for Fusion Science, and the Information Technology Center of Nagoya University for providing the computational resources. T. W. is supported by the Grant-in-Aid for Scientific Research No.17760139 in the Ministry of Education, Culture, Sports, Science and Technology of Japan.

References

1. Warhaft Z (2000) Annu Rev Fluid Mech 32: 203–240
2. Falkovich G, Gawedzki K, Vergassola M (2001) Rev Mod Phys 73: 913–975
3. Celani A, Lanotte A, Mazzino A, Vergassola M (2001) Phys Fluids 13: 1768–1783
4. Shraiman I, Siggia E D (2000) Nature London 405: 639–646
5. Watanabe T, Gotoh T (2006) Phys Fluids 18: 058105 (4pages)
6. Watanabe T, Gotoh T (2004) New J Phys 6: 40 (36pages)
7. Biferale L, Procaccia I (2005) Phys Rep 414: 43–164

Multifractal PDF Analysis of Energy Dissipation Rates in Turbulence

Toshihico Arimitsu[1], Naoko Arimitsu[2], Kyo Yoshida[1] and Hideaki Mouri[3]

[1] Grad Sch Pure and Applied Scis, Univ Tsukuba, Tsukuba 305-8571, Japan
`tarimtsu@sakura.cc.tsukuba.ac.jp`
[2] Grad Sch Environment and Information Scis, Yokohama Nat'l Univ,
Hodogaya-ku, Yokohama 240-8501, Japan
[3] Meteorological Research Institute, Tsukuba 305-0052, Japan

Abstract. The *multifractal PDF analysis*, an ensemble theory for those systems containing intermittent phenomena, is applied to the analyses of observed PDFs for energy dissipation rates extracted from experiments conducted in a wind tunnel. The generalized dimension is extracted and analyzed as well.

Keywords: multifractal analysis, energy dissipation rate, scale transformation, generalized dimension, Tsallis-type distribution function

1 Introduction

In this paper, with the help of the formula within *multifractal probability density function analysis* (MPDFA) [1, 2, 3, 4], we analyze PDFs for energy dissipation rates extracted out from time series data of wind velocity measured in the experiments conducted in a wind tunnel, and show how to get the generalized dimension from the time series date.

It is believed that the velocity field $\mathbf{u} = \mathbf{u}(\mathbf{x}, t)$ of fully developed turbulence can be described by the Navier-Stokes (N-S) equation $\partial \mathbf{u}/\partial t + (\mathbf{u} \cdot \boldsymbol{\nabla})\mathbf{u} = -\boldsymbol{\nabla} p + \nu \nabla^2 \mathbf{u}$ for incompressible fluids satisfying $\boldsymbol{\nabla} \cdot \mathbf{u} = 0$ with $p = \check{p}/\rho$ and the kinematical viscosity $\nu = \check{\eta}/\rho$, where $\check{p} = \check{p}(\mathbf{x}, t)$, $\rho = \rho(\mathbf{x}, t)$ and $\check{\eta}$ are, respectively, the pressure of fluid, the mass density and the viscosity. N-S equation is invariant under the scale transformation: [5, 6, 7]

$$\mathbf{x} \to \mathbf{x}' = \lambda \mathbf{x}, \quad \mathbf{u} \to \mathbf{u}' = \lambda^{\alpha/3}\mathbf{u}, \quad t \to t' = \lambda^{1-\alpha/3}t, \quad p \to p' = \lambda^{2\alpha/3}p \quad (1)$$

$$\nu \to \nu' = \lambda^{1+\alpha/3}\nu \quad (2)$$

for an arbitrary real number α. Because of the invariance under the scale transformation (1), the velocity derivative, the fluid particle acceleration and the energy transfer rate have "singularities", respectively, for $\alpha < 3$, $\alpha < 1.5$ and $\alpha < 1$. The exponent α plays the role of an index representing the degree of singularities within the description based on N-S equation.[1]

---

[1] As the experimental resolution in observing hydrodynamic quantities and the mesh size of lattice in conducting numerical simulations of N-S equation are finite,

117

Y. Kaneda (ed.), *IUTAM Symposium on Computational Physics and New Perspectives in Turbulence*, 117–123.
© 2008 *Springer*.

Although the energy transfer rate $\delta\epsilon_n$ is a convenient quantity for the phenomenological description of turbulence such as the energy cascade model [8], its definition in terms of the multi-point functions is not clear nor unique [9]. Here, we introduced the velocity difference (fluctuation) $\delta u_n = u(\bullet + \ell_n) - u(\bullet)$ of a component u of the velocity field $\mathbf{u}$ between two points separated by the distance $\ell_n = \ell_{\text{in}}\delta_{n-n_{\text{in}}}$, with $\delta_n = \delta^{-n}$ ($n = 0, 1, 2, \cdots$), $\delta > 1$. We put $\delta = 2$ throughout this paper. The length $\ell_{n=n_{\text{in}}} = \ell_{\text{in}}$ is the diameter of the largest eddies.[2] Note that the energy transfer rates are related to the velocity difference by the relation $|\delta\epsilon_n| = |\delta u_n|^3/\ell_n$.

As a substitute for the energy transfer rate, one introduces the formal energy dissipation rate ε_n within the inertial range through $|\delta\epsilon_n| \sim \varepsilon_n \equiv \nu_n \left(\delta u_n/\ell_n\right)^2$ which takes the values in the range $[0, \Lambda_n]$ where Λ_n is a positive real number, and can be ∞ depending on theoretical models. Here, ν_n is the effective dissipation coefficient, associated with the eddies having the diameter ℓ_n, satisfying $\nu_n/\nu_{\text{in}} = \delta_n^{1+\alpha/3}$ with $\nu_{\text{in}} = \nu_{n=n_{\text{in}}}$, hence ε_n is related to α by $\varepsilon_n/\epsilon = \delta_n^{\alpha-1}$. It is divergent in the large n limit for $\alpha < 1$, which is the same as $|\delta\epsilon_n|$. Note that ν_n satisfies the scale transformation (2) *with* (1) under which N-S equation remains unchanged. Since the dissipation rate ν_n is the quantity defined in the inertial range, it may be interpreted as the turbulent viscosity whose origin is attributed to the nonlinearity of convection term. ν_n reduces to ν at the Kolmogorov length η [10]. Note that it behaves as $\nu_n \to 0$ for the limit $n \to \infty$ as far as α satisfies $\alpha > -3$.

The energy dissipation rate ε_n within the inertial range is a convenient quantity since it is positive semi-definite, and therefore it allows the introduction of its log-normal distribution [11, 12, 13]. However, the direct observation or the extraction of the scaling property of ε_n is not possible yet. The substitution for ε_n is usually given by the average of the microscopic dissipation rate $\breve{\varepsilon}(\mathbf{r})$ per unit mass, whose origin is the effect of the dissipation term in N-S equation, over the space in a volume element ΔV_n, i.e., $\varepsilon_n \sim \int_{\Delta V_n} d^d r \, \breve{\varepsilon}(\mathbf{r})/\ell_n^d$. Here, the volume of the element ΔV_n is chosen to be ℓ_n^d with d being the dimension of real space. One expects that the values for $\breve{\varepsilon}(\mathbf{r})$ are quite close to the dissipation rate $\breve{\varepsilon}_\eta$ at about the Kolmogorov length, and are fluctuating as the same way as the dissipation rate ε_n in the inertial range, since they are connected through the energy cascade process.[3]

---

one cannot take the limit producing the singularities in practice. Therefore, it may be appropriate to interpret the term "singularity" here meaning to take abnormally large values.

[2] We are assuming that energy is put into turbulent system at about this length scale with the energy input rate ϵ which is considered to be constant.

[3] Since we are considering a stationary state of forced turbulence, we assume that $\epsilon = \langle \delta\epsilon_n \rangle = \langle \varepsilon_n \rangle$ is satisfied. The ensemble average $\langle \cdots \rangle$ is taken by $P^{(n)}(\alpha)$ given in (3) below.

2 Formulation

MPDFA starts with the assignment of the probability, to find a singularity specified by the strength α within the range $\alpha \sim \alpha + d\alpha$, in the form [2]

$$P^{(n)}(\alpha)d\alpha = \sqrt{|f''(\alpha_0)||\ln \delta_n|/2\pi} \; \delta_n^{1-f(\alpha)} \, d\alpha. \tag{3}$$

Here, $f(\alpha)$ represents an appropriate multifractal spectrum defined in the range $\alpha_{\min} \leq \alpha \leq \alpha_{\max}$, which is related to the mass exponent $\tau(\bar{q})$ through the Legendre transformation $f(\alpha) - \alpha\bar{q} = \tau(\bar{q})$ with $\bar{q} = df(\alpha)/d\alpha$ and $\alpha = \alpha_{\bar{q}} = -d\tau(\bar{q})/d\bar{q}$. The generalized dimension (the Rényi dimension) $D_{\bar{q}}$ is introduced by the relation $\tau_d(\bar{q}) = (1-\bar{q})D_{\bar{q}}$. Note that $f(\alpha)$ does not depend on n because of the scale invariance of the system under consideration.

Let us derive expressions of PDFs for the fluctuation $\delta x_n = x(\bullet + \ell_n) - x(\bullet)$ of a physical quantity related to α by the relation $|x_n| \equiv |\delta x_n/\delta x_0| = \delta_n^{\phi\alpha/3}$. Its spatial derivative defined by $|x'| = \lim_{\ell_n \to 0} \delta x_n/\ell_n \propto \lim_{n \to \infty} \ell_n^{\phi\alpha/3-1}$ diverges when $\alpha < 3/\phi$. The quantity x' reduces to the velocity derivative and fluid particle acceleration for $\phi = 1$ and $\phi = 2$, respectively, and formally to the energy transfer rate and the energy dissipation rate for $\phi = 3$. Within MPDFA, it is assumed that the probability $\Pi_\phi^{(n)}(x_n)dx_n$ to find the physical quantity x_n taking a value in the domain $x_n \sim x_n + dx_n$ can be, generally, divided into two parts as $\Pi_\phi^{(n)}(x_n)dx_n = \Pi_{\phi,\mathrm{S}}^{(n)}(x_n)dx_n + \Delta\Pi_\phi^{(n)}(x_n)dx_n$. Here, the first term describes the contribution from the abnormal part of the physical quantity x_n due to the fact that its singularities distribute themselves multifractal way in real space. This is the part given by $\Pi_{\phi,\mathrm{S}}^{(n)}(|x_n|)d|x_n| \propto P^{(n)}(\alpha)d\alpha$ [2] through the variable translation between $|x_n|$ and α. On the other hand, the second term $\Delta\Pi_\phi^{(n)}(x_n)$ represents the contributions from the dissipation term in N-S equation and so on (for the details of $\Delta\Pi_\phi^{(n)}(x_n)$, see [3, 4]). The dissipation term violates the invariance based on the scale transformation (1) without (2).[4] The second term is the correction term to the first one in the case of turbulent fluids. The values of $|x_n|$ representing the part originated from the singularities are describing the large deviations due to intermittency. Whereas the values of $|x_n|$ for the part contributing to the correction term is smaller than or the order of its standard deviation.

For those PDFs with variables taking both positive and negative values, we symmetrize the PDFs before we start analyses, under the assumption that the important characteristics of intermittency manifest themselves in the large deviations from the mean value of the quantity under consideration, i.e., the fat-tail behavior of PDFs. Then, the normalization of PDF is given by $\kappa \int_0^{\Lambda_n} dx_n \Pi_\phi^{(n)}(x_n) = 1$ with $\kappa = 2$ for the variable whose domain is $[-\Lambda_n, \Lambda_n]$

---

[4] In experiments both ordinary and numerical, turbulence is measured in a fluid with a fixed ν. Therefore, we utilize the scale transformation (1) without (2) in order to extract those variables having singular properties that are responsible to the intermittency in turbulence [6, 7].

120   T. Arimitsu et al.

and $\kappa = 1$ for $[0, \Lambda_n]$. Here, Λ_n is a positive real number that can be ∞. The formula for the mth order structure function (moments) of the variable $|x_n|$ is given by

$$\langle\!\langle |x_n|^m \rangle\!\rangle_\phi \equiv \kappa \int_0^{\Lambda_n} dx_n |x_n|^m \Pi_\phi^{(n)}(x_n) = \kappa \gamma_{\phi,m}^{(n)} + \left(1 - \kappa \gamma_{\phi,0}^{(n)}\right) a_{\phi m}\, \delta_n^{\zeta_{\phi m}} \quad (4)$$

with $\gamma_{\phi,m}^{(n)} = \int_0^{\Lambda_n} dx_n |x_n|^m \Delta\Pi_\phi^{(n)}(x_n)$, $a_{\phi m} = \sqrt{|f''(\alpha_0)|/|f''(\alpha_{\phi m/3})|}$ and $\zeta_{\phi m} = 1 - \tau(\phi m/3)$. Note here again that the independence of $\zeta_{\phi m}$ on the multifractal depth n is a manifestation of the invariance under the scale transformation.

Within A&A model, a harmonious representation of MPDFA, we adopt the Tsallis-type distribution function for $P^{(n)}(\alpha)$ [14], i.e.,

$$P^{(n)}(\alpha) = [1 - ((\alpha - \alpha_0)/\Delta\alpha)^2]^{n/(1-q)}/Z_\alpha^{(n)} \quad (5)$$

with $(\Delta\alpha)^2 = 2X/(1-q)\ln 2$. Here, q is the entropy index introduced in the definition of the Rényi entropy [15] and/or the Tsallis entropy [16, 17]. The range of α is $\alpha_{\min} \leq \alpha \leq \alpha_{\max}$ with $\alpha_{\min} = \alpha_0 - \Delta\alpha$, $\alpha_{\max} = \alpha_0 + \Delta\alpha$. The multifractal spectrum $f(\alpha)$ is given by $f(\alpha) = 1 + \log_2[1 - ((\alpha - \alpha_0)/\Delta\alpha)^2]/(1 - q)$ which, then, produces the mass exponent

$$\tau(\bar{q}) = 1 - \alpha_0 \bar{q} + 2X\bar{q}^2/(1 + \sqrt{C_{\bar{q}}}) + [1 - \log_2(1 + \sqrt{C_{\bar{q}}})]/(1 - q) \quad (6)$$

with $C_{\bar{q}} = 1 + 2\bar{q}^2(1 - q)X\ln 2$. The parameters α_0, X and q are determined as functions of μ through the relations $\tau(1) = 0$ (the energy conservation law, i.e., $\langle \varepsilon_n \rangle = \epsilon$), $\tau(2) = \mu - 1$ (the definition of the intermittency exponent μ, i.e., $\langle \varepsilon_n^2 \rangle = \epsilon^2 \delta_n^{-\mu}$) and the scaling relation $1/(1 - q) = 1/\alpha_- - 1/\alpha_+$ with $f(\alpha_\pm) = 0$.

Since, within the treatment of K41 [10], $|\delta\epsilon_n|$ is assumed to be constant and independent of n, we see that it is the case corresponding to $\alpha = 1$, i.e., $\alpha_0 = 1$ and $\Delta\alpha = 0$. If we look at this way, the arbitrariness of α, appeared in the scale transformation (1), indicates that $\delta\epsilon_n$ can be viewed as a stochastic variable, i.e., one can introduce fluctuations in $\delta\epsilon_n$. Actually, within the log-normal model [11, 12, 13], Gaussian distribution function was assigned to $P^{(n)}(\alpha)$ because of the central limit theorem under the assumption that $\ln(\varepsilon_n/\varepsilon_{n-1})$ are independent stochastic variables having a common distribution.

3 Turbulence in a Wind Tunnel

We display in Fig. 1 the time dependence of the quantity $(du/dt)^2$, closely related to the energy dissipation rates ε_n,[5] extracted out from the time-series data of the wind velocities measured in a wind tunnel of the Meteorological Research Institute in Tsukuba. The Reynolds number for this experiment is $\mathrm{Re} = 1.582 \times 10^5$ that is estimated from the microscale Reynolds number $\mathrm{Re}_\lambda = 1258$. The experimental details are found in [18].

---

[5] In replacing the time derivative to the spatial derivative, we used Taylor's frozen hypothesis, i.e., there is a given relation between the time and spatial differences since the turbulent velocity field moves downstream riding on the mean flow.

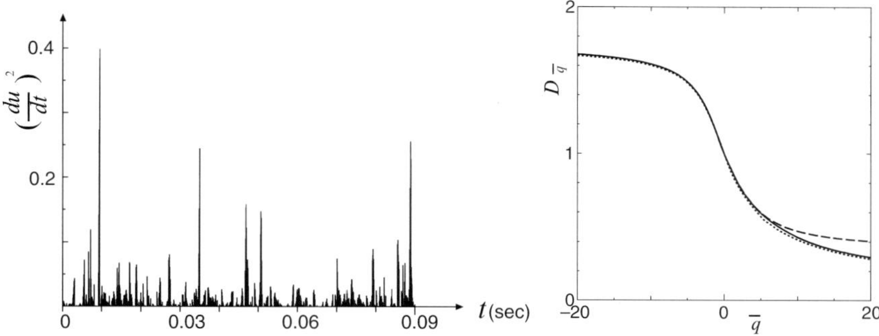

Fig. 1. (Left) Time dependence of the quantity $(du/dt)^2$ $(\mathrm{cm}^2/\mathrm{sec}^4)$.

Fig. 2. (Right) Generalized dimension $D_{\bar{q}}$ extracted out from PDFs of energy dissipation rates measured in the wind tunnel at $R_\lambda = 1255$.

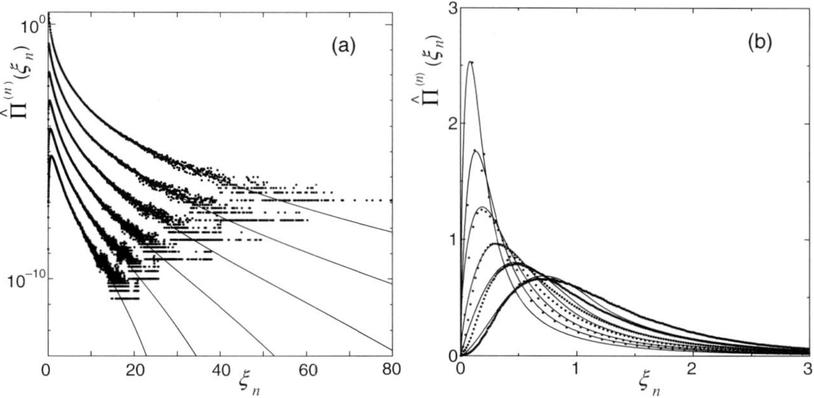

Fig. 3. PDFs of energy dissipation rates measured in the wind tunnel at $R_\lambda = 1255$ on (a) log and (b) linear scale. Open circles are the experimental data points. Solid lines represent the curves given by the present theory with $\mu = 0.230$ ($q = 0.368$). For better visibility, each PDF is shifted by -1 unit along the vertical axis in (a).

In Fig. 3 displayed on (a) log and (b) linear scale are PDFs $\hat{\Pi}^{(n)}(\xi_n)$ of the energy dissipation rates ξ_n scaled by the standard deviations in x_n. Dots are the experimental data points. The lengths r/η of the region in which energy dissipation rates are averaged to produce PDF are, from the top to bottom: 32.6, 65.1, 130, 260, 521 and 1040 with the Kolmogorov length $\eta = 1.06 \times 10^{-2}$ cm. The estimated inertial range is the region $67.3 < r/\eta < 642$. Solid lines represent the curves given by the present theoretical PDFs with $\mu = 0.230$ that gives $q = 0.368$, $\alpha_0 = 0.273$ and $X = 1.13$, self-consistently. For the theoretical curves, from top to bottom, $n = 10.0,\ 9.00,\ 8.15,\ 7.20,\ 6.20,\ 5.20$. Other parameters are found in [4].

Extracting $\zeta_{3\bar{q}}$ for $-\infty < \bar{q} < \infty$ through the formula (4) with $\kappa = 1$ out of PDFs of energy dissipation rates ($\phi = 3$), we can obtain the generalized dimension $D_{\bar{q}}$. The result is given in Fig. 2. The solid line represents the generalized dimension derived via observed PDFs of energy dissipations with the help of (4) by the new route in which we make up the lack of observed data for larger x_n by the substitution of theoretical PDF curve, $\Pi_{\phi,S}^{(n)}(x_n)$. The dotted line represents the generalized dimension given by the present theory with $\mu = 0.230$, which almost overlaps with solid line. Dashed line is the result obtained by the formula (4) *without* making up the lack of observed data for larger x_n. Note that, for $\bar{q} < 0$, contribution of the first term in (4) becomes conspicuous, therefore, we should subtract the contributions originated from the term violating the invariance of the scale transformation.

4 Discussion

By the success of MPDFA, it can be said that one has gotten a clue to search for the fundamental process of intermittency, i.e., the origin of singularities and the reason why the singularities distribute themselves *multifractally*, etc., which may provide us with a fruitful insight to produce something for the dynamical approach. We also expect that since MPDFA can reproduce observed PDF with very high accuracy, the analysis of the center part makes it possible to extract the local structure and the local dynamics of the system. We need, however, more precise investigation on the center part of PDF, smaller than 1 (the standard deviation), to explain the observed PDF as can be seen from Fig. 3(b). Anyway, it is one of the attractive future problems to find out two different dynamics one of which determines the tail part of PDF, and the other of which determines the central part.

References

1. Arimitsu T, Arimitsu N (2000) Phys Rev E 61:3237–3240
2. Arimitsu T, Arimitsu N (2001) Physica A 295:177–194
3. Arimitsu T, Arimitsu N (2005) J Phys: Conf Series 7:101–120
4. Arimitsu T, Arimitsu N, Yoshida K, Mouri H (2006) Multifractal PDF analysis for intermittent systems. In: Riccardi C, Roman H (eds) Anomalous fluctuation phenomena in complex systems: Plasma physics, bio-science and econophysics (Special Review Book for Research Signpost), Transworld Research Network, Kerala, India, in press
5. Moiseev S, Tur A, Yanovskii V (1976) Sov Phys JETP 44:556–561
6. Frisch U, Parisi G (1985) On the singularity structure of fully developed turbulence. In: Ghil M, Benzi R, Parisi G (eds) Turbulence and Predictability in Geophysical Fluid Dynamics and Climate Dynamics. North-Holland, New York
7. Meneveau C, Sreenivasan K (1987) Nucl Phys B (Proc. Suppl.) 2:49–76
8. Richardson L (1922) Wheather Prediction by Numerical Process. Cambridge Univ Press, Cambridge
9. Aoyama T, Ishihara T, Kaneda Y, Yokokawa M, Itakura K, Uno A (2005) J Phys Soc Japan 74:3202–3212

10. Kolmogorov A (1941) Dokl Akad Nauk SSSR 30:301–305; 31:538–540
11. Oboukhov A (1962) J Fluid Mech 13:77–81
12. Kolmogorov A (1962) J Fluid Mech 13:82–85
13. Yaglom A (1966) Sov Phys Dokl 11:26–29
14. Tsallis C (1999) Braz J Phys 29:1–35
15. Rényi A (1961) Proc 4th Berkeley Symp Maths Stat Prob 1:547
16. Tsallis C (1988) J Stat Phys 52:479–487
17. Havrda J, Charvat F (1967) Kybernatica 3:30–35
18. Mouri H, Hori A, Kawashima Y (2004) Phys Rev E 70:066305

Scale Interactions and Non-Local Flux in Hydrodynamic Turbulence

Pablo D. Mininni, Alexandros Alexakis, and Annick Pouquet

NCAR, PO Box 3000, Boulder, CO 80307, USA
mininni@ucar.edu, alexakis@ucar.edu, pouquet@ucar.edu

Direct numerical simulations of hydrodynamical turbulence are performed in the presence of a large-scale forcing up to Taylor-based Reynolds numbers of 800 for a Taylor-Green flow and 1000 for ABC forcing. Statistics on the triadic interactions responsible for the transfer of energy, and on their integrated counterparts are obtained from simulations with a turbulent steady state integrated for more than ten eddy turn-over times. The maximum resolution of the runs performed on regular grids are 1024^3 points. It is shown that the energy transfer is local but at the highest resolution as much as 20% of it comes from non-local interactions involving widely separated scales, with a particular emphasis on the forcing scale k_0^{-1}. The scaling of the ratio of nonlocal to local energy flux with Reynolds number is found to be a power law.

Keywords: turbulence, high-Reynolds-number flows, direct numerical simulations, isotropic and homogeneous turbulence, scale interactions

1 Introduction

The Kolmogorov phenomenology is based on the assumption of local interactions in Fourier space, most of the energy transfer occurring between eddies of similar sizes. The numerical study of such an assumption has received attention both numerically [1] and theoretically as well [2]. Evidence of non-local interactions and slower than expected recovery of isotropy and homogeneity in the small scales have been observed in the atmosphere [3] and in experiments [4, 5]. The failure to recover isotropy has been linked to the presence of small scale strong events (vortex filaments) in the flow [6], with a memory of how the flow is driven in the large scales. Recently, it was observed in experiments that small scales are driven by two processes [5], a fast process local in Fourier space, and a slow process which is non-local in Fourier space.

We revisit this problem using high-resolution runs that allow for some scale separation in the inertial range. Numerical investigations of turbulent flows are often done in a classical framework, that of three-dimensional rectangular periodic boundary conditions. This allows for using one of the best

125

Y. Kaneda (ed.), *IUTAM Symposium on Computational Physics and New Perspectives in Turbulence*, 125–130.

existing numerical techniques for investigating turbulence, namely the Orszag-Patterson pseudospectral method. Another reason is that the results of such computations appear in a form that compares easily with homogeneous and isotropic turbulence theory predictions in the Kolmogorov mold. However, in many of these studies artificial initial conditions or forcing functions are used. In the case of forced runs, phases are changed randomly in time and coefficients are picked to ensure statistical isotropy and homogeneity in the large scales. This is not the case in turbulent flows observed in nature, where generally turbulent fluctuations arise as a result of an instability of a (often anisotropic and inhomogeneous) large scale flow.

We discuss results from simulations of incompressible hydrodynamic turbulence (see Figs. 1 and 2 for rendering of the enstrophy displaying the well-known myriad of vortex filaments) with Taylor Reynolds numbers R_λ up to 1000 [7, 8] ($R_\lambda = U\lambda/\nu$; U is the rms velocity, ν the kinematic viscosity, $\lambda = 2\pi(E/\Omega)^{1/2}$ the Taylor scale, E the energy, and Ω the enstrophy). We use random forcing or coherent forcing functions (the latter giving a well defined flow in the large scales) with the intent of revising some of the prevailing assumptions about the locality of interactions between spatial scales in Kolmogorov theory. The study is carried out by studying the nonlinear triadic interactions and transfer functions in Fourier space. Sharp filters are used to define shells. The forcing functions are either a Taylor-Green (TG) vortex [10], the ABC flow [11], or isotropic random forcing. While TG forces only the x and y components of the flow, ABC is the superposition of Beltrami flows in three different directions. In all cases, a single shell k_0 is forced.

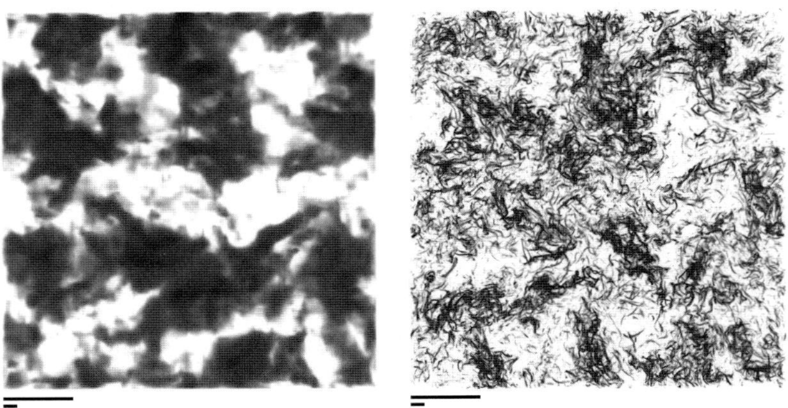

Fig. 1. Rendering of energy density filtered to preserve only the large scales (left), and enstrophy density at full resolution (right) using VAPOR [9] in a $1024\times64\times1024$ slice of a 1024^3 simulation with helical forcing. The black bars at the bottom indicate respectively the integral, Taylor, and dissipation scales. Clusters of vortex tubes can be observed, and the separation between these regions is close to the integral scale of the flow and corresponding to energy patches as observed on the left.

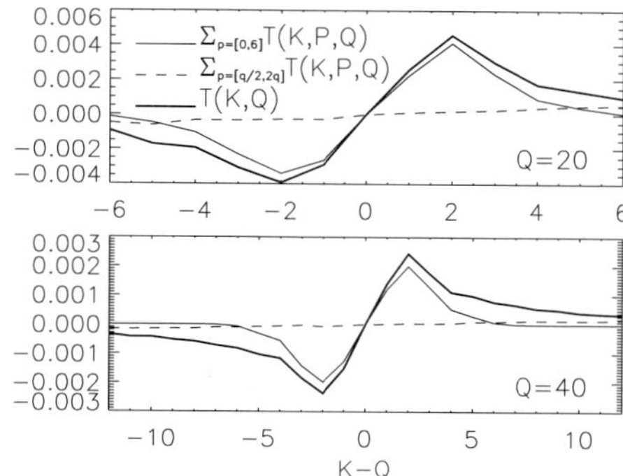

Fig. 2. Left: rendering of enstrophy density in a subvolume of a 1024^3 helical simulation (for the three black bars, see Fig. 1). The length of the vortex tubes ranges from the Taylor scale to the integral scale, while the thickness is close to the dissipation scale. Right: shell-to-shell transfer in a 1024^3 simulation at $Q = 20$ and 40. The dashed and dotted lines indicate respectively the contribution to $T_2(K,Q)$ due to triadic interactions with the large scale flow, and within octaves in Fourier space.

2 Triadic Interactions and Energy Transfer

Triadic interactions are defined as usual, with $\mathbf{v}$ the velocity and $\mathbf{k} = \mathbf{p} + \mathbf{q}$: $T_3(K,P,Q) = -\int \mathbf{v}_K \cdot (\mathbf{v}_P \cdot \nabla)\mathbf{v}_Q \mathrm{d}\mathbf{x}^3$; the subindex K denotes the velocity field filtered to preserve modes only with wave vectors $K \leq |\mathbf{k}| < K + 1$. If we sum over P, we obtain the shell-to-shell energy transfer $T_2(K,Q)$:

$$T_2(K,Q) = \sum_P T_3(K,P,Q) = -\int \mathbf{v}_K \cdot (\mathbf{v} \cdot \nabla)\mathbf{v}_Q \mathrm{d}\mathbf{x}^3. \qquad (1)$$

Positive transfer implies that energy is transferred from shell Q to K, and negative from K to Q. Finally, the energy transfer function $T_1(K)$ and the energy flux through the wavenumber k are respectively

$$T_1(K) = \sum_Q T_2(K,Q), \qquad \Pi(k) = -\sum_K^k T_1(K). \qquad (2)$$

3 Local Versus Non-Local Interactions

For hydrodynamic turbulence, we find that nonlinear triadic interactions are dominated by their non-local components, involving widely separated scales, even though the nonlinear transfer itself is local. Figure 2 shows the shell-to-shell energy transfer $T_2(K,Q)$ in a 1024^3 simulation forced at $k_0 \sim 3$, computed at two different scales in the inertial range. Although the transfer is local (energy is transferred between nearby shells), triadic interactions with the large scale flow ($0 \leq P \leq 6$) give a significant contribution to $T_2(K,Q)$, and are responsible for the peaks at $K - Q \sim \pm k_0$. Local triadic interactions (i.e. interactions between wavenumbers K, P, and Q in the same octave) are responsible for the tails of $T_2(K, Q)$ for large values of $K - Q$.

Performing simulations changing the wavenumber k_0 at which the energy input occurs in the flow, it is clear from the position of the peaks in the shell-to-shell transfer that the dominant triadic interactions are with the forcing scale. In fact, a study of $T_3(K, P, Q)$ in 1024^3 simulations with ABC and TG forcing shows that individual triadic interactions with $P \sim k_0$ are two orders of magnitude larger than individual triadic interactions involving modes within the local octave. Similar results for T_2 and T_3 were obtained when isotropic random forcing was used, albeit at a lower resolution (256^3).

In spite of the strong triadic interactions, the simulations discussed present a scaling in the energy spectrum close to the classical Kolmogorov law, and constant flux in the inertial range. The reasons for this can be twofold. On one hand, it was shown [7] that the multiscale character of the vortex tubes (see Fig. 2) permits non-local interactions to give a scaling consistent with Komogorov law. On the other hand, although local triadic interactions are weaker, they are more numerous. As modes are summed to obtain the energy flux in (1) and (2), the local interactions start to dominate.

We can quantify the relevance of local vs. non-local interactions in the energy flux by means of $\Pi_{LS}(k)/\Pi(k)$, the ratio of energy flux due to interactions with the large scale flow [i.e., doing the sums in (1) and (2) only for $0 \leq P \leq 6$], to the total energy flux. At high Reynolds numbers ($R_\lambda \sim 800$), non-local interactions with the large scale flow give $\sim 20\%$ of the total flux, while local interactions within triads are responsible for most of the remaining $\sim 80\%$ (note there are also non-local interactions with modes outside the octave but with $P > 6$). In the 1024^3 simulations, the ratio $\Pi_{LS}(k)/\Pi(k)$ is approximately constant for wavenumbers k in the inertial range, while for smaller Reynolds numbers $\Pi_{LS}(k)/\Pi(k)$ decreases with increasing k in the inertial range.

As the Reynolds number is decreased, the fraction $\Pi_{LS}(k)/\Pi(k)$ increases. Figure 3 shows the ratio $\Pi_{LS}(k)/\Pi(k)$ evaluated at the wavenumber $k_\lambda = 2\pi/\lambda$, where λ is the Taylor scale, as a function of the Reynolds numbers based on the integral scale of the flow (R_e), and based on the Taylor scale (R_λ). The best fit gives a scaling $\Pi_{LS}(k)/\Pi(k) \sim R_e^{-0.7}$, indicating that as the Reynolds number is increased the contribution of the non-local interactions with the large scale flow to the total flux decreases slowly. The same scaling was observed for all forcing functions explored, although the prefactor is forcing dependent. The scaling of $\Pi_{LS}(k)/\Pi(k)$ with R_λ is consistent with the well-known scaling $R_\lambda \sim R_e^{1/2}$. From dimensional grounds $\Pi_{LS}(k_\lambda) \sim U_L u_\lambda^2/L$ (U_L is a characteristic velocity at the large scale L, and u_λ is a characteristic velocity at the Taylor scale; note that this relation does not take into account that structures are in fact multiscale [7]), while for $\Pi_{LS}/\Pi \ll 1$, $\Pi(k_\lambda) \sim u_\lambda^3/\lambda$. As a result, $\Pi_{LS}(k_\lambda)/\Pi(k_\lambda) \sim R_e^{-1/2}$. The

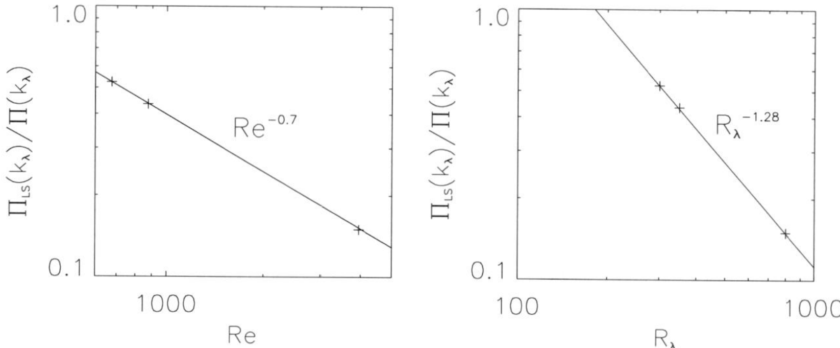

Fig. 3. Ratio of the energy flux due to interactions with the large scale flow to the total energy flux, evaluated at the Taylor scale λ, as a function of the Reynolds number R_e based on the integral scale (left) and the Reynolds number R_λ based on the Taylor scale (right). The three crosses correspond to simulations with $R_\lambda \sim 300$ (resolution of 256^3 grid points), $R_\lambda \sim 350$ (512^3), and $R_\lambda \sim 800$ (1024^3), while the straight lines indicate the best fit, *viz.* $R_e^{-0.7}$ and $R_\lambda^{-1.28}$ respectively (TG forcing).

condition $\Pi_{LS}/\Pi \ll 1$ is notsatisfied in the simulations with $R_\lambda \sim 300$ and 350. A phenomenological understanding of such power laws may have to further await results from simulations at higher Reynolds numbers, in order to study the scaling for different wavenumbers in the self-similar inertial range.

4 Discussion

We reported the results of analyzing the energy transfer, triadic interactions, and energy flux in several high resolution DNS with periodic boundaries. Several forcing functions were studied, and spatial resolutions up to 1024^3 grid points were considered.

In all cases we confirmed that the cascade of energy is local between Fourier shells, although it is strongly mediated by individual triadic interactions which are non-local in Fourier space (corresponding to widely separated scales). As a result, the energy cascades from one shell to the next with a fixed step proportional to the forcing wavenumber k_0. The forcing function leaving a trace of its influence on the small scales, a feature that can be related to the slow return to isotropy in the small scales (see e.g. [12]). As more modes are summed to define the energy flux, the larger number of small scale modes dominate, and in the energy flux local interactions within triads are responsible for the larger contribution.

However, the relevance of non-local interactions is non negligible, and decays slowly with the Reynolds number. As a result, the large scale flow can play an important role in the development and the statistical properties of the small scale turbulence. The link between these findings and the intermittency of the small scales, and their consequences for modeling of turbulent flows are discussed in Ref. [8]. As a last remark, we note that stronger non-locality is observed in magnetohydrodynamics [13], both for a traditional Ohm's law and when the Hall current is included, as relevant to the magnetosphere [14]; this type of analysis also allows for an understanding of the role of magnetic helicity [15]. Note that the scaling of the degree of nonlocality of nonlinear transfer in turbulent flows with Reynolds numbers will have to be confirmed using high-resolution numerical simulations for which a more substantial scale separation can be achieved.

NSF grant CMG-0327888 at NCAR is acknowledged. Computer time was provided by NCAR and by the NSF Computing System at PSC.

References

1. Domaradzki JA (1988) Phys Fluids 31:2747–2749; Yeung PK, Brasseur JG (1991) Phys Fluids A 3:884–897; Brasseur JG, Wei CH (1994) Phys Fluids 6:842–870; Kishida K et al (1999) Phys Rev Lett 83:5487–5490
2. Waleffe F (1992) Phys Fluids A 4:350–363; Eyink GL (2005) Physica D 207:91–116; Verma MK et al (2005) Pramana J Phys 65:297–310
3. Stewart RW (1969) Radio Sci 4:1269–1278
4. Shen X, Warhaft Z (2000) Phys Fluids 12:2976–2989; Staicu A, van de Water W (2003) Phys Rev Lett 90:094501
5. Poulain C, Mazellier N, Chevillard L, Gagne Y, Baudet C (2006) Eur Phys J B 53:219–224
6. Pumir A, Shraiman BI (1995) Phys Rev Lett 75:3114–3117
7. Alexakis A, Mininni PD, Pouquet A (2005) Phys Rev Lett 95:264503
8. Mininni PD, Alexakis A, Pouquet A (2006) Phys Rev E 74:016303
9. Clyne J, Rast M (2005) A prototype discovery environment for analyzing and visualizing terascale turbulent fluid flow simulations. In: Erbacher RF et al (eds) Visualization and data analysis 2005 SPIE, Bellingham Wash (URL: http://www.vapor.ucar.edu)
10. Taylor GI, Green AE (1937) Proc Roy Soc Lond Ser A 158:499–512; Brachet ME et al (1983) J Fluid Mech 130:411–452
11. Arnol'd VI (1965) C R Acad Sci Paris 261:17–21 (1965); Galloway D, Frisch U (1984) Geophys Astrophys Fluid Dyn 29:13–18
12. Pumir A, Shraiman BI (1995) Phys Rev Lett 75:3114–3117; Shen X, Warhaft Z (2000) Phys Fluids 12:2976–2989
13. Alexakis A, Mininni PD, Pouquet A (2006) Phys Rev E 72:046301; Mininni PD, Alexakis A, Pouquet A (2006) Phys Rev E 72:046302
14. Mininni PD, Alexakis A, Pouquet A (2006) J Plasmas Phys (in press), arXiv:physics/0510053
15. Alexakis A, Mininni PD, Pouquet A (2006) Astrophys J 640:335–343

Control over Multiscale Mixing
in Broadband-Forced Turbulence

Arkadiusz K. Kuczaj[1] and Bernard J. Geurts[1,2]

[1] Department of Applied Mathematics, J.M. Burgers Center for Fluid Mechanics,
University of Twente, P.O. Box 217, 7500 AE Enschede, The Netherlands
[2] Department of Applied Physics, Eindhoven University of Technology,
P.O. Box 513, 5600 MB Eindhoven, The Netherlands

Abstract. The effects of explicit flow modulation on the dispersion of a passive scalar field are studied. Broadband forcing is applied to homogeneous isotropic turbulence to modulate the energy cascading and alter the kinetic energy spectrum. Consequently, a manipulation of turbulent flow can be achieved over an extended range of scales beyond the directly forced ones. This modifies transport processes and influences the physical-space turbulent mixing of a passive scalar field. We investigate by direct numerical simulation the stirring-efficiency associated with turbulence modified by forcing. This is quantified by monitoring the surface-area and wrinkling of a level-set of the passive scalar field. We consider different forcing to manipulate the quality and rate of mixing. The instantaneous mixing efficiency measured in terms of surface-area or wrinkling is found to increase when additional energy is introduced at the smaller scales. The increased intensity of small scales significantly influences the small-scale mixing characteristics depicted by wrinkling, while the forcing of large scales primarily affects the surface-area. Evaluation of geometrical statistics in broadband-forced turbulence indicates that the self-amplification process of vorticity and strain is diminished. This leads generally to smaller extremal values of the velocity gradients but higher average values as a result of the competition between the natural cascading processes and the explicit small-scales forcing.

Keywords: modulated turbulence, multi-scale forcing, passive scalar mixing

1 Introduction

Turbulent mixing of embedded scalar fields is important in a diverse range of fluid mechanics problems, from process-engineering, environmental issues to non-premixed combustion. The efficiency of mixing is governed by a number of aspects. Nowadays, growing computational capabilities allow the determination of statistics of turbulent flows at quite high Reynolds and Schmidt numbers [1]. Simultaneously, the engineering approach is directed towards control of mixing by modulation of the driving velocity fields [2]. Recently, the use of multiscale forcing methods was proposed to model turbulent flows that are disturbed at various spatial scales [3] as may arise in case of flows through

Y. Kaneda (ed.), *IUTAM Symposium on Computational Physics and New Perspectives in Turbulence*, 131–136.

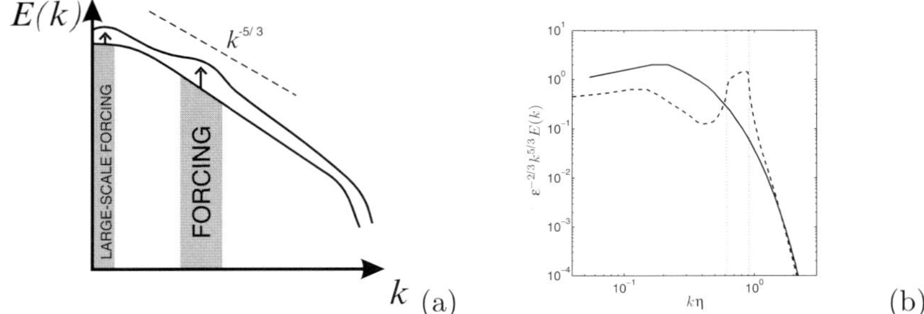

Fig. 1. (a) Broadband forcing in spectral space. (b) Time-averaged compensated energy spectrum for the large-scale ($\mathbb{K}_{1,1}$: solid) and additional broadband ($\mathbb{K}_{17,24}$: dashed) forced turbulence (ε - energy-dissipation rate, η - Kolmogorov scale).

complex geometrical structures such as metal foams, or forest canopies. These numerical experiments indicate that turbulent mixing properties, e.g., expressed by surface-area and surface-wrinkling growth-parameters of scalar level-sets, can be significantly influenced by external agitation.

We consider the incompressible Navier–Stokes equations with broadband forcing working as a complex stirrer in which a spectrum of length-scales is simultaneously perturbed. Traditionally, only large-scale forcing was included in a simulation. This induces an average flow of energy toward smaller scales. The additional forcing in a high wavenumber band agitates a specific range of spatial scales as depicted in Fig. 1(a). We also investigated the case in which forcing is applied to the high-k range only (see Ref. [3]). We found that such forcing leads to a strongly reduced turbulence intensity, e.g., expressed in a significantly reduced Taylor-Reynolds number. In this paper we decided to always include the large-scale energy injection as driving force and consider the effects of additional small-scale forcing bands. We focus on the control over basic mixing-properties that may be obtained from such explicit broadband forcing and concentrate on the consequences *(i)* for the time needed to reach a perfectly-mixed state and *(ii)* the accumulated large- and small-scale mixing.

We investigate the dispersion of strongly localized initial scalar concentrations. Direct numerical simulation of the forced turbulence shows that the maximal surface-area and wrinkling as well as the time at which such a maximum is achieved can be controlled by variation of forcing parameters. The time-integrated surface-area and wrinkling are indicators of the accumulated effect. The simulations show that at small Schmidt numbers, a higher emphasis on small-scale flow agitation yields a significant increase in the mixing.

2 Computational Flow Model

We solve the incompressible Navier–Stokes equations in spectral space:

$$\left(\partial_t + \mathrm{Re}^{-1} k^2\right) u_\alpha(\mathbf{k}, t) = M_{\alpha\beta\gamma} \sum_{\mathbf{p+q=k}} u_\beta(\mathbf{p}, t) u_\gamma(\mathbf{q}, t) + F_\alpha(\mathbf{k}, t), \quad (1)$$

where $u_\alpha(\mathbf{k}, t)$ is the velocity field coefficient at wavevector $\mathbf{k}$ ($k = |\mathbf{k}|$) and time t, and Re is the computational Reynolds number. The tensor $M_{\alpha\beta\gamma} = (k_\beta D_{\alpha\gamma} + k_\gamma D_{\alpha\beta})/(2\imath)$, where $D_{\alpha\beta} = \delta_{\alpha\beta} - k_\alpha k_\beta/k^2$, accounts for the pressure and incompressibility effects. We adopted a forcing procedure motivated by flow through a fractal gasket [4]. This particular forcing has a constant energy input rate ε_w for the entire system:

$$F_\alpha(\mathbf{k}, t) = \frac{\varepsilon_w k^\beta}{\sum_{\mathbf{k} \in \mathbb{K}_{m,p}} \sqrt{|\mathbf{u}(\mathbf{k}, t)|} k^\beta} \, e_\alpha(\mathbf{k}, t). \tag{2}$$

Each forced band $\mathbb{K}_{m,p}$ ($m \le p$) consists of $p - m + 1$ adjacent spherical shells $\mathbb{S}_n = 2\pi(n - 1/2)/L_b < |\mathbf{k}| \le 2\pi(n + 1/2)/L_b$: $m \le n \le p$. Here L_b is the size of the computational domain. We force the first shell $\mathbb{K}_{1,1}$ with a constant energy injection rate $\varepsilon_{w,1}$ and a single high-k band $\mathbb{K}_{m,p}$ with $\varepsilon_{w,2}$. The vector $\mathbf{e}(\mathbf{k}, t) = \mathbf{u}(\mathbf{k}, t)/|\mathbf{u}(\mathbf{k}, t)| + \imath \mathbf{k} \times \mathbf{u}(\mathbf{k}, t)/(|\mathbf{k}||\mathbf{u}(\mathbf{k}, t)|)$ has the general form proposed in [4] and the complexity of the stirring object is parameterized by the exponent $\beta = D_f - 2$ related to the fractal dimension D_f.

The scalar concentration C evolves in a velocity field $\mathbf{v}$ by:

$$\partial_t C(\mathbf{x}, t) + (\mathbf{v}(\mathbf{x}, t) \cdot \nabla)C(\mathbf{x}, t) = (\mathrm{Re\,Sc})^{-1}\nabla^2 C(\mathbf{x}, t), \tag{3}$$

where Sc is the Schmidt number. We adopt a level-set integration method to quantify basic mixing-properties of the evolving scalar fields [5]. Geometric properties of a level-set $S(a, t) = \{\mathbf{x} \in \mathbb{R}^3 \mid C(\mathbf{x}, t) = a\}$ may be evaluated by integrating a corresponding 'density function' g over this set. In fact, we have:

$$I_g(a, t) = \int_{S(a,t)} dA \, g(\mathbf{x}, t) = \int_V d\mathbf{x} \, \delta(C(\mathbf{x}, t) - a)|\nabla C(\mathbf{x}, t)|g(\mathbf{x}, t), \tag{4}$$

where the volume V is the flow-domain. Setting $g(\mathbf{x}, t) = 1$ we may determine the surface-area A of S. In case $g(\mathbf{x}, t) = |\nabla \cdot \mathbf{n}(\mathbf{x}, t)|$, where $\mathbf{n}(\mathbf{x}, t) = \nabla C(\mathbf{x}, t)/|\nabla C(\mathbf{x}, t)|$ is a unit normal vector, we can determine the wrinkling W of S. We focus on the evolution of the surface-area and wrinkling, monitoring the instantaneous value as well as the accumulated effect:

$$\vartheta_Z(a, t) = \frac{I_Z(a, t)}{I_Z(a, 0)} \; ; \; \zeta_Z(a, t) = \int_0^t \vartheta_Z(a, \tau)d\tau \; ; \; Z \in \{A, \, W\}. \tag{5}$$

By determining ϑ_A and ϑ_W we may quantify the rate at which surface-area and wrinkling develop, the maximal values that are obtained and the time-scale at which these are achieved. The cumulative measures $\zeta_A(a, t)$ and $\zeta_W(a, t)$ express the total surface-area and wrinkling that has developed in the course of time. The measures (5) express 'mixing efficiency' in terms of the progress *relative* to the initial state. It is not straightforward to relate this progress-interpretation to the optimal mixed state that could maximally be achieved. However, (5) does allow a clear interpretation of changes in the turbulent transport arising from differences in the forcing procedure.

3 Mixing Efficiency

The broadband forcing is observed to modify the kinetic energy spectrum in a strongly non-local manner as seen in Fig. 1(b). This illustrates the deviations from the classical Kolmogorov picture that is characterized by a $-5/3$ slope in the spectral energy distribution (for more details see [3]).

To establish the influence of forcing on mixing properties we simulated the spreading of a passive tracer at Schmidt number $Sc = 0.7$. The length and time scales of the simulations are chosen by picking $L_b = 1$ and $\varepsilon_{w,1} = 0.15$ for the large-scale energy injection rate as a reference case. Choosing the computational Reynolds number $\mathrm{Re} = 1061$ results in a Taylor-Reynolds number $R_\lambda \approx 50$. The simulations started from a spherical tracer distribution C of radius $r = 3/16$ scaled to be between 0 and 1 and the level-set $a = 1/4$ was considered. The resolution requirements were satisfactorily fulfilled: $k_{\max}\eta$ ranges from 2.3 to 3.5 using a resolution in the range $128^3 - 192^3$ grid-cells. Here $k_{\max}$ is the highest wave-number that is resolved in the simulation. For the passive scalar these resolutions correspond to $k_{\max}\eta_{OC}$ in the range from 3 to 4.5, where η_{OC} is the Obukhov-Corrsin scale [6]. The characterization of the mixing-efficiency was based on an ensemble of 20 simulations, each starting from an independent fully-developed realization of the velocity field. Individual velocity fields were separated by two eddy-turnover times.

In Fig. 2 we compare the instantaneous and accumulated mixing properties for a number of forcing parameters. We include large-scale forcing in $\mathbb{K}_{1,1}$ at $\varepsilon_{w,1} = 0.15$ as well as forcing of the band $\mathbb{K}_{5,8}$ at various $\varepsilon_{w,2}$. We observe that the large-scale forcing mainly governs the development of the surface-area, while forcing in the second band has a larger influence on the wrinkling. An increase in the strength of the forcing in $\mathbb{K}_{5,8}$ leads to a slight increase in ϑ_A and a considerable reduction in the time at which ϑ_A reaches its maximum. The final cumulative surface area, however, decreases with increasing $\varepsilon_{w,2}$. In contrast, an increase in $\varepsilon_{w,2}$ quite strongly influences the instantaneous wrinkling; the maximal value increases and the time of maximal mixing decreases. The cumulative effect on W increases notably with increasing $\varepsilon_{w,2}$. These results give an overall view of the changes in turbulence characteristics and transport properties arising from small-scale forcing. Similar results were obtained when forcing the band $\mathbb{K}_{13,16}$ or even $\mathbb{K}_{17,24}$ instead.

Evaluation of geometrical statistics [7] shows that broadband forcing considerably changes the general characteristics of turbulence. The self-amplification process of vorticity and strain is diminished and the statistical flow-structure altered. This can be seen in Fig. 3 where the pdfs of the alignment between the vorticity $\boldsymbol{\omega} = \nabla \times \mathbf{u}$ and eigenvectors of the rate of strain and the vortex stretching vector, which is given by $W_i \equiv \omega_j S_{ij}$ in terms of the vorticity and the rate of strain tensor $\mathbf{S}$, are plotted for various forcing-strengths in the second band. Increased forcing of the small scales leads to less pronounced alignment.

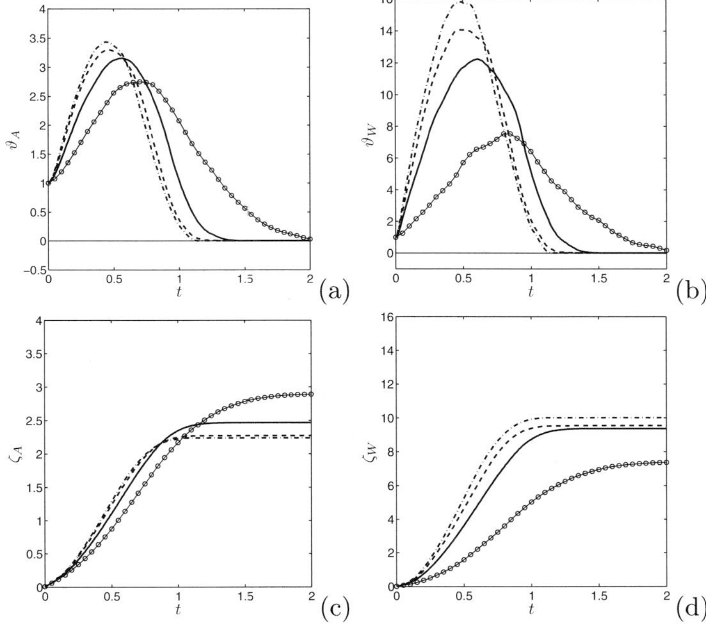

Fig. 2. Evolution of passive scalar dispersion parameters: a) surface-area ϑ_A, b) wrinkling ϑ_W, c) accumulated surface-area ζ_A, d) accumulated wrinkling ζ_W. Large-scale forcing $\mathbb{K}_{1,1}$ with $\varepsilon_{w,1} = 0.15$ and additional forcing in the band $\mathbb{K}_{5,8}$ at $\varepsilon_{w,2} = 0, 0.30, 0.45, 0.60$ ($\bullet$, solid, dash, dash-dotted).

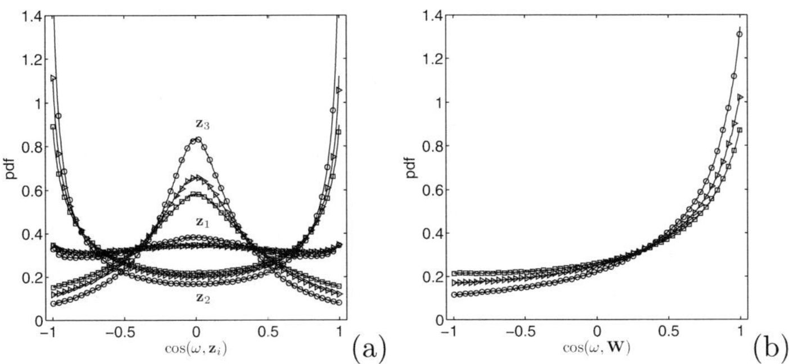

Fig. 3. PDFs of the cosine of the angle between vorticity $\boldsymbol{\omega}$ and (a) the eigenvectors $\mathbf{z}_i$ of the rate of strain tensor, and (b) the vortex stretching vector $\mathbf{W}$, for large-scale forcing $\mathbb{K}_{1,1}$ at $\varepsilon_{w,1} = 0.15$ and additional broadband forcing with energy input $\varepsilon_{w,2} = 0, 0.15, 0.45$ to the second band $\mathbb{K}_{17,24}$ ($\circ$, $\triangleright$, $\square$).

4 Conclusions

Forcing methods agitating a flow in a wide range of scales induce significant differences compared to the case obtained classically in which only the large scales are forced. In this study we devoted attention to a recently proposed multiscale forcing that models a flow under the influence of an additional perturbation by a complex stirrer [4]. We performed numerical simulations of the dispersion of a passive scalar field in a turbulent flow that is driven by such forcing. By monitoring global properties of level-sets of the evolving passive scalar we could quantify the modification of the mixing that results from broadband forcing. It was found that broadband forcing causes additional production of smaller scales in the flow. This is directly responsible for the enhancement of wrinkling. In contrast, the surface-area of a level-set of the tracer is found to be mainly governed by convective sweeping by the larger scales in the flow. Hence, the surface-area is controlled to a greater extent by the energy injected at the largest scales. The additional energy introduced by forcing at small scales competes with processes that govern Kolmogorov-type turbulence, e.g., expressed by the self-amplification of vorticity and strain and vortex stretching. The forcing also changes the structure of turbulence; it modifies the alignment of vorticity with eigenvectors of the rate of strain tensor. Future study will be devoted to clarifying this role of the small-scale forcing by evaluating the geometrical statistics of the turbulent flow as function of forcing parameters.

Acknowledgments

This work is part of the FOM research program. AKK would like to thank Arkady Tsinober (Imperial College, London) for many fruitful comments regarding geometrical statistics in turbulence.

References

1. Yeung PK, Donzis DA, Sreenivasan KR (2005) Phys Fluids 17:081703
2. Kearney M Engineered fractal cascades for fluid control applications In Proc of Fractals in Engineering Conference, Arcachon, France INRIA, June, 1997
3. Kuczaj AK, Geurts BJ (2006) J Turbul 7(67):1–28
4. Mazzi B, Vassilicos JC (2004) J Fluid Mech 502:65
5. Geurts BJ (2001) J Turbul 2:1–24
6. Watanabe T, Gotoh T (2004) New J Phys 6(40):1
7. Tsinober A, Galanti B (2003) Phys Fluids 15(11):3514–3531

Coherent Structures in Marginally Turbulent Square Duct Flow

Markus Uhlmann[1], Alfredo Pinelli[1], Atsushi Sekimoto[2]
and Genta Kawahara[2]

[1] Modeling and Numerical Simulation Unit, CIEMAT, Spain
[2] Division of Nonlinear Mechanics, Osaka University, Japan
`kawahara@me.es.osaka-u.ac.jp`

Abstract. Direct numerical simulation of fully developed turbulent flow in a straight square duct was performed in order to determine the minimal requirements for self-sustaining turbulence. It was found that turbulence can be maintained for values of the bulk Reynolds number above approximately 1100, corresponding to a friction-velocity-based Reynolds number of 80. The minimum value for the streamwise period of the computational domain measures around 190 wall units, roughly independently of the Reynolds number. Furthermore, we present a characterization of the marginal state, where coherent structures are found to have significant relevance to the appearance of secondary flow of Prandtl's second kind.

Keywords: square duct, marginally turbulent flow, coherent structure, secondary flow

1 Introduction

The appearance of secondary flow of Prandtl's second kind is a well-known phenomenon in fully developed turbulent duct flow. The intensity of the secondary flow is two orders of magnitude smaller than the mean axial velocity; however, it plays an important role in the cross-streamwise momentum transfer. To the authors' knowledge the mechanism of generation of the secondary flow has not been elucidated, albeit a number of investigations which were focussed upon the budget of statistical quantities, such as the mean streamwise vorticity component [1, 2].

We shall consider the secondary flow from the viewpoint of coherent structures, i.e. streamwise vortices and streaks, which are observed in the near-wall region. Coherent vortices and streaks are believed to be key ingredients of the self-sustaining process of near-wall turbulence, which has been extensively studied in the plane channel [3]. The main objective of our present study is the characterization of the relation between the coherent structures in duct flow and the generation of secondary flow. For very low Reynolds numbers the coherent structures have a cross-streamwise length scale comparable to

137

Y. Kaneda (ed.), *IUTAM Symposium on Computational Physics and New Perspectives in Turbulence*, 137–142.
© 2008 *Springer*.

the duct width and the length scale of the secondary flow. Therefore, we expect that they have significant relevance to the appearance of the secondary flow under marginal conditions. Since the critical values for the sustenance of turbulence are not known in this configuration, we first conduct a systematic minimization study similar to reference [4]. Then we proceed by characterizing the secondary flow pattern under marginal conditions quantitatively. Finally we present the p.d.f. of the location of the centers of the coherent vortices to be compared with the corresponding secondary flow pattern.

2 Numerical Methods

We consider an incompressible viscous fluid flowing through an infinite straight square duct. The Navier-Stokes equations are solved by means of a pressure correction method. We use an implicit scheme for the viscous terms, and a three-step Runge-Kutta method for the non-linear terms [5]. For the spatial discretization, Fourier expansion is employed in the streamwise (x) direction, while Chebyshev-polynomial expansions are used in the cross-streamwise (y, z) directions. The non-linear terms are evaluated pseudo-spectrally with full de-aliasing in the streamwise direction. The Helmholtz and Poisson problems for each Fourier coefficient are solved by the fast diagonalisation technique of [6].

Turbulent statistics obtained with our implementation of this spectral method have been confirmed to be in good agreement with results from finite-difference-based DNS of Gavrilakis [1] as well as experimental measurements of Kawahara, *et al.* [7].

In the following we present results from simulations performed at several Reynolds numbers, $Re_b \equiv u_b h/\nu < 2600$, where u_b is the bulk mean velocity, h is half the duct width, and ν is the kinematic viscosity of the fluid. In our simulations a constant mass flow rate is imposed. Furthermore, for each case the number of Fourier modes was chosen such that the streamwise grid spacing was below 15 wall units, and the number of Chebyshev polynomials was adjusted such that the maximum cross-streamwise grid spacing was less than 6 wall units.

3 Marginal Reynolds Number and Streamwise Period

In order to determine marginally turbulent states, Re_b is gradually reduced from $Re_b = 2200$ or the streamwise period is shortened from $L_x/h = 4\pi$. Figure 1(a) shows the friction factor f as a function of Re_b for several values of L_x/h. It can be seen that the friction factor increases with decreasing Re_b and then drastically reduces to the lower level of a laminar state, indicated by a solid line, around the marginal Reynolds number $Re_b = 1100$. If we use the mean friction velocity u_τ as a velocity scale, the corresponding marginal Reynolds number is around $Re_\tau \equiv u_\tau h/\nu \approx 80$. At large Reynolds numbers, the minimal streamwise period for sustaining turbulence (lowest open circles

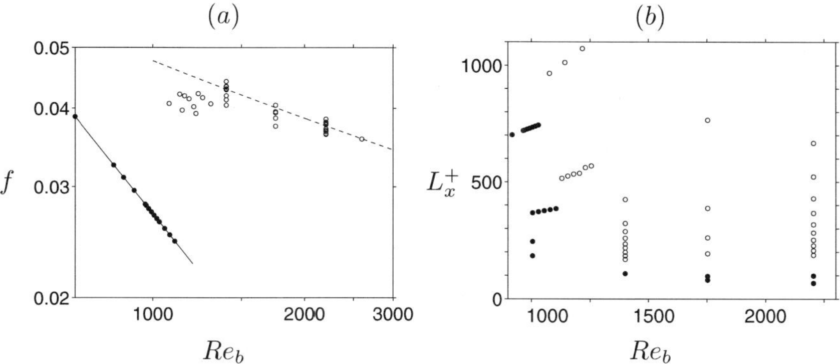

Fig. 1. (a) Variation of the friction factor as a function of the bulk Reynolds number for various values of the streamwise period of the computational domain. The friction factor for a square duct is given by $f = 8u_\tau^2/u_b^2$: •, present DNS (laminar); ∘, present DNS (turbulent); ——, laminar state represented by $f = 64/(2.25Re_b)$; $---$, turbulent state represented by $1/\sqrt{f} = 2\log_{10}(2.25Re_b\sqrt{f}) - 0.8$ (Jones' empirical correlation [8]). (b) Map of laminar/turbulent flow states in the plane defined by L_x^+ (the streamwise period of the computational domain in wall units) and Re_b. Symbols as in (a).

for fixed Re_b) is around $L_x^+ \equiv L_x u_\tau/\nu \approx 190$, roughly independent of Re_b (Fig. 1(b)). At relatively low Re_b close to the marginal value, however, the minimal streamwise period increases with decreasing Re_b.

4 Marginally Turbulent States in Square Duct Flow

In this section we shall examine flow structures just above the marginal Reynolds number determined in the preceding section. Figure 2 shows the temporally averaged velocity field for the streamwise-independent Fourier mode at $Re_b = 1100$ and $L_x/h = 4\pi$. It can be seen that the well-known secondary flow pattern appears, exhibiting eight longitudinal vortices which are visible in the longer temporal average (cf. Fig. 2(c)), while different patterns are observed during shorter temporal intervals (cf. Fig. 2($a - b$)). In these different patterns only four longitudinal vortices, in other words two pairs of counter-rotating vortices, are observed near two opposite walls; there are no secondary flow vortices located near the other pair of parallel walls. We refer to the state corresponding to these patterns (or the conventional pattern) as the 4-vortex (or 8-vortex) state.

In order to quantitatively identify the 4-vortex state, we consider the amplitude of the streamwise vorticity contained in each of the four triangular regions delimited by the diagonals, viz:

$$S_i \equiv \iint_{\Omega_i} \overline{\omega}_x^2 \mathrm{d}y\mathrm{d}z, \tag{1}$$

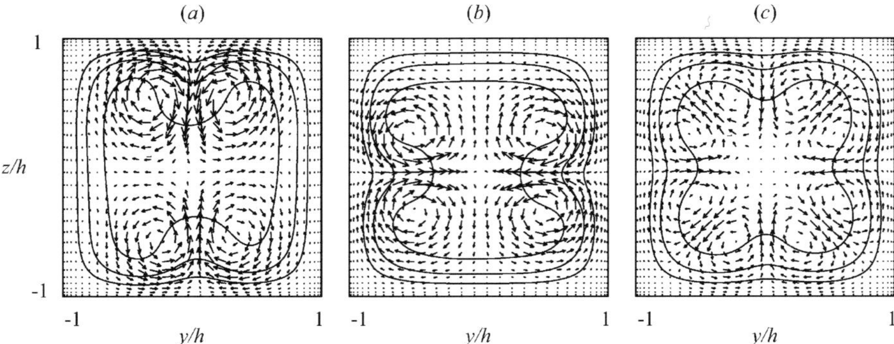

Fig. 2. Temporally averaged velocity field for the streamwise-independent Fourier mode at $Re_b = 1100$ and $L_x/h = 4\pi$. The cross-streamwise component is shown by vectors and the streamwise component is represented by iso-contours. (a), (b) temporal average over shorter interval $96h/u_b$; (c) temporal average over longer interval $1350h/u_b$.

where $\overline{\omega}_x$ is the streamwise vorticity of the streamwise-independent Fourier mode, and the regions are defined as:

$$\Omega_1 : \{(y,z)|z > -y \cap z < y\}, \quad \Omega_2 : \{(y,z)|z > -y \cap z > y\}, \tag{2}$$

$$\Omega_3 : \{(y,z)|z < -y \cap z > y\}, \quad \Omega_4 : \{(y,z)|z < -y \cap z < y\}. \tag{3}$$

Based upon S_i we introduce the following dimensionless indicator function I:

$$I \equiv \frac{S_1 + S_3 - S_2 - S_4}{S_1 + S_2 + S_3 + S_4}, \tag{4}$$

where $-1 \leq I \leq 1$. When I is close to zero, the streamwise vorticity is equipartitioned between the two triangular regions associated with the walls at $y = \pm 1$ and those associated with the walls at $z = \pm 1$, i.e. when the system exhibits the 8-vortex state. Conversely, large absolute values of I appear for the 4-vortex state, with the sign of I indicating the orientation of the arrangement of the secondary vortices. As an example, values of $I = -0.879, 0.866, -0.052$ are obtained for the averaged flow states in Fig. 2$(a - c)$, respectively.

Figure 3 shows the temporal variation of the indicator I for two cases with $Re_b = 1143$ and $Re_b = 2600$ (in both cases the length of the domain is approximately 4π). At the higher value for Re_b the indicator fluctuates around $I = 0$ with a low amplitude and a high frequency, implying that the flow is always in the conventional 8-vortex state. For lower Re_b, however, the indicator deviates largely from $I = 0$ and its sign changes over relatively long intervals of the order of $100h/u_b$. This latter behaviour indicates that the flow at this marginal Reynolds number exhibits the 4-vortex state with both orientations occurring during the observation interval. A statistical study of the transition between the 4- and 8-vortex state is currently underway.

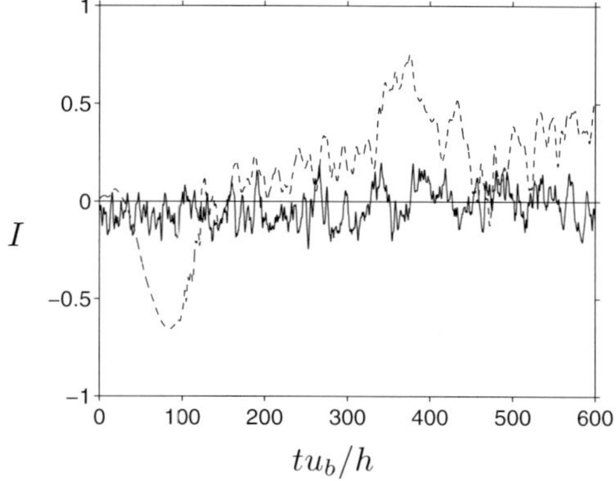

Fig. 3. Temporal evolution of the indicator function I for two cases: $---$, $Re_b = 1143, L_x/h = 4\pi$; ———, $Re_b = 2600, L_x/h = 11$.

5 Coherent Structures and Secondary Flow

In order to investigate the relation between coherent structures and the mean secondary flow we have determined the centers of the coherent vortices through the method of Kida and Miura [9]. In this approach vortex centers are identified by local minima of the pressure field and an additional swirl condition. The method is therefore free from any threshold value, which allows for the detection of vortices with any intensity. However, for our purposes the low-intensity vortices are not significant. Thus, we have eliminated weak vortices (with intensity below 1% of the maximum vorticity) a posteriori. Figure 4 shows data for one marginal case, accumulated over a limited time interval of $386h/u_b$ for which a 4-vortex state and no switching between different orientations were observed. The p.d.f.s of the location of the center of the vortices with positive (negative) sign have been accumulated separately and both are presented in Fig. 4(a) by dashed (solid) lines. Figure 4(b) shows the mean secondary flow vectors for the corresponding interval for comparison. It is observed that the most probable locations for the vortex centers with positive (negative) sign coincide with the centers of the corresponding secondary flow vortices. Statistics over much longer time intervals and for various parameter points (Re_b, L_x) are required before any definite conclusion can be reached. However, the present observations imply that – at least for marginal Reynolds numbers – the secondary flow is not due to any unique flow structure, but simply a consequence of predominant locations of the usual streamwise vortices in the duct geometry. For higher Reynolds numbers, the formation of secondary flow should involve the dynamics of larger structures (e.g. outer-layer structures) which still need to be clarified in the present geometry.

This work was supported in part by a Grant-in-Aid for Scientific Research (B) from Japan Society for the Promotion of Science and by the Center of

142   M. Uhlmann et al.

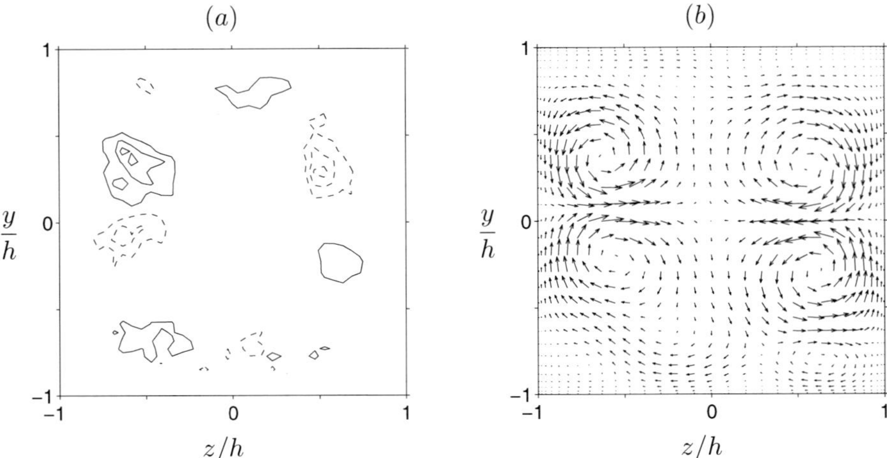

Fig. 4. (a) The p.d.f. of the position of vortex centers according to the criterion of [9] for a case with $Re_b = 1180$ and $L_x/h = 2\pi$, accumulated over an interval of $386h/u_b$. Contour levels indicate $0.5, 0.7, 0.9$ times the maximum probability; $---$, positive streamwise vorticity; ———, negative streamwise vorticity. (b) The corresponding mean flow in the cross-streamwise plane.

Excellence for Research and Education on Complex Functional Mechanical Systems (COE program of the Ministry of Education, Culture, Sport, Science, and Technology of Japan). The numerical computations were carried out at CIEMAT and on Altix3700 BX2 at YITP in Kyoto University.

References

1. Gavrilakis S (1992) J Fluid Mech 244:101–129
2. Gessner F (1973) J Fluid Mech 58:1–25
3. Hamilton MJ, Kim J, Waleffe F (1995) J Fluid Mech 287:317–348
4. Jiménez J, Moin P (1991) J Fluid Mech 225:213–240
5. Verzicco R, Orlandi P (1996) J Comput Phys 123:402–414
6. Haldenwang P, Labrosse G, Abboudi S, Deville M (1984) J Comput Phys 55: 115–128
7. Kawahara G, Ayukawa K, Ochi J, Ono F, Kameda E (2000) Trans JSME 66B: 95–102
8. Jones O (1976) Trans ASME J: J Fluids Engng 98:173–181
9. Kida S, Miura H (1998) J Phys Soc Japan 67(7):2166–2169

Extraction of Multi-Scale Vortical Structures from a Homogeneous Isotropic Turbulence

Waleed Abdel Kareem[1], Seiichiro Izawa[1], Ao-Kui Xiong[2] and Yu Fukunishi[1]

[1] Department of Mechanical Systems and Design, Graduate School of Engineering, Tohoku University, 6-6-01 Aramaki-Aoba, Aoba-ku, Sendai 980-8579, Japan
`fushi@fluid.mech.tohoku.ac.jp`
[2] Wuhan University of Technology, Wuhan, China. `akxiong@hotmail.com`

Abstract. An attempt to extract multi-scaled vortical structures from a turbulent field is carried out. The Lattice Boltzmann method with forcing is used to generate a statistically stationary velocity field in a resolution of 256^3. Two methods for filtering, namely the Fourier and wavelet decompositions, are used in the investigation and the results are compared. Three fields that contain vortical structures of different-scales area extracted from the original turbulent field. It is found that the results obtained by the two methods overlap at a high rate. Several universal characteristics among the filtered flow fields are found in the comparisons using the invariants of the velocity gradient, rotation and deformation tensors.

Keywords: lattice Boltzmann method, multi-scaled vortices, Fourier filter, wavelet filter, velocity gradient analysis

1 Introduction

Recent studies have revealed that coherent element eddies, sometimes called the "worms", are the building blocks of a turbulent flow field. They are small in size, with diameters only several times larger than the Kolmogorov scale. On the other hand, the energy spectrum [1] shows that there are motions of wide range of wavenumbers in the turbulent flow field. However, vortical structures of large-sizes corresponding to the low-wavenumber fluctuations cannot be recognized in the flow field. It is imagined that tiny eddies are contributing to the low-wavenumber fluctuations by forming groups or clusters. The groups of element eddies are difficult to handle, so it will be helpful if large-scale vortical structures that contribute to the low-wavenumber fluctuations can be directly extracted from the flow field. It may also help to understand the relations between the vortical structures of different-scales. Our motivation is to extract vortical structures of different sizes and to highlight the characteristics of the large-scale vortical structures and their interaction with the fine-scale structures. In this paper, the Lattice Boltzmann method (LBM)

Y. Kaneda (ed.), *IUTAM Symposium on Computational Physics and New Perspectives in Turbulence*, 143–148.

with a forcing scheme [2], which satisfies the continuity equation, is used to obtain a statistically stationary three-dimensional homogeneous and isotropic turbulence in a periodical box, 256^3.

2 Methods to Extract Multi-Scale Vortical Structures

Target flow filed is generated using LBM in a cubic box $2\pi^3$ with a resolution of 256^3. At the statistically stationary state, the characteristic scales of the flow field are given by as follows: the integral scale $l = 1.0$, the Taylor micro-scale $\lambda = 0.25$, and Kolmogorov sale $\eta = 0.01$. The equilibrium Taylor Reynolds number Re_λ is 179.6.

The Fourier decomposition is applied against the velocity field using sharp cutoff low-pass filters at different cutoff wavenumbers. This approach is basically the same as Tanahashi et al [3]. The three cut-off wavenumbers chosen in this study are $k_1 = 16$, $k_2 = 26$ and $k_3 = 64$, which are used to obtain the large-scale, the intermediate-scale and the fine-scale fields. The scales correspond to 39% of the integral length scale, Taylor micro-scale and 10 times of the Kolmogorov scale, respectively.

The vortex-identifying method using Q [4] is applied to the velocity fields after the filtering process in order to extract vortices of different scales from the flow field; the large-scale, intermediate-scale and fine-scale. The obtained fields are denoted as Q_F. The wavelet decomposition is applied against the Q field, where Q is computed from the unfiltered velocity field. The Morlet wavelet modulated by a Gaussian envelope of a unit width, defined as

$$\psi(\boldsymbol{x}) = e^{\mathrm{i}2\pi \boldsymbol{k}_\psi \cdot \boldsymbol{x}} e^{-\frac{|\boldsymbol{x}|^2}{2}}, \tag{1}$$

is used. Here, $\boldsymbol{k}_\psi$ is the central wavenumber vector of the mother wavelet. The real part of the definition is taken as the mother wavelet in our analysis. The wavelet transformation of the Q invariant is defined as;

$$Q_M = w(s) \int Q(\boldsymbol{x}) \psi^*\left(\frac{\boldsymbol{x} - \boldsymbol{b}}{s}\right) d\boldsymbol{x}, \tag{2}$$

where $w(s) = 1/s^{d/2}$ is the weighting function, and d is the dimension.

3 Results and Discussion

The flow field before filtering is visualized in Fig. 1 by Q, where Q is non-dimensionalized by the averaged enstrophy $< Q_W >$. The threshold value of Q^* is determined to be $Q^* = 5.981$, so that the total volume within the vortex structures will be 3% of the total volume. Here, the threshold value is chosen so that a large number of vortices could be visualized and at the same time different vortices could be distinguishable. In the figure, a large number of small thin vortex structures are found, whereas large-scale vortex structures are not. Figure 2 shows the energy spectra of the unfiltered field and

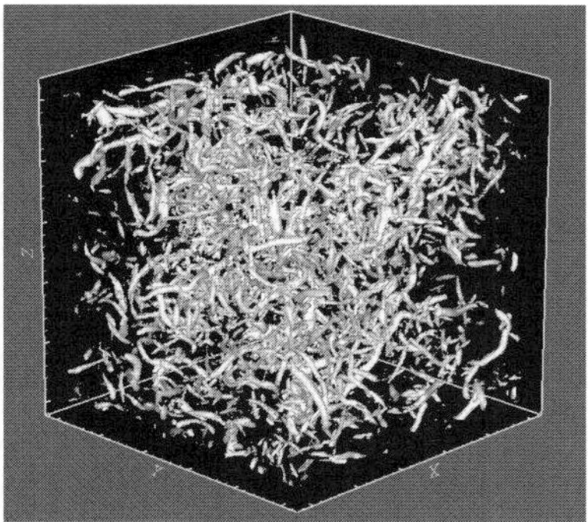 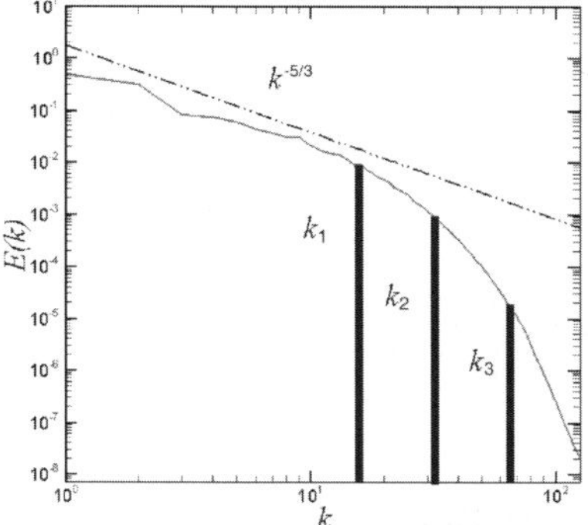

Fig. 1. Isosurfaces of Q in the non-filtered flow field with threshold $Q^* = 5.981$.

Fig. 2. Energy spectrum and the cutoff wavenumbers.

the filtered fields, namely the large-scale field, the intermediate-scale field and the fine-scale field. The vertical line shows the cut-off wavenumber of each filtered field. The non-filtered spectrum displays the Kolmogorov's $-5/3$ law with a Kolmogorov constant of 1.8.

3.1 The Structures Extracted by Fourier and Wavelet Methods

The invariants Q_S and Q_W are also non-dimensionalized by the averaged enstrophy of each flow field $< Q_W >$, where Q_S and Q_W are the half of the square of the strain and the vorticity tensors. The third invariant R of the velocity gradient tensor is non-dimensionalized by $< Q_W >^{3/2}$. It is important to point that the normalization is performed using the values of the filtered invariants at each scale. The extracted flow fields Q_F are shown in Fig. 3(a,b,c). From Fig. 3(a,b,c), it can be found that the vortices become thinner in the flow fields filtered at higher cutoff wave numbers.

By applying the wavelet decomposition to the Q field in the physical space, vortical structures of different scales can be captured. Figure 3(b) shows the cases for the wavelet coefficients Q_M at $s = 2^6$, 2^2 and $s = 2^1$, respectively. It can be found that as the spatial scale s increases, the captured vortices become larger especially at the cross-sections of the vortices. The results are similar to the results shown in Fig. 3(a), where the Fourier decomposition is used.

3.2 Similarity between the Results Using Two Methods

A contour map of the overlapping ratio $V_{ks} = \text{vol.}(Q_{F_k} \cap Q_{M_s})/\text{vol.}(Q_{M_s})$ between the regions marked as vortices through the Fourier filtering Q_{F_k} and the wavelet decomposition Q_{M_s} is shown in Fig. 4. The result shows that the vortical structures identified by the two methods are mostly overlapping,

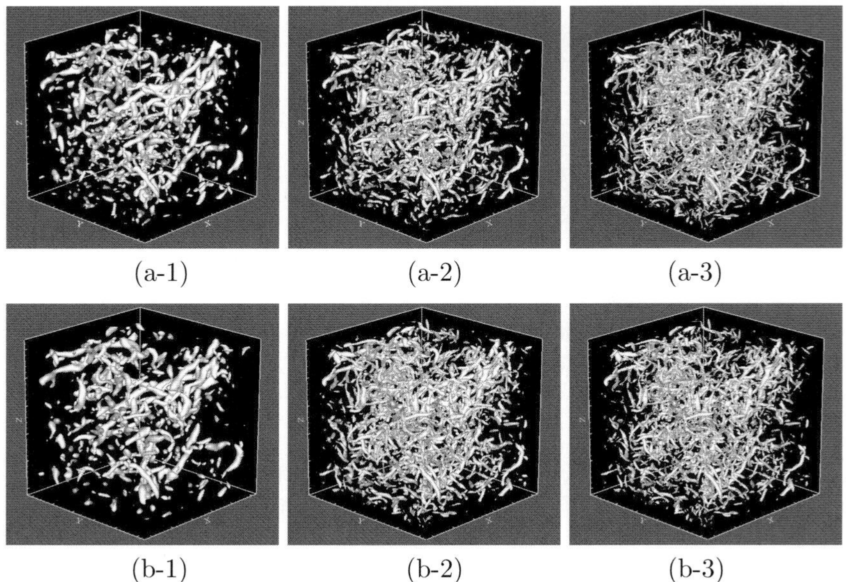

(a-1)   (a-2)   (a-3)

(b-1)   (b-2)   (b-3)

Fig. 3. Isosurfaces of filtered Q, Fourier fields (a-1) $k_1 = 16$, (a-2) $k_2 = 26$, (a-3) $k_3 = 64$ and wavelet fields (b-1) $s = 64$, (b-2) $s = 4$, (b-3) $s = 2$.

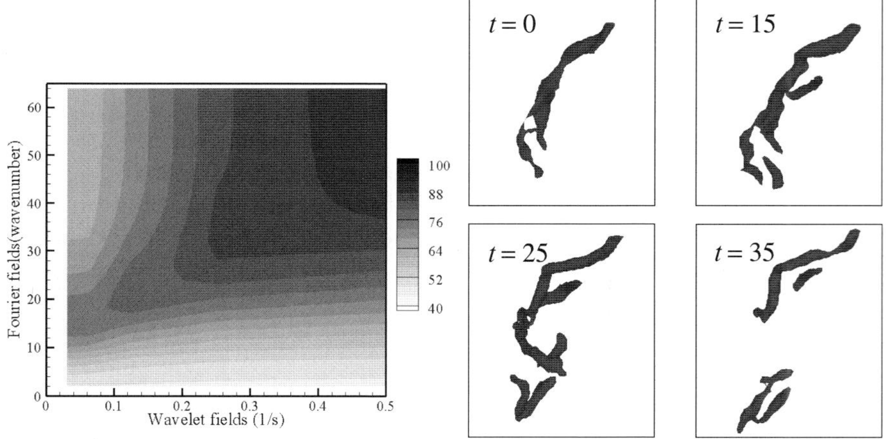

Fig. 4. Overlapping ratios between fields extracted by Fourier (Q_F) and wavelet (Q_M) decompositions.

Fig. 5. Extracted large structures (a), (c), (e) and the corresponding structures in the fine-scale field (b), (d), (f).

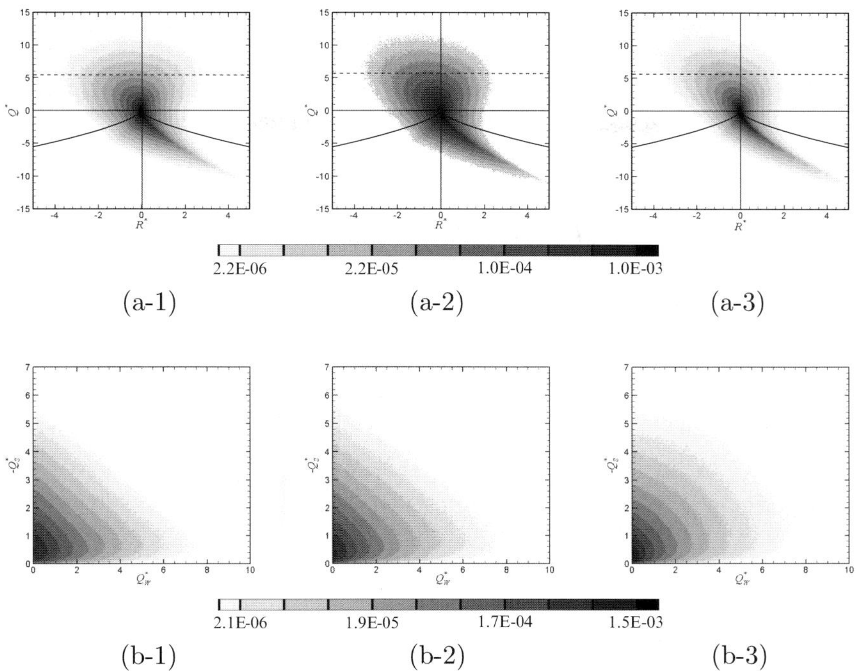

Fig. 6. Joint PDFs between the filtered invariants R and Q (a-1,2,3) and Joint PDFs between the filtered invariants Q_W and Q_S (b-1,2,3).

which imply that the results do not depend very heavily on the identification method used. This result does not agree with the results obtained by Farge et al. [5] where nonlinear wavelet filtering technique was used.

3.3 Interaction between Large and Fine-Scale Vortices

Three isolated structures are picked up from the large-scale flow field which are shown in Fig. 5(a,c,e). The structures in the fine-scale field that share the same grid points with the selected large-scale structures are shown in Fig. 5(b,d,f). This figure supports the idea that the fine-scaled vortical structures produce low wavenumber energy components of velocity by forming groups or clusters. The angles between the vorticity vectors for each large-scale structure and the corresponding fine-scale ones are measured for the three examples. They are 6.3, 4.3 and 6.3 degrees, respectively. The numbers of fine-scale structures which are found in the overlapping regions in the three cases are 5, 6 and 6, respectively.

3.4 Invariants (R, Q) in the Filtered Fourier Flow Fields

The joint PDFs of the third invariant R against the second invariant Q are studied for the filtered fields. The results are shown in Fig. 6(a), where the tent like curves represent $D = 0$. D is the discriminant of the velocity gradient tensor. The pattern displays a similar shape indicating that there are certain universalities in the invariant space (R, Q) among the filtered flow fields. It should be pointed that the regions below $D = 0$ are larger in larger-scales than in the fine-scale.

3.5 Invariants $(-Q_S, Q_W)$ in the Filtered Flow Fields

The $(-Q_S, Q_W)$ invariant space for the flow fields filtered by the Fourier decomposition is investigated. The joint PDFs for Q_W against $-Q_S$ of the filtered flow fields are shown in Fig. 6(b). The points close to the Q_W-axis imply the vortex-tube-like structure and those close to the $(-Q_S)$-axis correspond to the irrotational strain field. A self-similar shape of the contour lines can be observed regardless of the filtering scale. It can be observed that the highest local value of $-Q_S$ is smaller than the highest local value of Q_W in all flow fields.

4 Conclusion

An attempt to extract multi-scaled vortical structures from a turbulent flow field was carried out. Two filtering methods, the Fourier and wavelet decompositions, were tested and the results were compared. Three fields that contain vortical structures of different-scales were extracted from the original turbulent field. It was shown that by filtering the flow field, large-scaled coherent vortices could be identified, which could not be recognized in the original flow field. It was also shown that the vortical structures extracted by the two methods were overlapping at a high rate, which indicated the reliability of the two methods. Some examples of vortices in the large-scale field corresponding to several small vortices in the fine-scale field were presented. The result supports the idea that tiny eddies contribute to the low-wavenumber velocity fluctuation components by forming groups. Using the extracted fields of different scales, comparisons among the characteristics of the fields using the invariants of the velocity gradient, rotation and deformation tensors, were carried out. Some universal characteristics regardless of the scale of filtering were found.

References

1. Kaneda Y, Ishihara T, Yokokawa M, Itakura K, Uno A (2003) Phys of Fluids 15:L21–L24
2. Abdel KW, Izawa S, Xiong AK, Fukunishi Y (2006) Progress in Computational Fluid Dynamics 6:402–408
3. Tahanashi M, Midorikawa Y, Tsukamoto Y, Miyauchi T (2002) Proc 16th Symp Computational Fluid Dynamics (CD-ROM) E-28-1
4. Chong MS, Perry AE, Cantwell BJ (1990) Phys of Fluids A2:765–777
5. M Farge, K Schneider, G Pellegrino, A Wray and R Rogallo (2003) Phys of Fluids 15:2886–2896

Orthonormal Divergence-Free Wavelet Analysis of Nonlinear Energy Transfer in Rolling-Up Vortices

Keisuke Araki[1] and Hideaki Miura[2]

[1] Department of Intelligent Mechanical Engineering, Faculty of Engineering, Okayama University of Science, Okayama, Okayama 700-0005 JAPAN
`araki@are.ous.ac.jp`
[2] Theory and Computer Simulation Center, National Institute for Fusion Science, Toki, Gifu 509-5292 JAPAN

Abstract. Orthonormal divergence-free wavelet analysis of nonlinear transfer in some isolated vortices, which evolve from the instability of a thin shear layer, suggests that the rolling-up vortices are relevant to the energy cascade to smaller scales. A graphical representation method for the wavelet nonlinear interactions is developed so that the spatial scale and location information of the wavelets are depicted with enstrophy isosurfaces of vortices and vortex axis lines. It is found that the active nonlinear interactions are very closely distributed around the axis of rolling-up vortices irrespective of the forward or backward transfers and that the dominant nonlinear interactions are "local" in the sense of distance.

Keywords: isolated vortices, wavelets, nonlinear interactions

1 Introduction

Scale localness of nonlinear interactions in fully developed turbulence is one of the principal assumptions of Kolmogorov's phenomenology (cf. [1]). Many attempts and analyses have been made to relate this assumption to the dynamics of the Navier-Stokes equations (NSE) for an incompressible fluid

$$\frac{\partial \boldsymbol{u}}{\partial t} + (\boldsymbol{u} \cdot \nabla)\boldsymbol{u} = -\frac{1}{\rho}\nabla P + \nu \triangle \boldsymbol{u} \tag{1}$$

where ρ and P are the fluid density and the pressure, respectively. The Fourier analysis carried out in [2, 3, 4], which studied triadic interactions $T(k,p,q)$, shows the predominance of nonlocal interactions.[1] On the other hand, the analysis of the energy flux across the wavenumber k with respect to the scale disparity parameter s, $\Pi(k,s)$ appeals the importance of local interactions [5].

---

[1] In this paragraph we use the notations found in [5].

Y. Kaneda (ed.), *IUTAM Symposium on Computational Physics and New Perspectives in Turbulence*, 149–154.

Recently, Mininni et al. have carefully conducted numerical simulations, updated these knowledges and confirmed that the nonlocalness of T and the localness of Π are also true for the larger Reynolds numbers [6]. As is pointed out in [7], the disagreement between these conclusions is mainly due to the Fourier mode binning way and the wavelet analysis supports the dominance of local interactions.

By means of wavelet analysis Meneveau has shown that the spatial distribution of the wavelet energy flux is strongly fluctuated [8]. Using the coherent vortex simulation technique which is based on wavelet analysis, Farge et al. have shown that most energy transfers are due to coherent structures [9]. Scatter plots of the modulus of wavelet nonlinear interaction $|\langle j,\underline{l}|\boldsymbol{u}|k\rangle| := |-\int \boldsymbol{u}_{j\underline{l}} \cdot ((\boldsymbol{u} \cdot \nabla)\boldsymbol{u}_k) \mathrm{d}^3\underline{x}|$ and wavelet scale-location energy spectrum $E_{j\underline{l}} := \frac{1}{2}\int |\boldsymbol{u}_{j\underline{l}}|^2 \mathrm{d}^3\underline{x}$ for some assigned j's have been shown to be positively correlated [10]. How is this result related to the coherent structures such as vortices or shear layers? A detailed location-to-location wavelet nonlinear interaction analysis and comparison of its spatial features with that of vortex structures are needed.

2 Orthonormal Divergence-Free Wavelet Representation of the Navier-Stokes Equation

Orthonormal divergence-free wavelets, which we call "helical wavelets" and denote it by $\boldsymbol{\psi}_{j\epsilon\underline{l}\sigma}(\underline{x})$ hereafter, are obtained by the unitary transform of complex helical waves [7, 10]. Helical wavelets have four kinds of indices the implications of which are summarized in Table 1. Helical wavelets have the following properties: (1) each one defines a divergence-free vector field and (2) is real valued so that each expansion coefficient is a real number. (3) They constitute an orthonormal complete basis of the three-dimensional divergence-free vector fields. In the present study we use Meyer wavelet so that (4) each one has compact support in the wavenumber space.

Substituting the wavelet expansion of the velocity field $\boldsymbol{u} = \sum u_\lambda(t)\,\boldsymbol{\psi}_\lambda(\underline{x})$ where $u_\lambda(t) := \int \boldsymbol{u}(\underline{x},t)\cdot\boldsymbol{\psi}_\lambda(\underline{x})\,\mathrm{d}^3\underline{x}$ into (1), taking the inner product with $\boldsymbol{\psi}_\lambda$, we obtain the wavelet representation of the Navier-Stokes equation (W-NSE)

$$\frac{\mathrm{d}u_\lambda(t)}{\mathrm{d}t} = \sum \langle\boldsymbol{\psi}_\lambda|\boldsymbol{\psi}_\alpha|\boldsymbol{\psi}_\beta\rangle u_\alpha(t)u_\beta(t) + \nu \sum \langle\boldsymbol{\psi}_\lambda|\triangle|\boldsymbol{\psi}_\beta\rangle u_\beta(t), \quad (2)$$

Table 1. parameters of helical wavelets and their implications

index	range	implication
j	$0,1,2,3,\ldots$	spatial scale in physical space
$\epsilon = \xi + 2\eta + 4\zeta$	$1,2,\ldots,7$	anisotropy in wavenumber space
$\underline{l} = (l_x, l_y, l_z)$	$l_j = 0,\ldots,2^j - 1$	location in physical space
σ	$1, -1$	helicity

where λ, α and β stand for the set of wavelet indices (j,ϵ,l,σ) and the brackets are given by $\langle a|b|c\rangle := -\int a \cdot [(b \cdot \nabla)c]\,\mathrm{d}^3\underline{x}$, $\langle a|\triangle|c\rangle := \int a \cdot (\triangle c)\,\mathrm{d}^3\underline{x}$.

3 Numerical Results

In order to see the relation between coherent structures and nonlinear interactions clearly we analyzed isolated rolling-up vortices which are developed from instability of thin shear layers. A direct numerical simulation is carried out using the Fourier pseudospectral method. The number of grid points is 256^3. We analyzed a snapshot of the flow at $t = 17.5$. Its Taylor microscale Reynolds number and Kolmogorov wave number are $R_\lambda := \int |u|^2\mathrm{d}^3\underline{x}/\nu(\int |\mathrm{rot}u|^2\mathrm{d}^3\underline{x})^{\frac{1}{2}} = 1442$ and $k_d = 59.7$, respectively. The analyzed quadrant contains one large vortex and the kinked part of a neighbouring large vortex (Fig. 1).

In the present study we pay attention to the nonlinear interactions between different scales and locations. So we reduced the information on anisotropy and helicity of the bra- and ket-modes. We calculated the nonlinear interactions between the different scales and locations given by the integral

$$\langle j, \underline{k}|u|l, \underline{m}\rangle = -\int u_{j\underline{k}}(\underline{x}, t) \cdot \Big((u(\underline{x}, t) \cdot \nabla)u_{l\underline{m}}(\underline{x}, t)\Big)\,\mathrm{d}^3\underline{x} \tag{3}$$

where $u_{a\underline{b}}(\underline{x}, t) = \sum_{\epsilon,\sigma} u_{a\epsilon\underline{b}\sigma}(t)\,\psi_{a\epsilon\underline{b}\sigma}(\underline{x})$. We numerically verified for the scale parameters $j \geq 3$, with which each of the wavelets has comparable or smaller spatial scale than the diameter of the rolling-up vortices, the features of the nonlinear interactions are qualitatively similar. Due to the limitation of pages we will show here the results for the $j = 4$ and 5 resolution classes as a typical example. At these scales the disspation range characteristics are expected.

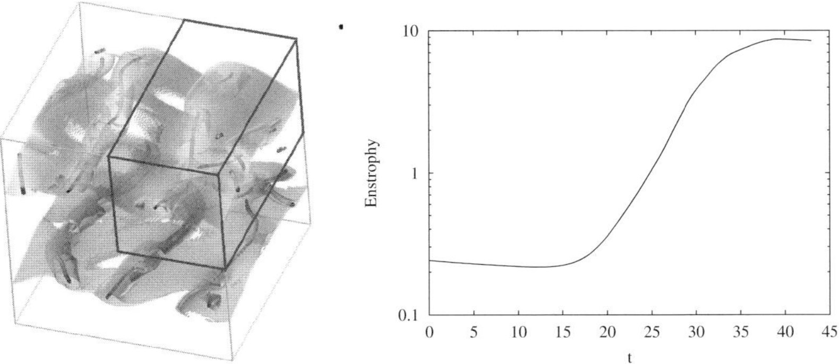

Fig. 1. Left: snapshot of the analyzed flow. The right top box is the analyzed quadrant. Isosurfaces of the enstrophy density and the vortex axes captured by the method described in [11] are shown simultaneously. Right: the time development of the enstrophy. About $t = 17$ the rapid enhancement of the enstrophy, which is due to the three-dimensional development of turbulence, begins.

Fig. 2. Color bar representation

In order to indicate the information on the scale and location of helical wavelets in (3) we introduce the color bar representation (see Fig. 2). The ends of a bar indicate the wavelet "site" in physical space (not exact but very close to the wavelet peak location). The color with respect to light gray versus dark gray/black depict the spatial scale of wavelets (light gray side indicates the larger scale wavelet site). The color with respect to dark gray versus black shows direction of energy transfer: the black bars represent the transfer to the smaller scales. In the present study the amplitude of each nonlinear interaction is not expressed.

In Figure 3 the nonlinear interactions are shown separately according to the sign of them, i.e. forward transfer or backward one. Not the whole nonlinear interactions but only such ones that have larger moduli are displayed. The thresholds are determined so that each of the depicted nonlinear interactions contains 30% of the net positive or negative transfer, respectively. The following features are seen from the figure and checked statistically;

1. The color bars are concentrated around the winding vortex tubes without any artificial selection of the nonlinear interactions except for the largest 30% picking-up.
2. The color bars are very short. Their typical lengths are about one grid of $j = 4$ mode, i.e. one-sixteenth of the side length of the whole domain. It is numerically verified that 99% transfer is due to the nonlinear interactions between the wavelets nearer than two grids of the $j = 4$ modes. This implies that the principal nonlinear interactions occur "locally" in the sense of distance. The interactions between distant wavelet sites are negligible though they are much larger than the expected numerical errors.
3. It is numerically verified that both the forward and the backward transfers are very active and the cancelled amount is very large though the energy is transferred to the smaller scales as a whole. Such strong spatial fluctuation and cancellation is consistent with Meneveau's findings [8]. Since both the positive and negative bars are concentrated around the vortex axes, our graphical representation suggests that such fluctuation and cancellation mainly occur in the vortex core regions. This size is comparable to the principal envelope size of each wavelet function of the $j = 4$ class.

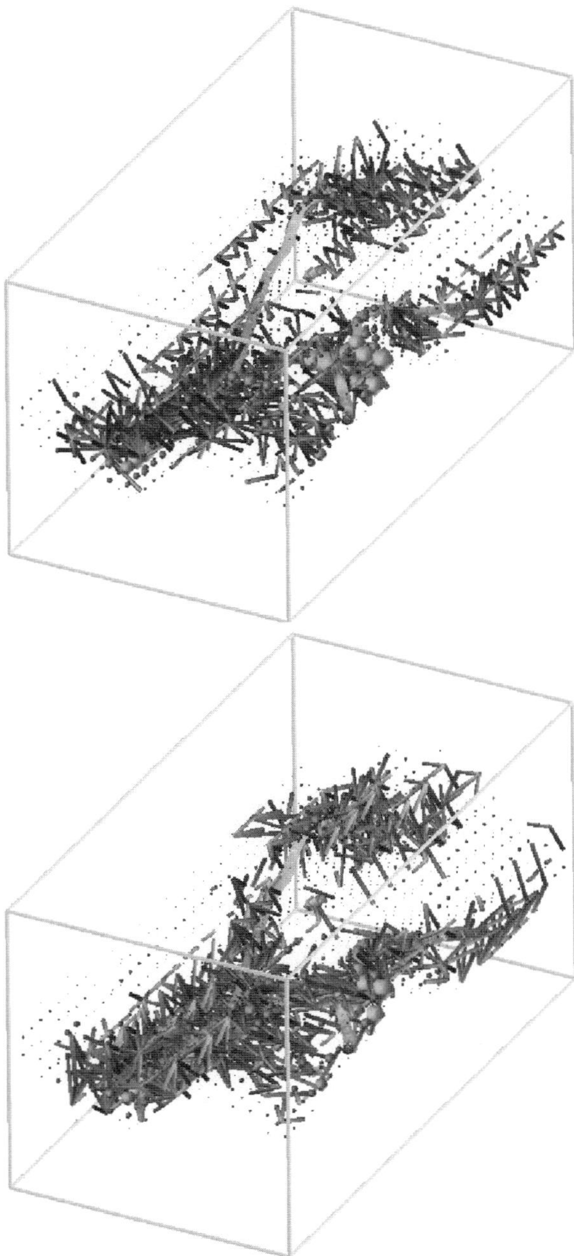

Fig. 3. Spatial distribution of the nonlinear interactions between the wavelets of $j = 4$ and 5 resolution classes; top: forward transfer, bottom: backward transfer. In order to identify the precise location of the rolling-up vortices vortex axes captured by the method described in [11] are also depicted. The balls are wavelet scale location spectrum of $j = 5$ class. The radii of balls are set to be proportional to the integral $E_{5\underline{l}} := \frac{1}{2} \int |\boldsymbol{u}_{5\underline{l}}|^2 \mathrm{d}^3\underline{x}$ and the center of each ball is located at $(l_x + \frac{1}{2}, l_y + \frac{1}{2}, l_z + \frac{1}{2})/(2^6\pi)$.

4 Concluding Remarks

In the present paper we investigated the features of the inter-scale, location-to-location nonlinear energy transfer by means of divergence-free wavelet analysis and developed a graphical representation method suitable for the analysis.

Application of the wavelet analysis and the representation method to the snapshot of rolling-up vortices clearly shows good spatial correspondence of the active nonlinear interactions with the coherent structures. Though we present here the result for the interactions between the $j = 4$ and 5 class wavelets, this seems to give a direct evidence that coherent structures are responsible for the nonlinear energy transfer to the smaller scale. This is consistent with the findings of Farge et al. [9].

We also confirmed the existence of strong spatial fluctuation of nonlinear interactions and their cancellation which is consistent with Meneveau's findings [8]. As is pointed out in his paper, such large spatial fluctuations may be due to the sweeping of large eddies. As well as the sweeping, the dense clustering of color bars around the vortex cores may suggest that the vortex core size itself affects the strong fluctuation of the wavelet nonlinear interactions if the principal envelope size of each analyzing wavelet function is comparable to that of the coherent structure.

References

1. Frisch U (1995) Turbulence. Cambridge Univ Press, Cambridge
2. Domaradzki JA, Rogallo RS (1990) Phys Fluids A 2:413–426
3. Yeung PK, Brasseur JG (1991) Phys Fluids A 3:884–897
4. Ohkitani K, Kida S (1992) Phys Fluids A 4:794–802
5. Zhou Y (1993) Phys Fluids A 5:1092–1094
6. Mininni PD, Alexakis A, Pouquet A (2006) Phys Rev E 74:016303
7. Kishida K, Araki K, Kishiba S, Suzuki K (1999) Phys Rev Lett 83:5487–5490
8. Meneveau C (1991) J Fluid Mech 232:469–520
9. Farge M, Schneider K, Pellegrino G, Wray AA, Rogallo RS (2003) Phys Fluids 15:2886–2896
10. Kishida K, Araki K (2002) Orthonormal divergence-free wavelet analysis of spatial correlation between kinetic energy and nonlinear transfer in turbulence. In: Kaneda Y, Gotoh T (eds) Statistical Theories and Computational Approaches to Turbulence, Modern Perspectives and Applications to Global-Scale Flows. Springer, Tokyo
11. Miura H, Kida S (1997) J Phys Soc Japan 66:1331–1334

Wavelet-Based Statistics of Energy Transfer in High Reynolds Number Three-Dimensional Homogeneous Isotropic Turbulence

Naoya Okamoto and Katsunori Yoshimatsu

Department of Computational Science and Engineering, Graduate School
of Engineering, Nagoya University, Chikusa-ku, Nagoya 464-8603, Japan
okamoto@fluid.cse.nagoya-u.ac.jp, yosimatu@fluid.cse.nagoya-u.ac.jp

Abstract. We examine small-scale statistics of energy transfer from a scale within grid scales to subgrid scales localized in space and scale in high Reynolds number fully developed turbulence. An orthonormal wavelet decomposition method is applied to data of three-dimensional incompressible homogeneous isotropic turbulence with Taylor microscale Reynolds number 471. We find that the energy transfer and the kinetic energy localized in space and scale have strong correlation in the inertial subrange though they are almost uncorrelated in the dissipation range. The correlation is enhanced by the positive values for the transfer, while it is reduced by the negative values for the transfer.

Keywords: wavelet-based statistics, energy transfer, homogeneous isotropic turbulence, high Reynolds number

1 Introduction

In fully developed turbulence, eddies of a wide range of scales coexist and exhibit strong spatial intermittency. Keeping track of both location and scale to represent fully developed turbulence may play a key role in revealing nature of turbulence and in exploration for its small-scale universality. An orthonormal discrete wavelet expansion method is one of the most powerful tools to resolve information on both scale and location of turbulence. The reader interested in reviews on wavelet applications to turbulence may refer to [1, 2].

Energy transfer from grid scales to subgrid scales in high Reynolds number turbulence is one of the most important phenomena for understanding small-scale statistics of turbulence and for developing turbulence models, such as Large Eddy Simulation. Statistics of the energy transfer in homogeneous isotropic turbulence with Taylor microscale Reynolds number up to approximately 1130 has been studied by the use of the spectral cut and Gaussian filters [3]. Wavelet representation of the energy transfer from a scale within grid scales to subgrid scales has been proposed, and wavelet-based statistics of the transfer in low Reynolds number homogeneous turbulence was examined

Y. Kaneda (ed.), *IUTAM Symposium on Computational Physics and New Perspectives in Turbulence*, 155–160.

in [4]. Properties of methods separating between large and small scales are summarized in the subsection 2.2 of [5].

In this paper, we examine wavelet-based statistics of the energy transfer in high Reynolds number fully developed turbulence. A three-dimensional orthonormal wavelet analysis is applied to the DNS data of incompressible homogeneous isotropic turbulence at resolution $N = 1024^3$ grid points and Taylor microscale Reynolds number $R_\lambda = 471$ [6, 7].

2 Three-Dimensional Orthonormal Wavelet Analysis

We consider a scalar function $f(\mathbf{x})$ in $\mathbf{T}^3 = [0, 2\pi]^3$. Three-dimensional orthonormal wavelet transform unfolds $f(\mathbf{x})$ into scales, positions and directions by three-dimensional mother wavelets $\Psi_m(\mathbf{x})$ ($m = 1, 2, \cdots, 7$). The mother wavelets are constructed by tensor product of one-dimensional scaling function $\psi_0(x_i)$ and mother wavelet $\psi_1(x_i)$ as $\Psi_m(\mathbf{x}) = \psi_\xi(x_1)\psi_\eta(x_2)\psi_\zeta(x_3)$ ($\xi, \eta, \zeta = 0, 1$, and $m = \xi + 2\eta + 4\zeta$), where x_i is the component of $\mathbf{x}$ of the i-th Cartesian direction. Here, $i = 1, 2, 3$. We use the Meyer wavelet [8]. The field $f(\mathbf{x})$ sampled on 2^J equidistant grid points in each space direction of the Cartesian coordinates can be decomposed into an orthonormal wavelet series,

$$f(\mathbf{x}) \simeq \bar{f} + \sum_{\alpha=1}^{J} f_\alpha(\mathbf{x}), \quad f_\alpha(\mathbf{x}) = \sum_{\iota_1,\iota_2,\iota_3=0}^{2^{\alpha-1}-1} \mathcal{W}^\alpha_{m,\boldsymbol{\iota}}[f] \, \Psi^\alpha_{m,\boldsymbol{\iota}}(\mathbf{x}), \tag{1}$$

where $\bar{f} = \int_{\mathbf{T}} f(\mathbf{x}) \, d\mathbf{x}/(2\pi)^3$, $\mathcal{W}^\alpha_{m,\boldsymbol{\iota}}[f] = \int_{\mathbf{T}} f(\mathbf{x})\Psi^\alpha_{m,\boldsymbol{\iota}}(\mathbf{x}) \, d\mathbf{x}/(2\pi)^3$, $\Psi^\alpha_{m,\boldsymbol{\iota}}(\mathbf{x}) = 2^{3\alpha/2}\Psi_m(2^\alpha\mathbf{x} - 2\pi\boldsymbol{\iota})$, and $\boldsymbol{\iota} = (\iota_1, \iota_2, \iota_3)$. The scripts α, $\boldsymbol{\iota}$ and m denote the scales, the positions and the seven directions of the wavelets, respectively. Hereafter, the summation convention is used for the alphabetical repeated indices but not for the Greek ones. Readers interested in details on orthonormal wavelet transform may refer to, e.g. [9].

3 Wavelet-Based Statistics of Energy Transfer

3.1 DNS Data and Wavelet-Based Quantities of Turbulence

We apply the three-dimensional orthonormal wavelet expansion method to the data with $k_{\max}\eta \simeq 1$ at $N = 1024^3$ and $R_\lambda = 471$, which were obtained by DNS of three-dimensional homogeneous isotropic turbulence performed on the Earth Simulator [6, 7]. Here $k_{\max}$ is the maximum wavenumber retained in the DNS, and η is the Kolmogorov length scale. The DNS field obeys the Navier-Stokes equations for incompressible fluid in a periodic box with side length 2π. Negative viscosity is introduced in the range $1 \le k < 2.5$ so that the total energy is kept almost constant, where k is the wavenumber. The minimum wavenumber of the DNS is 1. The details of the DNS may be referred to [6, 7, 10].

Wavelet-based quantities, local energy $e_\alpha[\boldsymbol{\iota}]$ at scale α ($\alpha = 1, 2, \cdots$, $\log_2 N^{1/3}$) and location $\boldsymbol{\iota}$, local energy flux of kinematic energy to scale α

at location $\boldsymbol{\iota}$, $\pi_\alpha[\boldsymbol{\iota}]$ and energy transfer from a scale α within grid scales to subgrid scales at location $\boldsymbol{\iota}$, $t_{\alpha,\beta}[\boldsymbol{\iota}]$ $(\alpha \le \beta)$ are defined by

$$e_\alpha[\boldsymbol{\iota}] = \frac{1}{2}\{\mathcal{W}^\alpha_{m,\boldsymbol{\iota}}[u_l]\}^2, \tag{2}$$

$$\pi_\alpha[\boldsymbol{\iota}] = \sum_{\beta=1}^{\alpha} 2^{3(\beta-\alpha)}\mathcal{W}^\alpha_{m,\boldsymbol{\kappa}}[u_l]\mathcal{W}^\alpha_{m,\boldsymbol{\kappa}}[u_j\partial_j u_l + \partial_l p] \ \text{ with } \boldsymbol{\kappa} = 2^{\beta-\alpha}\boldsymbol{\iota}, \tag{3}$$

$$t_{\alpha,\beta}[\boldsymbol{\iota}] = \mathcal{W}^\alpha_{m,\boldsymbol{\iota}}[u_i]\,\mathcal{W}^\alpha_{m,\boldsymbol{\iota}}\left[\partial_j\left(u_i u_j - u_i^{>\beta}u_j^{>\beta}\right) + \frac{1}{\rho}\partial_i p^{<\beta}\right], \tag{4}$$

respectively [4]. Here, u_l $(l = 1,2,3)$ is the l-th component of velocity, $\partial_l = \partial/\partial x_l$, p the pressure and ρ fluid density. The grid and subgrid scale contributions to a scalar function $f(\mathbf{x})$ are expressed by $f^{>\beta}(\mathbf{x}) = \sum_{\alpha=1}^{\beta} f_\alpha(\mathbf{x})$, and $f^{<\beta}(\mathbf{x}) = f(\mathbf{x}) - f^{>\beta}(\mathbf{x})$, respectively.

3.2 Mean Wavelet Spectra of Energy and Its Flux

Figure 1 shows the mean wavelet spectrum of the local energy $E(k_\alpha)$ and that of the local energy flux $\Pi(k_\alpha)$. They are written as $E(k_\alpha) = N_\alpha \langle e_\alpha[\boldsymbol{\iota}]\rangle /\Delta k_\alpha$ and $\Pi(k_\alpha) = N_\alpha \langle \pi_\alpha[\boldsymbol{\iota}]\rangle$, respectively [4]. Here, k_α is equal to $2^\alpha/3$, which is the centroid wavenumber of the Meyer wavelet at scale α, $\Delta k_\alpha = (k_{\alpha+1} - k_\alpha)\ln 2$ and $N_\alpha = 2^{3(\alpha-1)}$. $\langle\cdot\rangle$ denotes the mean value of $\cdot$ at each scale. The spectrum of the flux is not divided by Δk_α, because the summation from $\beta = 1$ to $\beta = \alpha$ has been already performed in (3). A bump is observed for $k\eta = k_b\eta(\sim 0.1)$ in the compensated mean wavelet energy spectrum. Figure 1(b) shows $\Pi(k_\alpha)/\langle\epsilon\rangle \simeq 1$ in the range that $0.01 \lesssim k\eta \lesssim 0.03$. We call here the range the inertial subrange. The scales 4 and 5 are the representative scales in the inertial subrange. For readers' convenience, the Fourier energy spectrum and the Fourier energy transfer rate are shown in Fig. 1.

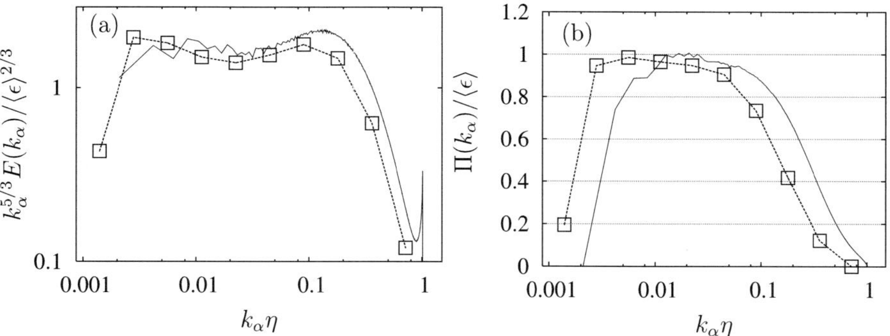

Fig. 1. Mean wavelet spectra and Fourier spectra of (a) the local energy and (b) the local energy flux. ——, Fourier spectra; $\square$, mean wavelet spectra. Here, $\langle\epsilon\rangle$ is the mean energy dissipation rate per unit mass.

3.3 Statistics of Energy Transfer from a Scale within Grid Scales to Subgrid Scales

Statistics of the local energy transfer $t_{\alpha,\beta}[\boldsymbol{\iota}]$ is examined. We call the positive (negative) values for $t_{\alpha,\beta}[\boldsymbol{\iota}]$ the forward (backward) energy transfer. We here focus on two typical cases of $\beta = 6$ ($k_6\eta \simeq 0.045$) and 9 ($k_9\eta \simeq 0.36$). The scale 6 is adjacent to the inertial subrange and satisfies $k_6\eta < k_b\eta$, while the scale 9 does $k_9\eta > k_b\eta$.

Figure 2 shows dual spectra of $t_{\alpha,\beta}[\boldsymbol{\iota}]$ normalized by $(\langle\epsilon\rangle\nu)^{3/4}$, where ν is the kinematic viscosity. The dual spectra consist of the mean wavelet spectrum and the mean plus and minus one standard deviation spectra. The mean wavelet spectra $T(k_\alpha|k_\beta)$ and the standard deviation spectra $\sigma_t(k_\alpha)$ which are the spatial fluctuations at scale α are defined as $T(k_\alpha|k_\beta) = N_\alpha \langle t_{\alpha,\beta}[\boldsymbol{\iota}]\rangle /\Delta k_\alpha$ and $\sigma_t(k_\alpha) = N_\alpha\{\langle (t_{\alpha,\beta}[\boldsymbol{\iota}])^2\rangle - \langle t_{\alpha,\beta}[\boldsymbol{\iota}]\rangle^2\}^{1/2}/\Delta k_\alpha$, respectively [4]. In Fig. 2, the forward transfer is dominant in $t_{\alpha,6}[\boldsymbol{\iota}]$ ($\alpha \geq 2$) and in $t_{9,9}[\boldsymbol{\iota}]$. We observe slightly negative $T(k_\alpha|k_9)$ for $k_\alpha\eta \lesssim k_7\eta$ and almost constant σ_t for $k_4\eta \lesssim k_\alpha\eta \lesssim k_6\eta$. We also find that $t_{\alpha,\alpha}[\boldsymbol{\iota}]$ plays an important role of the energy transfer from the grid scales to the subgrid scales.

We consider the link between the spatial distributions of $t_{\alpha,\beta}[\boldsymbol{\iota}]$ and $e_\alpha[\boldsymbol{\iota}]$. Figure 3 shows three-dimensional snapshots of the intense forward (backward) transfer in $t_{5,6}[\boldsymbol{\iota}]$ and the local energy at scale 5, respectively. The intense region of the forward transfer overlaps the intense region of $e_5[\boldsymbol{\iota}]$ better than that of the backward transfer does.

To examine the dependence of the link on the scale α quantitatively, we introduce the scale-by-scale correlation between $e_\alpha[\boldsymbol{\iota}]$ and $t_{\alpha,\beta}[\boldsymbol{\iota}]$ defined as

$$C_{\alpha,\beta} = \frac{\langle (e_\alpha[\boldsymbol{\iota}] - \langle e_\alpha[\boldsymbol{\iota}]\rangle)(t_{\alpha,\beta}[\boldsymbol{\iota}] - \langle t_{\alpha,\beta}[\boldsymbol{\iota}]\rangle)\rangle}{\sqrt{\langle (e_\alpha[\boldsymbol{\iota}] - \langle e_\alpha[\boldsymbol{\iota}]\rangle)^2\rangle}\sqrt{\langle (t_{\alpha,\beta}[\boldsymbol{\iota}] - \langle t_{\alpha,\beta}[\boldsymbol{\iota}]\rangle)^2\rangle}}. \qquad (5)$$

Figure 4(a) shows that the correlation is strong (about 0.6) in the case that α is in the inertial subrange for $\beta = 6$. It is almost uncorrelated for $\beta = 9$. We

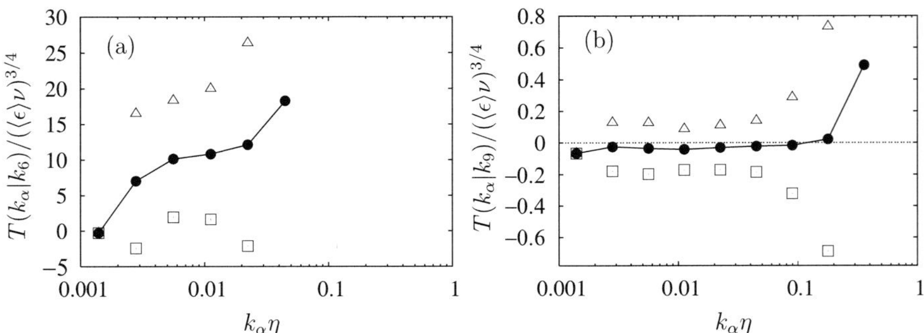

Fig. 2. Dual spectra of the local energy transfer to smaller scales than (a) $\beta = 6$ and (b) $\beta = 9$. $\bullet$, mean spectra; $\triangle$, mean plus one standard deviation; $\square$, mean minus one standard deviation.

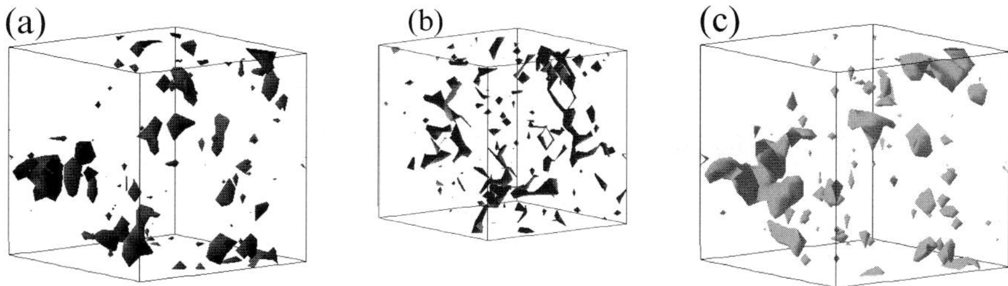

Fig. 3. Isosurfaces of (a) forward transfer of $t_{5,6}[\iota]$, and (b) backward transfer of negative $t_{5,6}[\iota]$ and (c) $e_5[\iota]$. The values of the isosurfaces are $t_{5,6}[\iota] = \langle t_{5,6}[\iota] \rangle + 2\hat{\sigma}_t(>0)$, $t_{5,6}[\iota] = \langle t_{5,6}[\iota] \rangle - \hat{\sigma}_t(<0)$, and $e_5[\iota] = \langle e_5[\iota] \rangle + 2\hat{\sigma}_e$, where $\hat{\sigma}_t$ and $\hat{\sigma}_e$ are the standard deviations of $t_{5,6}[\iota]$ and $e_5[\iota]$, respectively.

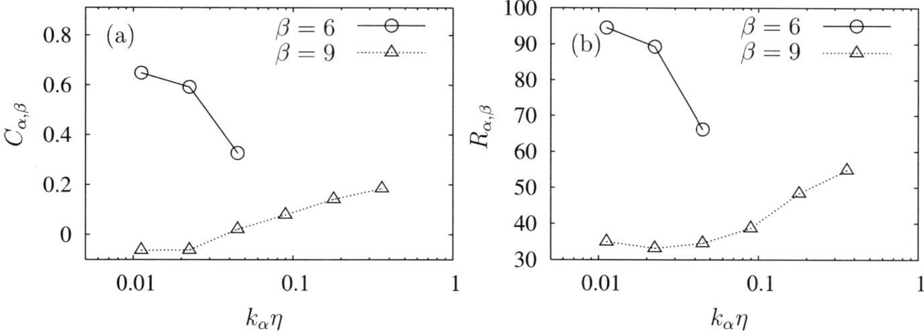

Fig. 4. (a) Correlation $C_{\alpha,\beta}$ vs. $k_\alpha\eta$. (b) Volume ratio $R_{\alpha,\beta}$ of the region of forward transfer vs. $k_\alpha\eta$.

plot the volume ratio of the region of the forward transfer denoted by $R_{\alpha,\beta}$ vs. $k_\alpha\eta$ in Fig. 4(b). The ratio $R_{\alpha,\beta}$ is given by $R_{\alpha,\beta} = 100N_{\alpha,\beta}^+/N_\alpha$. Here, $N_{\alpha,\beta}^+$ is the number of grid points where $t_{\alpha,\beta}[\iota] > 0$ at scale α. Comparison of Fig. 4(a) with Fig. 4(b) shows that the dependence of $C_{\alpha,\beta}$ on $k_\alpha\eta$ is similar to that of $R_{\alpha,\beta}$. The forward transfer enhances the correlation, while the backward transfer reduces it. In Fig. 4, only the results for the smaller scales than the inertial subrange, i.e. $\alpha \geq 4$ are presented.

4 Conclusions

The wavelet-based statistics of the energy transfer $t_{\alpha,\beta}[\iota]$ from a scale α within grid scales to subgrid scales at location ι has been examined in homogeneous isotropic turbulence at $R_\lambda = 471$. We focused on two typical cases, the scale $\beta = 6$ ($k_6\eta \simeq 0.045$) adjacent to the inertial subrange and the scale $\beta = 9$ ($k_9\eta \simeq 0.36$). We find that the forward transfer is dominant for $\beta = 6$ and that the backward transfer occurs in the range $k_\alpha\eta \lesssim 0.1$ more frequently than in $k_\alpha\eta \gtrsim 0.1$ for $\beta = 9$. The scale-by-scale correlation between $t_{\alpha,\beta}[\iota]$

and the local energy is strong in the case that α is in the inertial subrange for $\beta = 6$, though it is almost uncorrelated for $\beta = 9$.

In [11], a positive correlation between kinetic energy and magnitude of nonlinear interactions of a velocity field between adjacent scales is observed for homogeneous isotropic turbulence at $R_\lambda = 100$. Relevance between the results and the present ones remains to be pursued. Results on wavelet-based statistics of homogeneous isotropic turbulence with R_λ up to 732 will be reported elsewhere.

The computations were carried out on HPC2500 system at the Information Technology Center of Nagoya University. The authors would like to express their thanks to Y. Kaneda and T. Ishihara, M. Yokokawa, K. Itakura and A. Uno for providing us with the DNS data. They thank Y. Kaneda for fruitful discussions on turbulence and T. Ishihara for support of the data handling. They are also grateful to M. Farge and K. Schneider for valuable discussions on wavelets and their applications to turbulence and for providing us with their wavelet decomposition code for single processor. This work was supported by a Grant-in-Aid for the 21st Century COE "Frontiers of Computational Science".

References

1. Farge M (1992) Ann Rev Fluid Mech 24:395–457
2. Addison P S (2002) The illustrated wavelet transform handbook. Institute of Physics Publishing, Bristol and Philadelphia
3. Aoyama T, Ishihara T, Kaneda T, Yokokawa M, Itakura K, Uno A (2005) J Phys Soc Jpn 74:3202–3212
4. Meneveau C (1991) J Fluid Mech 232:469–520
5. Meneveau C, Katz J (2000) Ann Rev Fluid Mech 32:1–32
6. Yokokawa M, Itakura K, Uno A, Ishihara T, Kaneda Y (2002) Proc IEEE/ACM SC2002 Conf, Baltimore; http://www.sc-2002.org/paperpdfs/pap.pap273.pdf
7. Kaneda Y, Ishihara T, Yokokawa M, Itakura K, Uno A (2003) Phys Fluids 15:L21–L24
8. Daubchies I (1992) Ten lectuers on wavelets. SIAM, Philadelphia
9. Mallat S (1999) A wavelet tour of signal processing Second Edition. Academic Press
10. Kaneda Y, Ishihara T (2006) J Turbul, 7 20
11. Kishida K, Araki K (2003) Proc Int Workshop on Statistical Theories and Computational Approaches to Turbulence ed. Kaneda Y and Gotoh T. Springer, Tokyo 248–259

Part II

Experimental Approach to Fundamental Problems in Turbulence

Visualization of Quantized Vortex Dynamics

Gregory P. Bewley[1], and Matthew S. Paoletti[1], and Daniel P. Lathrop[1], and Katepalli R. Sreenivasan[1,2]

[1] Universityof Maryland, College Park, MD 20742
[2] International Center for Theoretical Physics, Trieste, Italy 34014
   krs@ictp.it

Abstract. We discuss an experimental technique developed to visualize quantized vortices in helium II. We illuminate micron-sized solid particles of hydrogen suspended in the fluid, and mark quantized vortex cores within the volume observed by a camera. While, under some circumstances, the particles modify vortex dynamics by the action of viscous drag and by pinning vortex intersections, we do capture basic vortex dynamics such as vortex ring decay, vortex reconnection, and superfluid turbulence.

Keywords: turbulence, quantized turbulence, superfluid helium, vortex dynamics and visualization

1 Introduction

At the Symposium, one of the authors (KRS) of this article presented a talk describing the gains made by working with cryogenic helium in the experimental studies of fluid dynamics. Among these advantages is the ability to generate flows at very high Reynolds and Rayleigh numbers and to span many decades of these parameters in a single apparatus. These aspects have been described in adequate detail in a recent review article [1], and need not be restated here. We describe briefly the second part of the talk dealing with the fluid dynamics of helium II. Even though certain properties of helium II such as superfluidity have no counterparts in classical fluid dynamics, we shall be concerned here with the dynamics of vortices associated with the superfluid part of helium II. These vortices are of the order of an angstrom in diameter and the circulation around them is quantized. However, the properties of these vortices beyond a few diameters from the core are classical in nature and can be understood largely in those terms.

Above $T_\lambda \approx 2.17\ K$, liquid helium behaves like a classical fluid and the Navier-Stokes equations describe its dynamics. As the fluid temperature is lowered, a phase transition at T_λ introduces more complicated behavior, and the fluid acquires superfluid properties. At these lower temperatures, we must appeal to quantum mechanics to describe even the macroscopic motions of

163

Y. Kaneda (ed.), *IUTAM Symposium on Computational Physics and New Perspectives in Turbulence*, 163–170.

the fluid [2]. One consequence of this is that vortices can have quantized circulation [3], rather than arbitrary circulation as in a classical fluid. We discuss an experimental technique presented by Bewley et al. [4] for marking the cores of these quantized vortices in the superfluid phase of liquid helium. We show that the vortices trap small particles of solid hydrogen, and observe the behavior of filaments formed by the collection of many such particles along the vortex core. For the first time, it is possible to observe directly in this way the interactions of quantized vortices and to locally probe superfluid turbulence.

There is a long tradition of studying fluid behavior by observing the motions of tracer particles (for examples, see [5]), and several groups have used small particles to study the motions of liquid helium (e.g., [6] [7] [8]). In liquid helium, the challenge is to generate particles that are small enough to trace accurately the flow [9], and to understand the response of the particles to the motions of the superfluid [10].

Other researchers have detected quantized vortices indirectly by a variety of means, including using the vortex attenuation of second sound in the superfluid [11]. Until now, individual vortices have been resolved only in two-dimensional projections, where the vortices are straight and parallel to each other [12]. This technique was perfected by Yarmchuk et al. [13], who observed vortices at their intersection with the top boundary of the fluid, and determined that in solid body rotation, the superfluid formed a triangular lattice of vortices. The uniqueness of our method is that it allows the study of the dynamics of quantized vortices in the bulk of the fluid.

2 Apparatus

We generate small solid hydrogen particles in liquid helium contained in a cryostat with optical access provided by four windows [14]. We inject a room-temperature mixture of hydrogen gas diluted with helium gas into normal liquid helium, at a temperature above T_λ. The injector tube has a $3 \ mm$ inner diameter, and opens below the free surface of the liquid helium. The procedure yields particles with a wide size distribution, though most are probably about $2 \ \mu m$ in diameter or smaller. We estimate the characteristic particle size by dividing the known mass of hydrogen injected among the number of particles in the volume of liquid, which we find by counting particle images. Our experience has been that the observations described below are possible only if the particles form in the warmer phase of liquid helium. The suspension of particles is subsequently cooled through the transition temperature to a desired temperature below T_λ. We cool the liquid helium by evaporation using a mechanical vacuum pump. The pumping rate is controlled manually using a diaphragm valve, and we measure temperature by determining the resistance of a calibrated semiconductor probe.

Some of our images are acquired when the cryostat is rotating, others are not. In the absence of system rotation, we acquire images of the particle suspension during the cooling process, and record the time and temperature for each image. We use a digital movie camera with 16 $\mu m/pixel$ resolution that is focused on a laser-illuminated sheet approximately 100 μm thick. We do not disturb the fluid, except by the thermally driven flows caused by the cooling process.

The cryostat and camera are mounted on an air bearing that allows them to rotate freely about the vertical axis. The laser beam is passed to the rotating apparatus in such a way that they rotate in unison. Rotation complicates the method for acquiring images because the cryostat is initially tethered to a vacuum pump in order to cool the liquid helium. In the absence of cooling, the temperature of the fluid rises due to heat leaks into the cryostat. We therefore cool the fluid to below the phase transition temperature before spin-up, and disconnect the vacuum pump in order to spin the cryostat independently of the pump. We cool the fluid by an amount sufficient to keep the fluid below the transition temperature throughout the spin-up of the cryostat and fluid, and for the period thereafter during which measurements are made. The steady state behavior of superfluid helium at a fixed rotation rate does not depend on the temperature [15].

3 Results

3.1 Evidence

Images acquired above T_λ show that particles are randomly distributed (see Fig. 1a). As we cool the fluid without rotation, the images show that a fraction of the suspended particles collect onto slender filaments, often several millimetres in length (see Fig. 1b). These appear when the temperature is from a few millikelvin below T_λ down to 1.9 K, which is the lowest temperature we have

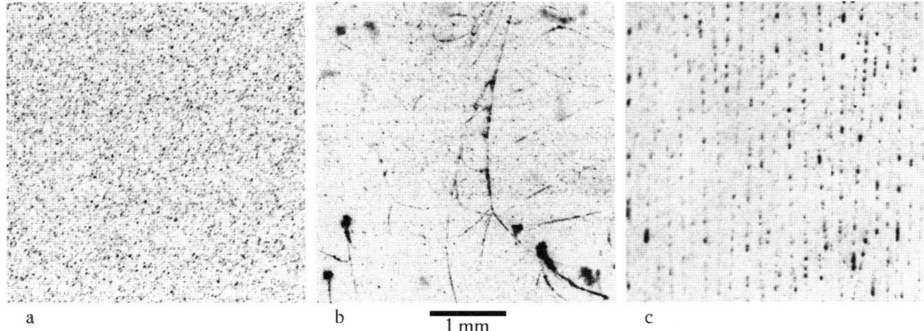

Fig. 1. (a) shows hydrogen particles in liquid helium at a temperature above T_λ; (b) shows similar particles after the fluid is cooled to a temperature below T_λ. Some of the particles have collected along branching filaments. Panel (c) shows particles arrayed along vertical lines in liquid helium that is steadily rotating at $\Omega = 0.3\ Hz$ at a temperature slightly less than T_λ.

explored. The filaments evolve slowly as they drift upward through the observation volume at a rate of roughly 1 mm/s. The remaining particles are randomly distributed, as is the case above T_λ (see Fig. 1a). In Bewley et al. [4], we presented briefly the evidence that suggests the observed filaments are formed by particles trapped on quantized vortex cores. Here, we consider this evidence and our assumptions in greater detail.

The first piece of evidence that the particles mark the cores of the quantized vortices is that the filaments appear only in images taken at temperatures when quantized vortices exist—that is, at temperatures below T_λ. Parks and Donnelly [16] describe a mechanism by which quantized vortices trap ions. Ions are drawn to the cores of quantized vortices by the steep pressure gradient supporting the circulating superfluid. Hydrogen particles are hydrodynamically comparable to ions, though the ions are much smaller. Since both the ions and the hydrogen particles are substantially larger than a quantized vortex core, the mechanism is equally applicable to both entities.

A second piece of evidence is that when the liquid helium cell is set in steady rotation, the particles arrange themselves along uniformly spaced lines parallel to the rotation axis, as shown in Fig. 1c. This observation agrees with the expectation that quantized vortices form a rectilinear lattice aligned with the axis [2], since the vorticity of rotation resides in the quantized vortices.

Feynman [2] showed that the population of quantized vortices aligned with the axis of rotation in steadily rotating liquid helium should have at equilibrium a number density equal to

$$n_o = 2\Omega/\kappa \approx 2000\Omega \ vortices/cm^2 \qquad (1)$$

where Ω is the angular velocity of the system in $radians/s$ and $\kappa = h/m$ is the quantum of circulation, h being the Planck's constant and m the mass of the helium atom. Additional evidence that the columns of particles described above are vortices is that the number density of lines per unit area normal to the axis of rotation is consistent with Feynman's rule (1) for a series of rotation rates. For system rotation rates from 0.15 Hz to 0.46 Hz, we calculate the number density of vortex lines observed in each state, according to assumptions described below. The value we measure is typically 25% larger than is predicted by (1).

3.2 Discussion of Assumptions

From the images of particles in rotating superfluids helium, we calculate the number density of lines under several assumptions. We argue that the measurement of line density is not sensitive to these assumptions, and that the conclusion is favorable even when the conditions hold only approximately.

The array of vortices appears rectilinear in the plane of observation, but we cannot know in what way the vortices are arranged in the normal plane. In (1), we have assumed that the intersections of the vortex cores with the light sheet are at the corners of the theoretical triangular array (see [17]) and that the laser light sheet illuminates a section of such a triangular array. If, for example, the lines in the array were on the corners of squares instead of triangles, we err by about 14% in computing the vortex line density, which nudges the corresponding theoretical line toward the data.

We assume that the light sheet is aligned with the lattice in such a way that the vortices appear equally spaced in the image, and that the separation between them is the minimum spacing of vortices in the lattice. Deviations from these assumptions lead to a calculation of line density that is in excess by about 33%. We also assume that the thickness of the illuminating sheet is roughly equal to the lattice spacing, so that it illuminates only one row of lines in the lattice. As mentioned above, the thickness of the illuminating sheet is about 100 μm. For the range of rotation rates explored we expect from Feynman's rule (1) that the minimum separation between vortices is between 140 and 250 μm, and so our assumption is reasonable.

Finally, we assume that every vortex in the fluid is decorated with particles, and will be observed if illuminated. In making our estimates, we have assumed that all vortices are singly quantized, as favored by energy considerations. Our vortex count provides only a lower bound for the total circulation they produce, if there exist multiply quantized vortices. We take our results to be consistent with the view that each of the lines revealed by particles is singly quantized.

3.3 Vortex Dynamics

Using the technique for marking quantized vortex cores, we are able to study fundamental vortex dynamics. We observe, for example, the decay of vortex rings and the reconnection of pairs of vortices (see Fig. 2). In addition, we have observed phenomena which indicate that under certain conditions, particles are not passive tracers of vortex dynamics. For example, continuously decorated filaments form branches rather than reconnecting, and a vortex ring decays more slowly than predicted by the theory when its core is completely covered with particles.

It is evident in Fig. 1b that the continuously decorated filaments form networks with stable forks, whereas simulations show vortex filaments as smooth curves with brief dynamic intersections through reconnections [18]. We are unaware of any discussion in the literature of the possibility that particles in superfluid helium transform the topology of a vortex tangle. This phenomenon may be related to ability of a particle to trap more than one quantized vortex, analogous to a condition examined by Tsubota in computer simulations [19].

We attribute the retarded decay time of a vortex ring (see Fig. 2, row (a)) to viscous drag on the particles, and conjecture that this drag is important when the particles are spaced by less than about 10 particle diameters according to the following argument. We compare the drag per unit length on a vortex line due to friction with the normal fluid [15], which is a normal part of quantized vortex dynamics, to the Stokes drag on a spherical particle on

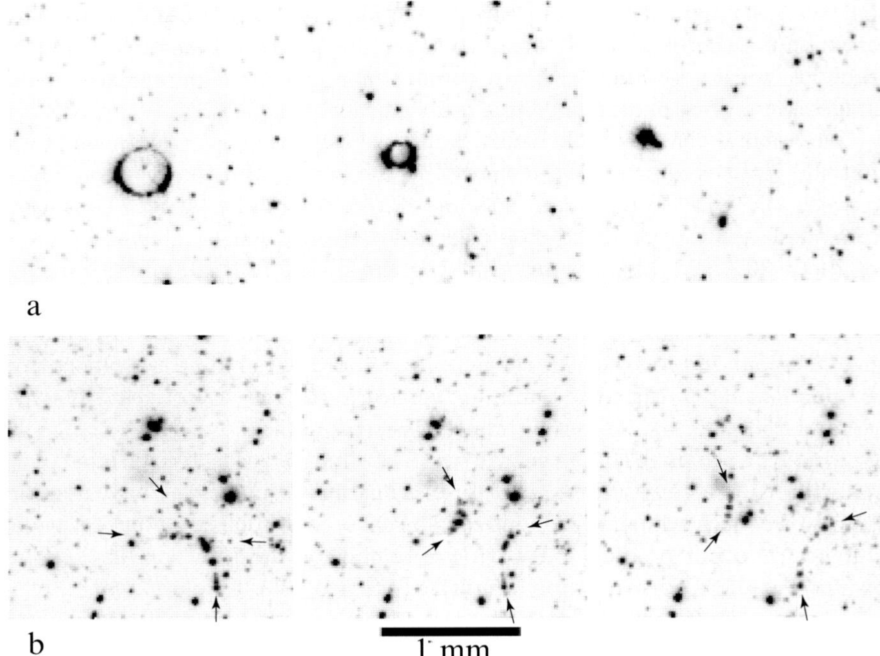

Fig. 2. Row (a) shows images taken 1.5 s apart of a ring whose diameter decays until the particles form an amorphous aggregate. The theory [15] predicts that the ring should decay more than 3 times faster than it does. Row (b) shows what we believe is the reconnection of two vortices that have crossed. The vortices are marked by particles that have space between them, rather than by a continuous coating of particles. The left-most frame shows the two vortices just before the moment of reconnection. We have indicated where we believe the vortices leave the thin plane of illumination, and become invisible to the camera, with arrows. The subsequent two frames are taken 0.25 and 0.50 s after the first, and show that the vortices have exchanged ends and are recoiling.

the line per sphere diameter, which could modify the behavior of the quantized vortex. To a rough approximation, this ratio is $\rho_s \kappa / 3\pi\mu$, and its value is always less than one. This result suggests that the drag of particles trapped on the line dominates, as long as the line is continuously decorated with particles. However, for a vortex line with particles every 10 particle diameters, the drag contributions are comparable for a range of temperatures between 1.5 K and 2.1 K.

Under the condition that quantized vortices are decorated with particles spaced by about ten diameters, we observe vortex reconnections (see Fig. 2, row (b)). The dotted lines indicated in the figure consist of particles that are roughly 10 μm in diameter and spaced by about 100 μm. In the first frame, the dotted lines ap pear crossed in projection, but they are probably separated in the direction normal to the image plane. At some time between the first and second frames, the vortices move toward one another and intersect. Their evolution after reconnection is in qualitative agreement with simulations [20] [18] in that the vortices draw away from each other, and their curvature decreases. Details regarding observations such as this one will appear in forthcoming publications.

4 Summary

We find that particles collect along filaments in superfluid helium and present evidence that these filaments mark the cores of quantized vortices. In this way, we have found a method for tracking the vortices, so that their cores can be viewed in the bulk for the first time. We observe that the filaments behave in some respects as is expected of quantized vortices. They align with the axis when the fluid is rotating, and the diameter of a ring decays with time due to interaction with the normal fluid. Because of this, we identify the filaments with quantized vortices. However, we propose that the hydrogen particles modify the behavior of the vortices in two ways. First, the viscous drag on the hydrogen particles is more important than mutual friction when the particles completely cover the line, and second, the hydrogen particles stabilize vortex intersections, which might not otherwise exist. Our observation of what appears to be a decaying vortex ring further validates an explanation of dissipation in quantized vortex turbulence, and confirms that the hydrogen spines do not immobilize the quantized vortices, or render them rigid. We predict that if particles are spaced along a vortex line by more than ten diameters, the vortex may behave as if bare of particles, but still be observable. In fact, we have recorded images of what we believe are vortices with widely and evenly spaced particles, and captured such dotted lines reconnecting with each other.

References

1. Niemela JJ, Sreenivasan KR (2006) J Low Temp Phys 143:163–212
2. Feynman RP (1955) In: Gorter CJ (ed) Progress in low temperature physics vol 1. North Holland, Amsterdam
3. Onsager L (1953) Proc Intern Conf Theor Phys, Kyoto and Tokyo, Science Council of Japan, Tokyo:877–880
4. Bewley GP, Lathrop DP, Sreenivasan KR (2006) Nature 441:588
5. Batchelor GK (1967) An introduction to fluid dynamics. Cambridge University Press, Cambridge
6. Chopra KL, Brown JB (1957) Phys Rev 108:157
7. Murakami M, Ichikawa N (1989) Cryogenics 29:438–443
8. Zhang T, Van Sciver, SW (2005) Nature Physics 1:36–38
9. White CM, Karpetis AN, Sreenivasan KR (2002) J Fluid Mech 452:189–197
10. Poole DR, Barenghi CF, Sergeev YA, Vinen WF (2005) Phys Rev B 71:064514-1–16

11. Vinen WF (1957) Proc Rot Soc A 240:114–127
12. Williams GA, Packard RE (1974) Phys Rev Lett 33:280–283
13. Yarmchuk EJ, Gordon MJV, Packard RE (1979) Phys Rev Lett 43:214–218
14. Bewley GP, Sreenivasan KR, Lathrop DP (2007) submitted to Expts in Fluids
15. Donnelly RJ (1991) Quantized Vortices in Helium II. Cambridge University Press, Cambridge
16. Parks PE, Donnelly RJ (1966) Phys Rev Lett 16:45–48
17. Campbell LJ, Ziff RM (1979) Phys Rev B 20:1886–1902
18. Schwarz KW (1985) Phys Rev B 31:5782–5804
19. Tsubota M (1994) Phys Rev B 50:579–581
20. Koplik J, Levine H (1993) Phys Rev Lett 71:1375–1378

Lagrangian Measurements of Fluid and Inertial Particles in Decaying Grid Generated Turbulence

Sathyanarayana Ayyalasomayajula[1], Armann Gylfason[2], and Zellman Warhaft[1]

[1] Sibley School of Mechanical and Aerospace Engineering, Cornell University, Ithaca, NY 14850, USA
zw16@cornell.edu
[2] Reykjavik University, Iceland
armann@ru.is

Abstract. We present preliminary measurements of the Lagrangian acceleration probability density function (pdf) of fluid particles in decaying grid-generated turbulence and show that they are in very good agreement with direct numerical simulations determined at the same Reynolds number. We contrast these pdf's with inertial particle acceleration pdf's done in the same apparatus.

Keywords: turbulence experiments, Reynolds number, wind tunnels, Lagrangian measurements, probability density functions

It has been nearly a decade since the first quantitative Lagrangian measurements of fluid particles at realistic Reynolds numbers began to appear [9, 7, 5]. These results have shown the extremely intermittent nature of the small scale structure, manifested by the highly stretched tails of the probability distribution (pdf) of the particle acceleration. More recently measurements of inertial particles, particles with mass significantly greater mass than the surrounding fluid, have been carried out (Ayyalasomayajula et al (2006) [2], from here on referred to as AGCBW). These experiments show that the acceleration variance is reduced and, although the particle trajectories are still highly intermittent, they are less so than the inertia-less fluid particles. This is due to the clustering and filtering effects that occur because of the particle inertia. These are discussed in detail in Ayyalasomayajula et al 2007 [1] (see also Bec et al 2006 [3]). The inertial particles were measured in decaying, grid generated turbulence, while all of the existing the fluid particle measurements were made in a cylindrical tank stirred from above and below (the so called French washing machine). It is important to determine the sensitivity of the pdf's to the boundary conditions of the flow. Here we present a preliminary measurement of the probability density function of fluid particles in decaying

171

Y. Kaneda (ed.), *IUTAM Symposium on Computational Physics and New Perspectives in Turbulence*, 171–175.
© 2008 *Springer*.

Fig. 1. (left) The wind tunnel showing the camera (far left in the picture, at the beginning of its trajectory), the sled and the laser sheet. The active grid and spray system are at the tunnel entrance (just above the camera lens). The copper strips (right foreground) are the magnetic braking system for the camera sled. (upper right) Close-up of the active grid with array of spray nozzles. (lower right) Close-up of the mist generated by the spray nozzles.

grid turbulence. It is our purpose to compare these measurements with those of the stirred tank, and with the inertial particle pdf's done in the decaying grid turbulence.

The experiments were conducted in a large open circuit wind tunnel (1 m x 0.9 m x 20 m) with an active grid at its entrance section. For the inertial particle measurements water sprays were placed slightly downstream of an active grid (Figure 1). For the fluid particles of approximately 7 micron diameter (see below), it was necessary to locate their source further down stream, closer to the detector, since evaporation and low seeding presented a problem. Thus we placed a household acoustic humidifier on the floor of the tunnel, approximately 1 meter from the region where we tracked them using the high speed camera mounted on a sled beside the tunnel (AGCBW). The mean speed of the flow was 1.9 m/s and the Taylor scale Reynolds number was 250. The Stokes number of the particles was approximately 0.01, i.e. inertial effects were negligible. (The Stokes number is defined as $St = \tau_p/\tau_\eta$, where τ_p is the time scale defined as $(1/18)[\rho_p/\rho_f]d^2/\nu$ where ρ_p, ρ_f, d and ν are the particle density, fluid density, particle diameter, and fluid kinematic viscosity.) By contrast, the St of our inertial measurements was approximately 0.1, AGCBW.

Figure 2 shows the particle size distribution of the fluid particles measured by a phase Doppler particle sizer. The distribution is close to lognormal. We have also plotted the particle size distribution of the inertial particles, from AGCBW.

In Fig. 3 we present the pdf of the fluid particle accelerations and compare it with the pdf of fluid particle accelerations measured in the stirred tank as

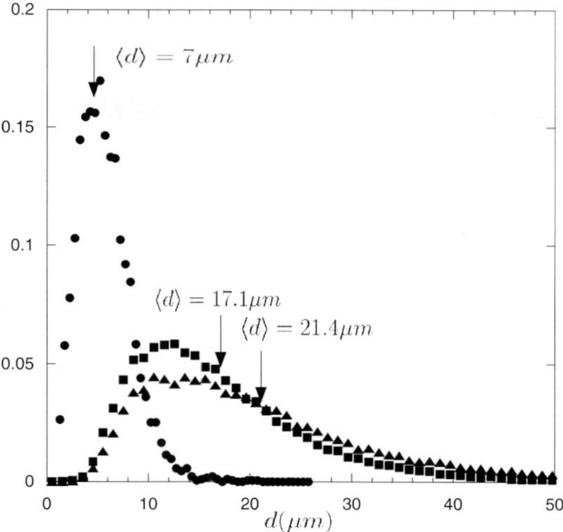

Fig. 2. Particle size distributions at the measurement station. The vertical arrows are the mean diameters calculated from the distributions. The resulting mean Stokes numbers for the three cases are 0.01 ± 0.005 (circles), 0.09 ± 0.03 (squares) and 0.15 ± 0.04 (triangles).

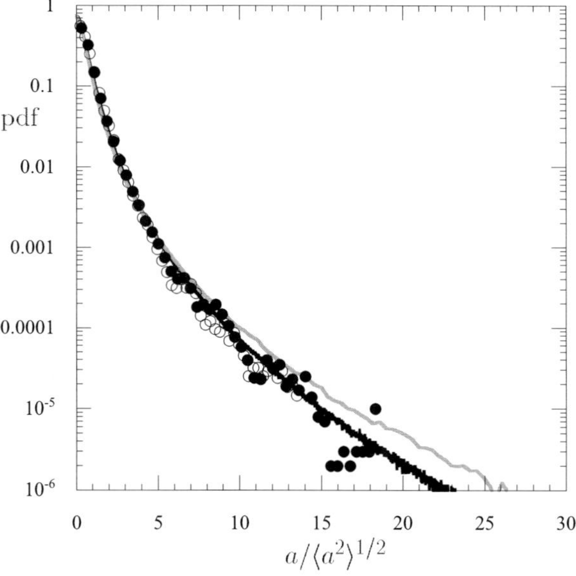

Fig. 3. The acceleration probability density functions (pdf). Symbols - fluid particles ($\langle St \rangle = 0.01 \pm 0.005 \ll 1$, filled and open circles are pdfs for acceleration measured in x and y directions). Solid black line - fluid particles from Bec et al. (2006) [3] at comparable R_λ to the experiments ($R_\lambda \approx 185$). Solid gray line - fluid particles from Mordant et al. (2004) [4] at a higher R_λ of 690. The pdfs are of the acceleration normalized by the rms, $\langle a^2 \rangle^{1/2}$.

well as the pdf of fluid particle acclerations determined by direct numerical simulations (DNS) by Bec et al (2006) [3]. The Reynolds number for the DNS is similar to those of our experiment. The Reynolds number for the stirred tank was significantly higher, and this is reflected in the slightly broader tails of the pdf. (Intermittency effects increase with Reynolds number. This is well established in the Eulerian framework, e.g. Mydlarski and Warhaft, 1996 [6]. Evidence from Lagrangian measurements is less clear ([8]) indicating that further experimental work is needed.) There is remarkably good agreement between the DNS and our preliminary measurement. Nevertheless our measurement is not well resolved and further measurements are under way. In Fig. 4 we show fluid particle pdf's compared with the inertial particle pdf's. The systematic decrease in the spread of the tails with increasing Stokes number is clearly evident. Note, that at least for moderate Stokes numbers, the departure from the fluid particle pdf occurs quite deep in the tails (below 0.001). This is also evident in the DNS of Bec et al. 2006 [3].

In summary, we have presented preliminary measurements of the pdf of fluid particles in decaying grid generated turbulence. The results compare well with direct numerical simulations at the same Reynolds number.

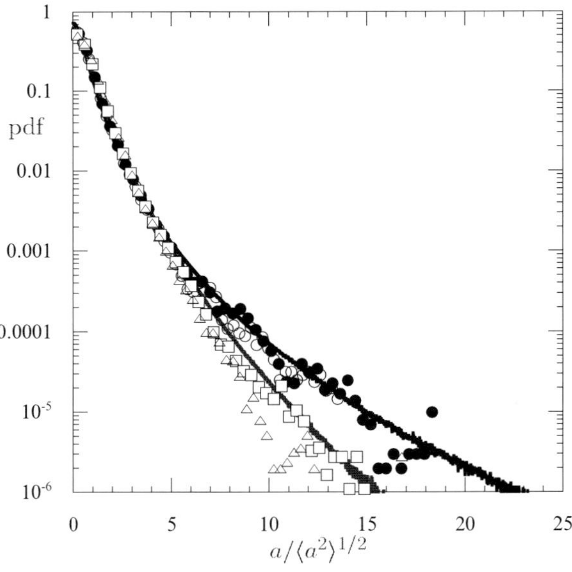

Fig. 4. The acceleration probability density functions (pdf). Inertial particles, squares $\langle St \rangle = 0.09 \pm 0.03$ and triangles, $\langle St \rangle = 0.15 \pm 0.04$ from Ayyalasomayajula et al. (2006) [2] and circles - passive particles ($\langle St \rangle = 0.01 \pm 0.005 \ll 1$, filled and open circles are pdfs for acceleration measured in x and y directions) at $R_\lambda = 250$. Also shown are the DNS results of comparable R_λ and St, from Bec et al. (2006) [3]; black line - fluid particles; gray line - $St = 0.16$ at $R_\lambda = 185$. The pdfs are of the acceleration normalized by the rms, $\langle a^2 \rangle^{1/2}$.

We thank Eberhard Bodenschatz, Haitao Xu & Nicholas Ouellette. This work was done in association with International Collaboration for Turbulence Research (ICTR) and was supported by the U.S National Science Foundation.

References

1. Ayyalasomayajula S, Collins LR, Warhaft Z (2007) In preparation
2. Ayyalasomayajula S, Gylfason A, Collins LR, Bodenschatz E, Warhaft Z (2006) Phys Rev Lett 97:144507
3. Bec J, Biferale L, Boffetta G, Celani A, Cencini M, Lanotte A, Musacchio S, Toschi F (2006) J Fluid Mech 550:349–358
4. Mordant N, Crawford AM, Bodenschatz E (2004) Physica D 193:245–251
5. Mordant N, Metz P, Michel O, Pinton J-F (2001) Phys Rev Lett 87:214501
6. Mydlarski M, Warhaft Z (1996) J Fluid Mech 320:331–368
7. Ott S, Mann J (2000) J Fluid Mech 422:207–223
8. Voth GA, La Porta A, Crawford AM, Alexander J, Bodenschatz E (2002) J Fluid Mech 469:121–160
9. Voth GA, Satyanarayan K, Bodenschatz E (1998) Phys Fluids 10 (9):2268–2280

Shear Effect on Pressure and Particle Acceleration in High-Reynolds-Number Turbulence

Yoshiyuki Tsuji[1], Jens H. M. Fransson[2], and P. Henrik. Alfredsson[2], and Arne V. Johansson[2]

[1] Department of Energy Engineering and Science, Nagoya University, Nagoya 464-8603, Japan `c42406a@nucc.cc.nagoya-u.ac.jp`
[2] Royal Institute of Technology (KTH), Stockholm, Sweden

Abstract. Pressure fluctuations are an important ingredient in turbulence, however the variation of the pressure fluctuations inside a turbulent boundary layer is not well known and its generation not fully understood. One reason for this is the difficulty inherent in measuring the fluctuating pressure and, consequently, a lack of published experimental results of this quantity. In the present study, the new methods and improved measurement techniques are developed. Fluctuating pressure has been measured in turbulent boundary layer up to $R_\theta \simeq 20000$ and in mixing layer up to $R_\lambda \simeq O(10^3)$. In this report, the shear effect on the Kolmogorov's $-7/3$ power-law scaling of pressure spectrum is discussed and the preliminary measurement of particle acceleration is reported.

Keywords: pressure statistics, particle acceleration, Kolmogorov scaling

1 Introduction

There is an immense body of literature on the behavior, distribution and scaling of velocity fluctuations in turbulent boundary layers, however so far very little is known about similar behavior of pressure fluctuations. The main reason for the lack of such results is that no measurement technique so far has been able to actually measure pressure fluctuations inside the boundary layer. Pioneering experiments to measure pressure fluctuations were performed by [1], who measured the static pressure fluctuation behind the wake of cylinder using the microphone with ratio-discriminator. This technique was succeeded by [2] for eduction of vortical structures in a noncircular jet.

Although there were several attempts to measure the pressure fluctuations in turbulence, they were not enough to confirm the accuracy of measured data until [3] compared the experimental results with direct numerical simulation (DNS). They reported that the probability density function of pressure matched that of DNS well, and a power law close to $-7/3$ was realized in the range of $R_\lambda \geq 600$. These Reynolds numbers are much larger than those for

177

Y. Kaneda (ed.), *IUTAM Symposium on Computational Physics and New Perspectives in Turbulence*, 177–182.

which velocity fluctuations achieve Kolmogorov scaling. They concluded that the spectral constant is not universal but depends on Reynolds number.

The present study is the first attempt, as far as we know, to investigate the detailed pressure statistics in high-Reynolds-number turbulent boundary layers. The accuracy of measured data is examined carefully and compared with DNS by [4], who computed the zero pressure gradient flow up to $R_\theta = 716$. The main results were reported in [5] and they are summarized as a full paper [8]. Here, we focus on the pressure spectra. It is discussed from the point of Kolmogorov scaling.

2 Experimental Conditions

The measurement of pressure fluctuation is accomplished with a small piezo-resistive transducer and a standard condenser microphone (1/4 and 1/8-inch). The pressure probe is a standard Pitot-static tube measuring 0.5 mm in outside diameter and 0.05 mm in thickness. Four static-pressure holes (0.15 mm in diameter) are spaced 90° apart and located at a distance of 22 tube diameters from the tip of the probe to minimize sensitivity to cross-flow error. The detailed explanations are given in the reference [3].

Experiments are performed in the MTL wind tunnel at KTH [5]. This is a high quality flow tunnel with a stream-wise turbulence intensity of less than 0.02% and a total pressure variation less than 0.01% across the test section. The data are measured in the zero-pressure-gradient boundary layer, which covers the Reynolds number range $2600 \le R_\theta \le 26700$. Stream-wise velocity, wall pressure and pressure fluctuations inside the boundary layer are measured simultaneously [5, 8].

3 Results and Discussions

3.1 Shear Effect on Pressure Spectra

Recent theoretical studies, numerical simulations, and experiments have revealed that the large-scale anisotropy caused by mean shear significantly affects the inertial range statistics [6]. In the case of two-dimensional boundary layers, the essential parameter characterizing the shear effect is the anisotropy or shear parameter (S^*) defined as follows,

$$S^* = (\nu/\langle\varepsilon\rangle)^{1/2}\, S\,, \tag{1}$$

where S is the mean velocity gradient, $S \equiv dU/dy$. S^* is the ratio of mean shear time scale to the smallest eddy time scale $\tau_\eta \equiv (\nu/\langle\varepsilon\rangle)^{1/2}$. Although S varies like $\simeq y^{-1}$ in the overlap region, S^* behaves in a different way. Since $\langle\varepsilon\rangle$ behaves approximately as y^{-1} in the log-layer, S^* will decrease more slowly than S itself. We should, hence, expect S^* to vary approximately as $y^{-1/2}$ with increasing distance from the wall (in the log-layer). If S^* is small, the level of anisotropy created by the mean shear might also be expected to be

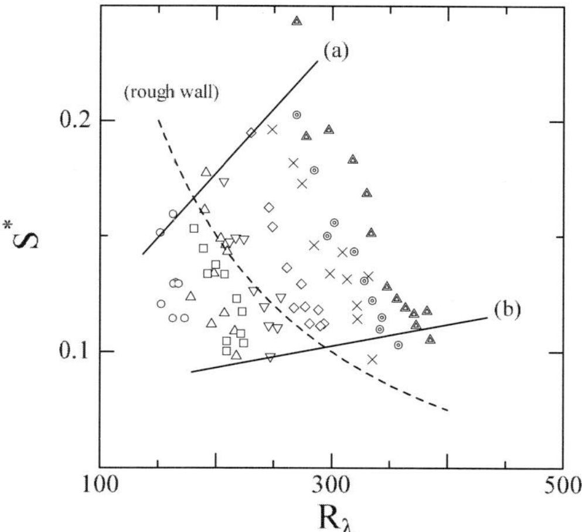

Fig. 1. Relation between the shear parameter and the Reynolds number in the log-region. Symbols indicate different Reynolds numbers. $\circ$: $R_\theta = 5870$, $\triangle$: $R_\theta = 7420$, $\square$: $R_\theta = 8920$, $\triangledown$: $R_\theta = 10500$, $\diamond$: $R_\theta = 12100$, $\odot$: $R_\theta = 13600$, $\times$: $R_\theta = 15200$, $\triangle$: $R_\theta = 16700$. The solid line (a) indicates the relation between S^* and R_λ at the lower end of the log-region ($y^+ \simeq 180$) and solid line (b) is the one at the outer end of the log-region ($y/\delta \simeq 0.15$). Dashed line is the case of rough wall boundary layer [6], in which the relation $S^* \propto R_\lambda^{-1}$ is satisfied in the overlap region for $4925 \leq R_\theta \leq 13060$.

small. In the log-region, both S^* and R_λ are plotted in Fig. 1, where the different symbols indicate the different R_θ. Away from the wall, S^* decreases but R_λ increases. Or in other words, S^* is a decreasing function of R_λ for each R_θ. The solid line (a) indicates the relation between S^* and R_λ at the lower end of log-region ($y^+ \simeq 180$) and solid line (b) is the one at the outer end of the log-region ($y/\delta \simeq 0.15$). As the Reynolds number R_θ increases, R_λ becomes larger and S^* increases accordingly. Therefore, as plotted by the solid lines in the figure, large R_λ is equivalent to large S^*, and we cannot keep S^* small for large R_λ. This trend is contrary to the case of rough wall boundary layers [6], as plotted by the dashed line, in which S^* is a decreasing function of R_λ, and the relation $S^* \propto R_\lambda^{-1}$ is satisfied in the overlap region for $4930 \leq R_\theta \leq 13100$.

Kolmogorov presented hypotheses for small-scale statistics based on the idea of local isotropy. The one-dimensional pressure spectrum E_{pp} exhibits an inertial subrange (for $k_1 \eta \ll 1$), when the Reynolds number becomes large, with a simpler form independent of ν:

$$E_{pp}(k_1) = K_p \rho^2 \langle \varepsilon \rangle^{4/3} k_1^{-7/3}, \tag{2}$$

where k_1 is the stream-wise wave number, $\langle \varepsilon \rangle$ is the energy dissipation rate per unit mass on average. And K_p is a universal constant. The $-7/3$ power-law scaling was supported theoretically with various assumptions in the 1950's.

Recently, Tsuji & Ishihara [3] have examined the pressure spectrum in fully developed nearly isotropic turbulence in the centre of a plane jet for the Reynolds number range of $200 \leq R_\lambda \leq 1200$. A power-law exponent of the pressure spectrum was systematically obtained by fitting the relation

$$E_{pp}(k_1) = K_p{}' \rho^2 \, \langle \varepsilon \rangle^{3/4} \, \nu^{7/4} (k_1 \eta)^{\gamma_p} \,, \tag{3}$$

to the measured spectrum where here k_1 is normalized by the Kolmogorov length scale η defined as $\eta = \left(\nu^3 / \langle \varepsilon \rangle \right)^{1/4}$ and $K_p{}'$ is a non-dimensional quantity.

In the log-region, we may expect the Kolmogorov scaling with a $-7/3$ power law behavior. However, the measured spectra for several R_θ in the range of $150 < R_\lambda < 400$ did not follow the Kolmogorov scaling (graph is not shown here). There is no evidence that the power-law exponent approaches the $-7/3$, nor that the spectra collapse, so the Kolmogorov scaling is not satisfied. Compared with the result of the centre line of a turbulent jet [3], which is close to the homogeneous isotropic condition, the pressure spectra in the shear flow are widely different from (2). This is likely to be due to the shear effect, which is present in the boundary layer case.

The exponent γ_p evaluated by the relation of equation (3) is plotted against R_λ and S^* in Fig. 2. The slope γ_p is obtained by computing the local slope of

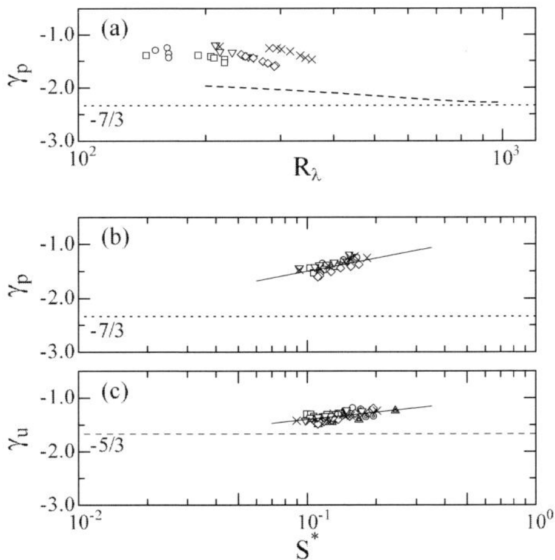

Fig. 2. (a) The power-law exponent of the pressure spectrum in the log- region is plotted against the Reynolds number. The dashed line is the result of $S^* = 0$ measured in the centreline of a turbulent jet. (b c) The power-law exponent of the pressure and velocity spectra in the log-region are plotted against the shear parameter. Solid line is the relation given in equation (5) obtained by least-square fit to the data. Symbols indicate the different Reynolds numbers. $\circ$: $R_\theta = 5870$, $\triangle$: $R_\theta = 7420$, $\square$: $R_\theta = 8920$, $\triangledown$: $R_\theta = 10500$, $\diamond$: $R_\theta = 12100$, $\odot$: $R_\theta = 13600$, $\times$: $R_\theta = 15200$.

the spectrum (detailed procedure, such as the wave number at which the local slope is computed, is mentioned in appendix of [8]). The symbols correspond to the different R_θ. The dashed line is the result of $S^* = 0$ measured on the centreline of a turbulent jet. γ_p is a decreasing function of R_λ for a fixed value of R_θ. As R_λ increases, however, γ_p deviates substantially from the isotropic value of $-7/3$, and it also differs from the value obtained from the turbulent jet. This trend is better understood, with the help of Fig. 1, as a function of S^*. γ_p is less dependent of R_θ and it is an increasing function of S^*. Velocity spectra, on the other hand, have the scaling exponent γ_u defined by

$$E_{uu}(k_1) = K_u{}' \, \langle \varepsilon \rangle^{5/4} \, \nu^{1/4} (k_1 \eta)^{\gamma_u} \,, \tag{4}$$

and in the log-law region the variation of γ_u and γ_p as a function of S^* is given by the least square fit,

$$\gamma_p = 0.8 \log_{10} S^* - 0.7 \quad, \quad \gamma_u = 0.45 \log_{10} S^* - 0.95 \,. \tag{5}$$

Thus, the exponents indeed depart from the isotropic values; $\gamma_p = -7/3$ and $\gamma_u = -5/3$ for large S^*. The data indicate that the shear effect is more significant for the pressure, that is, the pressure exponent changes more rapidly than that of the velocity. Extending these solid lines, local isotropy is expected to become realized for $S^* \simeq O(10^{-2})$. Hence, we have to conclude that local isotropy is not restored in the inertial range even if R_θ is $O(10^4)$. In the present experiment we believe that the observed $\gamma_p > -7/3$ is mostly due to the fact that Reynolds number is too small for a given S^*, which make the anisotropic contribution to dominate the pressure fluctuation. For the smooth wall it is not possible to increase R_λ while keeping S^* small in the overlap region. Therefore, local isotropy conditions may not be achieved in this region. One possibility to achieve local isotropy is to use a rough wall, thereby increasing the turbulence intensity near the wall.

3.2 Acceleration Measurement and Shear Effect on It

The motion of fluid particles as they are pushed along the trajectories by fluctuating pressure gradient is fundamental to transport and mixing in turbulence. In principle, fluid particle trajectories are easily measured by seeding a turbulent flow with small tracer particles and following their motions with an imaging system. But, in practice, this can be a very challenging task because we must fully resolve particle motions which take place on the order of Kolmogorov time scale. This kind of measurements was recently achieved by [7] with using a specially designed particle tracking system in high energy physics. In a usual notation, acceleration vector is given by N.S. equation as follows; $\mathbf{a} = D\mathbf{u}/Dt = -\nabla(p/\rho) + \nu\nabla^2\mathbf{u}$. This means that acceleration is decomposed into the contribution from pressure gradient and viscous force while the fluid density is constant. In a fully developed turbulence, the viscous damping term is small compared with the pressure gradient term, therefore, the acceleration is closely related to the pressure gradient.

Pressure gradient measurement was performed by using two pressure probes in a mixing layer up to $R_\lambda \simeq 10^3$. The pressure difference measured at a distance Δy(or Δx) becomes pressure gradient $\mathrm{d}p/\mathrm{d}y$(or $\mathrm{d}p/\mathrm{d}x$) as far as Δy (or Δx) is sufficiently small ($\simeq 5\eta$). Although the graph is not shown here, the measured pressure gradient distribution has a stretched exponential shape, which is close to the previous result [7], in which the tails extend much further than a Gaussian distribution. This indicates that acceleration is an extremely intermittent variable. When the flatness of velocity gradient is compared with that of pressure, it is clearly understood that the acceleration field is more intermittent. In the acceleration spectra, the expected power-law exponent $-1/3$ is well observed in the present measurements. It is noted that the acceleration spectrum is hard to obtain as far as pursuing the particle trajectories. Following the Kolmogorov's idea, acceleration is scaled by the energy dissipation rate and kinematic viscosity as $\langle a_i a_j \rangle = a_0 \varepsilon^{3/2} \nu^{-1/2} \delta_{ij}$, where $a_i = (1/\rho)\partial p/\partial x_i$. The constant a_0 is expected to be universal. But the recent DNS and La Porta's experiment do not show that a_0 is not constant but a function of Reynolds number. The present measurement in a shear flow (mixing layer) shows that a_0 is much smaller than those of DNS (isotropic flow). As R_λ increases, a_0 increases and approaches the DNS values. This indicates that the local isotropy is realized at $R_\lambda \simeq 2000$ in the inertial range. But more detailed discussions are necessary for the shear effect (large scale anisotropy) on the acceleration.

References

1. Kobashi Y (1957) J Phys Soc Jpn 12:533–543
2. Toyoda K, Okamoto T, Shirahama Y (1993) Fluid Mechanics and its Applications 21:125–136
3. Tsuji Y, Ishihara T (2003) Phys Rev E 68: 026309-1–026309-5
4. Skote M (2001) TRITA-MEK Tech Rep 2001:01, Dept Mech, KTH, Stockholm, Sweden
5. Tsuji Y, Fransson JHM, Alfredsson PH, Johansson AV (2005) Proc Fourth Int Symp on Turbulence and Shear Flow Phenomena, Williamsburg, VA USA, 27–29 June : 27–32
6. Tsuji Y (2003) Phys Fluids 12: 3816–3828
7. La Porta A, Voth GA, Crawford AM, Alexander J, Bodenschatz E (2001) Nature 409:1017–1019
8. Tsuji Y, Fransson JHM, Alfredsson PH, Johansson AV (2007) Pressure statistics and its scaling in high Reynolds number turbulent boundary layers, accepted for publication in J Fluid Mech

On the Development of Wall-Bounded Turbulent Flows

Kapil A. Chauhan[a] and Hassan M. Nagib[b]

Illinois Institute of Technology, 10 W. 32nd St., Chicago, IL 60616, USA
[a]chaukap@iit.edu, [b]nagib@iit.edu

Abstract. Three mechanisms leading to non-equilibrium behavior in boundary layers are identified, and examined using mean-flow characteristics such as the Coles' wake parameter Π. Well-understood criteria are realized for design and analysis of experiments and computations of "fully-developed" zero pressure gradient turbulent boundary layers in equilibrium.

Keywords: development, equilibrium, self-similar, transition and pressure-gradient

1 Introduction

Among the three canonical wall bounded turbulent flows, boundary layers differ fundamentally from pipes and channels because of their streamwise inhomogeneity. Experimental turbulent boundary layers typically originate from a leading edge, or from an upstream flow with history, and will at least initially also have some "artificial" streamwise inhomogeneity. Such inhomogeneity can be due to perturbations at the leading edge, post-transition growth, and structures originating from trips or transition triggers; or even from an upstream contraction for boundary layers on test-section sidewalls. All of these "artificial" effects should vanish at far downstream distances, in order to achieve a high-quality canonical boundary layer. Hence, it is critical in any fundamental study of fully-developed turbulent boundary layers (TBLs) to detect the presence of such artificial inhomogeneities and to distinguish them from equilibrium flow behavior. The present study identifies such evidence in turbulent boundary layers under zero pressure gradient (ZPG) using data from three recent and several classical experiments. The analysis presented is based on fitting the experimental data of U vs y to a recently refined composite velocity profile; see [1]. This composite velocity profile was developed based on the classical theory incorporating a logarithmic law in the overlap region with $\kappa = 0.384$. The inner function uses an integrated Padé 45 expression, U^+_{inner}, while an exponential-type wake function, $W(y/\delta)$ is added to it to represent the outer flow behavior:

Y. Kaneda (ed.), *IUTAM Symposium on Computational Physics and New Perspectives in Turbulence*, 183–189.

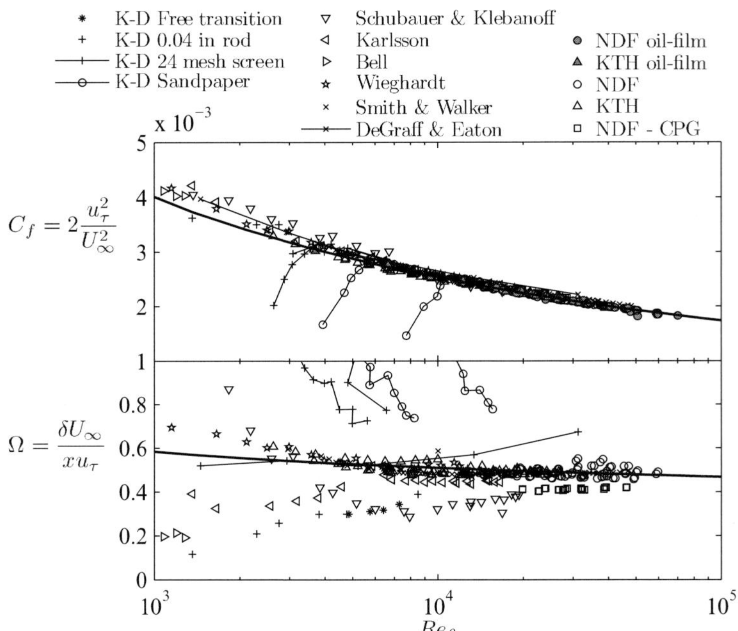

Fig. 1. The variation of C_f and Ω with Re_θ for ZPG TBL's. Solid line represents prediction of the classical theory using Clauser scaling.

$$U^+ = U^+_{inner}\left(\frac{yu_\tau}{\nu}\right) + \frac{2\Pi}{\kappa}W\left(\frac{y}{\delta}\right), \quad 0 \le y \le \delta \tag{1}$$

The mean velocity data are fitted to find δ, u_τ and Π based on least-square errors. The "development" of a boundary layer is then examined as a function of Reynolds number utilizing parameters like C_f, Ω $[= (U_\infty\delta)/(u_\tau x)]$, Π, Δ/δ $[= (U_\infty\delta^*)/(u_\tau\delta)]$ and x/δ.

Figure 1 displays the development of the skin-friction coefficient, C_f, with Reynolds number, Re_θ. The C_f is determined using the u_τ obtained from the composite fit and experimental values of U_∞. The resulting C_f is compared with the direct and independent oil-film interferometry measurements from the recent NDF and KTH experiments. Comparisons are also made with the Coles-Fernholz skin-friction relation of the logarithmic type, with $\kappa = 0.384$, which is represented by a solid grey line (see [6]). The experimental data were acquired using different measurement techniques in different physical setups, and even the number of points available for fitting the profile is different between the various experiments. Therefore, it is remarkable to find that these varying conditions do not seem to ultimately affect the skin-friction data, and that all the experiments collapse very well with each other in Fig. 1. Such an agreement implies that the near wall region, which contributes the most to the wall-shear stress, is not sensitive to gradual changes in "outer" flow conditions, and that it achieves self-similarity readily. The lower part of Fig. 1

includes the development of the parameter Ω with Reynolds number Re_θ. The parameter Ω represents the ratio of a turbulent time scale to a mean flow time scale, as estimated by δ/u_τ and x/U_∞, respectively. For fully-developed ZPG TBLs in equilibrium, Ω exhibits a self-similar behavior with Reynolds number independent of initial conditions (see [6, 7]). In Fig. 1, we see a well correlated monotonic behavior of Ω for the data of Smith & Walker [9], KTH [8] and NDF [7]. However, for the other experiments, distinct trends are observed, and can be attributed to the non-equilibrium local effects. In general, we find that the behavior of Ω for these other experiments has trends that converge towards equilibrium at high Reynolds numbers. Consequently, the local non-equilibrium effects can be classified in three categories as: flows with different transition triggers; flows with history of pressure gradient; and flows with inadequate development length.

2 Flows with Different Transition Triggers

In the experiments of Klebanoff & Diehl [5], three different transition mechanisms were used to thicken the boundary layer as seen in the figure. The experiments were repeated for mesh screen and sandpaper with two different freestream velocities, and are represented as distinct lines on the plot. Consistent with their findings, the composite fitting yields a relatively lower boundary layer thickness, and results in a low Ω, which subsequently increases with Re for the cases of natural transition and the 0.04 in rod. Transition due to a mesh screen and sandpaper leads to higher boundary layer thickness and Ω; again subsequently decreasing with Re towards equilibrium. In both cases, different transition triggers have direct effects on the physical development of the flow. Hence, the outer structure of the mean flow itself would be different, which is also reflected in the parameters Π and Δ/δ. Figure 2 depicts the development of Π, Δ/δ and x/δ with Re_θ for various experiments. Consistent with classical theory, Π and Δ/δ asymptote to a constant at high Re. The magnitude of Π is proportional to the mean velocity defect in the outer part when scaled with y/δ. Hence, the changes in Π are directly proportional to changes in δ^*, and for the same U_∞^+, behavior of Π and Δ/δ complement each other as seen in Fig. 2. The data of Klebanoff & Diehl show distinct trends for Π and Δ/δ for different transition mechanisms as also reflected in Ω. For transition with sandpaper and mesh screen, the over-developed boundary layer has high values of Π and Δ/δ. The values of Π and Δ/δ quickly drop and then show constant behavior with Re_θ. However, they are relatively lower because the boundary layer is substantially thickened. The free transition and 0.04in rod, on the other hand, show opposite behavior and develop towards asymptotically consistent values of Π and Δ/δ.

3 Flows with History of Pressure Gradient

Relative to a ZPG TBL, a boundary layer under favorable pressure gradient (FPG) should have a smaller boundary layer thickness, while under adverse pressure gradient (APG) it would have a larger thickness. The experiments of Schubauer & Klebanoff (data from [2]), and the complex pressure gradient (CPG) case from the NDF, were designed to have an initial FPG which relaxes to a ZPG. For both these cases, Ω decreases under FPG, complemented with a decrease in Π and Δ/δ. However, once a ZPG is established in the test section, these parameters start tending towards the consistent equilibrium behavior. The measurements of Karlsson [4] were acquired under ZPG, although the transition trip, and hence the origin of the boundary layer, was positioned in the tunnel contraction, which imposes a FPG on the initial development. This history of FPG is also well represented in Karlsson's data at low Re_θ, where we see lower values of Ω, Π and Δ/δ. In all three experiments the flow appears to ultimately overcome the local pressure gradient effects. In contrast, for flows with a history of adverse pressure gradient, increased values of Ω, Π and Δ/δ will be seen. The non-equilibrium effects can also be detected in the parameter x/δ, which is representative of the non-dimensional growth of a boundary layer and is plotted in the bottom of Fig. 2 against Re_θ. In all the figures, the predicted equilibrium growth of a boundary layer based on

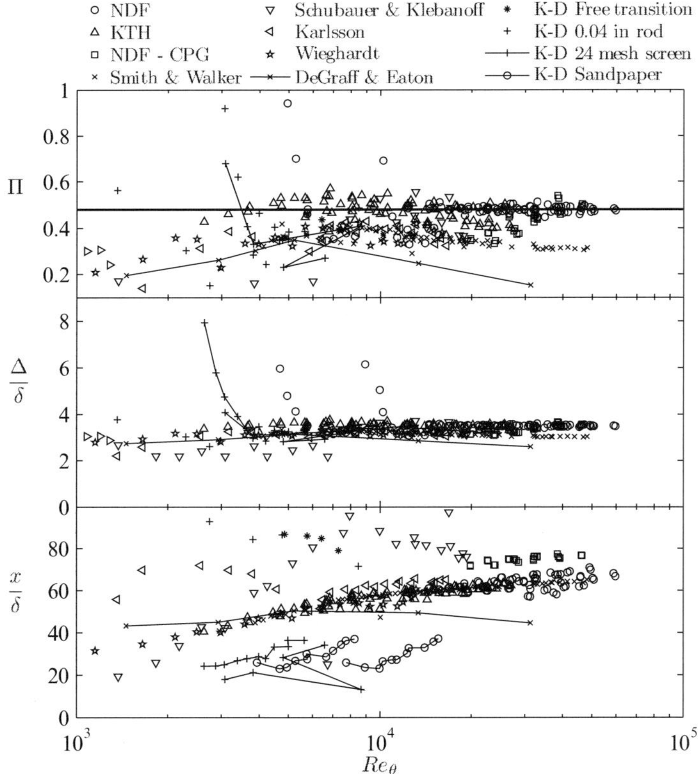

Fig. 2. The variation of Π and Δ/δ with Re_θ for ZPG TBL's.

classical theory is shown by a solid grey line. For flows with history of FPG, x/δ is higher due to decreased δ, while for those with history of APG, x/δ is lower due to increased δ.

4 Flows with Inadequate Development Length

Finally, we examine evidence of flows with inadequate development length or time; a typical example is when a high Re is achieved by large U_∞ at a small streamwise distance x. Such flows will have a fully-developed near wall region, while the outer region is still developing with post-transition growth; i.e., they will lack an adequate amount and appropriate size of outer turbulent structures. While the Reynolds number is accepted as a fundamental parameter in many flows including boundary layers, there is no general agreement if Re_θ or Re_x is more appropriate for TBLs. Here, we will focus on Re_x and utilize a nontraditional exercise by splitting Re_x into two parts in order to reveal the separate effects of development distance and time on the outer flow. Figure 3

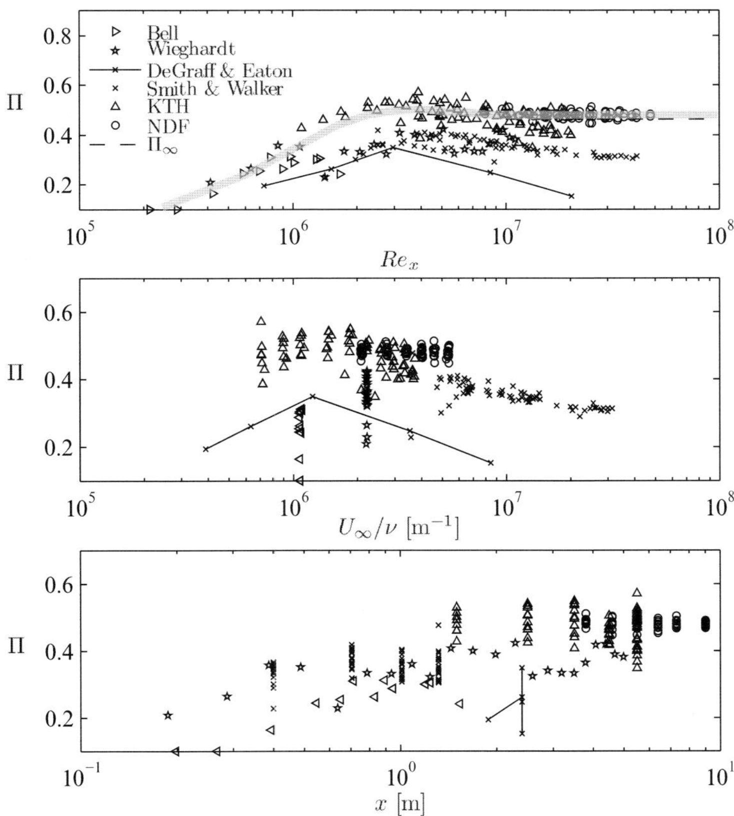

Fig. 3. The variation of Π with Re_x, U_∞/ν and x for ZPG TBL's.

shows the variation of Π with Re_x, U_∞/ν, and x, where U_∞/ν is traditionally called the unit Reynolds number. In the top part of the figure we find, similar to plots using Re_θ, that Π initially increases at low Reynolds numbers but reaches its asymptotic value at high Re, which is clearly seen for NDF data. However, a very interesting trend of an initial increase and subsequent decline in value of Π with Re_x is seen for some experiments. This anomalous trend is explained by plotting Π against U_∞/ν and x separately. We find that Π is lower for experiments with measurements at small x, which implies that the development fetch is not sufficient for the boundary layer to fully mature. This feature is clearly found in data of Wieghardt and Bell (from [2]) and some data of Smith & Walker [9] at lowest x. We also see low Π values for flows with moderate x positions, but they correspond to high Re_x achieved by very high U_∞/ν. In the experiments of DeGraff & Eaton [3], high Re_x was achieved by decreasing the viscosity, and we see that Π decreases as U_∞/ν is increased. A similar effect is clearly demonstrated by the data of Smith & Walker for which, even at the last measurement station, low Π values are reached with the aid of very high U_∞ required to achieve high Re. On the other hand, the lack of adequate data points in the outer part of Österlund's measurements results in underestimating some of his Π values. However, in general the trend of his data is consistent with those of the NDF and the "overall" trend shown with a grey line. Boundary layers developed in long narrow test sections can also exhibit lower Π values, but at least the cases of Smith & Walker [9] and DeGraff & Eaton [3] require a different explanation as revealed here.

5 Summary

Non-equilibrium mean flow effects in many of the classical experiments for ZPG TBLs have been examined, and adequate "development" criteria have been quantified for future design of quality experiments and computational domains. These findings have helped us better understand the physics of boundary layer development and outer/inner flow interactions, as well as clarified a number of unexplained trends and discrepancies associated with several of the measurement sets frequently used in the literature.

References

1. Chauhan KA, Nagib HM, Monkewitz PA (2007) 45th AIAA Aerospace Sciences Meeting and Exhibit Reno NV
2. Coles DE and Hirst EA (1968) Computation of turbulent boundary layers-1968 AFOSR-IFP-Stanford Stanford University
3. DeGraff DB, Eaton JK (2000) J Fluid Mech 422:319–346
4. Karlsson RI, Ph.D. thesis (1980) Chalmers University of Technology Sweden
5. Klebanoff PS and Diehl ZW (1952) NACA Report no 1110

6. Nagib HM, Chauhan KA, Monkewitz PA (2007) Philosophical Transactions of the Royal Society A 365:755–770
7. Nagib HM, Chauhan KA, Monkewitz PA (2005) 4th AIAA Theoretical Fluid Mechanics Meeting Toronto Canada AIAA 2005–4810 paper
8. Österlund JM, Ph.D. thesis (1999) Kungl Tekniska Högskolan Sweden
9. Smith DW, Walker JH (1958) NACA Technical Note 4231

Scaling Law of the Near Wall Flow
Subjected to an Adverse Pressure Gradient

Shinsuke Mochizuki[1], Keisuke Ohta[2], and Takatsugu Kameda[3],
and Hideo Osaka[4]

[1] Department of Mechanical Engineering, Graduate School of Engineering,
Yamaguchi University, 2-16-1 Tokiwadai, Ube 755-8611, Japan
`shinsuke@yamaguchi-u.ac.jp`
[2] Kyocera Co. Ltd., Fushimiku, Kyoto 612-8501, Japan
[3] Yamaguchi University `kameda@yamaguchi-u.ac.jp`
[4] Yamaguchi University `ohsaka@yamaguchi-u.ac.jp`

Abstract. Detailed experiments have been conducted to investigate the effect of
adverse pressure gradients on the log-law in turbulent boundary layers. The wall
shear stress was measured by a direct measurement device and the scaling law of the
mean velocity was discussed based on high-accuracy experimental data. Considering
the significant contribution of the inertia term in the equations of motion, a new
velocity scale is defined and a similarity law was obtained for the mean velocity
profile subjected to an adverse pressure gradient.

Keywords: turbulent boundary layer, law of the wall, log-law, adverse pressure gradient, skin friction measurement

1 Introduction

According to Ludwieg-Tillmann's experiment [1], the log-law in the wall layer
is believed to be independent of adverse pressure gradients. However, recent
experiments indicate that the numerical constants involved in the log-law
depend on pressure gradients [2]. Carefully obtained experimental data using well-designed facilities and techniques are desired for the investigation
of similarity under adverse pressure gradients. Even without pressure gradients, experimental studies should satisfy five requirements: 1) the momentum
thickness Reynolds number should be greater than 10,000, 2) the mean velocity and wall shear stress should be measured independently, 3) the uncertainty of the mean velocity normalized with the friction velocity should be
less than 1-2%, 4) the mean velocity gradients should be able to be calculated
with high accuracy, and 5) if possible, higher-order moments should be involved [3]. In addition, the development of the boundary layer should remain
two-dimensional and should avoid the upstream history in the outer layer.
The present experiment was conducted in the boundary layer at $R_\theta = 10{,}200$
under the power law variation of the free stream velocity. The wall shear stress

Y. Kaneda (ed.), *IUTAM Symposium on Computational Physics and New Perspectives in
Turbulence*, 191–196.

was measured directly by the drag balance measurement method, which was newly developed for low dynamic pressure flows. The similarity of the mean velocity profile was examined in the wall layer with the experimental data.

2 Experimental Set-Up

The experiment was conducted in a low-turbulence wind tunnel having a width of 910 mm and a length of 6,000 mm (see Fig. 1). The standard zero pressure gradient boundary layer develops beyond 1,500 mm from the leading edge of the test plate. The free stream is adjusted to vary with equation (1) over the adverse pressure gradient section.

$$U_e \propto (x - x_0)^{-0.188} \tag{1}$$

Velocity and other parameters were measured at $x = 4,500$ mm, where the flow approached the equilibrium state. The pressure gradient parameter of $\beta \equiv (\delta^*/\tau_w)dP_e/dx$ is 1.3, and that of $p^+ \equiv (\nu/u_\tau^3)dP_e/dx$ is 0.0026. Departure from the two-dimensional condition is evaluated with the momentum integral equation. The imbalance in the equation at $x = 4,500$ mm is approximately 3.5% of C_f determined by drag balance measurement. The local wall shear stress was measured by the drag balance device shown in Fig. 2. The diameter and lip size of the floating element were chosen such that the unanticipated lip force would be less than 1.5% of the total measured force. Velocity fluctuation measurements were performed by hot-wire probe and constant temperature anemometers. By taking account of uncertainties in drag balance and hot-wire measurements, the resultant uncertainty was estimated to be 1.8% for U^+.

3 Results and Discussion

Figure 3 shows the local skin-friction coefficient as a function of R_θ. The adverse pressure gradient reduces the skin-friction coefficient, which is

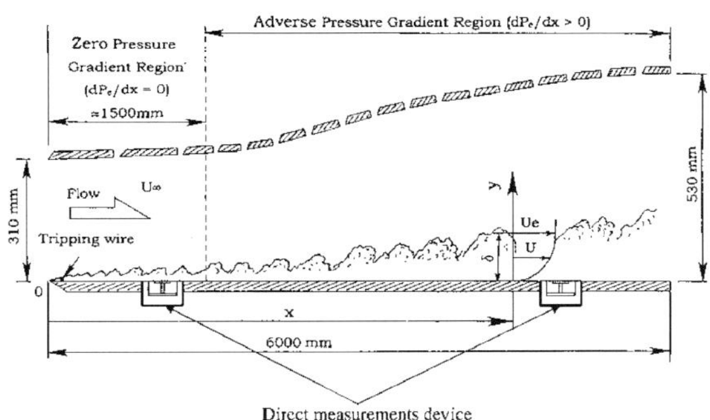

Fig. 1. Schematic of flow field, coordinate system and nomenclature.

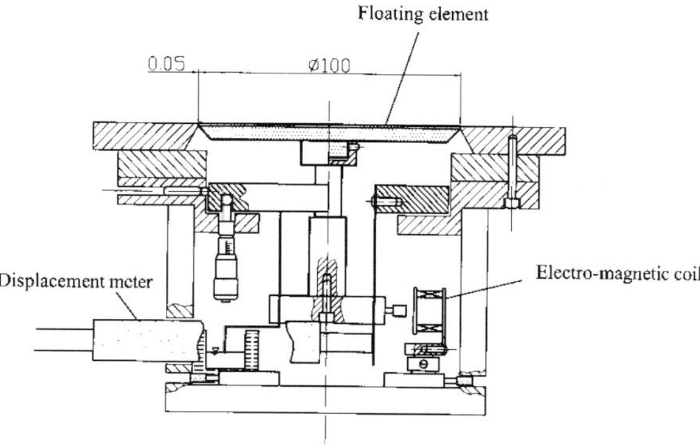

Fig. 2. Direct measurement device for wall shear stress measurement.

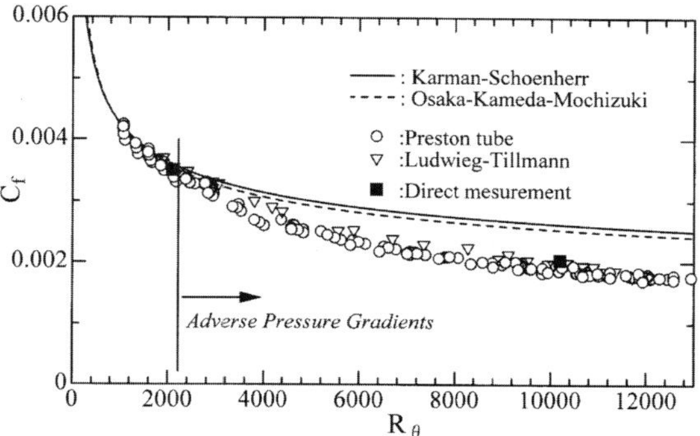

Fig. 3. Local skin friction coefficient.

$C_f = 0.00204$ at $R_\theta = 10,200$. The skin-friction value obtained by direct measurement is larger than those of the other methods by approximately 5%. Terms in the equations of motion were evaluated based on experimental data and were examined with respect to their contributions to the shear stress distribution in Fig. 4. Experimental data and calculated profiles from equation (2) are compared in the figure.

$$\frac{-\overline{uv}}{u_\tau^2} = 1 + p^+ y^+ - \frac{1}{u_\tau^2} \int_0^y \left(U \frac{\partial U}{\partial x} + V \frac{\partial U}{\partial y} \right) dy' \tag{2}$$

The last term in equation (2) is denoted by I_{IT}. The wall layer is defined over $y^+ = 200$ - 350, where the skewness values of fluctuating velocities are constant [4]. In contrast to the measurement uncertainty for the Reynolds shear stress, the contribution of the inertia term is considerable in the equation of motion. Townsend [5] proposed a velocity scale containing the pressure gradient effect on the shear stress distribution. The finite contribution of the inertia term suggests that the velocity scale should contain the inertia term

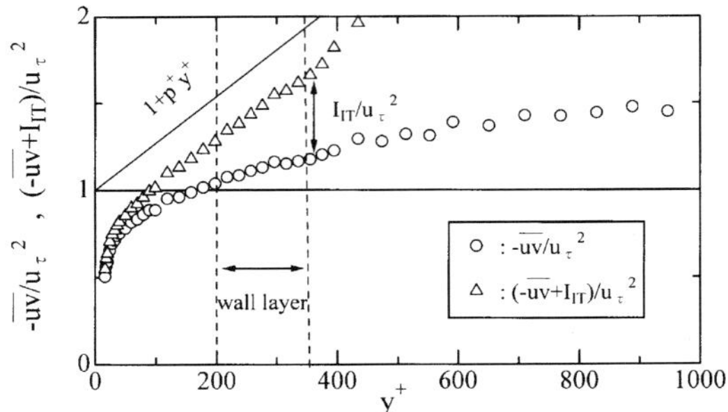

Fig. 4. Reynolds shear stress distributions.

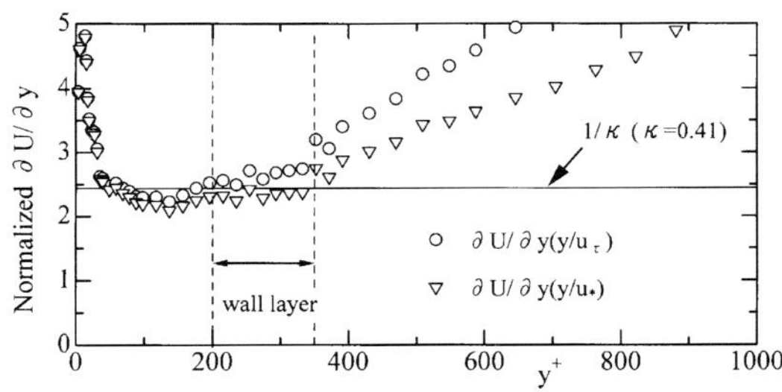

Fig. 5. Normalized mean velocity gradients.

effect as well as the pressure gradient force. Here, we propose the velocity scale as follows:

$$u_* \equiv u_\tau \sqrt{1 + \alpha p^+ y^+}. \tag{3}$$

The coefficient α determined empirically from experimental data is 0.40 for the present flow and 0.58 for Bradshaw's experiment [6] with $\beta = 0.82$. The mean velocity gradients are normalized with the friction velocity and the new velocity scale u_* in Fig. 5. The normalized mean velocity gradient with u_* remains approximately constant over the wall layer. Considering the contribution of the inertia term in the momentum balance, the mean velocity gradient in the wall layer should be scaled as follows:

$$\frac{\partial U}{\partial y} = \frac{u_*}{\kappa y}. \tag{4}$$

Integration of equation (4) yields the modified logarithmic velocity profile:

$$U^+ = \frac{1}{\kappa} \ln y^+ + C - \frac{2}{\kappa} \ln \left(\frac{1 + \sqrt{1 + \alpha p^+ y^+}}{2} \right) + \frac{2}{\kappa} \left(\sqrt{1 + \alpha p^+ y^+} - 1 \right). \tag{5}$$

Setting $\alpha = 0.4$, $\kappa = 0.43$ and $C = 4.9$ based on the available experimental data, the results obtained by the modified log-law are confirmed to agree

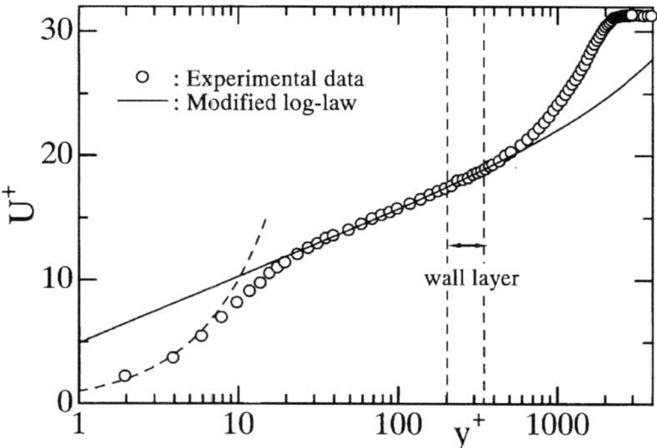

Fig. 6. Comparison between the modified lag-law calculated by equation (5) and experimental data.

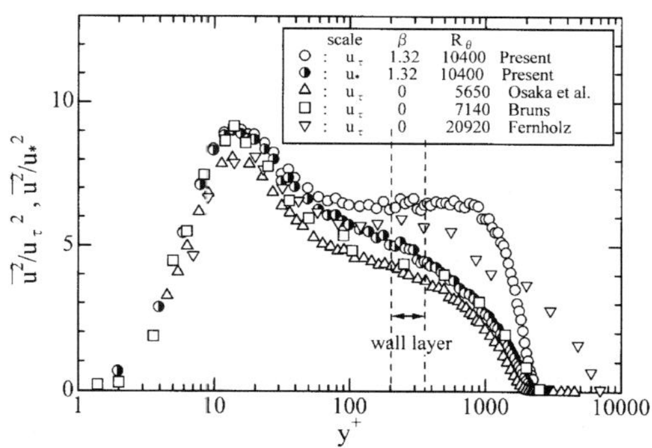

Fig. 7. Streamwise normal Reynolds stress normalized with u_* and u_τ.

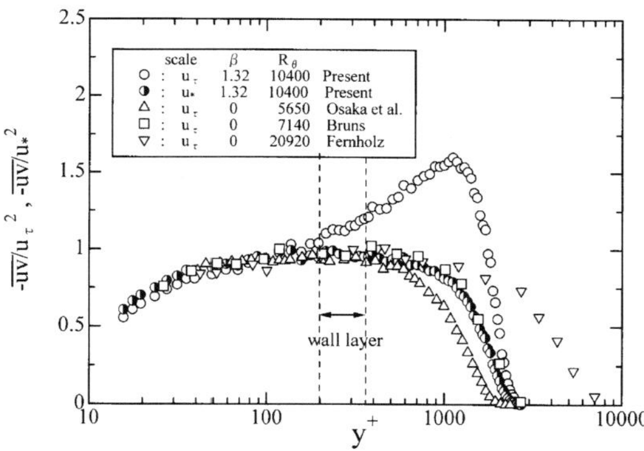

Fig. 8. Reynolds shear stress normalized with u_* and u_τ.

with the experimental data, as shown in Fig. 6. The modified profile agrees well with the present experimental data. The streamwise Reynolds normal stress and Reynolds shear stress are normalized with the velocity scale u_* in Figs. 7 and 8, respectively. The velocity scale u_* shows better agreement in the profiles of the normalized Reynolds normal stress components with and without pressure gradient. The Reynolds shear stress component normalized with the velocity scale u_* collapse on the single profile in the wall layer.

4 Summary

Considering the fact that the inertia term has a finite contribution to the equation of motion, the velocity scale $u_* \equiv u_\tau \sqrt{1 + \alpha p^+ y^+}$ is proposed for the similarity in the wall layer. The modified log-law well represents the experimental data in the wall layer. The velocity scale provides good similarity with respect to the Reynolds normal stress and shear stress profiles in the wall layer.

References

1. Ludwieg H, Tillmann W (1949) NACA TM1285
2. Skåre PE, Krogstad P-Å (1994) J Fluid Mech 272:319–348
3. Buschmann MH, Gad-el-Hak M (2003) AIAA J 41-4:565–572
4. Tsuji Y, Lindgren B, Johansson AV (2005) Fluid Dynamic Research 37:293–316
5. Townsend AA (1976) The structure of turbulent shear flow, Cambridge University Press, 2nd ed., 273
6. Bradshaw P (1967) J Fluid Mech 29:625–645

Non-Equilibrium and Equilibrium Boundary Layers without Pressure Gradient

Takatsugu Kameda[1], Shinsuke Mochizuki[2] and Hideo Osaka[3]

[1] Department of Mechanical Engineering, Graduate School of Science
and Engineering, Yamaguchi University, 2-16-1 Tokiwadai, Ube 755-8611, Japan
`kameda@yamaguchi-u.ac.jp`
[2] Yamaguchi University, `shinsuke@yamaguchi-u.ac.jp`
[3] Yamaguchi University, `ohsaka@yamaguchi-u.ac.jp`

Abstract. The effect of the friction parameter ω, defined as the ratio of the friction velocity to the free stream velocity, has been investigated on mean velocity fields for non-equilibrium and equilibrium boundary layers developing under zero pressure gradient. The wall shear stress was measured by a drag balance using a floating element device with a zero displacement mechanism. For the equilibrium boundary layer, the local skin friction coefficient is independent of two parameters, both the streamwise distance and the Reynolds number, based on the momentum thickness, and the boundary layer thickness is proportional to the streamwise distance. On the other hand, for the non-equilibrium boundary layer, the local skin friction coefficient depends on the above two parameters. The wake parameters for both boundary layers approach constant values, which depend on the surface condition, for high Reynolds numbers. From analysis using both the momentum integral equation and Coles's wake law, the wake parameter for the equilibrium boundary layer is uniquely expressed as a function of the friction parameter. However, for the non-equilibrium boundary layer, the wake parameter depends on the friction parameter as well as the growth rate of the boundary layer thickness.

Keywords: equilibrium boundary layer, rough surface, drag balance, friction parameter, wake parameter

1 Introduction

In the turbulent boundary layer, the statistical distributions normalized with the local velocity and length scales become approximately self-similar (e.g., Tennekes H, Lumley JL [1]) at high Reynolds numbers far distant from the leading edge because of small streamwise variations in the local skin friction coefficient c_f. Such a flow is referred to as a non-equilibrium boundary layer. On the other hand, in an equilibrium boundary layer (e.g., Rotta JC [2]), the statistical distributions are completely similar by virtue of a constant c_f value. In both cases, the partial differential equation of motion can be reduced to an ordinary differential equation and can yield a self-similar solution if

197

Y. Kaneda (ed.), *IUTAM Symposium on Computational Physics and New Perspectives in Turbulence*, 197–202.
© 2008 *Springer*.

the Reynolds stresses are modeled after the gradient-diffusion model. For the turbulent boundary layer described with the ordinary differential equation, we can understand the effects of external forces and boundary conditions on the flow.

In the present study, we investigate the effect of the friction parameter on the mean velocity profiles for the turbulent boundary layers. First, the characteristics of the mean velocity field will be shown for the non-equilibrium and the equilibrium boundary layers. Second, the flow characteristics of the mean velocity profile will be examined analytically and experimentally with respect to the friction parameter.

2 Flow Fields and Experimental Conditions

The equilibrium boundary layer in the present study corresponds to the flow of No.2 condition that is one of the six equilibrium boundary layers proposed by Rotta JC [2]. The wall shear stress is only the external force in the equilibrium boundary layer. Figure 1 gives the flow field for the equilibrium boundary layer. The rough surface consists of two-dimensional square ribs, and the roughness height k is proportional to streamwise distance x with $dk/dx = 0.00125$. The cavity width b between roughness elements has a size of $b = 3k$ and the roughness width w is $w = k$, so that the roughness pitch ratio $p(= (b+w)/k)$ is 4. The coordinate system is chosen as the streamwise distance x from the upstream edge of the rough surface and normal distance y_T from the crest of the roughness element. In addition, the normal distance y is $y = y_T + d_0$, where d_0 is the error in origin, which is the distance between the crest and the vertical origin of the log law profile. Finally, d_0 was estimated by Furuya-Fujita's method [3].

The streamwise velocity component was measured with a single hot wire probe and a constant temperature anemometer. Measurement was performed under zero pressure gradient. The Reynolds number based on the momentum

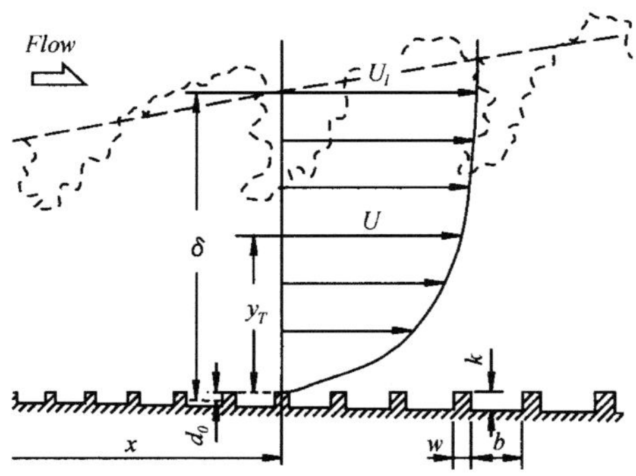

Fig. 1. Flow field, nomenclature and coordinate system for the equilibrium boundary layer

thickness R_θ ($=U_1\theta/\nu$, where U_1 is the free stream velocity, and ν is the kinematic viscosity) is within R_θ=2990 to 9080 in the velocity measurement. The wall shear stress was directly measured by a drag balance using a floating element device of a zero displacement mechanism.

The non-equilibrium boundary layers were selected turbulent boundary layers developing over smooth and k-type rough surfaces [4, 5]. The k-type rough surface, which has a constant roughness height and a roughness pitch ratio of 4, is known as the error in origin is proportional to the roughness height.

3 Characteristics of the Boundary Layers

Figure 2 shows the variation of c_f with respect to R_θ. In Fig. 2, all of the data except for the empty square symbols were measured using the drag balance. The empty square symbol was estimated with the velocity fit method proposed by Krogstad et al. [6]. The value of c_f for the equilibrium boundary layer is independent of both R_θ and x, and is constant at 0.00826. For the non-equilibrium boundary layers over the smooth and the k-type rough surfaces, the value of c_f depends on two parameters, and decreases with the increase of the parameters.

Figure 3 shows the boundary layer development. The horizontal axis is the streamwise distance X ($=x$-x_0) from the virtual origin x_0 of the boundary layer. The lines shown in Fig. 3 were estimated as a power function using the least squares method. The boundary layer thickness evolves as X^1 for the equilibrium boundary layer but as $X^{0.79}$ and $X^{0.92}$ for the non-equilibrium boundary layers. Figure 4 shows the mean velocity profiles normalized with the outer variables (U_1 and δ). The profiles of the flow over the two rough surfaces for the equilibrium and the non-equilibrium boundary layers take a large deficit compared with that of the flow over the smooth surface of the non-equilibrium boundary layer. Next, in order to recognize the difference between the equilibrium and the non-equilibrium boundary layers, the wake parameter Π is shown in Fig. 5. Here, Π is defined as $\Pi = \kappa/2 \times \Delta U_1/u_\tau$, where $\Delta U_1/u_\tau$

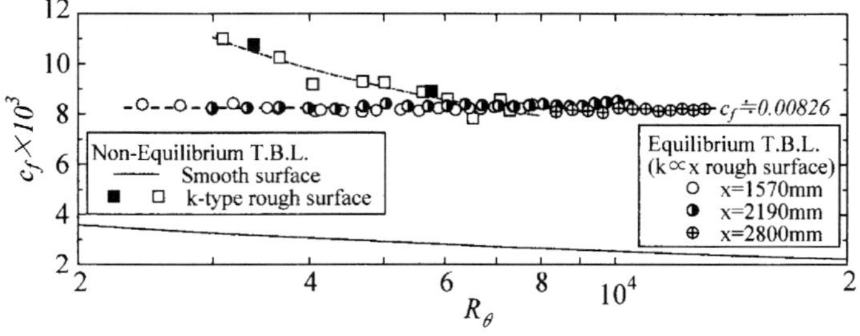

Fig. 2. Local skin friction coefficient

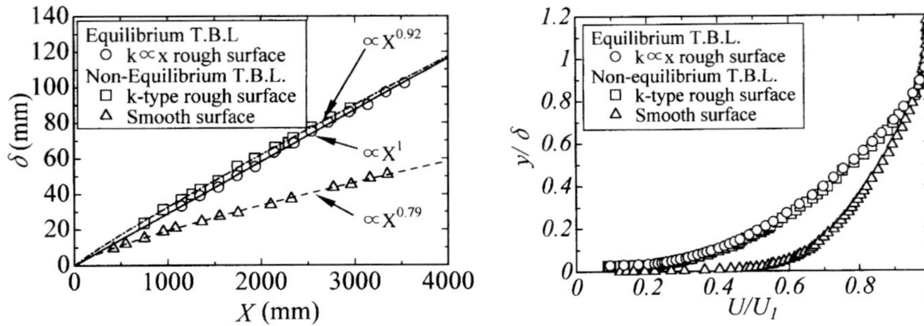

Fig. 3. Boundary layer thickness

Fig. 4. Mean velocity profiles

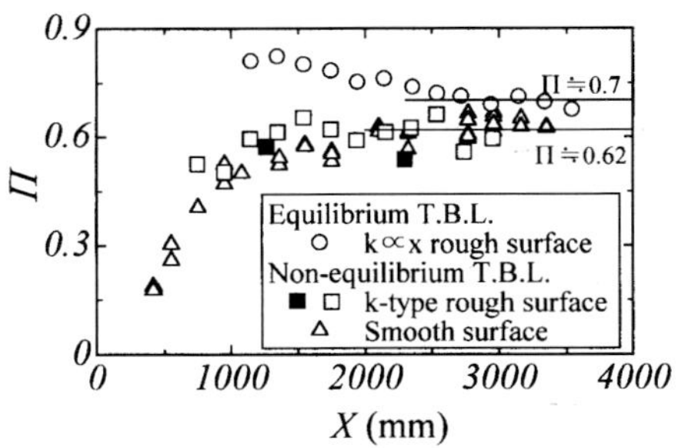

Fig. 5. Variation of wake parameter to streamwise distance

is a maximum departure from the log law in the outer layer, and κ is Kármán constant ($\kappa = 0.41$). It appears that Π is approximately constant for both boundary layers in the far downstream region (i.e., at high Reynolds number). However, the constant value depends on the flow conditions and is 0.70 for the equilibrium boundary layer and is approximately 0.62 for the non-equilibrium boundary layers. In the next section, we will discuss the effect of the flow characteristics on the wake parameter both analytically and experimentally.

4 Relation between the Flow Characteristics and the Wake Parameter

In the present analysis, we analyzed momentum integral equation (1) using Coles's wake law (2), assuming the constant or approximately constant c_f value.

$$\omega^2 = \frac{c_f}{2} = \frac{d\theta}{dx}. \tag{1}$$

$$\frac{U}{u_\tau} = \frac{1}{\kappa} \ln\left(\frac{y u_\tau}{\nu}\right) + A - \frac{\Delta U}{u_\tau} + \frac{\Pi}{\kappa} W\left(\frac{y}{\delta}\right). \tag{2}$$

Here, A is the additive constant ($A = 5.0$), $\Delta U/u_\tau$ is roughness function and $W(y/\delta)$ is the wake function. $W(y/\delta)$ was used in the empirical formula (3) proposed by Lewkowicz [7].

$$W(\eta) = 2\eta^2(3 - 2\eta) - \frac{1}{\Pi}\eta^2(1 - \eta)(1 - 2\eta). \tag{3}$$

We can derive the momentum thickness θ as follows based on Eqs. (2) and (3), correcting the contribution a in the viscous or roughness sublayer.

$$\theta = \frac{\omega\delta}{\kappa}\left(\Pi + \frac{59}{60} + a\right) - \frac{\omega^2\delta}{\kappa^2}\left(\frac{8437}{4200} + \frac{667}{210}\Pi + \frac{52}{35}\Pi^2\right). \tag{4}$$

Using the experimental data of the three flow fields, the correct term a can be expressed as a first approximation as follows.

$$a = -0.165\Pi + 0.126. \tag{5}$$

Next, we substitute Eqs. (4) and (5) into Eq. (1) and assume that the variations of ω and Π with respect to x are neglected, so that we can obtain the following quadratic equation with respect to Π.

$$\frac{52}{35}\Pi^2 + \left(\frac{667}{210} - \frac{\kappa}{\omega} + 0.165\frac{\kappa}{\omega}\right)\Pi + \left(\frac{8437}{4200} - \frac{59}{60}\frac{\kappa}{\omega} - 0.126\frac{\kappa}{\omega} + \frac{\kappa^2}{d\delta/dx}\right) = 0. \tag{6}$$

For an actual turbulent boundary layer observed in physical space, the solution must be real. The condition that the value is real will be given from the quadratic discriminant. In the solution of the multiple root of Eq. (6), semi-empirical relation (8) can be obtained for Π as a function of ω.

$$\Pi = 0.281\frac{\kappa}{\omega} - 1.069. \tag{7}$$

This relation indicates that the equation of motion must have a unique solution at a given ω. Figure 6 shows the variation of Π with respect to ω and

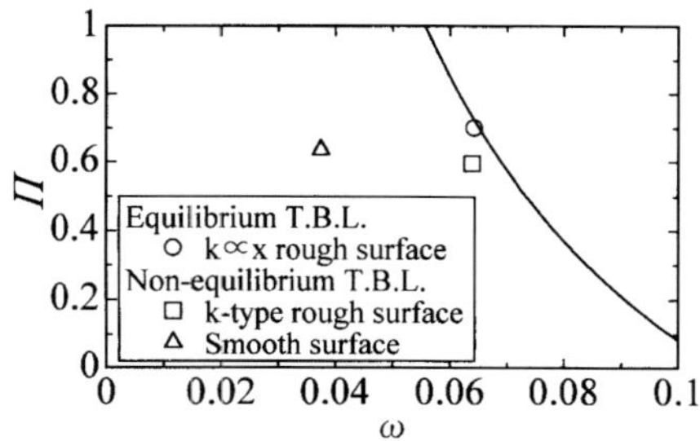

Fig. 6. Variation of the wake parameter to the friction parameter

compares the experimental data. The solid line represents Eq. (7). The data of the equilibrium boundary layer collapses to the solid line. However, for the non-equilibrium boundary layer, Π cannot be determined by ω because it is a function of ω as well as $d\delta/dx$.

5 Summary

For the equilibrium boundary layer, the local skin friction coefficient c_f is independent of both streamwise distance x and Reynolds number R_θ, and the boundary layer thickness δ evolves as X^1. However, for the non-equilibrium boundary layer, the local skin friction coefficient c_f depends on two parameters. The wake parameter Π for both boundary layers approaches a constant for a high Reynolds number. For the analytical method, the wake parameter Π for the equilibrium boundary layer is expressed as a function of only the friction parameter ω. However, for the non-equilibrium boundary layer, the wake parameter Π depends on the friction parameter ω as well as growth rate of the boundary layer thickness $d\delta/dx$.

References

1. Tennekes H, Lumley JL (1972) A First Course in Turbulence The MIT Press
2. Rotta JC (1962) Turbulent Boundary Layer in Incompressible Flow Second Edition Pergamon Press
3. Furuya Y, Fujita H (1966) Trans JSME 32-237:725–733
4. Osaka H, Kameda T, Mochizuki S (1998) JSME Int JB 41-1:123–129
5. Kameda T, Osaka H, Mochizuki S (1998) 13th Australasian Fluid Mechanics Conference:357–360
6. Krogstad P-A, Antonia RA, Browne LWB (1992) J Fluid Mech 245:599–617
7. Lewkowicz AK (1982) Z Flugwiss Weltramforsch 6:261–266

Experimental Study of Laminar Turbulent Boundary Layer Transition Influenced by Anisotropic Free Stream Turbulence

Toshiaki Kenchi[1] and Masaharu Matsubara[2]

[1] Department of Mechanical Systems Engineering, Graduate School of Engineering, Shinshu University, Nagano-shi, Nagano 380-8553, Japan
t04h410@amail.shinshu-u.ac.jp
[2] Department of Mechanical Systems Engineering, Shinshu University,
mmatsu@shinshu-u.ac.jp

Abstract. The purpose of this study is to clarify the boundary layer transition process influenced by anisotropic and low-level free stream turbulence generated by a turbulence grid mounted upstream from a contraction. Flow visualization experiments in the boundary layer revealed the appearance of wave packets with streamwise wave numbers that transform into 'Λ' shaped structures and immediately break down into turbulent spots. The wave packets were mapped out around the upper branch of the neutral curve in a linear stability diagram, suggesting the modal disturbance is dominant and triggers transition to turbulence in the boundary layer subject to relatively low-level free stream turbulence on the boundary layer scale.

Keywords: boundary layer, transition, instability, streaky structure, wave packet

1 Introduction

In boundary layer transition subjected to moderate intensity free stream turbulence, streaky structures, which greatly elongate in the streamwise direction, develop and then begin vibrating in the spanwise direction. This waviness breaks down to turbulence within a few wavelengths. This local turbulence forms an arrow-head shaped turbulent spot and propagates until the boundary layer is filled with turbulence [1] [2]. This streak generation caused by the free stream turbulence has been confirmed by DNS results [3], and the fluctuation profiles and the energy growth of the disturbances of the streaky structures have been accurately predicted by the non-modal theory [4]. However a modal disturbance, i.e. a Tollmien-Schlichting wave, still has the possibility of growing and triggering the transition to turbulence in the case that the intensity of the free stream turbulence is relatively weak. In addition, the effects of the scale length of free stream turbulence have not been thoroughly explored.

203

Y. Kaneda (ed.), *IUTAM Symposium on Computational Physics and New Perspectives in Turbulence*, 203–208.
© 2008 *Springer*.

In this study, the transition process in a boundary layer subject to characteristic free stream turbulence generated by a turbulence grid upstream of the contraction is investigated.

2 Experimental Setup

The experiments were performed in a closed wind tunnel which has a 4000 mm long test chamber with a 400 × 600 mm cross-section preceded by a three-dimensional contraction with a 9:1 ratio. A test plate was mounted vertically in the test chamber, with its leading edge located at 1500 mm from the exit of the contraction, as shown in Fig. 1. The leading edge is a 10:1 ellipse with a 100 mm major axis. The counter wall to the test plate was carefully adjusted to have a pressure gradient of zero along the plate. A trailing edge flap allows for adjustments in the stagnation line on the leading edge for prevention of separation of the boundary layer from the surface.

The coordinate system is set so that x is the streamwise direction, y for the wall-normal direction and z for the direction normal to both the x and y directions. The x origin is at the leading edge, and y is measured from the surface of the test plate.

Two different grids, grid A and grid B, were used. Grid A has a 30 mm mesh size and consists of 6 mm square bars and was set downstream of the contraction as a typical turbulence grid. Grid B was mounted 500 mm upstream of the contraction with a distance from the grid to the leading edge of 3800 mm. Grid B has a 120 mm mesh size and consists of 13 mm diameter round pipes each with 2 mm diameter holes bored oriented upstream. These holes are to increase the free stream turbulence intensity by blowing jets from these holes with 1.0 kPa pressurized air in the pipe.

An X-constant-temperature hot-wire anemometer consisting of 2.5μm platinum wire with a length of 1.5 mm was used. A three axes robot arm, which was designed to minimize flow resistance, traverses the hot wire probe in the y-z plane, and for the streamwise movement of the arm, a 2100 mm length rail was mounted over the test section roof.

Flow visualization was created with an alcohol mist that was allowed to seep in slowly through a spanwise slot. The slot is located at $x = 210$ mm and is 2 mm wide and has a 500 mm spanwise length. In the absence of the turbulence grid, the smoke forms a homogeneous layer on the surface of the

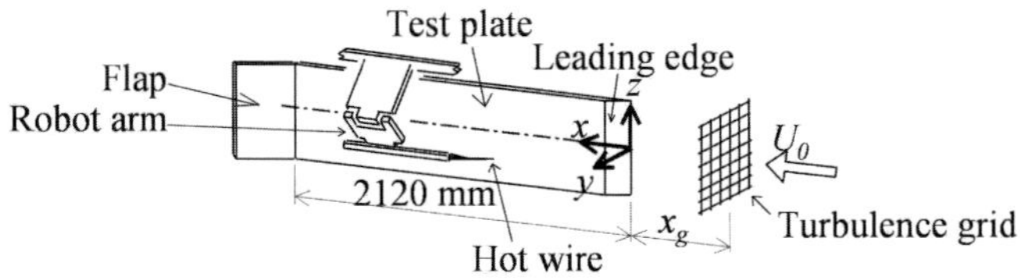

Fig. 1. Experimental set-up.

test plate. Both video and photographic recordings of the flow visualization were made by cameras mounted on the outside of the transparent counter wall. The smoke was illuminated by a flashlight and two 500 W floodlights from under the bottom wall of the test section.

This wind tunnel has a 0.3% turbulence intensity at a free stream velocity of 10 m/s. Energy from the free stream turbulence without a turbulence grid concentrates in the low frequency range of less than 10 Hz. Turbulence intensity is 0.08% with components larger than 1.0 m filtered, which is on the order of the length of the test chamber section.

3 Decay of the Free Stream Turbulence and Its Spectra

The streamwise variation of the lateral(u) and vertical(v) velocity fluctuations in the free stream direction and their spectra are shown in Fig. 2. The free stream velocity at $x = 0$, U_0, was 6 m/s for grid A and 10 m/s for grid B. x_g is the position of the grid and X is the distance from the grid normalized by the mesh size M. Figure 2 (a) is the case of grid A with the fitting curves of the typical power-law decay of grid turbulence,

$$\frac{u_{rms,\infty}}{U_0} , \frac{v_{rms,\infty}}{U_0} = C(X - X_0)^b, \qquad (1)$$

where X_0 is the virtual origin of the grid. The curves are in good agreement with the measurements of both the lateral and vertical components with an exponent $b = -0.54$. The ratio of the lateral component to the vertical

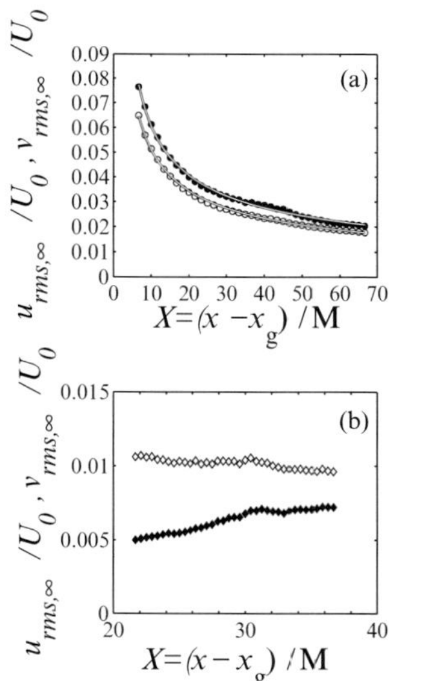
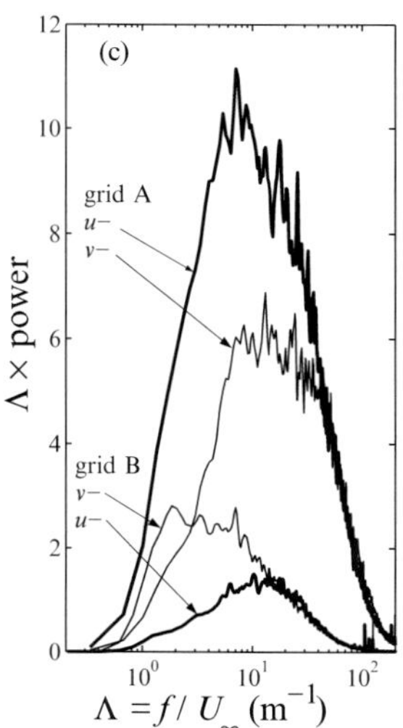

Fig. 2. Streamwise variations of the lateral and vertical velocity fluctuations and their spectra. (a) in the grid A case, (b) in the grid B case, solid marks for $u_{rms,\infty}/U_0$ and open marks for $v_{rms,\infty}/U_0$. (c) Spectra of the free stream turbulence at $x = 0$.

component is about 1.2 independent of the streamwise position except immediately downstream of the grid. The values of the exponent b and the ratio of the component u_{rms}/v_{rms} values are adequate compared with other investigations of grid turbulence [5]. On the other hand, in the case of grid B, the lateral component increases downstream while the vertical component decays, as shown in Fig. 2 (b). This phenomenon was previously observed by Uberoi [6], who generated turbulence with a grid mounted upstream of a contraction and observed the redistribution of the directional component energies.

The spectral density distribution of each directional component for both grids A and B are compared in Fig. 2 (c). Λ indicates the inverse of the length scale estimated by the Taylor frozen flow hypothesis with the free stream velocity U_0. The vertical axis is multiplied by the horizontal axis value so that the area represents the energy of the free stream turbulence. In the case of grid A, the lateral component is higher than the vertical component $\Lambda \leq 40\,\mathrm{m}^{-1}$, while in the case of grid B, the vertical component is higher than the lateral component $\Lambda \leq 20\,\mathrm{m}^{-1}$. However, in both cases both directional components become almost equal for large Λ values. These scales corresponded to the order of the boundary layer thickness. It is noticeable that, if compared on the scale of the boundary layer thickness, the free stream turbulence generated by grid B is much weaker than in the case of grid A.

4 Smoke Visualization

Figure 3 shows the transitional boundary layer in the grid A case. The left and right edges of the photo correspond to $x = 240\,\mathrm{mm}$ and $x = 575\,\mathrm{mm}$, respectively. There exist stripes of the spanwise scale of the boundary layer thickness, which are extremely elongated in the streamwise direction. These strips were recognized as streaky structures due to the non-modal growth [1]. As seen in Fig. 3, some parts of the streaks start a spanwise waviness with relatively short wavelengths. Using video observation, this waviness immediately breaks down to a turbulent spot that grows downstream until the boundary layer is filled with turbulence.

Figure 4 (a) and (b) show the case of grid B with the $10\,\mathrm{m/s}$ free stream velocity. The areas of the photographs are the same as in Fig. 3. Unlike the case of grid A, the streaky structures do not exist. Instead, Λ shaped structures

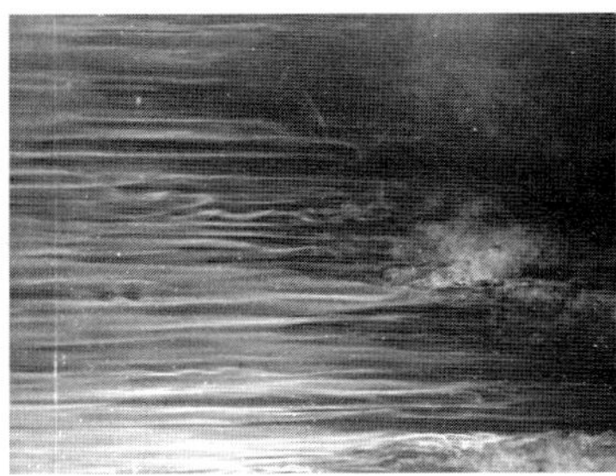

Fig. 3. Flow visualization in the grid A case.

Fig. 4. Flow visualization in the grid B case. (a) and (b) are different shots in time.

aligned in the streamwise direction are observed as seen under the center of Fig. 4 (a). In another shot of the area shown in Fig. 4 (b), a turbulent spot is seen on the right side. In terms of the generation and growth of the turbulent spots, the final stage of the transition is the same as in the grid A case. It comes into question what causes the generation of the turbulent spots in the absence of the streaky structures and the resulting waviness break down. One possibility is that the free stream turbulence directly generates the turbulent spots even though the intensity of the free stream turbulence is relatively low compared with the grid A case. Another hypothesis is that the turbulent spots originate from the Λ shaped structures.

In order to clarify this, detailed video observations were made. Figure 5 shows the sequence of development of a Λ shaped structure. The time duration between the pictures of the sequence is 8.3 msec. The streamwise position of the left and right sides of the photograph are the same as in Fig. 4. In Fig. 5 (a), weak waves with their wave fronts directed in the spanwise direction appear. In Fig. 5 (b), these waves deform in an arch shape, as is evident by the higher contrast in the photo. In Fig. 5 (c), they form Λ shapes and immediately break down, as seen Fig. 5 (d). Though it is not shown in this sequence, this point-like turbulence propagates via an arrow-head shape that can be identified as a turbulent spot. The transition process in the grid B case clearly differs in terms of 1) no streaky structure before the transition, and 2) the appearance and break down of the wave packets with the formation of the Λ shaped structure.

The two dimensional waves seen in Fig. 5 (a) could be expected as localized Tollmien-Schlichting waves. To confirm this, a comparison between the neutral stability curve and the appearance of the wave packets was made in Fig. 6. The horizontal axis is the Reynolds number based on the displacement boundary layer thickness $R = 1.72\sqrt{Re_x}$, and the vertical axis is the normalized frequency $F = 2\pi f \nu 10^6 / U_0^2$. This frequency is estimated by the wavelength and the phase velocity of the waves from the video sequences. The plots are close to the upper branch of the neutral curve indicating that the wave packets were in an unstable state before they appeared. This result

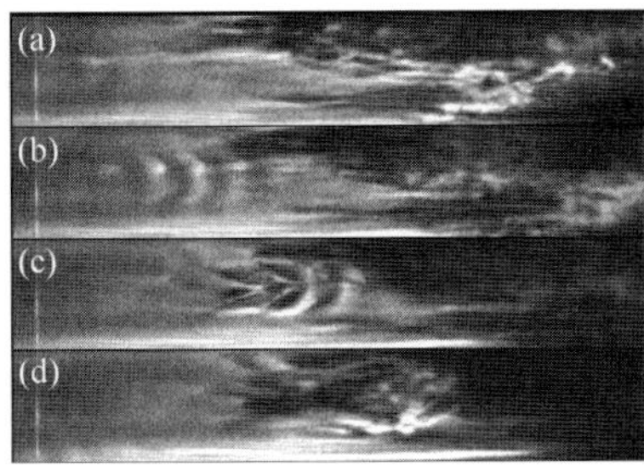
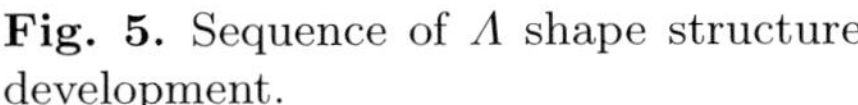

Fig. 5. Sequence of Λ shape structure development.

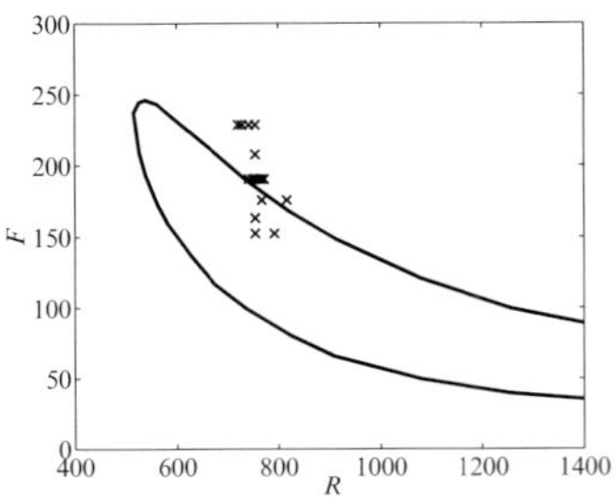

Fig. 6. Linear stability diagram. The neutral curve is by Klingmann [7].

strongly suggests that these wave packets are localized modal disturbances. The wave packet is influenced by its edge in the spanwise direction so that three dimensionalization quickly proceeds in the short distance of a few wavelengths as seen in the video sequence.

The reason for the remarkable difference in the transition scenario between the cases of grids A and B could be explained by the weakness of the free stream turbulence as compared with the scale of the boundary layer thickness. To clarify the cause of the different transition scenario, further detailed investigations with precise control of the scales and components of the free stream turbulence are required.

5 Conclusion

In this experiment, the boundary layer transition influenced by the characteristics of free stream turbulence generated by a turbulence grid mounted upstream from a contraction was investigated. Free stream turbulence has strong anisotropy on a large scale, while free stream turbulence on the boundary layer scale is isotropic and much weaker such that the streaky structures do not appear in the boundary layer. Flow visualization reveals the appearance of wave packets with a streamwise wave number that immediately transform to Λ shaped structures and break down to a turbulent spot. The Reynolds number of the wave packet appearance and its wave length correspond to an unstable modal disturbance predicted by linear stability theory, suggesting localized Tollmien-Schlichting waves are dominant and trigger the transition to turbulence in the boundary layer subjected to weak free stream turbulence.

References

1. Matsubara M, Alfredsson PH (2001) J Fluid Mech 430:149–168
2. Fransson JHM, Matsubara M, Alfredsson PH (2005) J Fluid Mech 527:1–25
3. Jacobs RG, Durbin PA (2001) J Fluid Mech 428:185–212
4. Luchini P (2000) J Fluid Mech 404:289–309
5. Grant HL, Nisbet ICT (1957) J Fluid Mech 2:263–272
6. Uberoi MS (1957) J Appl Phys 28:1165–1170
7. Klingmann BGB, Boiko AV, Westin KJA, Kozlov VV, Alfredsson PH (1993) Eur J Mech 12:493–514

Statistical Features of Scalar Flux
in a High-Schmidt-Number Turbulent Jet

Yasuhiko Sakai[1], Kenji Uchida[1], Takashi Kubo[2] and Kouji Nagata[1]

[1] Department of Mechanical Science and Engineering, Nagoya University,
Furo-cho, Chikusa-ku, Nagoya, 464-8603, Japan
ysakai@mech.nagoya-u.ac.jp, kenji-u@sps.mech.nagoya-u.ac.jp,
nagata@nagoya-u.jp

[2] EcoTopia Science Institute, Nagoya University, Furo-cho, Chikusa-ku, Nagoya,
464-8603, Japan
t-kubo@nagoya-u.jp

Abstract. In this paper, we investigate the statistical characteristics of the mass
flux in an axisymmetric turbulent jet diffusion field of a high-Schmidt-number matter. The diffusing fluid is a dye solution (whose Schmidt number is about 3,800),
and the simultaneous measurements of two velocity components and concentration
at five different jet Reynolds numbers have been made by a combined probe of a
X-type hot-film and a fiber-optic concentration sensor. It is found that the mass
flux is hardly affected by the magnitude of Schmidt (or Prandtl) number. The axial
velocity-concentration cospectra on the jet centerline show $-7/3$ scaling law. And
their scaling range agree with the overlap range of the inertial subrange and the convective subrange. On the other hand, it seems that the radial velocity-concentration
cospectra at some distance off the jet centerline (at $r/b_U = 0.73$) show about -2
scaling law.

Keywords: turbulent jet, high Schmidt number, scalar flux, cospectrum,
scaling law

1 Introduction

The phenomena related to diffusion and mixing appear very often in the industrial plants, for example, the mixing of various fluids in chemical reactors and
the diffusion of fuel in combustion chambers. To understand these phenomena, the investigation of the joint statistics of velocity and scalar is important.
However, it is a very difficult task to obtain the accurate and stable data of
the joint statistics of velocity and scalar, in particular, high-Schmidt-number
scalar, and their Reynolds number dependence is also not clear yet. In this
study, a turbulent jet of a high-Schmidt-number dye solution is investigated
experimentally. The simultaneous measurements of velocity and concentration
are performed, then the statistical characteristics of the turbulent mass flux
are examined.

209

Y. Kaneda (ed.), *IUTAM Symposium on Computational Physics and New Perspectives in
Turbulence*, 209–214.

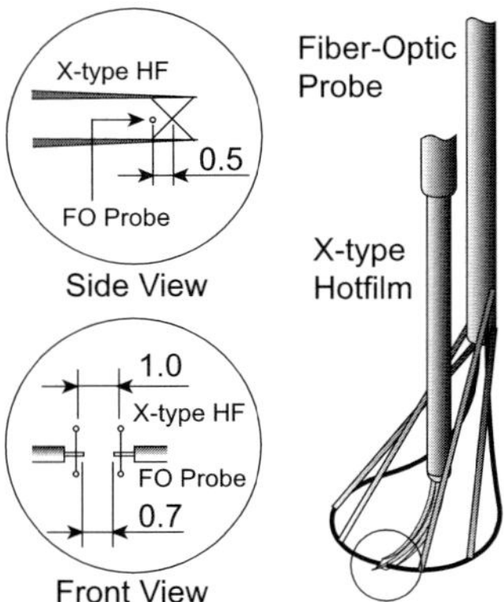

Fig. 1. The combined probe.

Table 1. Estimation of the turbulent Reynolds number at $x/d = 50$.

Re	R_λ $(r/b_U = 0.00)$	R_λ $(r/b_U = 0.73)$
6,600	167	113
9,900	228	169
13,300	269	218
15,500	282	241
17,700	296	263

2 Experimental Apparatus and Conditions

The flow field investigated in this study is a turbulent round jet of a high-Schmidt-number dye solution (Sc $\simeq$ 3,800). The origin of the coordinate system is at the nozzle exit (whose diameter $d = 2.0$ mm) and x and r are taken as streamwise and radial direction, respectively. The issuing Reynolds numbers, Re ($\equiv U_0 d/\nu$; U_0: initial velocity, ν: kinematic viscosity), are 6,600, 9,900, 13,300, 15,500 and 17,700. The combined probe of the X-type hot-film and fiber-optic concentration sensor based on the Lambert-Beer's law (see Fig. 1) is used for the simultaneous measurements of the axial velocity u, radial velocity v, and concentration c. The X-type hot-film sensor is a quartz-coated platinum wire whose diameter is 51 μm and the length of the sensing element is 1 mm. On the other hand, the fiber-optic concentration sensor consists of two facing optical fibers with the core diameter of 10 μm, and the spacing between two facing fibers is about 0.7 mm. The turbulent Reynolds numbers, $R_\lambda (\equiv u'\lambda/\nu$; u': the r.m.s. value of axial velocity, λ: Taylor microscale), at $x/d = 50$ are shown in Table 1. In the table, b_U denotes the half-width of radial profile of mean axial velocity U.

3 Results

3.1 Radial Profiles of Mass Flux

Figure 2 shows the radial profiles of the axial mass flux $\langle uc \rangle$ across the jet. The abscissa is the radial distance r normalized by b_U, and the ordinate is $\langle uc \rangle$ normalized by the mean value of axial velocity U_c and concentration C_c on the jet centerline. The profiles show a peak around $r/b_U \simeq 0.7$. In Fig. 2, the other results are also shown. P & L is the mass flux of helium gas (Sc $\simeq$ 0.7) obtained by the Panchapakesan and Lumley (1993) [1], Pa & Li is the mass flux of Rhodamine 6G (Sc $\simeq$ 1,000) obtained by Papanicolaou and List (1988) [2] and C & T is the heat flux (Pr $\simeq$ 0.7) obtained by Chevray

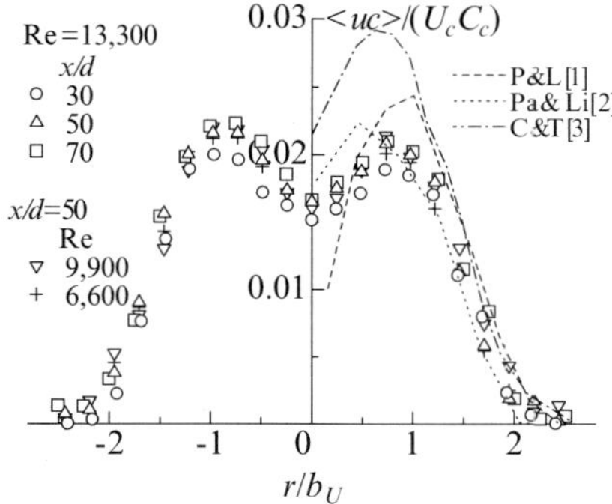

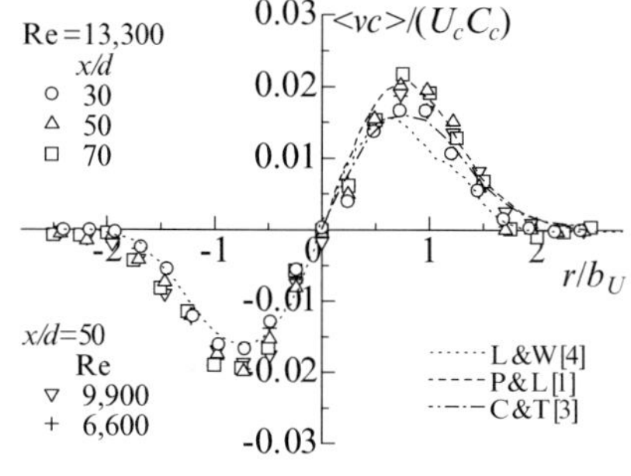

Fig. 2. Radial profiles of the axial mass flux.

Fig. 3. Radial profiles of the radial mass flux.

and Tutu (1978) [3]. In comparison with these results, it is found that our distributions show similar shapes to them.

On the other hand, the radial profiles of the radial mass flux $\langle vc \rangle$ across the jet are shown in Fig. 3. The ordinate is the normalized radial mass flux, $\langle vc \rangle / U_c C_c$. L & W in Fig. 3 is the mass flux of Rhodamine B (Sc $\simeq$ 1,000) obtained by Law and Wang (1998) [4]. From the figure, it is found that the profiles of each cross section show a good similarity and the influence of the Reynolds number hardly appears on them. Additionally, our distributions are in good agreement with the other researchers' results, although the Schmidt (or Prandtl) number is different significantly. This reason is, as shown in Fig. 9, most contribution to the scalar flux is by the larger-scale fluctuations than the integral scale. Therefore, it is found that the scalar flux is hardly affected by the magnitude of Schmidt (or Prandtl) number.

Although figures are not shown here, the value of correlation coefficient of u and c, $\rho_{uc} = \langle uc \rangle / u'c'$, is almost constant in the region of $-1.5 \lesssim r/b_U \lesssim 1.5$ and approaches asymptotically to 0.3 as the downstream distance is larger. On the other hand, the maximum and minimum value of the correlation coefficient of v and c, $\rho_{vc} = \langle vc \rangle / v'c'$, shows ± 0.3 at the far downstream region of $x/d \geq 50$.

3.2 PDFs of Mass Flux

Figure 4 shows the distributions of the probability density function (PDF) of axial mass flux uc, and radial mass flux vc, for five different Reynolds numbers at $x/d = 50$ on the jet centerline. The abscissa and ordinate are normalized by each r.m.s. value of instantaneous mass flux $(uc)'$ and $(vc)'$. ξ, ζ and η are the sample spaces of u, v and c, respectively. From the figure, the PDFs of uc and vc show a good similarity for each Reynolds number. The PDFs of uc skew to the positive direction, however, the PDFs of vc are distributed symmetrically. These suggest that the axial velocity and concentration have a positive

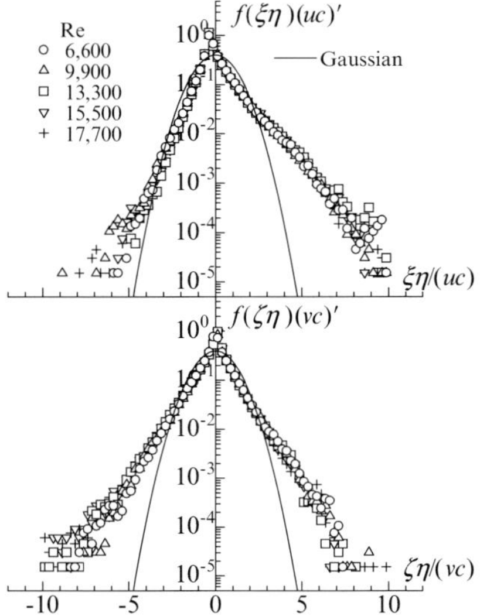

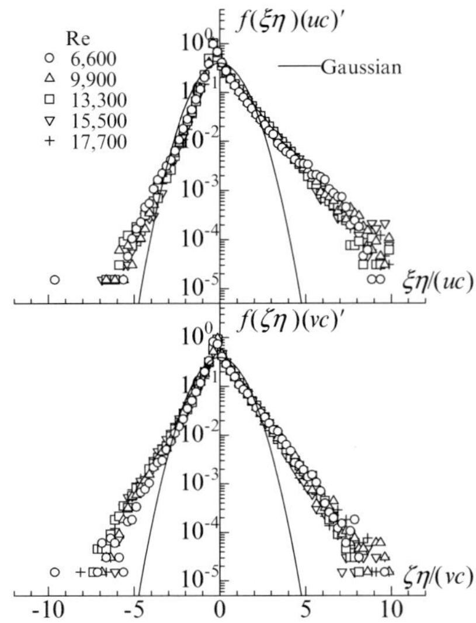

Fig. 4. Distributions of the axial- and radial-mass-flux PDF on the jet centerline at $x/d = 50$.

Fig. 5. Distributions of the axial- and radial-mass-flux PDF at $x/d = 50$, $r/b_U = 0.73$.

correlation, and the radial velocity and concentration have no correlation on the jet centerline.

The distributions of the PDF of uc and vc, for five different Reynolds numbers at $x/d = 50, r/b_U = 0.73$ are shown in Fig. 5. From the figure, it is found that the PDFs for both axial and radial mass flux skew to the positive direction.

3.3 Scaling Law of Velocity-Scalar Cospectrum

The velocity-scalar cospectrum $C_{u_i\theta}$ is defined so that the mean scalar flux is given by

$$\langle u_i\theta \rangle = \int_0^\infty C_{u_i\theta}(k)\,\mathrm{d}k, \tag{1}$$

where θ is the scalar fluctuation, u_i is the component of velocity in the i-direction, and k is the wavenumber. Lumley (1964) [5] applied a similarity hypothesis to predict the cospectrum of the velocity and potential temperature. For the case of passive scalar mixing, Lumley's result for the cospectrum in the inertial-convective range is

$$C_{u_i\theta}(k) \sim \mu_i\varepsilon^{1/3}k^{-7/3}, \tag{2}$$

where μ_i is the mean scalar gradient in the i-direction, and ε is the energy dissipation rate. Mydlarski & Warhaft (1998) [6] measured the temperature in the grid-generated turbulence for the values of R_λ up to 582, and found to be closer to k^{-2} scaling. Also, Mydlarski (2003) [7] evaluated the scalar flux structure function, $\langle \Delta u \Delta \theta \rangle(r)$, and found $\langle \Delta u \Delta \theta \rangle(r) \sim r^{1.02}$ which provides support to a $k^{-2.02}$ scaling for the cospectrum. In the laboratory boundary layer experiments of Antonia & Smalley (2000) [8], the velocity-scalar cospectrum showed the spectral slope of -1.80 at R_λ of 390.

In the next section, we show the velocity-concentration cospectrum $C_{u_i c}$ obtained in the present axisymmetric turbulent jet.

3.4 Velocity-Concentration Cospectra

The distributions of the axial velocity-concentration cospectrum C_{uc} for several Reynolds numbers at $x/d = 50$ on the jet centerline are shown in Fig. 6. The abscissa is the frequency f, normalized by the integral time scale of axial velocity t_{uI}, and the ordinate is C_{uc} normalized by the axial mass flux $\langle uc \rangle$. From the figure, it is found that C_{uc} shows the approximately $-7/3$ scaling for range of scales $2 \times 10^{-1} < ft_{uI} < 1 \times 10^0$. In the range of Reynolds number in this study, i.e. $167 \le R_\lambda \le 296$, the influence of Reynolds number doesn't appear. Here, the comparison of C_{uc} with the axial velocity spectrum E_u and concentration spectrum E_c at $Re = 13,300$ is shown in Fig. 7. The ordinate is an arbitrary unit. It is found that the scaling range of C_{uc} agrees with the overlap range of the inertial subrange and convective subrange.

Figures 8 and 9 show the axial velocity-concentration cospectra C_{uc} and the radial velocity-concentration one C_{vc} at $x/d = 50, r/b_U = 0.73$, respectively. C_{uc} shows the spectral slope of about $-7/3$ in the inertial-convective subrange. On the other hand, the spectral slope of C_{vc} is smaller than the one of C_{uc}, and its spectral slope seems to be about -2. Here, it is noted that the unit vector along the mean concentration gradient is described as $n = \nabla C/|\nabla C| = \mu_1 e_x + \mu_2 e_r$ (e_x: x-direction unit vector, e_r: r-direction unit vector), where μ_1 and μ_2 are -0.083 and -0.997, respectively, so that

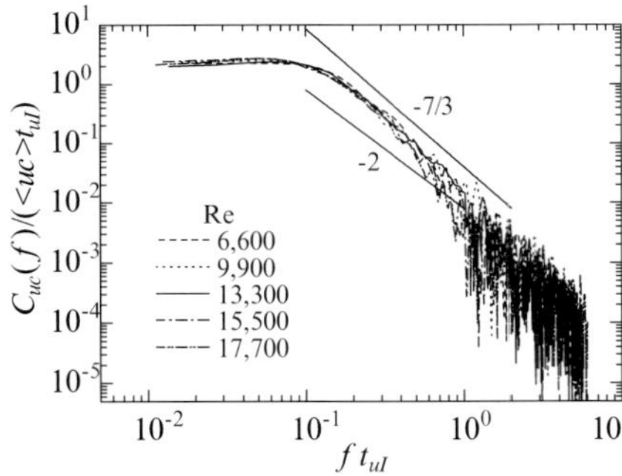

Fig. 6. Distributions of the axial velocity-concentration cospectrum at $x/d = 50$.

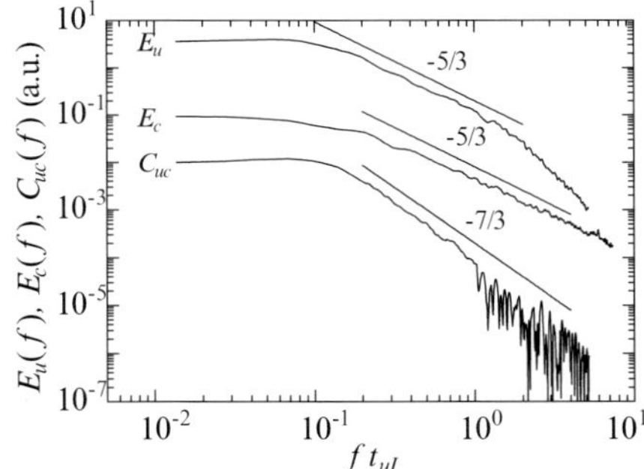

Fig. 7. Comparison of the cospectrum with the axial velocity and concentration spectra at $Re = 13,300$.

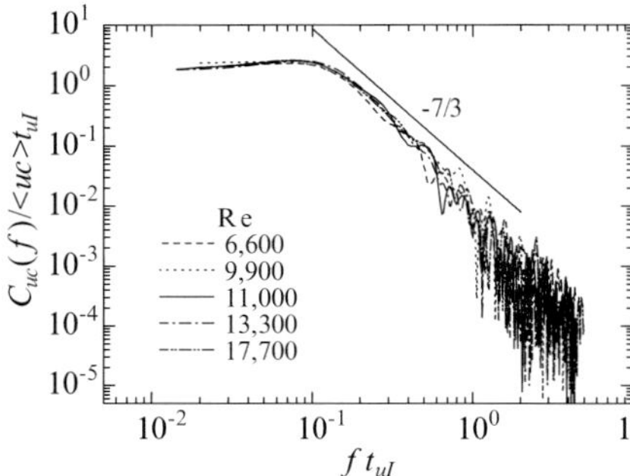

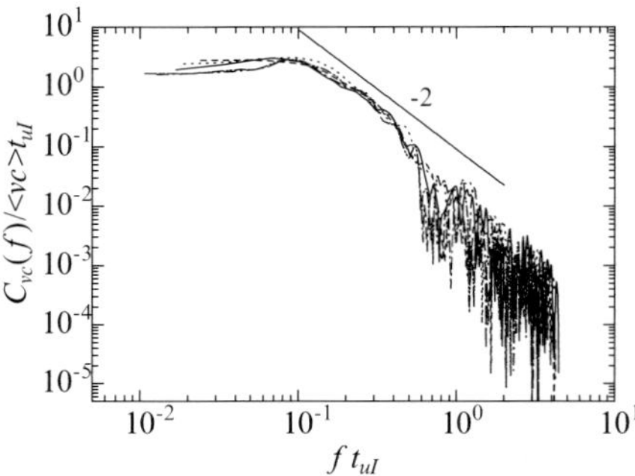

Fig. 8. Distributions of the axial velocity-concentration cospectrum at $x/d = 50, r/b_U = 0.73$.

Fig. 9. Distributions of the radial velocity-concentration cospectrum at $x/d = 50, r/b_U = 0.73$.

the angle between n and $-e_r$ is about 4.8°. In higher frequencies, the cospectra fluctuate violently because of losing the correlation between velocity and concentration.

4 Summary

To understand the statistical characteristics of mass flux, we performed the simultaneous measurements of two velocity components and concentration in a high-Schmidt-number turbulent jet diffusion field. It is found that the mass flux is hardly affected by the magnitude of Schmidt (or Prandtl) number. To investigate the relationship of the mass flux and frequency, we examine the distributions of velocity-concentration cospectrum. The axial velocity-concentration cospectra on the jet centerline show the $-7/3$ scaling law. And their scaling range agrees with the overlap range of the inertial subrange and the convective subrange. On the other hand, at some distance from the jet centerline (at $r/b_U = 0.73$), it seems that the radial velocity-concentration cospectra show about -2 scaling law, and then the axial velocity-concentration cospectra still show the $-7/3$ scaling law.

References

1. Panchapakesan NR, Lumley JL (1993) J Fluid Mech 246:225–247
2. Papanicolaou PN, List EJ (1988) J Fluid Mech 195:341–391
3. Chevray R, Tutu NK (1978) J Fluid Mech 88:133–160
4. Law AWK, Wang H (1998) Proc of 2nd Int Symp on Environ Hyd 211–216
5. Lumley JL (1964) J Atmos Sci 21:99–102
6. Mydlarski L, Warhaft Z (1998) J Fluid Mech 358:135–175
7. Mydlarski L (2003) J Fluid Mech 475:173–203
8. Antonia RA, Smalley RJ (2000) Phys Rev E 62:640–646

Large Scale Structure in Turbulent Plane Couette Flow

Masayuki Umeki[1] and Osami Kitoh[2]

[1] Trinity Industrial Corp.
Kamimotocho Toyota 471-0855, Japan
`m-umeki@trinityind.co.jp`
[2] Department of Engineering Physics, Electronics and Mechanics, Graduate
School of Engineering, Nagoya Institute of Technology, Showa-ku, Nagoya,
466-8555 Japan `kitoh.osami@nitech.ac.jp`

Abstract. Large-scale structures appearing in core region of turbulent Couette flow
have been studied in physical experiment. The behavior of low/high speed streaks
were measured by hot-wire arrays in x_1-x_3 plane. The structures having length
scale of $(40 - 60)h$ in streamwise and $(4 - 5)h$ in spanwise directions were detected
from spectrum pattern in accordance with DNS results reported so far. Since the
structures are disturbed randomly by the surrounding turbulent motion, the basic
pattern of the large-scale structure cannot be detected hitherto. By extracting large-
scale modes out of the wavelet energy spectrum and reconstructing the flow field
from them, the typical flow structure having inclined pattern along the streamwise
direction can be obtained. This pattern does not appear uniformly in space but a
sequence of it and void along the flow direction.

Keywords: Couette flow, low speed streak, large scale structure, wavelet
analysis

1 Introduction

In turbulent plane Couette flow, it has been known that large-scale structures (LSS), streamwise vortices and streaks, exist in the core region. From
power-spectrum study of the streamwise fluctuating velocity component u'_1,
developed by DNS adopting large computational domain, the characteristic
scales of the streaks are reported as about $4h$ in spanwise and $60h$ in streamwise direction, respectively [1], [2]. Since the turbulent motions disturb randomly these LSS, they are neither stationary nor fixed in position and it is
difficult to get the basic LSS from flow visualization. Although low/high speed
streaks were discerned in DNS-visualization, no particular streak patterns giving above mentioned scales have not been reported. Minimal Couette flow
simulations, in which the disturbance on LSS from random turbulent motion
is quite weak, enable us to see the basic LSS, i.e., the streamwise vortices and
low-speed streak in the core region appear sequentially and quasi-periodically

215

Y. Kaneda (ed.), *IUTAM Symposium on Computational Physics and New Perspectives in
Turbulence*, 215–220.
© 2008 *Springer*.

[3]. The characteristic streamwise scale (wavelength) of the wavy streak, estimated to be $5.5h$, is quite different from that of the full-scale simulations and need more work to understand the LSS in turbulent Couette flow.

The objective of this work is to extract the LSS, low/high speed streaks in particular, from physical experiment and identify the basic streak patterns that are responsible for the LSS reported so far in Couette flow.

2 Experimental Methods

Figure 1 shows the general arrangement of the Couette flow apparatus. The test channel consists of a stationary wall (top) and a moving-belt (bottom) that moves with a constant velocity U_b. The channel height $2h$ is 27 mm and the width and length are 880 mm, 5310 mm, respectively. The fully developed Couette flow can be obtained beyond $x_1/h = 144$. The mean velocity and the turbulence characteristics obtained with this apparatus are given in [4]. A pair of vortex generator (VG) is installed at $x_1/h = 152$ to produce robust low/high speed streaks in the downstream section [5]. No essential differences between Couette flows with and without VG are reported except for the robustness of the streak.

To study the behavior of the low and high speed streaks appearing in center plane ($x_2 = h$: h is half channel height) of the channel, sixteen I-type hot-wires arrayed in line in spanwise direction ($8h$ in width) were used to obtain the time series data of $u_1'(x_3, t)$, here x_2 and x_3 are wall-normal and spanwise axes, respectively. The Reynolds number based on the relative half wall velocity $U_b/2$ is $Re = hU_b/(2\nu) = 3750$. The time series data are translated into spatial

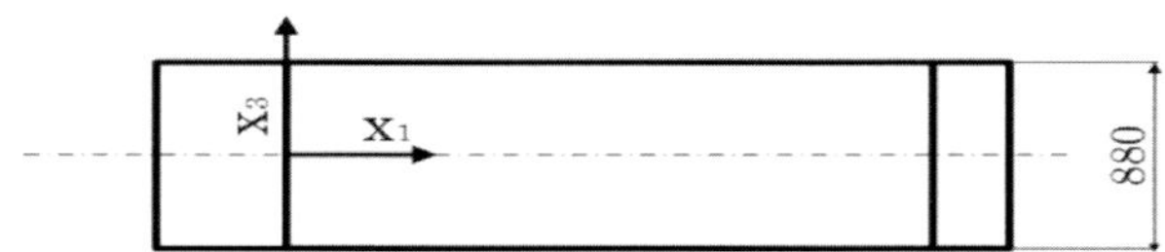

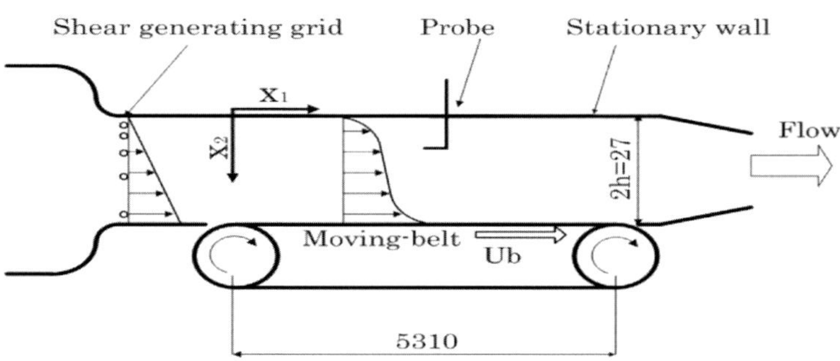

Fig. 1. Schematic of the experimental set up.

one $u_1'(x_1, x_3)$ using the relation $x_1 = -U_b t/2$. Hereafter, x_1 is a streamwise position from measuring point.

The wavelet transforms (1-D and 2-D) are adopted to separate the various scales in $u_1'(x_1, x_3)$ or to see the localized character of the particular scale. The 2-D discrete wavelet transform applied to $u_1'(x_1, x_3)$ is given as

$$\tilde{u}_{1j_1,m_1:j_3,m_3} = \int\int u_1'(x_1, x_3)\Psi_{j_1,m_1:j_3,m_3}(x_1, x_3)dx_1dx_3 \tag{1}$$

where $\Psi_{j_1,m_1:j_3,m_3}(x_1, x_3)$ is the Daubechies wavelet mother function with index $N = 20$, j_1, j_3 stand for the scale parameters and m_1, m_3 stand for the translation parameters.

3 Results and Discussions

Figure 2 shows a typical example of the streak behavior. The bold line areas are for low-speed streaks while the fine line areas are for high-speed ones. A long streak showing some waviness can be seen. We cannot see any particular basic pattern in there.

Figure 3 shows the wavelet energy spectrum $E_{j_1:j_3}$ distribution against the scale parameters j_1 (streamwise) and j_3 (spanwise). Here, $E_{j_1:j_3}$ is defined as

$$E_{j_1:j_3} = \sum_{m_1}\sum_{m_3}|\tilde{u}_{1j_1,m_1:j_3,m_3}|^2 \tag{2}$$

$E_{j_1:j_3}$ has two peaks, one is at around $j_1 = 11, 12$ (at relatively smaller scale) and the other is at $j_1 = 9$ (at larger scale). The length scale of the larger one corresponds roughly to the characteristic LSS reported so far ($(40\text{-}60)h$) while the smaller one corresponds to a peak scale of turbulent Poiseuille flow, not shown here, and is considered as that of ordinary turbulent motion. To extract LSS out of random turbulent flow, reconstruction the particular flow pattern, using the wavelet coefficients belonging the LSS ($j_1 = 8$, 9 and 10 with $j_3 = 3$) is introduced. u_1^L is a reconstructed velocity. Figure 4 shows the reconstructed flow-structure, high-speed and low-speed areas are indicated by

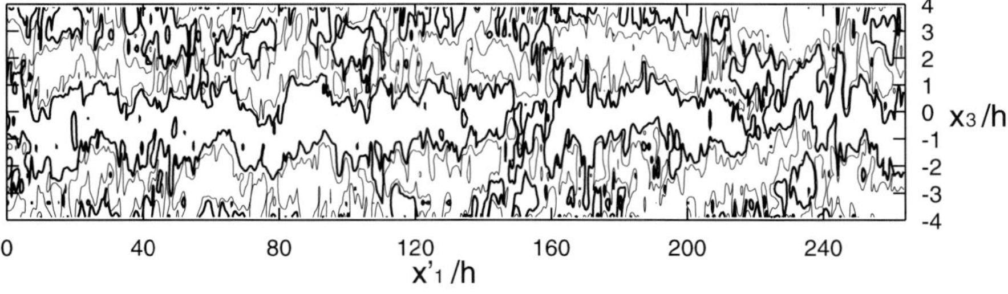

Fig. 2. Contours of $u_1'(x_1, x_3)$ at center plane: a bold line depicts $u_1'/(U_b/2) = -0.05$ and a thin line $u_1'/(U_b/2) = 0.05$.

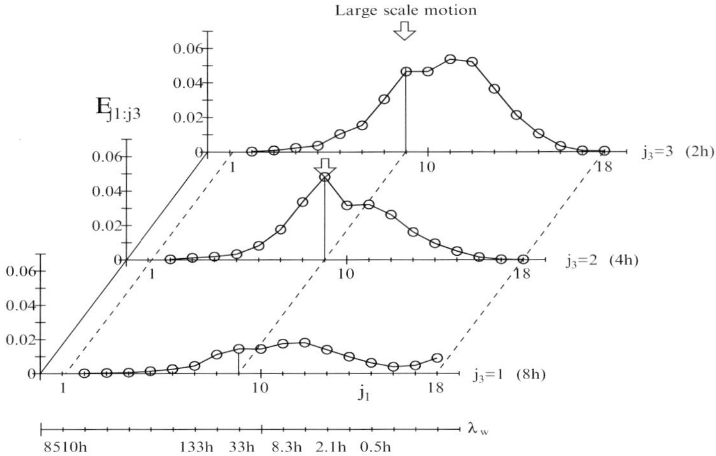

Fig. 3. Wavelet energy spectrum of Couette flow ($Re = 3750$).

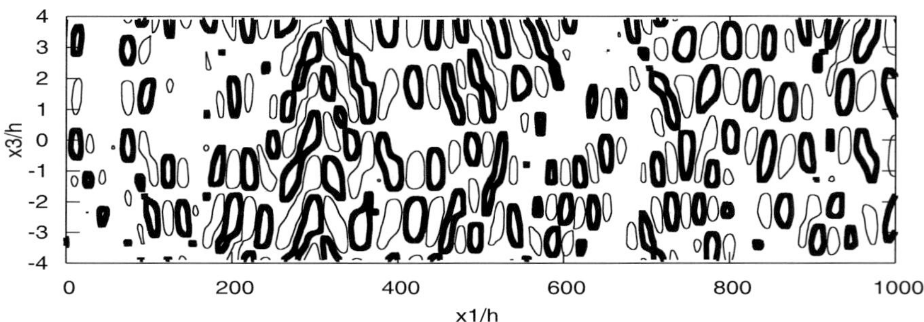

Fig. 4. Reconstructed flow-pattern: thin lines depict $u_1^L/(U_b/2) = 0.02$ and thick lines depict $u_1^L/(U_b/2) = -0.02$.

thin and thick lines, respectively. Note that the spanwise direction is largely stretched so that the particular structure can be seen easily. In addition to the plain patterns of the selected scales, large-scale structures (LSS), inclined along the streamwise direction and arrayed in parallel with one another, can be seen. The spanwise separation between two neighboring high (or low) speed inclined patterns is $(2.8 - 4)h$. The inclination-angle α with respect to x_1 axis is estimated to be $\pm(3° - 5°)$ similar to a coherent structure in wall turbulence ($\pm4°$ [6]). This configuration of LSS gives the typical streamwise length scale of $(45 - 75)h$. The angle α changes quasi-periodically along the streamwise direction, though not so regularly because the structures collide with each other or disappear from place to place. The irregularity would be caused by the random turbulent motion. The space-period of this angle variation is not clear but has an order of $300h$.

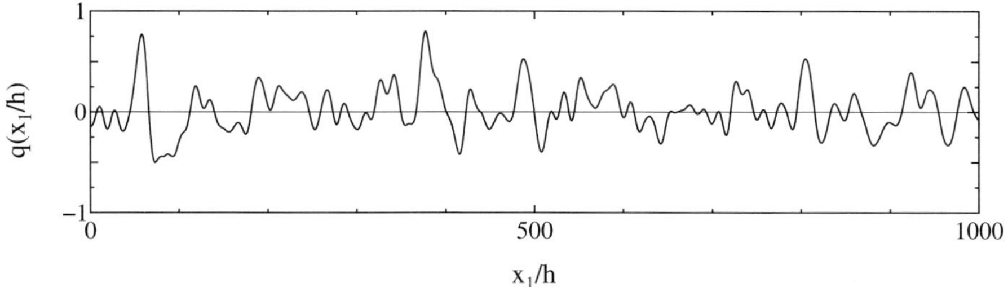

Fig. 5. Streak-line-function, $q(x_1/h)$.

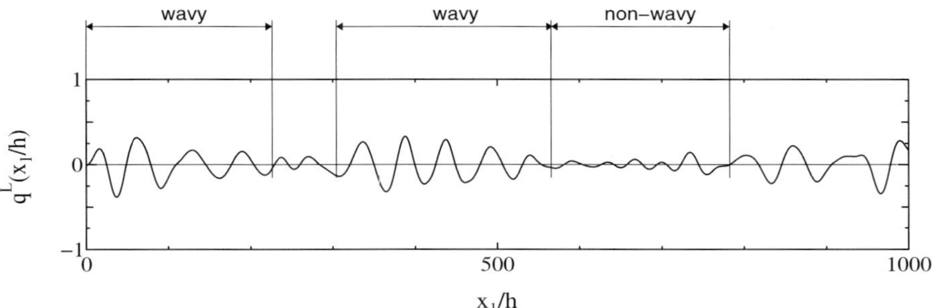

Fig. 6. The reconstructed streak-line-function, $q^L(x_1/h)$.

To study the streak pattern more quantitatively, a streak-line-function $q(x_1/h)$ defined as

$$q(x_1/h) = Z_s(x_1/h) - \overline{Z_s} \qquad (3)$$

is introduced, where $Z_s(x_1/h)$ is a spanwise position of minimum velocity within the low-speed streak and $\overline{Z_s}$ is its average position. Z_s is estimated from a least-square fitting curve of second degree using five consecutive discrete velocities including the lowest one. Figure 5 shows the typical variation of $q(x_1/h)$. The wavelet energy specrtum of the streak-line-function, $q(x_1/h)$, are also obtained. The spectrum has a peak at length scale of LSS. Figure 6 shows the reconstructed streak line, $q^L(x_1/h)$, using the peak wavelet coefficients. Wavy pattern does not appear uniformly but wavy and non-wavy patterns appear sequentially along the flow direction. On the average, the wavy patterns appear every $300h$.

The LSS in Couette flow reported so far is basically due to the existence of this structure found here. The mechanism of the LSS is not clear yet but, it might have something to do with the regeneration mechanism of the vortex-streak structure found in minimal-Couette simulation.

4 Concluding Remarks

Similar to DNS works reported so far, large-scale structures $((40\text{-}60)h$ in streamwise direction and $(4\text{-}5)h$ in spanwise direction) are confirmed to exist in physical experiment. The reconstructed pattern using the corresponding spectrum modes show large scale structures (LSS) having inclined patterns along the flow direction. LSS locates non-uniformly in space, resulting in a sequence of wavy and non-wavy streak pattern along the flow direction.

References

1. Komminaho J, Lundbladh A, Johansson A (1996) J Fluid Mech 320:259–285
2. Tsukahara T, Kawamura H, Shingai K (2006) J Turbul 7:19
3. Hamilton J, Kim J, Waleff F (1995) J Fluid Mech 287:317–348
4. Kitoh O, Nakabayashi K, Nishimura F (2005) J Fluid Mech 539:199–227
5. Umeki M, Kitoh O, Kameyama R (2003) Proc Turbulence Heat and Mass Trans 4:89
6. Jeong J, Hussain F, Schoppa W, Kim J (1997) J Fluid Mech 332:185–214

Vortex Tubes in Turbulence Velocity Fields at High Reynolds Numbers

Hideaki Mouri, Akihiro Hori, and Yoshihide Kawashima

Meteorological Research Institute, Nagamine, Tsukuba 305-0052, Japan
`hmouri@mri-jma.go.jp`

Abstract. The elementary structures of turbulence, i.e., vortex tubes, are studied from laboratory experiments at Reynolds numbers $Re_\lambda = 332$–1934. We find features of tube radius, circulation velocity, and spatial distribution that are independent of the Reynolds number Re_λ and of the configuration for turbulence production and are hence expected to be universal.

Keywords: experiment, small-scale coherent structures, duct flow, boundary layer

1 Introduction

Turbulence contains vortex tubes as the elementary structures [1]. Regions of strong vorticity are organized into tubes. They occupy a small fraction of the volume and are embedded in the random background flow.

The basic parameters of vortex tubes were derived from direct numerical simulations at microscale Reynolds numbers $Re_\lambda \le 200$ [1, 2, 3]. The radii are of the order of the Kolmogorov length η. The circulation velocities are of the order of the Kolmogorov velocity u_K or the rms velocity fluctuation $\langle u^2 \rangle^{1/2}$. However, the universality of such tube parameters has not been established because their behavior has not been known at high Reynolds numbers, $Re_\lambda > 200$, where the direct numerical simulation is difficult.

A more promising approach is velocimetry in laboratory experiments. For example, vortex tubes were studied in boundary layers at $Re_\lambda = 295$–1258 [4]. However, the Reynolds number could be increased still more. The dependence of tube parameters on the configuration for turbulence production has not been known. We accordingly study vortex tubes in velocity fields of duct flows at $Re_\lambda = 719$–1934 and boundary layers at $Re_\lambda = 332$–1304.

2 Experiments

The experiments were done in a wind tunnel. We use coordinates x, y, and z in the streamwise, spanwise, and floor-normal directions. The origin

Y. Kaneda (ed.), *IUTAM Symposium on Computational Physics and New Perspectives in Turbulence*, 221–226.

$x = y = z = 0$ m is on the tunnel floor at the entrance to the test section. Its size was $\delta x = 18$ m, $\delta y = 3$ m, and $\delta z = 2$ m. A hot-wire anemometer was used to measure the streamwise $(U + u)$ and spanwise (v) velocities. Here U is the mean while u and v are the fluctuations. To convert temporal variations into spatial variations, we used Taylor's frozen flow hypothesis.

2.1 Duct Flows

At $x = -2$ m, we placed a rectangular duct with width $\delta y = 1.3$ m and $\delta z = 1.4$ m. The duct center was on the tunnel axis. The measurement position was at $x = 15.5$ m and $z = 0.6$ m, where the flow was turbulent. We obtained $Re_\lambda = 719$–1934 by changing the duct-exit flow velocity from 11 to $55\,\mathrm{m\,s^{-1}}$.

2.2 Boundary Layers

Over the entire floor of the tunnel test section, we placed blocks as roughness elements. Their size was $\delta x = 0.06$ m, $\delta y = 0.21$ m, and $\delta z = 0.11$ m. Their spacing was $\delta x = \delta y = 0.5$ m. The measurement position was at $x = 12.5$ m, where the boundary layer was well developed, and $z = 0.25$–0.35 m in the log-law sublayer. We obtained $Re_\lambda = 332$–1304 by changing the incoming-flow velocity from 2 to $20\,\mathrm{m\,s^{-1}}$.

3 Model for Vortex Tubes

Using the Burgers vortex, an idealized model for vortex tubes, we discuss what information is available from a one-dimensional cut of the velocity field (see [5] for a similar discussion based on a direct numerical simulation). The Burgers vortex is an axisymmetric steady circulation in a strain field. In cylindrical coordinates, the circulation u_Θ and strain field (u_R, u_Z) are

$$u_\Theta \propto \frac{\nu}{a_0 R} \left[1 - \exp\left(-\frac{a_0 R^2}{4\nu} \right) \right] \quad \text{and} \quad (u_R, u_Z) = \left(-\frac{a_0 R}{2}, a_0 Z \right). \quad (1)$$

Here ν is the kinematic viscosity and a_0 (>0) is a constant. The circulation is maximal at $R = R_0 = 2.24(\nu/a_0)^{1/2}$. We regard R_0 as the tube radius.

Suppose that velocity data are obtained on a one-dimensional cut x of a flow containing vortex tubes, as illustrated in Fig. 1. The circulation flows of the vortex tubes intermittently enhance small-scale variations in the v signal. If we consider the enhancements above a high threshold, their scale and amplitude correspond to the radius and circulation velocity of strong vortex tubes. To demonstrate this, mean profiles are obtained along the cut x for the circulation flows u_Θ of the Burgers vortices with random positions (x_0, y_0) and orientations (θ_0, φ_0). The radii R_0 and maximum circulation velocities $V_0 = u_\Theta(R_0)$ are set to be the same. The threshold is $|\partial_x v| > |\partial_x v|_{x_0 = y_0 = \theta_0 = 0}/3$ at $x = 0$. When $\partial_x v$ is negative, the sign of the v signal is inverted before the averaging. The result is shown in Fig. 2. Around

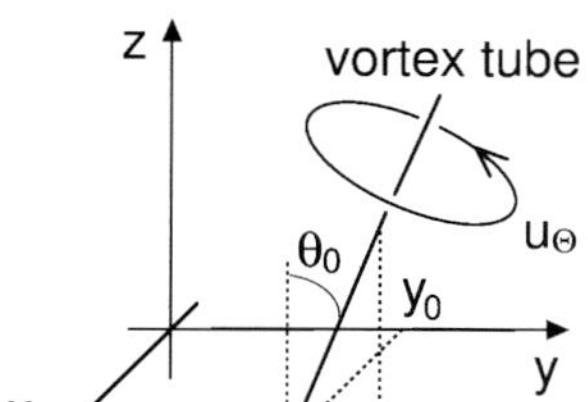

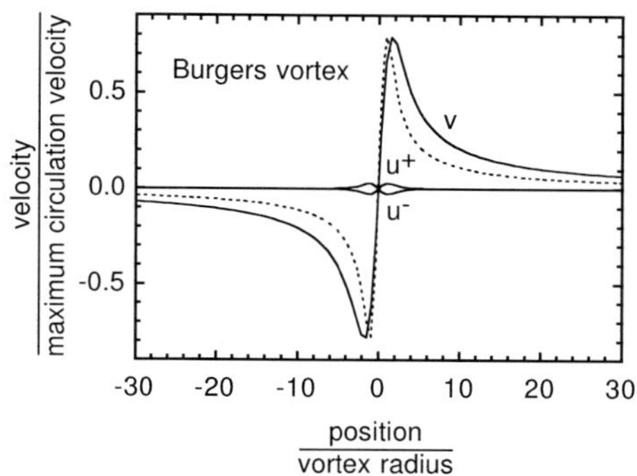

Fig. 1. Sketch of coordinates and a vortex tube.

Fig. 2. Mean velocity profiles of the circulation flows of the Burgers vortices.

the peaks, the mean v profile is similar to that of the Burgers vortex with $x_0 = y_0 = \theta_0 = 0$ (dotted line). The extended tails are due to the Burgers vortices with $|y_0| \gg R_0$ or $\theta_0 \gg 0$.

4 Mean Velocity Profile

Mean profiles for circulation flows of vortex tubes in the u and v velocities are extracted, by averaging signals centered at the position where the absolute spanwise-velocity increment $|v(x + r/2) - v(x - r/2)|$ is enhanced above a threshold. The scale r is set to be the sampling interval. The threshold is such that 0.1% or 1% of the increments are used for the averaging. Since these increments comprise the tail of the probability density distribution of all the increments as in Fig. 3, it follows that we study strong vortex tubes. An example of the results is shown in Fig. 4.

The v profile in Fig. 4 is similar to that in Fig. 2. We fit the v profile in Fig. 4 around its peaks by the v profile of the Burgers vortex with $x_0 = y_0 = \theta_0 = 0$ (dotted line), and thereby estimate the radius R_0 and maximum circulation velocity V_0. The radius R_0 is several times the Kolmogorov length η. The maximum circulation velocity V_0 is several tenths of the rms velocity fluctuation $\langle v^2 \rangle^{1/2}$ and several times the Kolmogorov velocity u_K [1, 2, 3, 4].

The u profile in Fig. 4 is separated for $\partial_x u > 0$ (u^+) and $\partial_x u \leq 0$ (u^-) at $x = 0$. These $u^\pm$ profiles have larger amplitudes than the $u^\pm$ profiles in Fig. 2. Hence the $u^\pm$ profiles in Fig. 4 are dominated by the circulation flow u_Θ of vortex tubes that passed the probe with some incidence angles relative to the mean flow direction, $\tan^{-1}[v/(U + u)]$. The radial inflow u_R of the strain field is not discernible, except that the u^- profile has a larger amplitude than the u^+ profile [4]. Unlike the Burgers vortex, a real vortex tube is not always oriented to the stretching direction [2, 3, 6].

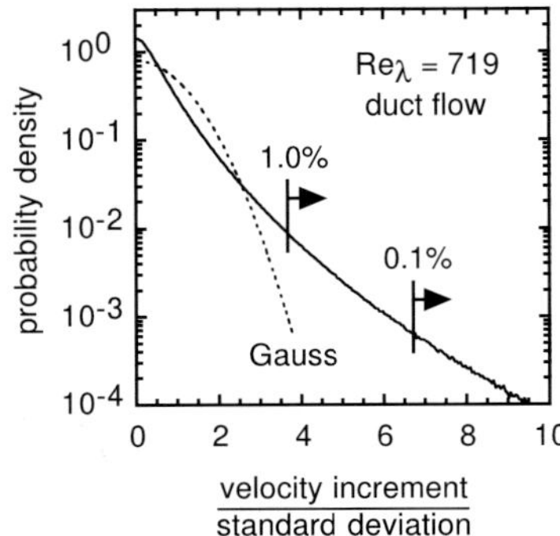

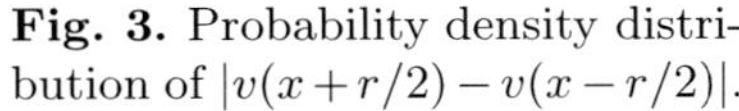

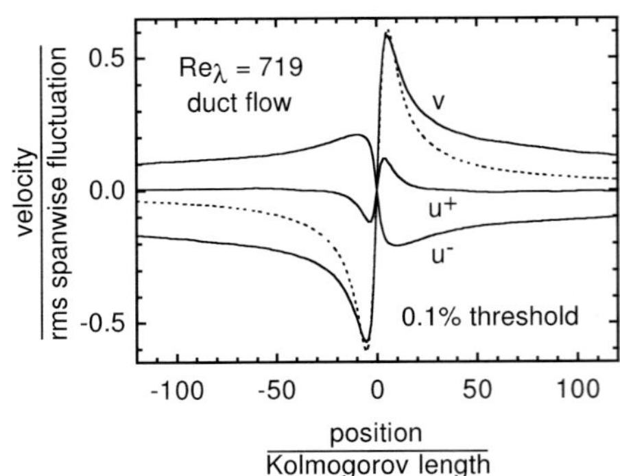

Fig. 3. Probability density distribution of $|v(x+r/2) - v(x-r/2)|$.

Fig. 4. Mean velocity profiles of vortex tubes.

5 Spatial Distribution

The spatial distribution of vortex tubes is studied using the distribution of interval δx_0 between successive enhancements of the absolute spanwise-velocity increment. We define the enhancement in the same manner as for the mean velocity profile. An example of the results is shown in Fig. 5.

Over small intervals (Fig. 5(a)), the distribution obeys a power law $\delta x_0^{-\mu_0}$ with $\mu_0 \simeq 1$ [7]. Thus, vortex tubes form self-similar clusters with no characteristic scale. In fact, a direct numerical simulation revealed that vortex tubes lie on borders of energy-containing eddies [2].

Over large intervals (Fig. 5(b)), the distribution obeys an exponential law, i.e., a characteristic of intervals for a Poisson process of random and independent events [4, 8]. The large-scale distribution of vortex tubes is random and independent.

6 Scaling Laws of Tube Parameters

The dependence of tube parameters on the microscale Reynolds number Re_λ and on the configuration for turbulence production, i.e., duct flow or boundary layer, is studied in Fig. 6. We normalize the quantities by their values in the duct flow at $Re_\lambda = 1934$: $R_0/\eta = 6.2$, $V_0/\langle v^2 \rangle^{1/2} = 0.63$, $V_0/u_K = 14$, $Re_0 = R_0 V_0/\nu = 87$, and $\mu_0 = 0.91$ for the 0.1% threshold, and $R_0/\eta = 7.0$, $V_0/\langle v^2 \rangle^{1/2} = 0.41$, $V_0/u_K = 9.2$, $Re_0 = 65$, and $\mu_0 = 1.1$ for the 1% threshold.

The radius R_0 scales with the Kolmogorov length η as $R_0 \propto \eta$ (Fig. 6(a)). Thus, the circulation flows of vortex tubes remain to be of smallest scales of turbulence.

The maximum circulation velocity V_0 scales with the rms velocity fluctuation $\langle v^2 \rangle^{1/2}$ as $V_0 \propto \langle v^2 \rangle^{1/2}$ (Fig. 6(b)). Although the rms velocity fluctuation is a characteristic of large-scale motions, vortex tubes could be formed via shear instability on borders of energy-containing eddies [2], where a velocity

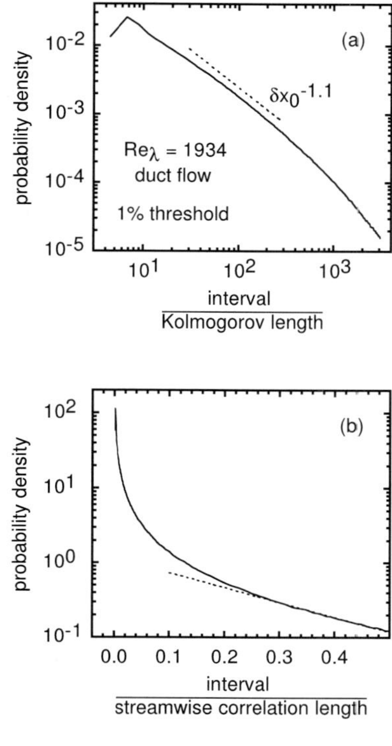

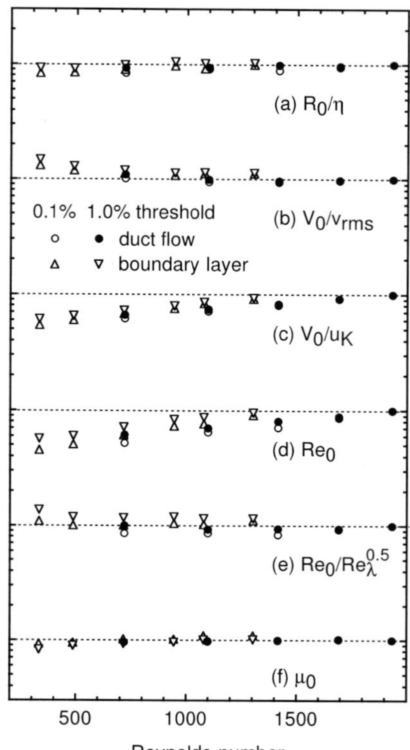

Fig. 5. Probability density distribution of interval between vortex tubes.

Fig. 6. Dependence of tube parameters on Re_λ.

variation over a small scale such as the tube radius could be comparable to the rms velocity fluctuation. The maximum circulation velocity does not scale with the Kolmogorov velocity u_K, a characteristic of smallest-scale motions, as $V_0 \propto u_K$ (Fig. 6(c)).

For strong vortex tubes, the scalings $R_0 \propto \eta$ and $V_0 \propto \langle v^2 \rangle^{1/2}$ were obtained from direct numerical simulations at $Re_\lambda \leq 200$ [2, 3] and laboratory experiments at $Re_\lambda \leq 1300$ [4]. We have found that the scalings exist up to $Re_\lambda \simeq 2000$, regardless of the configuration for turbulence production and of the threshold to identify the vortex tubes.

The scalings of the radius R_0 and circulation velocity V_0 lead to a scaling of the Reynolds number $Re_0 = R_0 V_0/\nu$ that characterizes the circulation flows of the vortex tubes [2, 3]:

$$Re_0 \propto Re_\lambda^{1/2} \quad \text{if } R_0 \propto \eta \text{ and } V_0 \propto \langle v^2 \rangle^{1/2}, \quad \text{or} \tag{2}$$

$$Re_0 = \text{constant} \quad \text{if } R_0 \propto \eta \text{ and } V_0 \propto u_K. \tag{3}$$

Our result favors the former scaling (Fig. 6(e)) rather than the latter (Fig. 6(d)). With an increase of Re_λ, strong vortex tubes progressively have higher Re_0 and are more unstable [2, 3]. Their lifetimes are shorter. It is known that the flatness factor $\langle(\partial_x v)^4\rangle/\langle(\partial_x v)^2\rangle^2$ scales with $Re_\lambda^{0.3}$ [9]. Since $\langle(\partial_x v)^4\rangle$ is dominated by strong vortex tubes, it scales with $\langle v^2\rangle^2/\eta^4$. Since $\langle(\partial_x v)^2\rangle^2$ is dominated by the background flow, it scales with u_K^4/η^4. If the number density of strong vortex tubes remains the same, we have $\langle(\partial_x v)^4\rangle/\langle(\partial_x v)^2\rangle^2 \propto \langle v^2\rangle^2/u_K^4 \propto Re_\lambda^2$. The difference from the real scaling implies that vortex tubes with $V_0 \simeq \langle v^2\rangle^{1/2}$ are less numerous at a higher Reynolds number Re_λ, albeit energetically more important.

The exponent μ_0 for the small-scale clustering, estimated in the range from $\delta x_0 = 30\eta$ to 300η, is constant (Fig. 6(f)). A similar result with $\mu_0 \simeq 1$ was obtained from laboratory experiments of the Kármán flow between two rotating disks at $Re_\lambda \simeq 400$–1600 [7]. The small-scale clustering of vortex tubes at high Reynolds numbers Re_λ is independent of the configuration for turbulence production.

7 Conclusion

For strong vortex tubes in duct flows at $Re_\lambda = 719$–1934 and boundary layers at $Re_\lambda = 332$–1304, we have obtained the scalings $R_0 \propto \eta$, $V_0 \propto \langle v^2\rangle^{1/2}$, and $Re_0 \propto Re_\lambda^{1/2}$. The exponent μ_0 for small-scale clustering is constant, $\mu_0 \simeq 1$. These features appear to have reached asymptotes at $Re_\lambda \simeq 2000$, regardless of the configuration for turbulence production, and are hence expected to be universal among strong vortex tubes at high Reynolds numbers Re_λ.

References

1. Makihara T, Kida S, Miura H (2002) J Phys Soc Jpn 71:1622–1625
2. Jiménez J, Wray AA, Saffman PG, Rogallo RS (1993) J Fluid Mech 255:65–90
3. Jiménez J, Wray AA (1998) J Fluid Mech 373:255–285
4. Mouri H, Hori A, Kawashima Y (2004) Phys Rev E 70:066305
5. Mouri H, Takaoka M, Kubotani H (1999) Phys Lett A 261:82–88
6. Kholmyansky M, Tsinober A, Yorish S (2001) Phys Fluids 13:311–314
7. Moisy F, Jiménez J (2006) Clustering of intense structures in isotropic turbulence. In: Kida S (ed) IUTAM symposium on elementary vortices and coherent structures: significance in turbulence dynamics. Springer, Dordrecht
8. La Porta A, Voth GA, Moisy F, Bodenschatz E (2000) Phys Fluids 12:1485–1496
9. Pearson BR, Antonia RA (2001) J Fluid Mech 444:343–382

Wavelet Analysis of Vortex Breakdown

Jori Elan Ruppert-Felsot[1], Marie Farge[1], and Philippe Petitjeans[2]

[1] Laboratoire de Météorologie Dynamique du CNRS, Ecole Normale Supérieure, Paris jori@lmd.ens.fr
[2] Laboratoire de Physique et Mécanique des Milieux Hétérogènes, UMR CNRS 7636, Ecole Supérieure de Physique et de Chimie Industrielles, Paris phil@pmmh.espci.fr

Abstract. We study the quasi-periodic turbulent bursting of a laboratory produced isolated vortex immersed in laminar flow. We analyze the experimentally measured flow field using orthogonal wavelets to observe the time evolution of the bursting. The discrete wavelet transform is used to separate the flow field into a coherent component, capturing the dynamics and statistics of the vortex during bursting, and an incoherent component, which is structureless and exhibits a different statistical behavior.

Keywords: wavelet, vortex, coherent structure, turbulence, bursting

1 Introduction

It remains an open question as to how the scaling of the classical energy spectrum is formed and what structures can be responsible for the $k^{-5/3}$ scaling in 3D turbulence. Recent experimental studies have focused on a solitary bursting vortex as a source of turbulence, leading to a transient buildup of a turbulent energy cascade [1, 2, 3]. The scaling of the energy spectrum was found to vary from k^{-1} to k^{-2} during the bursting with a $k^{-5/3}$ recovered in the time averaged spectra. The resulting vortex was found to be well approximated by a stretched spiral vortex following Lundgren's model [4], which also predicts a $-5/3$ time-averaged energy spectrum. However, the time evolution of the spectrum is not yet understood and depends upon the specific spatial structure assumed in the vortex model.

Previous studies were conducted using hot-film anemometry [1, 2]. These hot-film measurements have a good time resolution, but require a local Taylor hypothesis to obtain the spatial information necessary to calculate the energy spectrum. More recently particle image velocimetry (PIV) was used to measure the spatial distribution of the velocity field directly, without inferring it from a time series. Simultaneous hot-film probe measurements were used to synchronize the phase of the PIV with the bursting of the vortex. The PIV measurements were then phase averaged to obtain an ensemble average and

Y. Kaneda (ed.), *IUTAM Symposium on Computational Physics and New Perspectives in Turbulence*, 227–232.
© 2008 *Springer*.

to reconstruct an average time record of the bursting. The scaling of the energy spectra in the inertial range based upon the PIV measurements were in good agreement with the previous hot-film measurements [3]. However, the time resolution of the measurements was low and a single burst could not be followed in time. In the current study we use higher time resolution PIV to follow the time behavior of distinct individual bursts.

The vortex under study is a coherent structure that is well localized in space. It is therefore more natural to analyze this flow using a spatially localized set of basis functions rather than a Fourier basis. Wavelets consist of translations and dilations of a compact function and are well localized in physical and spectral space. Wavelets are thus an optimal choice to analyze such turbulent flows that contain features that are well localized in physical space [5]. Indeed, it has been found in simulation [6] and experiment [7] that the dynamics of turbulent flows are dominated by the contribution of a relatively small fraction of wavelet coefficients, the strongest of which correspond to the coherent structures.

2 Experiment

The vortex is produced in a laminar channel flow over a step, shown in Fig. 1. The vortex is intensified and stretched by suction of fluid through the channel walls, transverse to its axis. As the channel flow rate is increased, the vortex becomes increasingly strained. Above a critical channel flow rate the vortex detaches from the walls and eventually breaks down, resulting in a turbulent burst. The Reynolds number at the onset of the burst is estimated as 4000 in [2], based upon the circulation. The resulting turbulent flow is solely due to the bursting because the vortex is initially formed in laminar flow. A new vortex is formed shortly after the burst and this cycle repeats quasi-periodically at intervals of approximately 8 seconds.

We observe the vortex in a plane perpendicular to its axis at the center of the 12 cm $\times$ 7 cm cross section channel. Digital images are taken at a

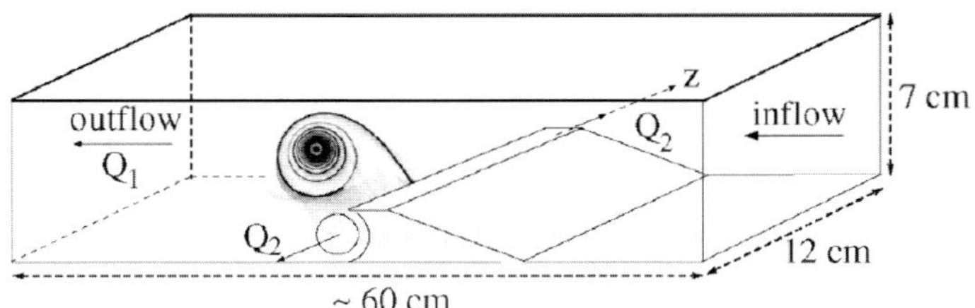

Fig. 1. Schematic of the experiment (from [2]). The vortex initiated by the step (5 mm high) is strained by the channel flow Q_1 and intensified and stretched by the axial suction Q_2 (i.e. the total flow rate through the channel $= Q_1 + 2Q_2$). The values $Q_1 = 12.5$ l min^{-1} and $Q_2 = 7.5$ l min^{-1} were chosen to produce an intense quasi-periodically bursting vortex

resolution of 1600×1200 pixels and 30 Hz frame rate. We use a pulsed laser to obtain successive exposures at separations of 1 ms. We then perform PIV on image pairs to measure the velocity field sampled at 15 Hz in a 6.4 cm $\times$ 4.8 cm region with 200×150 vector resolution. The vorticity component perpendicular to the plane is calculated from the measured 2D velocity field. The PIV measurements are repeated for many vortex breakdown cycles.

3 Wavelet Splitting

We apply wavelet analysis to the vorticity field calculated from the experiment. We follow the technique described in [6] to split the field into two orthogonal components.

The field is cropped to a size 128×128 for use with the discrete wavelet transform (DWT) which takes inputs of size $2^n \times 2^n$ (i.e. here $n = 7$). The DWT of a snapshot of the field is calculated using orthogonal wavelets. The optimal threshold is recursively computed as in [6] on the coefficients of the transform. The large amplitude coefficients above the threshold are taken as the coherent component of the field. We calculate their inverse discrete wavelet transform to obtain the coherent field in physical space. The remaining small amplitude coefficients correspond to the incoherent component of the field. Due to the orthogonality of the transform, the total (i.e. the original measured) field is the sum of the coherent and the incoherent fields. This splitting is repeated for each snapshot of the field.

4 Results

An example of splitting the measured vorticity field into coherent and incoherent components is shown in Fig. 2. The coherent field is comprised of a small number of the coefficients of the DWT, only 4% (i.e. 656 of $128 \times 128 = 16384$ total coefficients), and contains 83% of the enstrophy of the total field. The remaining 96% of the coefficients correspond to the incoherent field, containing 17% of the enstrophy of the total field. The coherent field preserves the same structures and features of the total field while the incoherent field is void of coherent spatial structure (see Fig. 2).

A scatter plot of the vorticity versus stream function indicating the spatial coherence of the fields is shown Fig. 3 (a). For a field that contains coherent structures, such as vortices, the distribution is organized along branches, each approximating a sinh function for a single vortex. This is evident in the long arm of the total and coherent fields observed prior to the vortex burst. As the bursting proceeds, this arm contracts and the scatter plot distribution becomes more compact and closer to the origin. The coherent field matches the behavior of the total field, while the incoherent field remains localized near the origin throughout the bursting due to its spatial incoherence.

A time trace of the statistics of the fields is shown in Fig. 3 (b). The field before the burst containing a solitary vortex is characterized by large

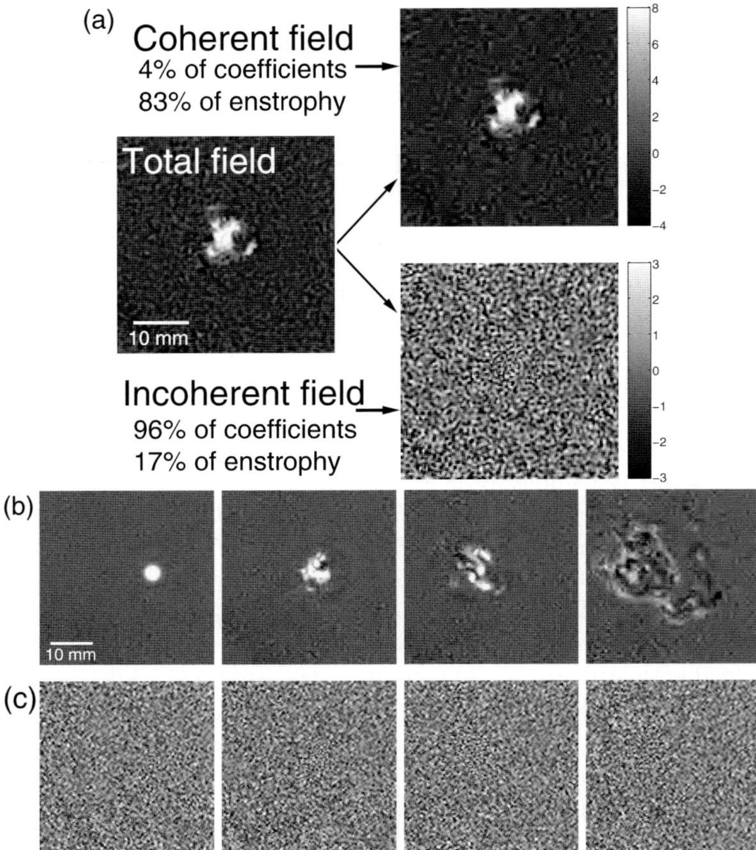

Fig. 2. (a) A split of the measured vorticity field into coherent and incoherent fields at the beginning of a burst. Each field snapshot has been renormalized by its standard deviation. (b) Time evolution of the coherent field and (c) incoherent field. Time proceeds from left to right in intervals of 0.33 seconds while the bursting vortex travels from right to left in the snapshots. The colormaps used for the coherent and incoherent fields are the same as in (a)

values of the variance, flatness, and skewness. A rapid decrease is observed during bursting as the vortex looses its coherence and breaks up. The moments return to their large values when a new vortex appears in the field. The statistics of the coherent field follows closely those of the total field, while the incoherent field remains close to Gaussian throughout the bursting. This can be seen in the probability density function (PDF) of the fields taken during the bursting as shown in Fig. 3 (c). The coherent and total fields have a PDF far from Gaussian with a broad and highly skewed distribution. The PDF of the incoherent fields is more symmetric and closer to Gaussian, as shown in the inset of Fig. 3 (c).

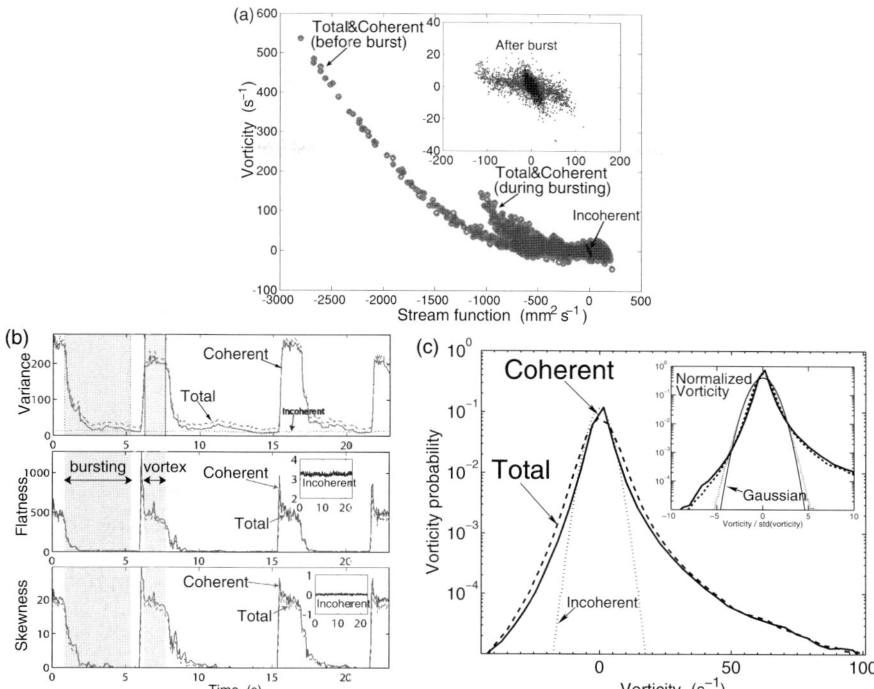

Fig. 3. (**a**) Scatter plot of the stream function versus vorticity showing the coherence of the vorticity field. The total and coherent distributions are nearly indistinguishable while the incoherent remains localized at the origin. The inset shows the fields after the burst. (**b**) Time evolution of the moments of the vorticity field. Plotted are the variance (top), flatness (middle) and skewness (bottom) of the total (dashed line), coherent (solid), and incoherent (dotted) fields during 24 seconds, capturing three bursting events. (**c**) Probability distribution functions of vorticity during the bursting. The inset shows the fields renormalized by their standard deviation

An example of the resulting enstrophy and energy spectra for a single field snapshot is shown in Fig. 4. The coherent spectra match that of the total field and dominate the contribution to the enstrophy in the large and intermediate scales. The incoherent field retains an enstrophy spectra scaling close to that of a random field (k^1 in 2D, corresponding to enstrophy equipartition) and contributes to the total field only in the small scales.

5 Summary

We have split the measured field of a bursting vortex into coherent and incoherent components following the algorithm in [6] using the discrete wavelet transform with orthogonal wavelets. We find that for our experimentally measured field, the coherent component captures the dynamics and statistics of the total field with a relatively small number of coefficients. The incoherent

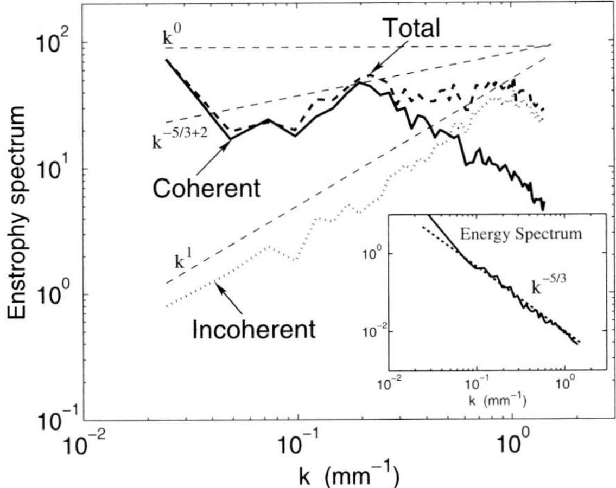

Fig. 4. Enstrophy and energy spectra in the inertial range for a single snapshot of the fields during the bursting. The scaling of the coherent field (solid line) matches the total (dashed) while the incoherent (dotted) remains close to the scaling of Gaussian white noise (k^1 in 2D). The energy spectrum scaling is approximately $k^{-5/3}$

field is void of structure, has near Gaussian statistics, and is relatively insensitive to the bursting.

Future studies will focus on probing the details of the dynamics of the vortex bursting process. We intend to utilize a high speed camera and the continuous wavelet transform to study the time evolution of the bursting and the buildup of the turbulent energy cascade.

References

1. Cuypers Y, Maurel A, Petitjeans P (2003) Phys Rev Lett 91:194502
2. Cuypers Y, Maurel A, Petitjeans P (2004) J Turbul 5:N30
3. Cuypers Y, Maurel A, Petitjeans P (2006) J Turbul 7:N7
4. Lundgren TS (1982) Phys Fluids 25:2193–2203
5. Farge M (1992) Ann Rev Fluid Mech 24:395–457
6. Farge M, Schneider K, Kevlahan N (1999) Phys Fluids 11:2187–2201
7. Ruppert-Felsot JE, Praud O, Sharon E, Swinney HL (2005) Phys Rev E 72:016311

Part III

Turbulence Modeling and Numerical Methods

Renormalized Numerical Simulation of Flow over Planar and 3D Fractal Trees

Stuart Chester and Charles Meneveau

Department of Mechanical Engineering and Center for Environmental and Applied Fluid Mechanics, Johns Hopkins University, Baltimore, MD 21218, USA
`meneveau@jhu.edu`

Abstract. We review the fundamentals of a new numerical modeling technique called Renormalized Numerical Simulation (RNS). It is applied to the calculation of drag forces produced by high Reynolds-number turbulent flow over tree-like fractals. The hallmark of RNS in this application is that the drag of the unresolved tree branches is modeled using drag coefficients measured from the resolved branches and unresolved branches (as modeled in previous iterations of the procedure). RNS results are used to study the influence of the tree fractal dimension on the tree drag. The procedure is generalized from trees in which all branches lie in a plane, to the more general case of 3D trees. Results illustrate that RNS enables numerical modeling of physical processes associated with fractal geometries using affordable computational resolution.

Keywords: fractals, large eddy simulation, turbulence, renormalized numerical simulation

1 Introduction

An inherent difficulty associated with simulating flow that occurs in many environmental systems, e.g., tree canopies [1] and coral reefs [2], is capturing the effects of the broad range of length scales involved in the fluid–boundary interactions. The study of drag produced by idealized fractal boundaries is a natural starting point in the study of drag produced by multi-scale boundaries encountered in environmental flows. However, little is known about the drag forces associated with fractal boundaries in high-Reynolds number (Re) flows. See [3] for a brief account of available literature. Here, we numerically study the drag force exerted by a three-dimensional, very high-Re flow on idealized fractal trees.

1.1 Flow Configuration

The trees to be studied are arranged in an infinite square lattice on a horizontal surface (or ground), resulting in a fully developed flow in the streamwise

235

Y. Kaneda (ed.), *IUTAM Symposium on Computational Physics and New Perspectives in Turbulence*, 235–242.
© 2008 *Springer*.

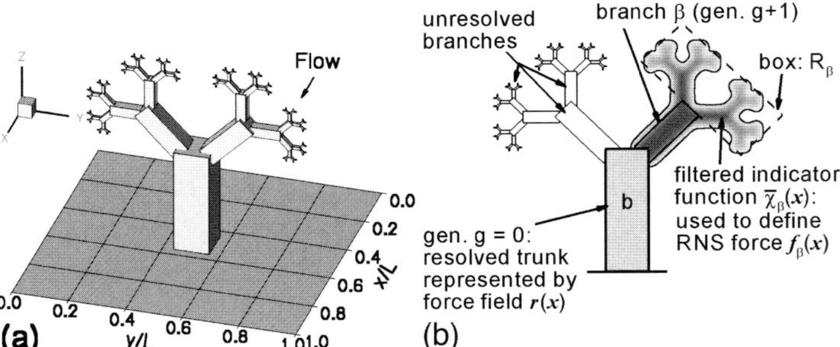

Fig. 1. Schematics of fractal tree considered. (a) Flow configuration. (b) RNS variables. The number of sub-branches per branch is $N_B = 2$, and the scale ratio between branch generations is $r = 0.53$. The similarity fractal dimension [4] is $D = \log N_B / \log r^{-1} \approx 1.09$. The trunk is a square cylinder with diameter (square-edge) $d_0 = L/8$ and height $l_0 = 3L/8$.

direction. For simplicity, we begin by considering trees where the fractal construction is confined to a plane perpendicular to the main flow direction, and we assume the trunk and branches have square cross-sections [3]. A schematic of the highly idealized tree geometry is shown in Figure 1(a). The mean flow is assumed to be along the x-direction, and the trees are spaced a distance L apart from each other in the streamwise (x) and spanwise (y) directions. The plane $z = 0$ defines the horizontal surface, or ground, upon which the trees stand. Our goal is to predict the drag that the trees apply to the flow.

Each tree has multiple branch generations, with generation 0 being the trunk. To each branch at each generation, N_B smaller branches representing the next generation are attached. The branches are square cylinders and all branches at a given generation have the same size. The trees are taken to be self-similar, so there is a constant scale ratio $r < 1$ between successive branch generations, i.e., the length l_g and diameter d_g of a branch at generation $g \geq 0$ are related to those at generation $g + 1$ by $l_{g+1} = rl_g$ and $d_{g+1} = rd_g$.

2 Renormalized Numerical Simulation (RNS)

The RNS technique is a computational tool for modeling the forces of unresolved geometry elements, e.g., drag, in simulations where the geometry is only partially resolved. In RNS, numerical simulation is used to solve the large-scale problem, using measurements made at resolved scales within this simulation to parameterize the unresolved scales.

2.1 Formulation

We consider high-Re flow over a solid geometry that can only be partially resolved by the computational mesh. In the context of the fractal trees from Section 1.1, this means that some branches are large enough to be represented directly in a numerical simulation, without a parameterization. The immersed boundary method [5] is used to represent these resolved branches, via a body force $\mathbf{r}(\mathbf{x})$ (per unit volume). From this point on, let the last resolved branch generation be denoted by g. The force field (per unit volume) applied by the tree on the flow is written as

$$\mathbf{f}(\mathbf{x}) = \mathbf{r}(\mathbf{x}) + \sum_{\beta} \mathbf{f}_{\beta}(\mathbf{x}), \tag{1}$$

where the sum is over all unresolved branches β at generation $g+1$, and $\mathbf{f}_{\beta}(\mathbf{x})$ is a force field (per unit volume) representing effects of these branches and all higher-generation descendant branches of the fractal attached to them. These force fields are defined so that they vanish outside the object they represent: $\mathbf{r}(\mathbf{x}) = 0$ for $\mathbf{x}$ falling outside the resolved trunk or branches, and $\mathbf{f}_{\beta}(\mathbf{x}) = 0$ for $\mathbf{x}$ falling outside the region covered by the branch β and its descendants (some numerical smoothing is involved, as will be discussed later), see Figure 1(b). The force $\mathbf{F}_{\beta}$ applied by a given branch β and its descendants through the distributed force field $\mathbf{f}_{\beta}(\mathbf{x})$ is

$$\mathbf{F}_{\beta} = \int \mathbf{f}_{\beta}(\mathbf{x}) \, \mathrm{d}^3\mathbf{x}. \tag{2}$$

Similarly, the force due to a generation-g branch b and its descendants is

$$\mathbf{F}_{b} = \int \left[\mathbf{r}_{b}(\mathbf{x}) + \sum_{\beta \in \mathrm{sub}(b)} \mathbf{f}_{\beta}(\mathbf{x}) \right] \mathrm{d}^3\mathbf{x}, \tag{3}$$

where $\mathbf{r}_{b}$ is the immersed boundary force field due to b, and $\mathrm{sub}(b)$ is the set of generation-$(g+1)$ sub-branches attached to b. The goal of RNS is the determination of the force fields $\mathbf{f}_{\beta}(\mathbf{x})$ at positions corresponding to the unresolved branches at generations $g+1$ and above. The essential feature of RNS is that it relates the total force $\mathbf{F}_{\beta}$ on the fluid due to the generation-$(g+1)$ branch β, and its descendants, to the total force due to the (resolved) generation g parent branch b and its descendants.

At time step m, the force $\mathbf{F}_{\beta}^{m}$ is modeled using the drag law

$$\mathbf{F}_{\beta}^{m} = -c_{\mathrm{D}}^{m}(g+1) \frac{\rho}{2} |\mathbf{V}_{\beta}^{m}| \mathbf{V}_{\beta}^{m} \, A_{\beta}, \tag{4}$$

where $c_{\mathrm{D}}^{m}(g+1)$ is the drag coefficient of all β (and their descendants). In the present implementation the drag coefficient is assumed to be the same at all β - more detailed implementation may distinguish between branches of different 'type' (such as branches closer to the ground, etc.). $\mathbf{V}_{\beta}^{m}$ is the spatially averaged velocity vector in a region R_{β} in the neighborhood of β and its descendants, and $A_{\beta} = ld/(1 - N_B r^2)$ is the projected area of branch β and its descendants as seen by the bulk streamwise flow. The region R_{β} is chosen to be the volume upstream of β shown as a dashed outline in Figure 1(b), containing fluid that is likely to come into contact with β or its descendants. The downstream face of R_β passes through the center of the tree (in the streamwise direction), and R_β extends upstream a distance $[d+2rl/(1-r)]/4$. The drag coefficient $c_D^m(g+1)$ in Eq. (4) is unknown because of the complexity of the flow around the branches. In incompressible flow, $c_D^m(g+1)$ can depend only on the geometric flow configuration and Re [6], and here we assume Re is high enough that $c_D^m(g+1)$ is Re-independent. Due to the self-similar geometry, we may thus invoke complete similarity and write that $c_D^m(g+1) = c_D^m(g)$, i.e., $c_D^m(g+1)$ can be measured from the branches at generation g. This is justified for $g \gg 1$, where 'outer' boundary conditions (such as proximity to the ground and imposed shear) are negligible. However in this work, as an approximation RNS will be applied even for low g.

To measure $c_D^m(g)$, the model in Eq. (4) is written for each branch b at generation g, and $c_D^m(g)$ is chosen to minimize the total square error ε:

$$\varepsilon = \sum_{b(g)} \left[\mathbf{F}_b^m(c_D^{m-1}) + c_D^m(g)\,\frac{\rho}{2}|\mathbf{V}_b^m|\mathbf{V}_b^m A_b \right]^2, \tag{5}$$

where $\sum_{b(g)}$ signifies that the sum includes every branch b at the last resolved generation g. In Eq. (5), $\mathbf{F}_b^m$ includes both the resolved force due to branch b as well as the modeled forces from its unresolved descendants. The latter are calculated using the value of $c_D^{m-1}(g+1)$ that is known from the previous time step. To make this dependence explicit, we have written $\mathbf{F}_b^m = \mathbf{F}_b^m(c_D^{m-1})$. Solving for the unknown average drag coefficient by setting $\partial\varepsilon/\partial c_D^m = 0$ yields

$$c_D^m(g) = -\frac{2\sum_{b(g)}\left[\mathbf{F}_b^m(c_D^{m-1})\cdot\mathbf{V}_b^m\right]|\mathbf{V}_b^m|A_b}{\rho\sum_{b(g)}|\mathbf{V}_b^m|^4 A_b^2}. \tag{6}$$

The iterations in Eq. (6) at time 0 are initialized using $c_D^0(g+1) = 0$.

We now determine how to distribute the force $\mathbf{F}_\beta$ on the computational mesh. We assume the force on the fluid at point $\mathbf{x}$ at time step m due to unresolved branches depends quadratically on the local velocity, and opposes it:

$$\mathbf{f}_\beta^m(\mathbf{x}) = -k_\beta^m\,|\tilde{\mathbf{u}}^m(\mathbf{x})|\,\tilde{\mathbf{u}}^m(\mathbf{x})\bar{\chi}_\beta(\mathbf{x}), \tag{7}$$

where k_β is a dimensional coefficient to be determined later, and where $\bar{\chi}_\beta$ is a Gaussian-filtered version of the branch indicator function χ_β (χ_β itself cannot be directly represented on the grid), see Figure 1(b). What remains to be done is to relate the unknown coefficient. k_β^m to the drag coefficient $c_D^m(g+1)$ such that the total given force from the distributed force of Eq. (7) equals the total force as implied by Eq. (4). Substitution of Eq. (7) into Eq. (2), using Eq. (4), and again using a least-squares error approach, leads to

$$k_\beta^m = c_D^m(g+1)\,\frac{\rho|\mathbf{V}_\beta^m|A_\beta(\mathbf{V}_\beta^m\cdot\mathbf{I}_\beta^m)}{2|\mathbf{I}_\beta^m|^2}, \tag{8}$$

where $\mathbf{I}_\beta^m = \int |\tilde{\mathbf{u}}^m(\mathbf{x})|\,\tilde{\mathbf{u}}^m(\mathbf{x})\bar{\chi}_\beta(\mathbf{x})\,\mathrm{d}^3\mathbf{x}$.

2.2 Simulation Details

High-Re turbulence in the bulk of the flow surrounding the trees is simulated using the LES technique. All simulations use the Smagorinsky SGS model [7] with prescribed coefficient $c_s = 0.16$. Details of the numerics and the immersed boundary implementation are given in [3]. Tests with our immersed boundary implementation show that a resolution of at least $d/h = 8$ is required to accurately capture the forces acting on a branch, and we consider a branch with this resolution to be resolved. The simulation domain is a cube with edge length L with a tree based at the center of the bottom wall. The top of the domain ($z = L$) is treated as a stress-free, impenetrable boundary, and periodic boundary conditions are applied at the sides of the domain to mimic the effect of an square infinite lattice of trees. Flow is forced through the domain parallel to the x-axis by applying a constant mean pressure gradient.

2.3 Results

RNS is used to perform simulations over the tree geometry using a resolution of only 64^3 grid points, resulting in a trunk diameter of $d = 8h$, which is just sufficient to resolve the tree trunk. All other branches are unresolved and modeled using the RNS technique, i.e., $g = 0$. To reduce CPU time, the RNS-determined $c_D(g = 0)$ is only updated once every five time steps (tests showed that results do not differ from those using updates at every time step). A snapshot of the RNS velocity field obtained using $r = 0.53$ is shown in Figure 2(a). The Y-shaped regions of slowed flow corresponding to positions of unresolved branches are due to the way the RNS force is non-uniformly distributed.

A series of RNS using different tree fractal dimensions was performed by varying the scale ratio r over the range $r = 0.20, 0.25, \ldots, 0.53$. As a global

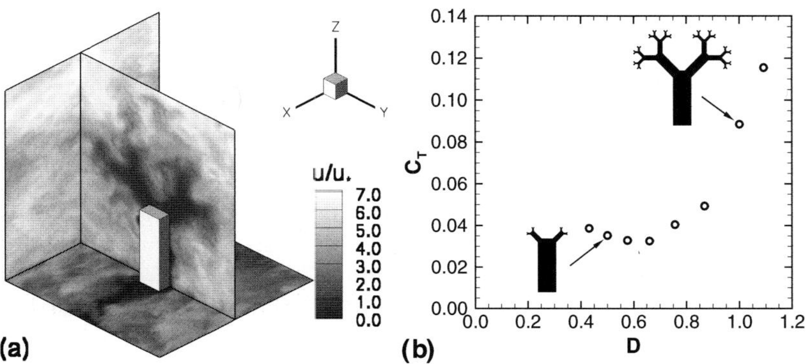

Fig. 2. (a) Resolved tree geometry, and instantaneous streamwise velocity-field slices from 64^3 RNS with $g = 0$ and $r = 0.53$. (b) Total drag coefficient C_T versus fractal dimension D for Y-trees, with schematics of the frontal tree geometry.

measure of the tree drag, the time-averaged effective total drag coefficient $C_\mathrm{T} = \bar{F}_T/(\frac{1}{2}\rho U^2 L^2)$, based on the frontal flow domain area L^2 is used, where $\bar{F}_T$ is the time-averaged total x-direction drag force the fluid applies on the tree, and U is the domain-mean streamwise velocity. The resulting C_T from this series are shown in Figure 2(b), plotted as a function of the tree fractal dimension D. The increase of C_T with D is significant for the larger values of D tested, as expected due to the increase in frontal area of the tree. However, RNS also predicts a slight decrease in C_T at low values of D, and further study is needed to determine the reason for this trend. More applications to other planar fractal trees are described in [3].

3 RNS For More General Geometries

We now explore a way to use the RNS methodology from Section 2 to simulate flow over idealized trees without restricting the branches to be perpendicular to the incoming flow. The starting point for performing RNS analysis on trees with arbitrary orientations is the crossflow principle [8]. In accordance with the crossflow principle, the drag model Eq. (4) underlying the RNS procedure is replaced by a more general model:

$$\mathbf{F}_\beta^m = -\frac{\rho}{2}\left(c_\mathrm{n}^m|\mathbf{V}_{\mathrm{n},\beta}^m|\mathbf{V}_{\mathrm{n},\beta}^m + c_\mathrm{a}^m|\mathbf{V}_{\mathrm{a},\beta}^m|\mathbf{V}_{\mathrm{a},\beta}^m\right)A_\beta, \qquad (9)$$

where the superscript m denotes the time step, c_n is the normal force coefficient of branch β and its descendants, c_a is the corresponding axial force coefficient, $\mathbf{V}_{\mathrm{n},\beta}$ and $\mathbf{V}_{\mathrm{a},\beta}$ are normal and axial projections of the velocity scale $\mathbf{V}_\beta$, and A_β is representative of the area that β and its descendants present to the flow. For simplicity, we now take $\mathbf{V}_\beta$ to be the average velocity in the entire domain. The area A_β is now taken as $\mathcal{V}_\beta^{2/3}$, where $\mathcal{V}_\beta = ld^2/(1 - N_B r^3)$ is the approximate volume of branch β and its descendants. The procedure for determining the model coefficients c_n and c_a, and the spatial force distribution is similar to that given in Section 2.

3.1 Application: Tripod Tree

The normal–axial RNS formulation is applied here to study flow over a the "tripod" geometry shown in Figure 3(a). The scale ratio is $r = 1/2$. Since resolving only the trunk would not provide robust resolved-scale information about the axial forces (they are nearly zero), this RNS is performed using $g = 1$. The computational mesh consists of $N_x \times N_y \times N_z = 256 \times 128 \times 128$ grid points in a $L_x \times L_y \times L_z = 2L \times L \times L$ flow domain. The tree is based on the lower wall at a distance $L/2$ from the upstream boundary, in the center of the domain in the cross-stream direction. A uniform inflow is prescribed upstream of the tree using a fringe region [9], while periodic boundary conditions are used for the lateral y boundaries. The top and bottom boundary conditions are the same as in Section 2.2.

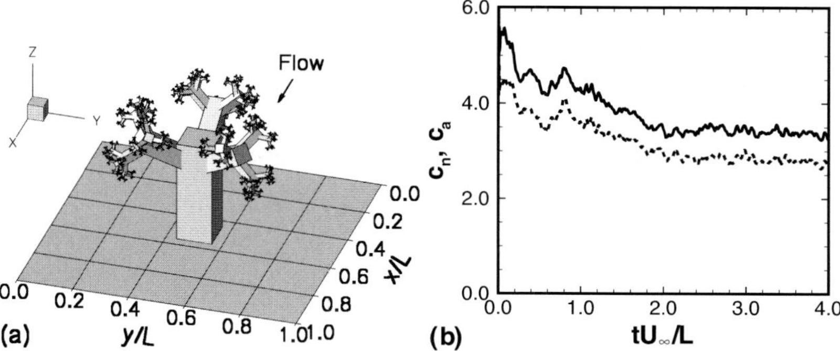

Fig. 3. (a) Tripod geometry studied using normal–axial RNS. The scale ratio between branch generations is $r = 1/2$, and there are $N_B = 3$ sub-branches per branch. (b) Time history of normal and axial force coefficients, dynamically obtained during RNS using the tripod geometry with $g = 1$ and the free-stream velocity scale. Solid line, normal force coefficient c_n; dashed line, axial force coefficient c_a.

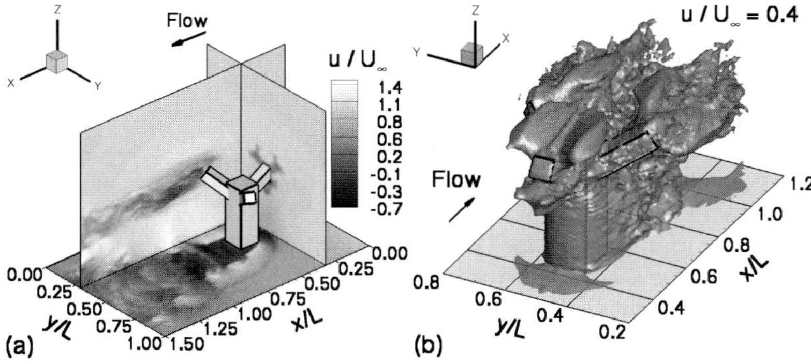

Fig. 4. Snapshots from normal–axial RNS using tripod geometry, free-stream velocity scale, and $g = 1$. The resolved tree geometry is outlined in black. (a) Planes showing instantaneous streamwise velocity. Note the horseshoe structure visible at the base of the trunk. (b) Isosurface of instantaneous streamwise velocity, $u/U_\infty = 0.4$.

The time history of the force coefficients, seen in Figure 3(b), shows that RNS produces statistically stable values (after the initial transient). RNS predicts different coefficients c_n and c_a for the two directions. An overall drag coefficient of $C_T = 0.16$ was obtained by averaging over the last half of the data. In this test, 42% of the total force on the tree comes from unresolved branches, underscoring the significance of accounting for the unresolved branches. Slices through the instantaneous velocity field are shown in Figure 4(a). The planes are positioned to show the slowing effect of unresolved branches. An isosurface of streamwise velocity (u) is shown in Figure 4(b). The resolved geometry is indicated with the black outline, and is visible through the translucent isosurface. The resolved branch closest to the viewer has three lobes of low speed fluid projecting from its tip that correspond to locations of unresolved branches. On these lobes, sub-lobes at positions of the second-generation sub-branches can be seen. Similar lobes are present around the two resolved branches at the rear, but they are less clear due to the turbulent flow around them. These results show that RNS can be applied to more realistic tree geometries than those in Section 2.3.

4 Summary

Turbulent flow over periodic arrangements of idealized fractal trees has been studied via a series of numerical simulations. To avoid great computational expense associated with directly simulating such multi-scale systems, a new technique, RNS, has been proposed. The technique was applied to study how the drag on fractal trees depends on their fractal dimension D. A more general formulation of RNS based on the crossflow principle was also tested. The results presented here suggest that RNS may be a useful technique for tackling other multi-scale flow problems, e.g., porous media flow and atmospheric flow over mountainous topography, at affordable computational cost.

Financial support: National Science Foundation (ATM-0222238).

References

1. Shaw RH, Patton EG (2003) Agr Forest Meteorol 115:5–17
2. Kaandorp JA, Koopman EA, Sloot PMA, Bak RPM, Vermeij MJA, Lampmann LEH (2003) Phil Trans R Soc Lond B 358:1551–1557
3. Chester S, Meneveau C, Parlange MB (2006) J Comput Phys (in press)
4. Mandelbrot BB (1982) The Fractal Geometry of Nature, WH Freeman, New York
5. Mittal R, Iaccarino G (2005) Annu Rev Fluid Mech 37:239–261
6. Landau LD, Lifshitz EM (1987) Fluid Mechanics Butterworth-Heinemann, Oxford, 2nd edition
7. Smagorinsky J (1963) Mon Weather Rev 91:99–164
8. Jones RT (1947) Technical Report 884, NACA
9. Bertolotti FP, Herbert T, Spalart PR (1992) J Fluid Mech 242:441–474

Wavelet-Based Extraction of Coherent Vortices from High Reynolds Number Homogeneous Isotropic Turbulence

Katsunori Yoshimatsu[1], Naoya Okamoto[1], Kai Schneider[2], Marie Farge[3], and Yukio Kaneda[1]

[1] Department of Computational Science and Engineering, Graduate School of Engineering, Nagoya University, Chikusa-ku, Nagoya 464-8603, Japan
yosimatu@fluid.cse.nagoya-u.ac.jp, okamoto@fluid.cse.nagoya-u.ac.jp, kaneda@cse.nagoya-u.ac.jp
[2] MSNM-CNRS & CMI, Université de Provence, 39 rue Frédéric Joliot-Curie, 13453 Marseille Cedex 13, France
kschneid@cmi.univ-mrs.fr
[3] LMD-IPSL-CNRS, Ecole Normale Supérieure, 24 rue Lhomond, 75231 Paris Cedex 05, France
farge@lmd.ens.fr

Abstract. A wavelet-based method to extract coherent vortices is applied to data of three-dimensional incompressible homogeneous isotropic turbulence with the Taylor micro-scale Reynolds number 471 in order to examine contribution of the vortices to statistics on the turbulent flow. We observe a strong scale-by-scale correlation between the velocity field induced by them and the total velocity field over the scales retained by the data. We also find that the vortices almost preserve statistics of nonlinear interactions of the total flow over the inertial range.

Keywords: wavelet, coherent vortices, high Reynolds number turbulence, energy transfer

1 Introduction

Wavelet techniques to analyze turbulence have been pioneered at the end of 1980ies [1, 2]. They have been developed and exploited to reveal nature of turbulence, to model it and to solve the Navier-Stokes equation directly in wavelet space [3, 4, 5, 6].

A wavelet-based coherent vortex extraction (CVE) method for two-dimensional flows have been introduced by the use of orthogonal wavelet decompositions [7]. This has been extended to three-dimensional flows [8, 9]. Coherent Vortex Simulation (CVS) proposed in [7] is based on deterministic simulation of the flow due to the coherent vortices by the use of an *adaptive* wavelet basis, while influence of incoherent background flow onto the coherent flow is

243

Y. Kaneda (ed.), *IUTAM Symposium on Computational Physics and New Perspectives in Turbulence*, 243–248.

neglected or statistically modeled. CVSs of two-dimensional flows and three-dimensional turbulent mixing layers is presented in [10] and [11], respectively.

In this paper, the wavelet-based CVE algorithm is applied to data obtained by DNS of three-dimensional incompressible homogeneous isotropic turbulence at resolution 1024^3 which corresponds to the Taylor micro-scale Reynolds number of 471 performed on the Earth Simulator [12, 13]. We examine coherent contribution to statistics of the turbulent flow, especially the energy transfer, as part of *a prior* test on a CVS of three-dimensional high Reynolds number turbulent flow.

2 Wavelet Analysis and CVE

2.1 Vector Valued Orthogonal Wavelet Decomposition

We consider a vector field $\mathbf{v}(\mathbf{x}) = (v_1(\mathbf{x}), v_2(\mathbf{x}), v_3(\mathbf{x}))$ in $\mathbf{T}^3 = [0, 2\pi]^3$. Three-dimensional orthonormal wavelet analysis unfolds $\mathbf{v}$ into scale, positions and directions using a mother wavelet Ψ_m constructed by tensor product of one-dimensional scaling function $\psi_0(x)$ and mother wavelet $\psi_1(x)$ as $\Psi_m(\mathbf{x}) = \psi_\xi(x_1)\psi_\eta(x_2)\psi_\zeta(x_3)$ ($\xi, \eta, \zeta = 0, 1$, and $m = \xi + 2\eta + 4\zeta$). We use the Coiflet 12, which is compactly supported, quasi-symmetric, defined with a filter of length 12, and has four vanishing moments. The field sampled on 2^J equidistant grid points in each space direction of the Cartesian coordinates can thus be decomposed into an orthogonal wavelet series:

$$\mathbf{v}(\mathbf{x}) = \overline{\mathbf{v}} + \sum_{\alpha=1}^{J} \mathbf{v}_\alpha(\mathbf{x}), \text{ and } \mathbf{v}_\alpha(\mathbf{x}) = \sum_{\iota_1,\iota_2,\iota_3=0}^{2^{\alpha-1}-1} \mathcal{W}_{m,\boldsymbol{\iota}}^\alpha [\mathbf{v}]\, \Psi_{m,\boldsymbol{\iota}}^\alpha(\mathbf{x}), \qquad (1)$$

where $\overline{\mathbf{v}} = \int_{\mathbf{T}} \mathbf{v}(\mathbf{x})\, d\mathbf{x}/(2\pi)^3$, $\Psi_{m,\boldsymbol{\iota}}^\alpha(\mathbf{x}) = 2^{3\alpha/2}\Psi_m(2^\alpha \mathbf{x} - 2\pi\boldsymbol{\iota})$, $m = 1, 2, \cdots, 7$ and $\boldsymbol{\iota} = (\iota_1, \iota_2, \iota_3)$. The l-th component of $\mathcal{W}_{m,\boldsymbol{\iota}}^\alpha [\mathbf{v}]$ is given by $\int_{\mathbf{T}} v_l(\mathbf{x})\Psi_{m,\boldsymbol{\iota}}^\alpha(\mathbf{x})\, d\mathbf{x}/(2\pi)^3$. The summation convention is used for repeated indices but not for the Greek indices. $\overline{\mathbf{v}_\alpha} = 0$ because $\overline{\Psi_{m,\boldsymbol{\iota}}^\alpha} = 0$. Readers interested in details on orthogonal wavelet transform may refer to, e.g. [2, 14].

2.2 CVE

A wavelet-based method to extract coherent vortices from two- and three-dimensional turbulent flows has been proposed [7, 8]. An orthogonal wavelet decomposition is applied to the vorticity field $\boldsymbol{\omega}$. A threshold based on denoising theory [15] splits the wavelet coefficients into two sets. The coherent vorticity $\boldsymbol{\omega}_\mathrm{C}$ is reconstructed from few wavelet coefficients whose moduli are larger than a given threshold depending on the enstrophy and resolution of the field. After applying the method, we obtain two orthogonal fields: the coherent vorticity $\boldsymbol{\omega}_\mathrm{C}$ and the incoherent vorticity $\boldsymbol{\omega}_\mathrm{I}$. We also reconstruct the coherent and incoherent velocity fields induced by the coherent and incoherent vorticity fields, respectively. Readers interested in the details may refer to [7, 8, 9, 16, 17].

3 CVE from Homogeneous Isotropic Turbulence

We apply the CVE method to data obtained by DNS of three-dimensional homogeneous isotropic turbulence performed on the Earth Simulator [12, 13]. The DNS fields obey the Navier-Stokes equations for incompressible fluid in a periodic box with sides 2π. We use data for $k_{\mathrm{max}}\eta \simeq 1$ at resolution $N = 1024^3$ and the Taylor miroscale Reynolds number $R_\lambda = 471$. Here k_{max} is the maximum wavenumber retained in the DNS, and η is the Kolmogorov length scale.

We find that the coherent vortices are represented by the $2.9\%N$ wavelet coefficients of the total vorticity field. The flow induced by the vortices retains 99.7% of the energy and 81.0% of the enstrophy. Figure 1 shows that the coherent vorticity well retains the vortex tubes observed in the total vorticity. In contrast, we observe no vortex tube in the incoherent vorticity field.

The probability density functions (PDFs) and energy spectra of velocity of the total, coherent and incoherent flows are shown in Fig. 2 (left). All velocity PDFs exhibit quasi-Gaussian distributions. The total and coherent velocity

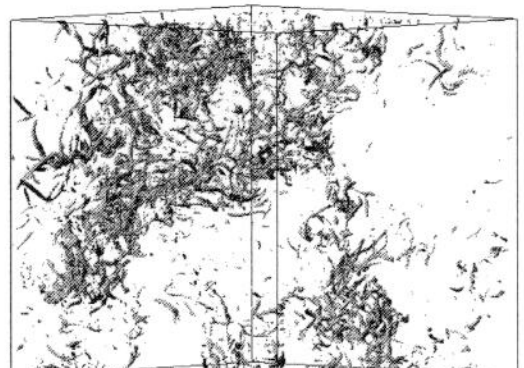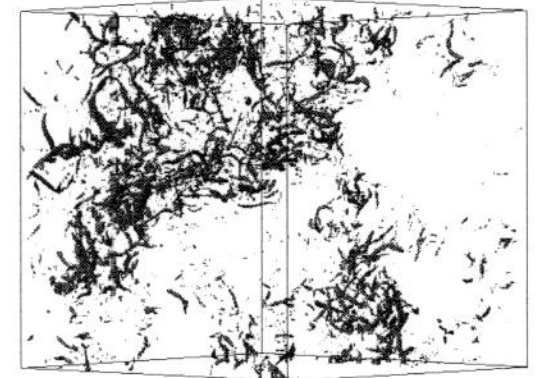

Fig. 1. Isosurfaces of total (left), coherent (middle) and incoherent (right) vorticity. The values of the isosurfaces are $|\boldsymbol{\omega}| = \omega_m + 3\sigma_\omega$ for the total and coherent vorticity and $2(\omega_m + 3\sigma_\omega)/5$ for the incoherent vorticity. ω_m and σ_ω are the mean value of $|\boldsymbol{\omega}|$ and the standard deviation of $|\boldsymbol{\omega}|$, respectively. Subcubes of size 256^3 are visualized.

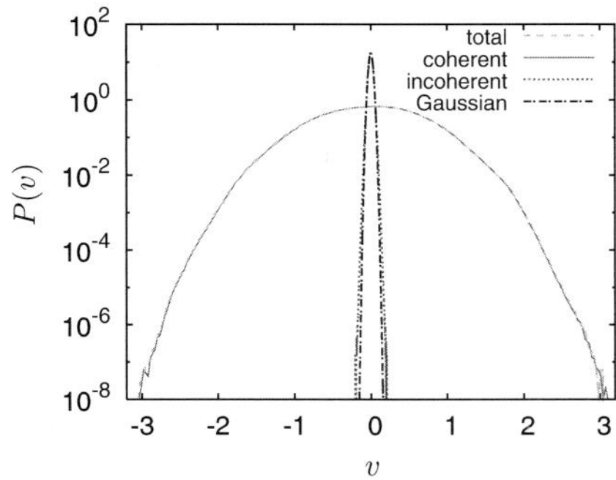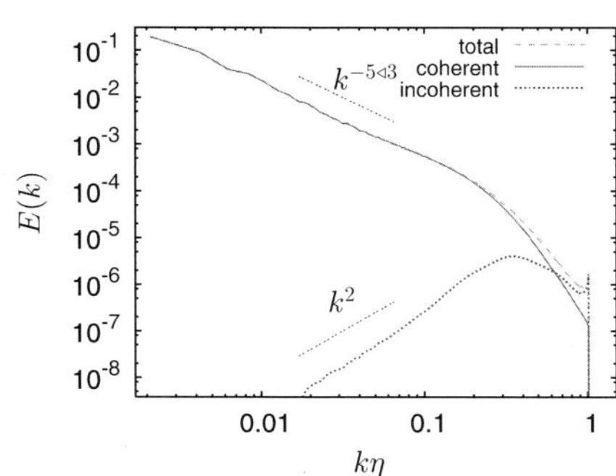

Fig. 2. Velocity PDFs (left) and energy spectra of the total, coherent and incoherent flow (right).

PDFs well coincide, while the incoherent one has a strongly reduced variance. The spectrum of the coherent energy is identical to that of the total velocity all along the inertial range, presenting a $k^{-5/3}$ scaling, and that the spectrum of the coherent energy only differs from the one of the total energy in the dissipative range, where k is the wavenumber. For the incoherent velocity, we observe that $E(k)$ scales as k^2, which corresponds to energy equipartition.

4 Contribution of Coherent Vortices to Energy Transfer

In order to examine contributions of the coherent vortices to energy transfer of the total flow, we consider scale-by-scale correlation between the vector fields due to the total flow and induced by the coherent vortices, and also the net energy transfer to scale α ($\alpha = 1, 2, \cdots, \log_2 N^{1/3}$) at location indexed by ι proposed in [18].

The scale-by-scale correlation between vector fields $\mathbf{A}(\mathbf{x})$ and $\mathbf{B}(\mathbf{x})$ is defined by

$$C_\alpha(\mathbf{A}, \mathbf{B}) = \frac{\overline{\mathbf{A}_\alpha \cdot \mathbf{B}_\alpha}}{\sqrt{\overline{|\mathbf{A}_\alpha|^2}} \sqrt{\overline{|\mathbf{B}_\alpha|^2}}}. \tag{2}$$

Figure 3 shows the correlation $C_\alpha(\mathbf{v}_\mathrm{T}, \mathbf{v}_\mathrm{C})$ and $C_\alpha(\mathbf{I}_\mathrm{T}, \mathbf{I}_\mathrm{C})$ vs. $k_\alpha \eta$, where $\mathbf{v}_\mathrm{T}$ ($\mathbf{v}_\mathrm{C}$) is the total (coherent) velocity field, $\mathbf{I}_\beta = (\mathbf{v}_\beta \cdot \nabla)\mathbf{v}_\beta + \nabla p_\beta/\rho$, $\beta \in (\mathrm{T}, \mathrm{C})$, p_β the pressure obtained from $p_\beta = -\rho \nabla^{-2}[\nabla \cdot \{(\mathbf{v}_\beta \cdot \nabla)\mathbf{v}_\beta\}]$, ρ fluid density. $k_\alpha = 2^{\alpha-1}/1.3$. $1/1.3$ for k_α is the centroid wavenumber of the Coifman 12. We find that the correlations are strong over all inertial scales. In the dissipation range, the latter is weaker than the former, and both of them decrease with increasing k_α.

The net energy transfer is given by $t_{\alpha,\beta}[\iota] = -\mathcal{W}^\alpha_{m,\iota}[\mathbf{v}_\beta] \cdot \mathcal{W}^\alpha_{m,\iota}[\mathbf{I}_\beta]$. Figure 4 shows mean and standard deviation wavelet spectra of the net energy transfers for the total and coherent flows. The spectra are normalized by $(\langle \epsilon \rangle \nu)^{3/4}$. Here $\langle \epsilon \rangle$ is the mean energy dissipation rate per unit mass of the total flow and ν is the kinematic viscosity. The mean wavelet spectrum $\tau_\beta(k_\alpha)$ is defined by $\tau_\beta(k_\alpha) = N_\alpha \langle t_{\alpha,\beta}[\iota] \rangle /\Delta k_\alpha$. Here, $N_\alpha = 2^{3(\alpha-1)}$ and

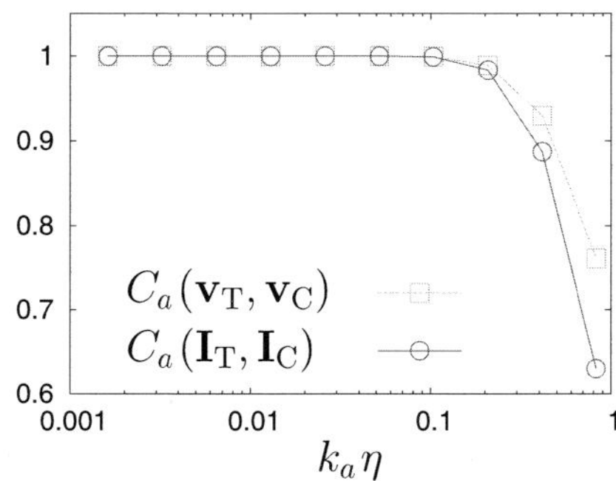

Fig. 3. $C_\alpha(\mathbf{v}_\mathrm{T}, \mathbf{v}_\mathrm{C})$ and $C_\alpha(\mathbf{I}_\mathrm{T}, \mathbf{I}_\mathrm{C})$ vs. $k_\alpha \eta$.

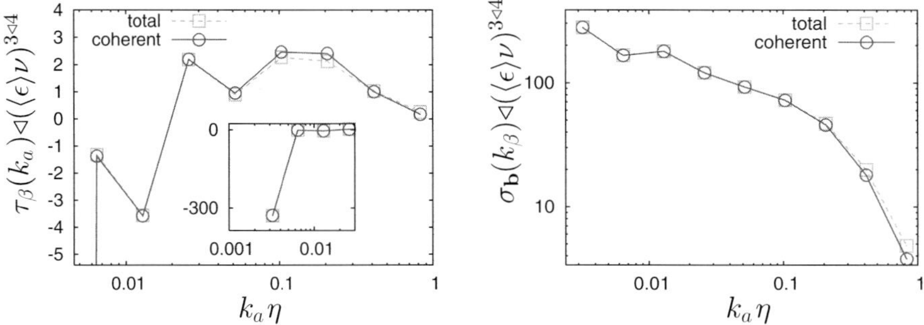

Fig. 4. Mean spectra (left) and standard deviation spectra (right) of the net energy transfer of the total and coherent flow. The standard deviation spectra are shown for $\alpha \geq 2$.

$\Delta k_\alpha = (k_{\alpha+1} - k_\alpha)\ln 2$. $\langle \cdot \rangle$ denotes the mean value of $\cdot$ at each scale. The spatial variability of the energy transfer at each scale is measured by the standard deviation wavelet spectrum defined by

$$\sigma_\beta(k_\alpha) = N_\alpha \sqrt{\langle \{t_{\alpha,\beta}[\boldsymbol{\iota}]\}^2 \rangle - \langle t_{\alpha,\beta}[\boldsymbol{\iota}] \rangle^2}/\Delta k_\alpha. \tag{3}$$

In Fig. 4 we find that $\tau_{\rm C}(k_\alpha)$ and $\sigma_{\rm C}(k_\alpha)$ well coincides with $\tau_{\rm T}(k_\alpha)$ and $\sigma_{\rm T}(k_\alpha)$ in the inertial range, while, for $k\eta \gtrsim 0.1$, we observe a small discrepancy between $\tau_{\rm C}(k_\alpha)$ $(\sigma_{\rm C}(k_\alpha))$ and $\tau_{\rm T}(k_\alpha)$ $(\sigma_{\rm T}(k_\alpha))$.

5 Conclusions

The wavelet-based method to extract coherent vortices has been applied to the data of three-dimensional incompressible homogeneous isotropic turbulence at resolution 1024^3 grid points and $R_\lambda = 471$. The energy spectrum of the coherent flow is in good agreement with that of the total flow all along the inertial range. We observe a strong scale-by-scale correlation between the velocity field induced by the coherent vortices and the total velocity field at all scales. The statistics of nonlinear interactions are almost preserved by the coherent flow all along the inertial range. The present results encourage further development of the CVS of high Reynolds number turbulent flows. Dependence of contribution of coherent vortices to total fields on Reynolds number will be reported in another paper.

The computations were carried out on HPC2500 system at the Information Technology Center of Nagoya University. This work was supported by a Grant-in-Aid for the 21st Century COE "Frontiers of Computational Science", Grant-in-Aids for Scientific Research (B)17340117 from the Japan Society for the Promotion of Science and IPSL, Paris. We would like to express their thanks to T. Ishihara, M. Yokokawa, K. Itakura and A. Uno for providing us with the DNS data. We also thank G. Pellegrino for providing his serial wavelet decomposition code (field16 written for a single processor) and T. Ishihara for support of the DNS data handling and for providing a FFT code that he parallelized in collaboration with T. Aoyama and M. Osada.

References

1. Farge M (1989) J Fluid Mech 206:433-462
2. Farge M (1992) Ann Rev Fluid Mech 24:395-457
3. van den Berg JC (Ed.) (1999) Wavelets in Physics. Cambridge University Press
4. Farge M, Schneider K (2002) New Trends in Turbulence. Les Houches 2000, Vol. 74 (Eds. Lesieur M, Yaglom A, David F), Springer: 449-503
5. Addison PS (2002) The Illustrated Wavelet Transform Handbook. Institute of Physics Publishing, Bristol and Philadelphia
6. Farge M, Schneider K (2006) In Encyclopedia of Mathematical Physics (Eds. Françoise JP, Naber G, Tsun TS), Elsevier:408-420
7. Farge M, Schneider K, Kevlahan N (1999) Phys Fluids 11:2187-2201
8. Farge M, Pellegrino G, Schneider K (2001) Phys Rev Lett 87:45011-45014
9. Farge M, Schneider K, Pellegrino G, Wray A, Rogallo B (2003) Phys Fluids 15: 2886-2896
10. Farge M, Schneider K (2001) Flow Turbul Comb 66: 393-426
11. Schneider K, Farge M, Pellegrino G, Rogers M (2005) J Fluid Mech 534: 39-66
12. Yokokawa M, Itakura K, Uno A, Ishihara T, Kaneda Y (2002) Proc IEEE/ACM SC2002 Conf, Baltimore; http://www.sc-2002.org/paperpdfs/pap.pap273.pdf
13. Kaneda Y, Ishihara T, Yokokawa M, Itakura K, Uno A (2003) Phys Fluids 15:L21-L24
14. Mallat S (1998) A Wavelet Tour of Signal Processing. Academic Press
15. Donoho D, Johnstone I (1994) Biometrika 81:425-455
16. Schneider K, Farge M, Azzalini A, Ziuber J (2006) J Turbul 7(44)
17. Azzalini A, Farge M, Schneider K (2005) Appl Comput Harm Anal 18: 177-185
18. Meneveau C (1991) J Fluid Mech 232:469-520

Decaying 2D Turbulence in Bounded Domains: Influence of the Geometry

Kai Schneider[1] and Marie Farge[2]

[1] MSNM–CNRS & CMI, Université de Provence, Marseille, France
   kschneid@cmi.univ-mrs.fr
[2] LMD–CNRS, Ecole Normale Supérieure, Paris, France
   farge@lmd.ens.fr

Abstract. We present direct numerical simulation of two-dimensional decaying turbulence in wall bounded domains. The Navier–Stokes equations are solved in a periodic square domain using the vorticity–velocity formulation. The bounded domain is imbedded in the periodic domain and the no–slip boundary conditions on the wall are imposed using a volume penalisation technique. The numerical integration is done with a Fourier pseudo–spectral method combined to a semi–implicit time discretization with adaptive time stepping. We study the influence of the geometry of the domain on the flow dynamics and in particular on the long time behaviour of the flow. We consider different geometries, a circle, a square, a triangle and a torus and we show that the geometry plays a crucial role for the decay scenario.

Keywords: 2D turbulence, penalisation, bounded geometry, long time decay

1 Introduction

Two–dimensional turbulence in wall bounded domains has many applications in geophysical flows, *e.g.* in oceanography and in planetary flows. Direct numerical simulations of two–dimensional turbulence in circular and square domains can be found, *e.g.* in [1, 2, 3]. The aim of the present paper is to study the influence of the geometry of the domain on the flow dynamics and in particular on the long time behaviour of the flow. Typically one observes the formation of stable large scale structures which persist for a long time before they are finally dissipated. Viscous dissipation is the dominant mechanism of these final states, as the nonlinear term in the Navier–Stokes equations is depleted when there is a functional relationship between the streamfunction and the vorticity. Late states of decaying two–dimensional flows in periodic boxes have for instance been investigted in [4]. Here we study the final states of wall bounded flows considering different geometries, a circle, a square, a triangle and a torus. Several theoretical predictions of the long time behaviour of two–

249

Y. Kaneda (ed.), *IUTAM Symposium on Computational Physics and New Perspectives in Turbulence*, 249–253.

dimensional flows can be found, *e.g.* in the book of Davidson [5]. Variational principles for predicting the final state are based on conservation of energy E and the decay of enstrophy Z. In this heuristic approach Z is minimized under constraint of conservation of E [6]. Another variational hypothesis is motivated by statistical mechanics. A measure of mixing can be introduced which leads to the definition of an entropy. The final states correspond to a maximum of entropy as turbulence maximizes mixing [7, 8, 9]. Another approach based on viscous eigenmodes of the Stokes flow has been used in [10] to predit the self-organisation of two-dimensional flows in a slip-free box, and for the no-slip case in [11].

2 Numerical Scheme

The Navier-Stokes equations are solved in a double periodic square domain of size $L = 2\pi$ using the vorticity–velocity formulation. The bounded domain is thus imbedded in the periodic domain and the no-slip boundary conditions on the wall $\partial\Omega$ are imposed using a volume penalisation method. The physical idea is to model the solid wall as a porous medium and to compute the flow in a larger domain with two regions of different permeability. A mathematical analysis of the method is given in [12], proving its convergence towards the Navier–Stokes equations with no–slip boundary conditions. The governing equations in vorticity-velocity formulation are,

$$\partial_t\omega + \mathbf{u}\cdot\nabla\omega - \nu\nabla^2\omega + \nabla\times(\frac{1}{\eta}\chi\,\mathbf{u}) = 0 \tag{1}$$

where $\mathbf{u} = (u,v)$ is the divergence-free velocity field, *i.e.* $\nabla\cdot\mathbf{u} = \partial_x u + \partial_y v = 0$, $\omega = \partial_x v - \partial_y u$ the vorticity, ν the kinematic viscosity and $\chi(\mathbf{x})$ a mask function which is 0 inside the fluid, *i.e.* for $\mathbf{x}\in\Omega$, and 1 inside the solid wall. The penalisation parameter η is chosen to be sufficiently small ($\eta = 10^{-3}$) [13]. The numerical technique we use here is based on a dealiased Fourier pseudospectral method with semi-implicit time discretization and adaptive time-stepping using a CFL condition for the maximum velocity. Details on the code together with its numerical validation can be found in [13].

The energy E, enstrophy Z and palinstrophy P of the flow can be defined as [14]

$$E = \frac{1}{2}\int_\Omega |\mathbf{u}|^2 d\mathbf{x}\,, \quad Z = \frac{1}{2}\int_\Omega |\omega|^2 d\mathbf{x}\,, \quad P = \frac{1}{2}\int_\Omega |\nabla\omega|^2 d\mathbf{x}, \tag{2}$$

respectively.

The energy dissipation is given by $d_t E = -2\nu Z$ and the enstrophy dissipation by

$$d_t Z = -2\nu P + \nu\oint_{\partial\Omega}\omega(\mathbf{n}\cdot\nabla\omega)ds, \tag{3}$$

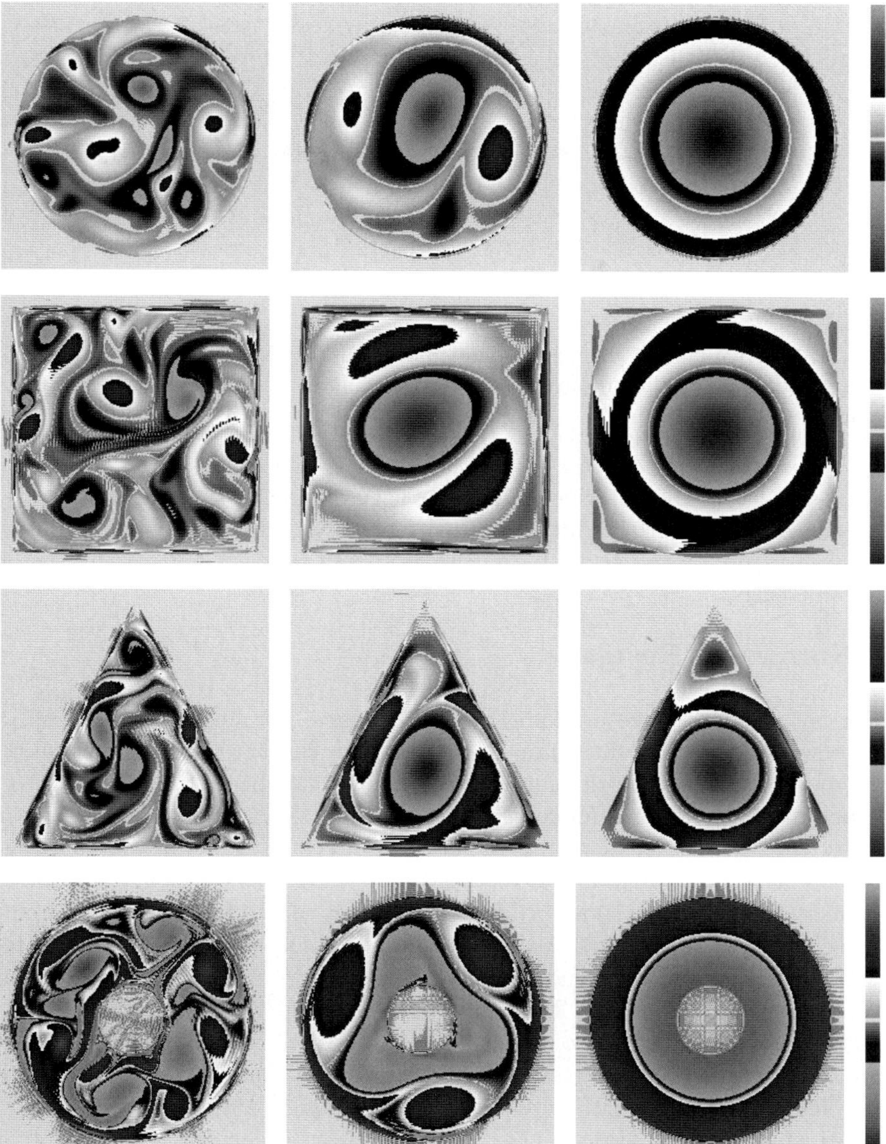

Fig. 1. 2d decaying turbulence in bounded domains. Vorticity fields at early (left), intermediate (middle) and late times (right). From top to bottom: circle, square, triangle and torus.

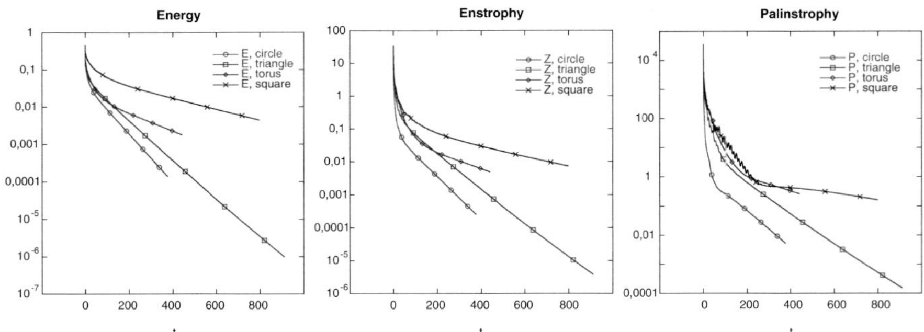

Fig. 2. 2d decaying turbulence in bounded domains. Decay of energy (left), enstrophy (middle) and palinstrophy (right) for the different geometries: circle, square, triangle and torus.

where **n** denotes the outer normal vector with respect to the boundary of the domain $\partial\Omega$. The surface integral in (3) reflects the enstrophy production at the wall involving the vorticity and its gradients which is not present in the periodic case.

3 Numerical Results

Starting with the same random initial conditions, *i.e.* a correlated Gaussian noise with an energy spectrum $E(k) \propto k^4$, we compute the flow evolution in four different geometries for initial Reynolds numbers, $Re = 2D\sqrt{2E}/\nu$, of about 1000 (where D denotes corresponds to the domain size). Figure 1 shows the vorticity fields at early, intermediate and late times for a circular, a square, a triangle and a torus geometry. All flows organize into large scale structures before more or less quasi–steady states form. For the circlular geometry (Fig. 1, top) we observe the transition via a tripole structure, before the final state, a negative circular vortex surrounded by a vortex ring of positive vorticty, forms. The final state of the toroidal geometry (Fig. 1, bottom) corresponds to two vortex rings, a positive one enclosed by a negative one. The transition phase shows a triangularly shaped vortical structure surrounded by three positive vortices. For the triangle and the square domain we see that the final state is note yet completely reached. However, we find also a negative circular vortex in the center which is encircled by a positive vorticity ring, which seems not yet be completely relaxed.

Figure 2 presents the decay of different integral quantities, energy (left), enstrophy (middle) and palinstrophy (right) for the four geometries. All quantities exhibit at early times a rapid monotonuous decay, except for the square and triangular geometry where we observe an oscillatory behaviour in the palinstrophy decay, however less pronouned for the latter case. These oscillations come from the enstrophy production at the boundary. At later times we find for all geometries and all quantities an exponential decay behaviour with slopes depending on the geometry. The square domain shows the slowest decay of all quantities. The decay is increasing in the following order: torus, triangle and the circle exhibits the fastest decay. The fact that the circular case decays fastest results from the smooth boundary, *i.e.* no corners are present and hence the enstrophy production at the wall is much less pronounced than for the other cases.

4 Conclusion

In conclusion, we have shown, by means of DNS of wall bounded flows in different domains, that no–slip boundaries play a crucial role for decaying turbulent flows. At early times we observe a decay of the flow which leads to self–organisation and the emergence of vortices in the bulk flow, similarly to flows in double periodic boxes. At later times larger scale structures form which depend on the domain geometry, and which finally relax to quasi-steady states. More details on the high Reynolds number simulation for shorter times can be found in [3]. Current work deals with comparison of the final states computed here with predictions of the viscous eigenmodes of the Stokes flow for the different geometries.

Acknowledgements:

We thankfully acknowledge financial support from Nagoya University and from the contract CEA/EURATOM n^0 V.3258.001.

References

1. Clercx HJH, Nielsen AH, Torres DJ, Coutsias EA (2001) Eur J Mech B - Fluids 20:557–576
2. Li S, Montgomery D, Jones B (1997) Theor Comput Fluid Dyn 9:167–181
3. Schneider K, Farge M (2005) Phys Rev Lett 95:244502
4. Segre E, Kida S (1998) Fluid Dyn Res 23:89–112
5. Davidson P (2004) Turbulence. An Introduction for Scientists and Engineers Oxford University Press
6. Leith CE (1984) Phys Fluids, 27(6):1388–1395
7. Joyce G, Montgomery D (1973) J Plasma Phys 10:107–121
8. Montgomery D, Joyce G (1974) Phys Fluids 17:1139–1145
9. Robert R, Sommeria J (1991) J Fluid Mech 229:291–310
10. Kondoh Y, Yoshizawa M, Nakano A, Yabe T (1996) Phys Rev E 54(3): 3017–3020
11. van de Konijnenberg JA, Flor JA, van Heijst GJF (1998) Phys Fluids 10(3): 595–606
12. Angot P, Bruneau CH, Fabrie P (1999) Numer Math 81:497–520
13. Schneider K (2005) Comput Fluids 34:1223–1238
14. Kraichnan RH, Montgomery D (1980) Rep Progr Phys 43:547–619
15. Montgomery D, Matthaeus WH, Stribling WT, Martinez D, Oughton S (1992) Phys Fluids A 4:3–6

New One- and Two-Point Scaling Laws in Zero Pressure Gradient Turbulent Boundary Layer Flow

Martin Oberlack and George Khujadze

Chair of Fluid Dynamics, Department of Mechanical Engineering, Technische Universität Darmstadt, Germany
oberlack@fdy.tu-darmstadt.de,khujadze@fdy.tu-darmstadt.de

Abstract. The Lie group or symmetry approach applied to turbulence as developed by Oberlack [1] is used to derive *new scaling laws* for various statistical quantities of a zero pressure gradient (ZPG) turbulent boundary layer flow. For this purpose the approach was applied to the two-point correlation (TPC) equations to find their symmetry groups and thereof to derive invariant solutions (scaling laws). For the verification of these new scaling laws three direct numerical simulations (DNS) at $Re_\theta = 810, 2240, 2500$ were performed using a spectral method with up to 538 million grid points.

Keywords: DNS, Lie groups, scaling laws, boundary layer flow, turbulence

1 Lie Group Analysis of TPC Equations

The present analysis is based on TPC equations with the mean velocity profile $\bar{u}_1 \equiv \bar{u}_1(x_2)$ i.e. the parallel flow assumption, where x_2 is the wall normal coordinate. Under this assumption and in the outer part of the boundary layer flow i.e. sufficiently apart from the viscous sublayer, and hence viscosity is negligible, we have

$$\frac{\bar{D}R_{ij}}{\bar{D}t} + R_{kj}\frac{\partial \bar{u}_i}{\partial x_k} + R_{ik}\frac{\partial \bar{u}_j}{\partial x_k}\bigg|_{x+r} + [\bar{u}_k(x+r) - \bar{u}_k(x)]\frac{\partial R_{ij}}{\partial r_k}$$

$$+ \frac{\partial \overline{p'u'}_j}{\partial x_i} - \frac{\partial \overline{p'u'}_j}{\partial r_i} + \frac{\partial \overline{u'_i p'}}{\partial r_j} + \frac{\partial R_{(ik)j}}{\partial x_k} - \frac{\partial}{\partial r_k}[R_{(ik)j} - R_{i(jk)}] = 0, \tag{1}$$

where

$$R_{ij} = \overline{u'_i(x,t)u'_j(x+r)}, \qquad \overline{p'u'_j} = \overline{p'(x)u'_j(x+r)},$$

$$R_{(ik)j} = \overline{u'_i(x)u'_k(\mathbf{x})u'_j(x+r)}$$

and $\frac{\bar{D}}{\bar{D}t} = \left(\frac{\partial}{\partial t} + \bar{u}_k\frac{\partial}{\partial x_k}\right)$ is the mean substantial derivative. x and $r = x' - x$ are coordinates in the physical and the correlation spaces respectively.

255

Y. Kaneda (ed.), *IUTAM Symposium on Computational Physics and New Perspectives in Turbulence*, 255–260.

In the present study Lie's procedure is used to find symmetry transformations and self-similar solutions of equation (1). Invoking the symmetry breaking constraint of an external length scale (details can be found in [1]), the exponential scaling law of the velocity in the wake region of a ZPG turbulent boundary layer flow has been obtained:

$$\bar{u}_1(x_2) = k_1 + k_2 e^{-k_3 x_2}, \tag{2}$$

where k_1, k_3 are group parameters and k_2 is a constant of integration. This scaling law was further validated in the different papers, e.g. in [2, 3, 4].

As an additional result we obtained the following scaling laws for the TPC functions:

$$R_{ij}(x_2, \boldsymbol{r}) = e^{-k_4 x_2} B_{ij}(\boldsymbol{r}), \tag{3}$$

$$\overline{u_i' p'}(x_2, \boldsymbol{r}) = e^{-(k_3+k_4)x_2} E_i(\boldsymbol{r}), \qquad \overline{p' u_i'}(x_2, \boldsymbol{r}) = e^{-(k_3+k_4)x_2} F_i(\boldsymbol{r}), \tag{4}$$

where k_4 is supposed to be universal and B_{ij}, E_i, F_i are only functions of the correlation distance $\boldsymbol{r}$;

Introducing equations (2 - 4) into equation (1) we obtain the reduced TPC equation:

$$
\begin{aligned}
k_2 k_3 [\delta_{i1} B_{2j}(\boldsymbol{r}) + \delta_{j1} B_{i2}(\boldsymbol{r})] + k_2(1 - e^{-k_3 r_2}) \frac{\partial B_{ij}(\boldsymbol{r})}{\partial r_1} + \\
(k_3 + k_4)\delta_{i1} F_j(\boldsymbol{r}) + \frac{\partial F_j(\boldsymbol{r})}{\partial r_i} - \frac{\partial E_i(\boldsymbol{r})}{\partial r_j} + \cdots = 0,
\end{aligned}
\tag{5}
$$

where the dots denote the higher order terms corresponding to the triple-correlation functions.

From the invariant solution (2) and (3) we may deduce the following important results:

- Mean velocity defect law:

$$\frac{\bar{u}_\infty - \bar{u}_1}{u_\tau} = \alpha \, \exp\left(-\beta \frac{x_2}{\Delta}\right), \tag{6}$$

where $\Delta \equiv \delta^* \bar{u}_\infty / u_\tau$ is the Rotta-Clauser length-scale, $\alpha = k_2/u_\tau$, $\beta = k_3 \Delta$, u_τ is the friction velocity, $\bar{u}_\infty$ is the free-stream velocity and δ^* is the boundary layer thickness.

- In non-dimensional form the Reynolds stresses have the form:

$$\frac{\overline{u_i' u_j'}(x_2)}{u_\tau^2} = b_{ij} \, \exp\left(-a \frac{x_2}{\Delta}\right), \tag{7}$$

where b_{ij} and $a = k_4 \Delta$ are universal constants that should be taken from DNS or experiments.

- The TPC correlation functions may be reduced to the correlation coefficients using the equations (3) and (7):

$$\mathcal{R}_{[ij]} \equiv \frac{R_{ij}(x_2, \boldsymbol{r})}{\overline{u_i' u_j'}(x_2)} = B'_{[ij]}(\boldsymbol{r}), \tag{8}$$

where $B'_{[ij]}$ is independent of x_2.

2 Verification of New Scaling Laws

For the verification of the scaling laws (4), (6), (7) and (8) DNS at three different Reynolds numbers $Re_\theta = 810, 2240, 2500$ (θ is the boundary layer momentum thickness and $Re_\theta = \bar{u}_\infty \theta / \nu$) were performed using a spectral method with up to 538 million grid points (see [2, 3]). The code for the DNS was developed at KTH, Stockholm [5]. We took great care for the accumulation of the statistics which in our DNS covered more then $60000\delta^*/\bar{u}_\infty$ time units for small Reynolds number ($Re_\theta = 810$), and more than $10000\delta^*/\bar{u}_\infty$ in the case of higher Reynolds numbers ($Re_\theta = 2240, 2500$). The mean velocity profile of the turbulent boundary layer data is plotted in the left plot in Fig. 1 for $Re_\theta = 2240$. The close-up of mean velocity profiles are presented in the right plot in Fig. 1 at different Reynolds numbers $Re_\theta = 1670, 1870, 2060, 2240$. Good collapse of the profiles in the exponential region is observed. Good agreement between DNS and the theoretical result (6) is visible in the region $x_2/\Delta \approx 0.01 - 0.16$. The left plot in Fig. 2 represents the Reynolds stresses in log-outer scaling showing exponential regions at Reynolds number $Re_\theta = 810$. The key result is that the constant in the exponent as anticipated by the invariant solution (7) is the same ($a = 7.2$) for all components of Reynolds stress tensor (see equation 9). The exponential region for for different components of Reynolds stress tensor $\overline{u_i' u_i'}$ has almost the same length. This is expressed

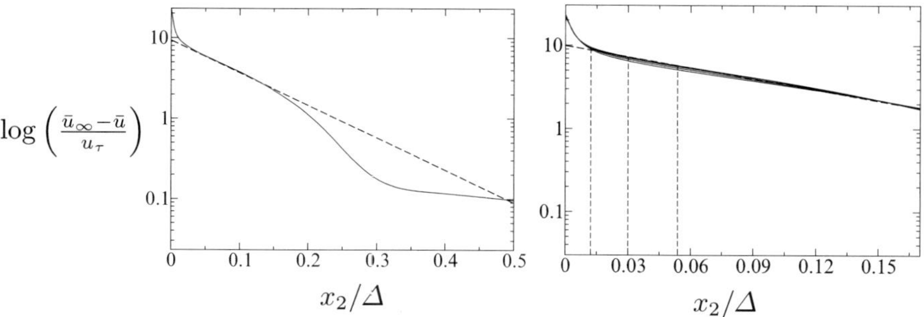

Fig. 1. Left Plot: Mean velocity profile at $Re_\theta = 2240$; **Right Plot:** Close-up plot of the mean velocity profiles in log-linear scaling at different Reynolds numbers ($Re_\theta = 1670, 1870, 2060, 2240$).

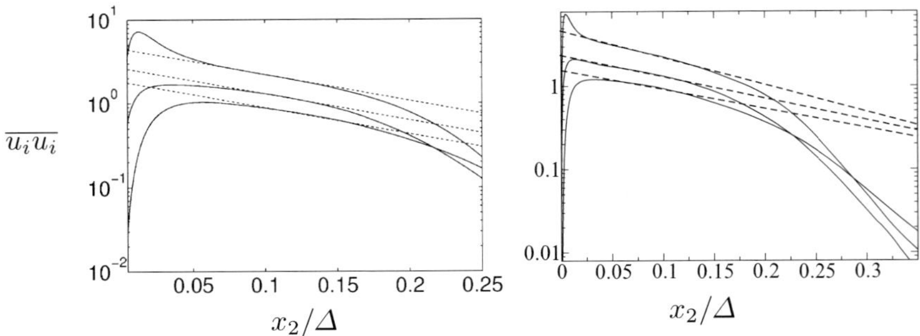

Fig. 2. Reynolds stresses in log-linear scaling. Upper, lower and middle curves are $\overline{u_1 u_1}$, $\overline{u_2 u_2}$, $\overline{u_3 u_3}$ respectively. - - - theoretical results (7), —— DNS. **Left plot:** $Re_\theta = 810$; **Right plot:** $Re_\theta = 2240$.

by the parallel dashed lines in Fig. 1. The exponential law covers the region $x_2/\Delta \approx 0.06 - 0.16$. Thus the equation (7) has the following form:

$$\overline{u_1 u_1}(x_2) = 4.55 \ e^{-7.2x_2/\Delta},$$
$$\overline{u_2 u_2}(x_2) = 2.7 \ e^{-7.2x_2/\Delta}, \tag{9}$$
$$\overline{u_3 u_3}(x_2) = 1.8 \ e^{-7.2x_2/\Delta}.$$

The right plot in Fig. 2 shows the flow case at $Re_\theta = 2240$. As one can see the lines on the plot are not fully parallel that means that the constants in the exponent of equation (7) is not the same for the different Reynolds stresses as it was in the case for $Re_\theta = 810$. The reason for the deviation of the DNS data from the theoretical result is the fact that the time of the accumulation of the turbulence statistics in the high Reynolds number case was smaller than in the low Reynolds number DNS. The high Reynolds number simulations require a considerable larger amount of computer resources that makes it so far impossible to get very large simulation times. However, the simulation time at high Reynolds number case is enough to get smooth statistics. A plot of $\overline{u_1' u_1'}$ at two different Reynolds numbers ($Re_\theta = 810, 2240$) is shown in Fig. 3. The exponential region is almost two times longer in the case of $Re_\theta = 2240$ and it increases towards the wall as the Reynolds number increases.

From the equations (3) and (8) at $r_1 = r_3 = 0$ we have:

$$\mathcal{R}_{[ij]}(r_2) \equiv \frac{R_{ij}(x_2, 0, r_2, 0)}{\overline{u_i' u_j'}(x_2)} = \frac{e^{-k_4 x_2} B_{ij}(0, r_2, 0)}{e^{-k_4 x_2} B_{ij}(0)} = B'_{[ij]}(0, r_2, 0). \tag{10}$$

Coefficients of the TPC functions are independent of the wall-normal coordinate x_2.

The left plot in Fig. 4 shows normalised TPC function $\mathcal{R}_{[22]}(r_2)$ for $Re_\theta = 810$ in the exponential region for the different initial points in wall-normal direction: $x_2/\Delta = 0.1, 0.12, 0.13, 0.144$. As predicted from the equation (10)

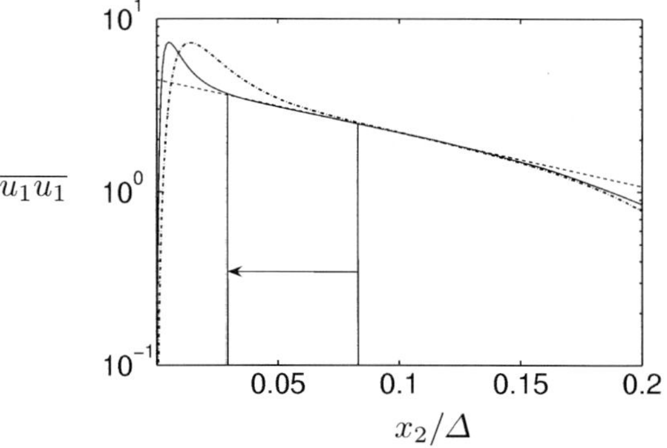

Fig. 3. $\overline{u_1 u_1}$ at —·— $Re_\theta = 810$, —— $Re_\theta = 2240$ Reynolds numbers; --- theoretical result (7).

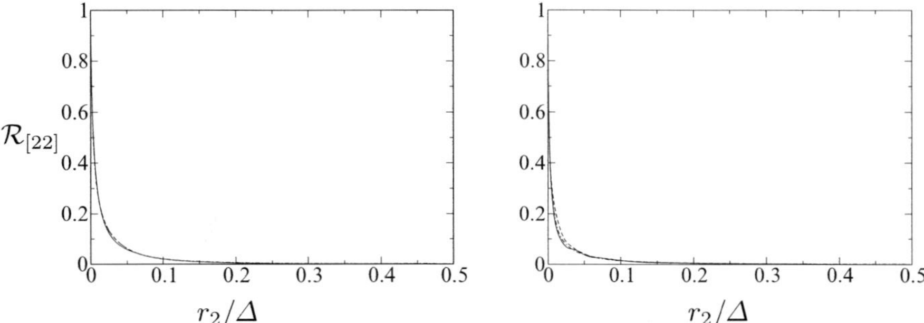

Fig. 4. Normalized TPC function $\mathcal{R}_{[22]}(r_2)$: **Left plot:** $Re_\theta = 810$ at different initial points in the exponential region: —— $x_2/\Delta = 0.1$, ········ $x_2/\Delta = 0.12$, --- $x_2/\Delta = 0.13$ and —·— $x_2/\Delta = 0.144$; **Right plot:** $Re_\theta = 2240$ at different initial points in the exponential region: —— $x_2/\Delta = 0.095$, --- $x_2/\Delta = 0.126$, —·— $x_2/\Delta = 0.16$.

the normalised TPC functions should collapse in one curve. We see that the results of the DNS nicely collapse on one curve. This fact shows the validity of the results of Lie group approach even for the low Reynolds number case. The right plot in Fig. 4 shows TPC functions calculated for the different initial points that are located in the exponential region ($x_2/\Delta = 0.095, 0.126, 0.16$) at $Re_\theta = 2240$. As one can see from the figure the agreement of DNS results with the theoretical one is of similar good quality as the low Reynolds number case.

In Fig. 5 DNS results are compared to the theoretical ones for the pressure-velocity correlations. The exponential region for $\overline{p' u_1'}$ is $x_2/\Delta \approx 0.04 - 0.18$, while the law for $-\overline{p' u_2'}$ is valid in the interval of $x_2/\Delta \approx 0.15 - 0.22$.

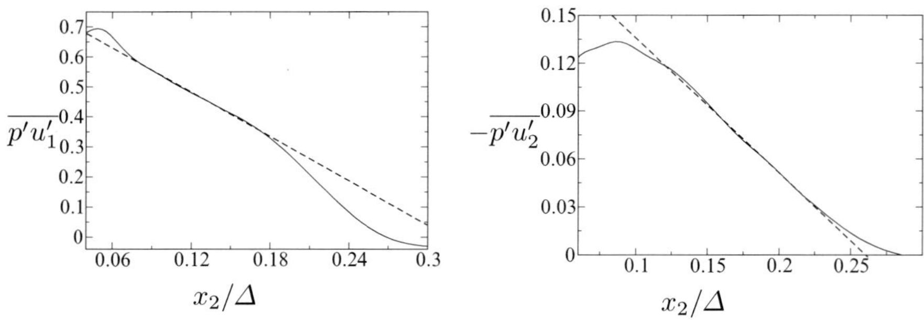

Fig. 5. Pressure-velocity correlations at $Re_\theta = 810$: **Left plot:** Pressure-streamwise velocity correlation, $\overline{pu_1}$; **Right plot:** Pressure-wall-normal velocity correlation, $-\overline{pu_2}$: - - - theoretical results (4); —— Results of the numerical simulations.

3 Summary

In the present work a Lie group analysis of the TPC equations is presented in the case of ZPG turbulent boundary layer flow using the parallel flow assumption $(\bar{u}_1 \equiv \bar{u}_1(x_2))$. It is shown that the Lie group analysis helps to gain deep insight into the description of the turbulent flow. This method provides a systematic procedure to derive symmetries from a set of differential equations and thereof obtain invariant solutions in turbulence called scaling laws. Several DNS of ZPG turbulent boundary layer flow were conducted for three different Reynolds numbers $(Re_\theta = 810, 2240, 2500)$ with up to 538 million grid points. The new scaling laws derived from Lie group analysis were verified using the DNS data.

References

1. Oberlack M (2001) J Fluid Mech 427:299-328
2. Khujadze G, Oberlack M (2004) Theor and Comp Fluid Dyn 18:391-411
3. Khujadze G (2006) DNS and Lie group analysis of zero pressure gradient turbulent boundary layer flow, PhD Thesis, TU Darmstadt, Germany
4. Lindgren B, Österlund JM, Johansson A (2004) J Fluid Mech 502:127-152
5. Lundbladh A, Berlin S, Skote M, Hildings C, Choi J, Kim J, Henningson D (1999) Tech Rep TRITA-MEK 1999:11, Royal Institute of Technology, Stokholm

Intrinsic Langevin Models for Turbulence

Hatem Touil[1], M. Yousuff Hussaini[2], Toshiyuki Gotoh[3,4],
Robert Rubinstein[5], and Stephen L. Woodruff[6]

[1] School of Computational Science, Florida State University, Tallahassee, FL
   32306, USA touil@scs.fsu.edu
[2] School of Computational Science, Florida State University, Tallahassee, FL
   32306, USA myh@scs.fsu.edu
[3] Department of System Engineering, Nagoya Institute of Technology, Gokiso-cho,
   Showa-ku, Nagoya 466-8555, Japan gotoh.toshiyuki@nitech.ac.jp
[4] CREST, Japan Science and Technology Agency, 4-1-8 Honcho,
   Kawaguchi, Saitama 332-0012, Japan
[5] Computational Aerosciences Branch, NASA Langley Research Center, Hampton,
   VA 23681 USA r.rubinstein@larc.nasa.gov
[6] Center for Advanced Power Systems, Florida State University, Tallahassee, FL
   32306 USA woodruff@caps.fsu.edu

Abstract. Large eddy simulation entails the projection of a large number of mode amplitudes onto a much smaller set of resolved modes. General considerations of non-equilibrium statistical mechanics [1] show that the interactions with the unresolved modes can be replaced by a damping and a forcing, that is, by a Langevin model. In the present work, we attempt to construct such Langevin models for turbulence directly from DNS data. Whereas LES modeling based on the Smagorinsky picture focuses exclusively on the construction of suitable damping, we will place equal emphasis on the random forcing. An important feature of the approach is that a 'universal' model is not sought; instead, following an important suggestion of Kraichnan [2], we emphasize that the analytical structure of the model depends crucially on what statistical quantities are to be predicted.

Keywords: large eddy simulation, subgrid model, Langevin model

1 Introduction

If the Navier-Stokes equations for homogeneous isotropic turbulence are written symbolically as $\dot{u}_i(\mathbf{k}, t) = N_i(\mathbf{k}, t) - \nu k^2 u_i(\mathbf{k}, t)$, then LES is defined by the decomposition

$$N_i(\mathbf{k}, t) = N_i^<(\mathbf{k}|k_c, t) + N_i^>(\mathbf{k}|k_c, t) \tag{1}$$

261

Y. Kaneda (ed.), *IUTAM Symposium on Computational Physics and New Perspectives in Turbulence*, 261–266.
© 2008 *Springer*.

where

$$N_i^<(\mathbf{k}|k_c,t) = \frac{\mathrm{i}}{2}P_{imn}(\mathbf{k})\int_{p,q\,\leq k_c} \mathrm{d}\mathbf{p}\,\mathrm{d}\mathbf{q}\,\,\delta(\mathbf{k}-\mathbf{p}-\mathbf{q})\,u_m(\mathbf{p},t)u_n(\mathbf{q},t)$$

$$N_i^>(\mathbf{k}|k_c,t) = \frac{\mathrm{i}}{2}P_{imn}(\mathbf{k})\int_{p\geq k_c\ \mathrm{or}\ q\geq k_c} \mathrm{d}\mathbf{p}\,\mathrm{d}\mathbf{q}\,\,\delta(\mathbf{k}-\mathbf{p}-\mathbf{q})\,u_m(\mathbf{p},t)u_n(\mathbf{q},t)$$

and k_c is an arbitrarily selected cutoff wavenumber. We are led to a Langevin-like representation by decomposing $N^>$ into parts correlated and uncorrelated with u : $N_i^>(\mathbf{k}|k_c,t) = \eta(k|k_c)u_i(\mathbf{k},t) + R_i(\mathbf{k}|k_c,t)$ where $\eta(k|k_c) = \langle N_i^>(\mathbf{k}|k_c,t)u_i(-\mathbf{k},t)\rangle/\langle u_i(\mathbf{k},t)u_i(-\mathbf{k},t)\rangle$ is the correlation coefficient and R_i is *defined* by

$$\dot{u}_i(\mathbf{k},t) = N_i^<(\mathbf{k}|k_c,t) - \eta(k|k_c)u_i(\mathbf{k},t) + R_i(\mathbf{k}|k_c,t) \tag{2}$$

Except for ignoring the small effect of viscosity on the resolved scales, this equation simply restates the governing equations (viscous effects on the unresolved scales can occur through R_i). But we can compare this equation to the Langevin representation of the test-field model [3] with the same decomposition

$$\dot{u}_i(\mathbf{k},t) = \underbrace{[-\eta^<(k|k_c)u_i(\mathbf{k},t) + f_i^<(\mathbf{k}|k_c,t)]}_{\text{resolved motion}} + \underbrace{[-\eta(k|k_c)u_i(\mathbf{k},t) + f_i(\mathbf{k}|k_c)]}_{\text{unresolved motion}}$$

where $\eta^<$, $f^<$, η and f are defined by the same integration regions that define $N^<$ and $N^>$ (for notational consistency, we do not write $\eta^>$ or $f^>$; it is understood that subsequently η and f always refer to the unresolved interactions).

Some basic properties of the test-field model are:

1. 'plateau-cusp' structure of the effective viscosity $\nu(k|k_c)$ defined by $\eta(k|k_c) = \nu(k|k_c)k^2$: $\nu(k|k_c) \sim k_c^{-4/3}$ for $k \ll k_c$ but $\nu(k|k_c) \approx \nu(k_c|k_c) - (k_c - k)^{1/3}$ for $k \uparrow k_c$.

2. k^4 'backscatter' at large scales: $E_f(k) \sim k^4$ for $0 \approx k \ll k_c$, where

$$E_f(k) = \oint_{|\mathbf{k}|=k} \langle f_i(\mathbf{k}|k_c,t)f_i(-\mathbf{k}|k_c,t)\rangle \mathrm{d}S \quad \text{for } 0 \approx k \ll k_c.$$

3. Gaussian character of the random forcing $f(\mathbf{k}|k_c)$.

2 Statistical Properties of the Eddy Viscosity

We first use DNS data to evaluate how well these properties of closure are satisfied by the data. The data is taken from 128^3 forced steady state DNS. In this problem, we are not interested in Kolmogorov scaling but in the properties of the decomposition itself, therefore this resolution is adequate.

The spectral eddy viscosity for different cutoffs k_c is shown in Figure 1. Following a very large number of investigations of the same type [4], we confirm the plateau-cusp structure of $\nu(k|k_c)$. Even given that we should not expect

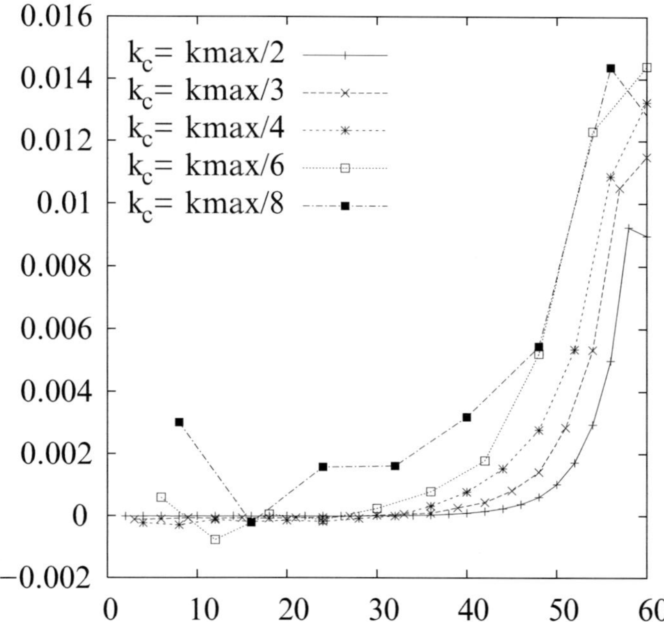

Fig. 1. Spectral eddy viscosity (ordinate) from DNS data for wavenumber cutoffs k_c between $k_{max}/2$ and $k_{max}/8$; abscissa represents shifted and rescaled wavenumbers so that all curves extend from 0 to 60.

the high Re plateau value $\propto k_c^{-4/3}$, the measured plateau value (≈ 0) is much smaller than the theoretical value, even using the measured spectrum in the test-field model transfer integral.

A possible explanation is that our value of the eddy viscosity includes the effects of interactions with the unresolved scales, whereas closure attempts to obtain the value from the resolved motion alone. This underscores a certain arbitrariness in the Langevin model. If we consider both the damping and the forcing to be adjustable, then the energy balance imposes only one condition, so that an infinite family of Langevin models will give the correct energy balance. This idea was used by Gotoh *et al.* [5] to develop an LES model consistent with the predictability spectrum.

3 Statistical Properties of the 'Remainder' Term $R(\mathbf{k}, t)$

Our main concern is analyzing the properties of the 'remainder' R defined in Eq. (2). We first compute the spectrum of R for different cutoffs k_c. A typical result is shown in Figure 2. R appears to follow the k^4 spectrum suggested by closure, but there is some uncertainty because this scaling should apply to the largest scales of motion, not to intermediate scales as the data seems to suggest. But in this case, it is important to consider the role of finite resolution,

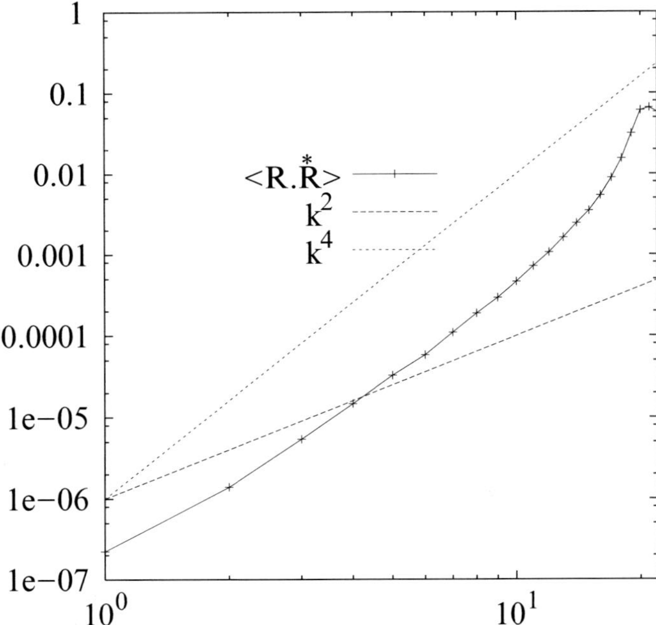

Fig. 2. Spectrum of remainder term $R(k)$ (ordinate) in Eq. (2) showing approximate k^4 scaling: abscissa is wavenumber k.

and in particular the inability of our DNS, which is forced at nearly the largest scales of motion, to treat asymptotically large scales.

Another quantity of interest is the pdf of R. Resolution requirements precluded obtaining this pdf at all wavenumbers, so instead, we evaluated the single-point pdf of R in physical space. The measured pdf (not shown) shows the 'heavy tails' often associated with 'intermittency' in turbulence; this feature is perhaps best characterized by the measured flatness of approximately 12., well in excess of the Gaussian value of 3.

Of course, this result does not resolve any wavenumber dependence. To this end, we evaluated the variance and flatness of the low- and high-pass filtered fields obtained by transforming $H(k^* - k)R_i(\mathbf{k}|k_c)$ and $H(k - k^*)R_i(\mathbf{k}|k_c)$ to physical space for various $k^* < k_c$ (Here, $H(x) = 1$ for $x \geq 0$ and $H(x) = 0$ otherwise). The results are shown in the table:

	variance $k \leq k^*$	$k \geq k^*$	flatness $k \leq k^*$	$k \geq k^*$
$k^* = k_c/2$	4.60×10^{-7}	1.147×10^{-4}	25.	5.2
$k^* = 3k_c/4$	5.58×10^{-6}	1.096×10^{-4}	16.	4.6
$k^* = 0.925k_c$	4.05×10^{-5}	7.47×10^{-5}	7.6	3.5

A tentative conclusion is that the fluctuations of R at large scales are more non-Gaussian than those near the cutoff, in disagreement with what might have been expected. This issue requires further study.

Langford and Moser [6] made the interesting suggestion to evaluate the ratio $\langle RR \rangle / \langle N^> N^> \rangle$, which measures how much of the variance of $N^>$ is not explained by the eddy viscosity model. Presumably, this ratio should be small. But instead, we confirm the finding of [6] that the variance of $N^>$ nearly equals the variance of R basically over the entire range of wavenumbers. Thus, fluctuations of $N^>$ are not well modeled by ηu.

There are two different ways to assess this finding. We might conclude that an LES model with $R = 0$ is inaccurate, or that accurate LES requires a model for the remainder R. But another view is that the results show that *most of the fluctuations of $N^>$ do not participate in mean energy transfer*. We advocate this second view, which is motivated by the results of Gotoh and Watanabe [7]. The pure eddy viscosity model with $R = 0$ should predict the mean energy transfer correctly, and it should therefore predict the resolved energy spectrum correctly, even if the fluctuations of the nonlinear term are poorly predicted. But we stress that the eddy viscosity in Eq. (2) is an 'exact' eddy viscosity that includes the effect of subgrid interactions. The eddy viscosity of practical LES models based on the resolved field alone may not predict the correct mean energy transfer.

4 Construction of General LES Models

It is a truism that the problem of LES is modeling the unresolved interactions. But the model depends on the goal: different models may be required to obtain the correct mean energy transfer and the correct energy transfer fluctuations. A different class of models may be required to obtain realistic time correlations [8].

From this viewpoint, we can say that LES modeling in general requires replacing $N^>$ by some estimate or approximation $\hat{N}$. The question then is how to construct $\hat{N}$. A powerful approach was suggested by Kraichnan [2], as part of a proposal for *decimated amplitude approximations*: namely, we can model $N^>$ by imposing statistical constraints on $\hat{N}$ based on the exact dynamics. For example, the constraint $\langle \hat{N}_i(\mathbf{k}) u_i(-\mathbf{k}) \rangle = \langle N_i^>(\mathbf{k}) u_i(-\mathbf{k}) \rangle$ preserves the energy flux balance $\langle \dot{u}_i(\mathbf{k}) u_i(-\mathbf{k}) \rangle = \langle N_i(\mathbf{k}) u_i(-\mathbf{k}) \rangle$. In this model, any part of $\hat{N}$ that is uncorrelated with u plays no role in the dynamics: the assumption that $\hat{N} \propto u$ simply recovers the eddy viscosity model without random forcing discussed in the previous section.

The stochastic constraint tells us the properties of $\hat{N}$ at each time, but constructing $\hat{N}$ as a stochastic process will require some model for its time correlations. The simplest model is to assume that $\hat{N}$ is white noise in time; then a suitable model is $\langle \hat{N}_i(\mathbf{k}, t) \hat{N}_i(-\mathbf{k}, s) \rangle = A(k)\delta(t - s)\delta_{ij}$ with $A(k) \propto \int_0^\infty d\tau \, \langle N_i^>(\mathbf{k}, t) N_i^>(-\mathbf{k}, t+\tau) \rangle$. We again stress that although higher order correlations between $N^>$ and u are ignored in this model, it should predict the correct energy flux, and the correct resolved energy spectrum. An important question is how the statistics of the random velocity field constructed by this model differ from those of the exact DNS field: we expect it to be less 'intermittent' and probably more Gaussian, but this is a topic for future research.

The usefulness of Kraichnan's approach is more evident if we consider nontrivial stochastic constraints. A different model can be constructed by requiring also $\langle (\hat{N}u)^2 \rangle = \langle (N^>u)^2 \rangle$ so that the variance of energy transfer fluctuations is preserved by the model. Kraichnan [2] discusses how to realize stochastic processes subject to any finite number of such constraints. In this way, we can construct an indefinite number of parodies of Navier-Stokes dynamics that share some, but not all of its properties.

5 Role of the PDF of the Forcing

We began by considering Eq. (2) as a kind of formal Langevin model for the velocity field u. But unlike the random force in a true Langevin model, the remainder R has a non-Gaussian pdf. Suppose then, following Kraichnan's ideas about realization that we replace R by an estimated $\hat{R}$ constrained only to have the same pdf at each wavenumber as R. The question is whether this pdf has any relevance for the dynamics of u.

Certainly, if u satisfied a Fokker-Planck equation, then only the second-order moment of $\hat{R}$ would be relevant. This is the situation in a standard Langevin model. However, if $\hat{R}$ has a non-Gaussian pdf, then its multiple time cumulants cannot vanish: for example, $\langle \hat{R}(1)\hat{R}(2)\hat{R}(3)\hat{R}(4) \rangle = \langle \hat{R}(1)\hat{R}(2) \rangle \langle \hat{R}(3)\hat{R}(4) \rangle + \langle \hat{R}(1)\hat{R}(3) \rangle \langle \hat{R}(2)\hat{R}(4) \rangle + \langle \hat{R}(1)\hat{R}(4) \rangle \langle \hat{R}(3)\hat{R}(2) \rangle$ implies $\langle \hat{R}^4 \rangle = 3\langle \hat{R}^2 \rangle^2$. Then u does not satisfy a Fokker-Planck equation [1]: the pdf of $\hat{R}$, not just its second-order moment, is relevant and the higher order Kramers-Moyal coefficients must be considered [9]. The random field generated by the model $\dot{u} = N^< - \eta u + \hat{R}$ with non-Gaussian $\hat{R}$ might have non-Gaussian joint mode amplitude pdfs [10].

References

1. Kubo R, Toda M, Hashitsume N (1991) Statistical physics II, non-equilibrium statistical mechanics. Springer, Berlin Heidelberg New York
2. Kraichnan RH (1985) Decimated amplitude equations in turbulence dynamics. In Dwoyer DL, Hussaini MY, Voigt RG (eds) Theoretical approaches to turbulence. Springer, Berlin Heidelberg New York
3. Kraichnan RH (1971) J Fluid Mech 47:513–524
4. Zhou Y (1993), Phys Fluids A 5:2511–2524
5. Gotoh T, Kakui I, Kaneda Y (2003) Subgrid models for two-dimensional turbulence based on Lagrangian statistical theory. In Kaneda Y, Gotoh T (eds) Statistical theories and computational approaches to turbulence Springer, Berlin Heidelberg New York
6. Langford JA, Moser R D (1999) J Fluid Mech 398:321–346
7. Gotoh T, Watanabe T (2005) J Turbul 6 N33:1–18
8. He G, Rubinstein R, Wang LP (2002) Phys Fluids 14:2186-2193
9. Renner C, Peinke J, Friedrich R (2001) J Fluid Mech 433:383–409
10. Brun C, Pumir A (2001) Phys Rev E 63:056313

Analysis of the Reynolds Stress Using the Green's Function

Fujihiro Hamba

Institute of Industrial Science, University of Tokyo, Meguro-ku, Tokyo 153-8505, Japan
hamba@iis.u-tokyo.ac.jp

Abstract. An exact nonlocal expression for the Reynolds stress in rotating frames is described using the Green's function for the velocity fluctuation. This expression is applied to the DNS of non-rotating and rotating channel flows. Spatially- and temporally-nonlocal and anisotropic properties are examined for better understanding and modeling the channel flows. The nonlocal eddy viscosity shows that spatially-nonlocal effect is stronger for the rotating channel flow than for the non-rotating channel flow. For the shear stress in the rotating channel flow, two terms related to the strain-rate and vorticity tensors are dominant and show long time correlation. This result supports a simple nonlinear eddy-viscosity model that reproduces the zero mean absolute vorticity state.

Keywords: Reynolds stress, Green's function, channel flow, nonlocal effect, nonlinear eddy viscosity

1 Introduction

Linear and nonlinear eddy-viscosity models are widely used to predict the mean velocity profile of turbulent flows. Although model expressions are local in space and time, it is not trivial whether this approximation is good. To examine nonlocal effect of the mean velocity gradient on the Reynolds stress, we pay attention to the Green's function for the velocity fluctuation in physical space. Kraichnan [1] derived an exact nonlocal expression for the Reynolds stress using the Green's function. However, the expression is an integral equation for the Reynolds stress and it is necessary to approximate it by expansion and truncation. By modifying the Green's function, we derived an explicit nonlocal expression for the Reynolds stress [2, 3]. We applied it to the non-rotating channel flow to examine its spatially-nonlocal properties. In the present work, we apply the nonlocal expression to the non-rotating and rotating channel flows and examine not only spatially-nonlocal property but also anisotropic and temporally-nonlocal properties of the channel flows.

Y. Kaneda (ed.), *IUTAM Symposium on Computational Physics and New Perspectives in Turbulence*, 267–272.

2 Formulation

The equation for the velocity fluctuation $u_i'(= u_i - U_i)$ in rotating frames can be written as

$$\frac{Du_i'}{Dt} + \Omega_{ij}u_j' + \frac{\partial}{\partial x_j}(u_j'u_i' - \langle u_j'u_i'\rangle) + \frac{\partial p'}{\partial x_i} - \nu\frac{\partial^2 u_i'}{\partial x_j\partial x_j} = -u_j'\left(S_{ij} + \bar{W}_{ij}\right), \quad (1)$$

where $D/Dt = \partial/\partial t + U_i\partial/\partial x_i$, $\Omega_{ij}(= \epsilon_{jik}\Omega_k)$ is the system rotation tensor, $S_{ij}[= (\partial U_i/\partial x_j + \partial U_j/\partial x_j)/2]$ is the strain-rate tensor, $W_{ij}[= (\partial U_i/\partial x_j - \partial U_j/\partial x_j)/2]$ is the vorticity tensor, and $\bar{W}_{ij}(= W_{ij} + \Omega_{ij})$ is the absolute vorticity tensor. The right-hand side of (1) can be formally considered as external forces for the velocity fluctuation. We then introduce the Green's function $g_{ij}(\mathbf{x}, t; \mathbf{x}', t')$; its equation is given by

$$\frac{Dg_{ij}}{Dt} + \Omega_{ik}g_{kj} + \frac{\partial}{\partial x_k}(u_k'g_{ij} - \langle u_k'g_{ij}\rangle) + \frac{\partial p_{Gj}}{\partial x_i} - \nu\frac{\partial^2 g_{ij}}{\partial x_k\partial x_k} = \delta_{ij}\delta(\mathbf{x}-\mathbf{x}')\delta(t-t').$$
$$(2)$$

Using this Green's function, a formal solution for the velocity fluctuation can be written as

$$u_i'(\mathbf{x}, t) = -\int d\mathbf{x}' \int_0^t dt' g_{ij}(\mathbf{x}, t; \mathbf{x}', t')u_k'(\mathbf{x}', t')\left(S_{jk}(\mathbf{x}', t') + \bar{W}_{jk}(\mathbf{x}', t')\right).$$
$$(3)$$

As a result, a nonlocal expression for the Reynolds stress is given by

$$\langle u_i'u_j'\rangle(\mathbf{x}, t) = -\int d\mathbf{x}' \int_0^t dt' \nu_{NLijkm}(\mathbf{x}, t; \mathbf{x}', t')\left(S_{km}(\mathbf{x}', t') + \bar{W}_{km}(\mathbf{x}', t')\right),$$
$$(4)$$

$$\nu_{NLijkm}(\mathbf{x}, t; \mathbf{x}', t') = (\langle u_i'(\mathbf{x}, t)g_{jk}(\mathbf{x}, t; \mathbf{x}', t')u_m'(\mathbf{x}', t')\rangle$$
$$+ \langle u_j'(\mathbf{x}, t)g_{ik}(\mathbf{x}, t; \mathbf{x}', t')u_m'(\mathbf{x}', t')\rangle)/2. \quad (5)$$

The nonlocal eddy viscosity $\nu_{NLijkm}(\mathbf{x}, t; \mathbf{x}', t')$ represents a contribution to the Reynolds stress at $(\mathbf{x}, t)$ from the mean strain-rate and vorticity tensors at $(\mathbf{x}', t')$.

We apply this nonlocal expression to theDNSof non-rotating and spanwise-rotating channel flows to examine nonlocal properties of the Reynolds stress. We carried out three runs with different rate of system rotation. The rotation number is set to $Ro_\tau(= 2\Omega h/u_\tau) = 0$, 2.5, and 7.5 for Cases 1, 2, and 3, respectively, where h is the channel half width and u_τ is the global friction velocity. The Reynolds number is set to $Re_\tau(= u_\tau h/\nu) = 180$ for Case 1 and $Re_\tau = 150$ for Cases 2 and 3.

3 Spatially-Nonlocal Properties of Channel Flows

First, we examine spatially-nonlocal properties in the wall-normal direction. Figure 1(a) shows profiles of the Reynolds shear stress for the non-rotating

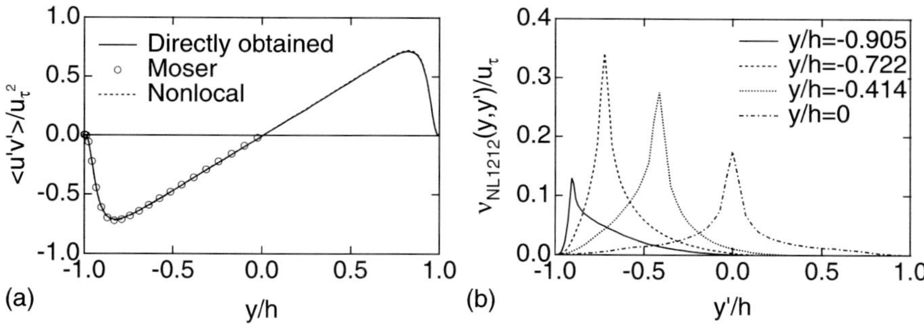

Fig. 1. Profiles of (a) shear stress $\langle u'v' \rangle$ and (b) nonlocal eddy viscosity ν_{NL1212} for the non-rotating channel flow in Case 1. ∘ : DNS data of Moser *et al.* [4].

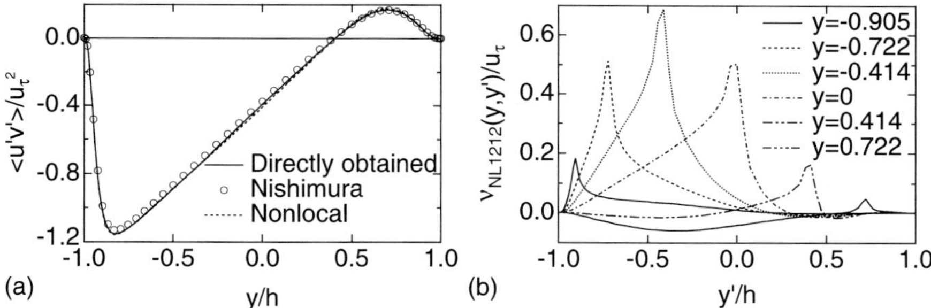

Fig. 2. Profiles of (a) shear stress $\langle u'v' \rangle$ and (b) nonlocal eddy viscosity ν_{NL1212} for the rotating channel flow in Case 2. ∘ : DNS data of Nishimura and Kasagi [5].

channel flow in Case 1. The solid line denotes the shear stress directly obtained from the present DNS whereas the dashed line denotes the shear stress evaluated using the nonlocal expression (4). The two curves agree well with each other; this agreement means that the nonlocal expression is accurate. Figure 1(b) shows profiles of the nonlocal eddy viscosity ν_{NL1212} as a function of y' for several y locations. For example, the dotted line represents that the Reynolds stress at $y = -0.414$ is affected by the velocity gradient at remote position within nearly half the channel width.

Figure 2 shows the result of the rotating channel flow in Case 2. The shear stress shows an asymmetric profile. The nonlocal eddy viscosity in Fig. 2(b) shows broader profiles compared to those for the non-rotating channel flow in Fig. 1(b). The largest convective motion in the rotating channel flow is as large as the channel height (not shown here) and corresponds to Görtler-like streamwise vortices observed in experiment and DNS [6, 7].

4 Anisotropic and Temporally-Nonlocal Properties of Channel Flows

In this section, we examine anisotropic and temporally-nonlocal properties of channel flows. A second-order nonlinear eddy viscosity model can be written as

$$\langle u'_i u'_j \rangle^* = -2\nu_T S_{ij} - \eta_1 (S_{ik}S_{kj})^* - \eta_2 (S_{ik}\bar{W}_{kj} - \bar{W}_{ik}S_{kj}) - \eta_3 (\bar{W}_{ik}\bar{W}_{kj})^*, \quad (6)$$

where $a^*_{ij} = a_{ij} - a_{kk}\delta_{ij}/3$. To validate such a model, we evaluate how the strain-rate and vorticity tensors affect the Reynolds stress using the nonlocal expression. The velocity fluctuation can be divided into three parts driven by S_{ij}, W_{ij}, and Ω_{ij}, respectively, as

$$u'_i = u'_{Si} + u'_{Wi} + u'_{\Omega i}, \quad (7)$$

where

$$u'_{Si}(\mathbf{x},t) = -\int d\mathbf{x}' \int_0^t dt' g_{ik}(\mathbf{x},t;\mathbf{x}',t')u'_m(\mathbf{x}',t')S_{km}(\mathbf{x}',t'), \quad (8)$$

$$u'_{Wi}(\mathbf{x},t) = -\int d\mathbf{x}' \int_0^t dt' g_{ik}(\mathbf{x},t;\mathbf{x}',t')u'_m(\mathbf{x}',t')W_{km}(\mathbf{x}',t'), \quad (9)$$

$$u'_{\Omega i}(\mathbf{x},t) = -\int d\mathbf{x}' \int_0^t dt' g_{ik}(\mathbf{x},t;\mathbf{x}',t')u'_m(\mathbf{x}',t')\Omega_{kn}. \quad (10)$$

The Reynolds shear stress can then be divided into nine terms as follows

$$\langle u'v' \rangle = \langle u'_S v'_S \rangle + (\langle u'_W v'_S \rangle + \langle u'_\Omega v'_S \rangle + \langle u'_S v'_W \rangle + \langle u'_S v'_\Omega \rangle)$$

$$+ (\langle u'_W v'_W \rangle + \langle u'_W v'_\Omega \rangle + \langle u'_\Omega v'_W \rangle + \langle u'_\Omega v'_\Omega \rangle). \quad (11)$$

For the non-rotating channel flow, terms involving $u_{\Omega i}$ vanish and the shear stress consists of four terms. Figure 3(a) shows the non-vanishing four terms in (11) in Case 1. Contribution from each term is very large. This large value is because the transport equations for $\langle v'^2_S \rangle$ and $\langle v'^2_W \rangle$ include non-zero production terms unlike that for $\langle v'^2 \rangle$.

Next, we examine temporally-nonlocal property of the Reynolds stress in the non-rotating channel flow. We introduce the velocity component u'_{ST} that is driven by the strain-rate tensor at time t'

$$u'_{STi}(\mathbf{x},t;t') = -\int d\mathbf{x}' g_{ik}(\mathbf{x},t;\mathbf{x}',t')u'_m(\mathbf{x}',t')S_{km}(\mathbf{x}',t'). \quad (12)$$

Using this velocity component, the shear stress can be written as

$$\langle u'_W v'_S \rangle = \int dt' \langle u'_W(\mathbf{x},t)v'_{ST}(\mathbf{x},t;t') \rangle, \quad (13)$$

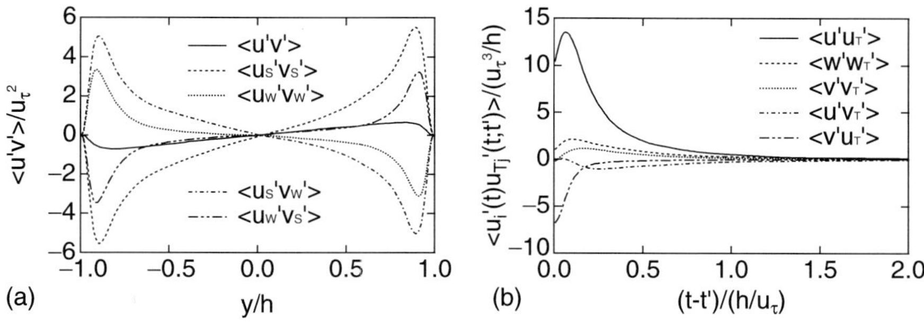

Fig. 3. Profiles of (a) terms in (11) and (b) integrands in (13) for the non-rotating channel flow in Case 1.

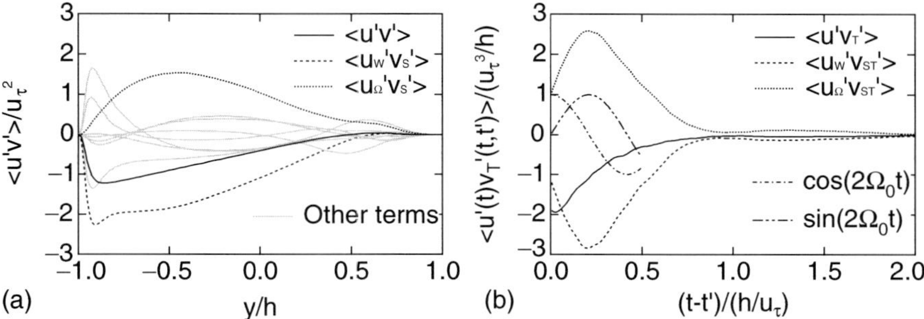

Fig. 4. Profiles of (a) terms in (11) and (b) integrands in (13) for the rotating channel flow in Case 3.

where

$$u'_{Si}(\mathbf{x}, t) = \int_0^t \mathrm{d}t' \, u'_{STi}(\mathbf{x}, t; t'). \tag{14}$$

The integrand of (13) is a kind of two-time velocity correlation and represents the effect of the strain rate at t' on the shear stress at t. Figure 3(b) shows the integrands for the Reynolds stress at $y = -0.8$ as functions of $t - t'$. The correlation time is comparable to the turbulent time scale $k/\varepsilon = 0.273$.

Finally, we examine the Reynolds stress in the rotating channel flow in Case 3. Figure 4(a) shows the nine terms in (11) for the rotating channel flow. It is clear that two terms $\langle u'_W v'_S \rangle$ and $\langle u'_\Omega v'_S \rangle$ are dominant and nearly cancelled to each other. This result suggests that the importance of the non-linear term involving $S_{ik}\bar{W}_{kj}$. Figure 4(b) shows the integrands of the two dominant terms and the shear stress itself at $y = -0.18$ as functions of $t - t'$. Compared to the integrand for the shear stress itself, those for the two dominant terms show longer time correlation. This long correlation suggests that temporally-nonlocal effect for the two terms can be important.

It is well known that the mean absolute vorticity nearly vanishes in the core region of spanwise rotating channel flow [6, 7]. The author previously proposed the following model to better understand the zero mean absolute vorticity state

$$\langle u_i' u_j' \rangle^* = -2\nu_T S_{ij} + \tau \eta_1 \left(\frac{\bar{D} S_{ik}}{Dt} \bar{W}_{kj} - \bar{W}_{ik} \frac{\bar{D} S_{kj}}{Dt} \right), \tag{15}$$

where $\bar{D} S_{ij}/Dt = DS_{ij}Dt + \Omega_{ik}S_{kj} - S_{ik}\Omega_{kj}$. This simple nonlinear eddy-viscosity model was shown to reproduce the nearly zero mean absolute vorticity in the core region. The present analysis of the Reynolds stress supports the importance of the second term on the right-hand side of (15) because this term involves S_{ij} and $\bar{W}_{ij}$ and $\bar{D} S_{ij}/Dt$ represents a temporally-nonlocal effect.

Since DS_{ij}/Dt vanishes in the rotating channel flow, $\bar{D} S_{ij}/Dt$ is replaced by $\Omega_{ik}S_{kj} - S_{ik}\Omega_{kj}$ in (15). The resulting expression suggests that the frame-rotation effect can be described by third-order nonlinear terms. Some first- and second-order nonlinear models can also account for the effect [8]. This is because the models involve a coefficient like $(1 + C\bar{W}_{ij}^2)^{-1}$ which can be approximated as $1 - C\bar{W}_{ij}^2$ and makes the models effectively higher order.

5 Conclusions

An exact nonlocal expression for the Reynolds stress derived using the Green's function is applied to the DNS of non-rotating and rotating channel flows. Spatially- and temporally-nonlocal and anisotropic properties of the Reynolds stress are examined. The nonlocal eddy viscosity shows that spatially-nonlocal effect is stronger for the rotating channel flow than for the non-rotating one. For the shear stress in the rotating channel flow, two terms related to the strain-rate and absolute vorticity tensors are dominant and show long time correlation. This result supports a simple nonlinear eddy-viscosity model that reproduces the zero mean absolute vorticity state of the rotating channel flow.

References

1. Kraichnan RH (1987) Complex Syst 1:805–820
2. Hamba F (2004) Phys Fluids 16:1493–1508
3. Hamba F (2005) Phys Fluids 17:115102
4. Moser RD, Kim J, Mansour NN (1999) Phys Fluids 11:943–945
5. Nishimura M, Kasagi N (1996) In: Proceedings of the Third KSME-JSME Thermal Engineering Conference. Kyongju, 3:77–82
6. Johnston JP, Halleen RM, Lezius DK (1972) J Fluid Mech 56:533–557
7. Kristoffersen R, Andersson HI (1993) J Fluid Mech 256:163–197
8. Nagano Y, Hattori H (2002) J Turbul 3:006

On Turbulence Characteristics Around the Two-Dimensional Symmetric Aerofoil: Kinematic Simulation and Experiments

Yasuhiko Sakai[1], Shinji Tagawa[1], Kouji Nagata[1], Takashi Kubo[2]
and Julian C. R. Hunt[3]

[1] Dept. of Mechanical Science and Engineering, Nagoya University, Furo-cho, Chikusa-ku, Nagoya 464-8603, Japan
ysakai@mech.nagoya-u.ac.jp, tagawa@sps.mech.nagoya-u.ac.jp, nagata@nagoya-u.jp

[2] EcoTopia Science Institute, Nagoya University, Furo-cho, Chikusa-ku, Nagoya 464-8603, Japan
kubo@nagoya-u.jp

[3] Dept. of Earth Sciences, University College London, U.K.
julian.hunt@cpom.ucl.ac.uk

Abstract. Turbulent field around a two-dimensional symmetric aerofoil in the grid turbulence has been measured and compared with the kinematic simulation results. The model used in this study is the combined model of random Fourier modes method and rapid distortion theory (RDT). For two cases of $\tau_{\max}/L_E$ ($\tau_{\max}$: the maximum aerofoil thickness, L_E: the integral length scale of incident turbulence), the effects of blocking and distortion on the wavenumber spectra and the r.m.s. velocity fluctuation have been mainly investigated. As a whole, the simulation results show good agreements with the experimental results, so the validity of the present kinematic simulation is confirmed.

Keywords: turbulence, aerofoil, rapid distortion theory, kinematic simulation, spectrum

1 Introduction

Understanding the turbulent field around a body is useful for engineering design, for example, for predicting the flow around structure or vehicle.

Numerical simulations of turbulent flow are classified roughly into dynamical simulation (DS) and kinematic simulation (KS). The typical ones of DS are the direct numerical simulation (DNS, e.g. Moin and Mahesh 1998 [1]) and the large eddy simulation (LES, e.g. Lesieur 1997 [2]), which are at present most extensively applied to the analysis of various flows. But in order to satisfy spatial and temporal resolution requirements and numerical stability for the turbulence simulation, both methods still require a large memory capacity

273

Y. Kaneda (ed.), *IUTAM Symposium on Computational Physics and New Perspectives in Turbulence*, 273–278.

and computational time so that the simulations inevitably become very expensive. On the other hand, KS makes it possible to simulate certain aspects of high Reynolds number turbulence at relatively low cost [3].

The purpose of this study is to develop the kinematic simulation method to predict the turbulence characteristics around a bluff body. The model used in this study is the combined model [4] of random Fourier modes method [3] and rapid distortion theory (RDT) [5].

In order to verify the validity of the kinematic simulation methods, a turbulent field around the two-dimensional symmetric aerofoil in the grid turbulence has been measured and compared with the simulation results.

2 Modeling

2.1 The Velocity Field by the Random Fourier Modes Method [3, 4]

In the method of random Fourier modes [3], a homogeneous isotropic turbulent field $\boldsymbol{u}_0$ is expressed as the sum of the finite Fourier modes as

$$\boldsymbol{u}_0(\boldsymbol{x}, t) = \sum_{n=1}^{N} (\boldsymbol{a_n} \times \hat{\boldsymbol{\kappa}}_n) \cos(\boldsymbol{\kappa}_n \cdot \boldsymbol{x} - \omega_n t)$$
$$+ (\boldsymbol{b}_n \times \hat{\boldsymbol{\kappa}}_n) \sin(\boldsymbol{\kappa}_n \cdot \boldsymbol{x} - \omega_n t). \tag{1}$$

The wavenumber vectors $\boldsymbol{\kappa}_n$ are randomly distributed on the surface of a sphere of a radius $|\boldsymbol{\kappa}_n|$ in each realization of the flow field. $\hat{\boldsymbol{\kappa}}_n = \boldsymbol{\kappa}_n/\kappa_n$ is the unit vector, where $\kappa_n = |\boldsymbol{\kappa}_n|$. The vectors $\boldsymbol{a}_n$, $\boldsymbol{b}_n$ are chosen randomly from the 3-dimensional isotropic Gaussian distribution. It is noted that the velocity field automatically satisfies the incompressibility condition, $\nabla \cdot \boldsymbol{u}_0 = 0$.

The angular frequency ω_n is modeled as follows [4],

$$\omega_n = \kappa_n \nu_n, \quad \nu_n^2 = \frac{2}{3}\gamma \int_{\kappa_{n-1}}^{\kappa_n} E(\kappa)d\kappa, \tag{2}$$

where γ is a factor to compensate for the energy contained beyond the maximum wavenumber κ_N and $E(\kappa)$ is the 3-dimensional energy spectrum. $E(\kappa)$ is specified by the modeled spectrum [6] defined by

$$E(\kappa) = C\epsilon^{2/3}\kappa^{-5/3} f_L(\kappa L_s) f_\eta(\kappa\eta), \tag{3}$$

where

$$f_L(\kappa L_E) = \left(\frac{\kappa L_E}{[(\kappa L_E)^2 + c_L]^{1/2}}\right)^{5/3+p_0}, \tag{4}$$

$$f_\eta(\kappa\eta) = \exp\left[-\beta\left\{[(\kappa\eta)^4 + c_\eta]^{1/4} - c_\eta\right\}\right]. \tag{5}$$

Here, $p_0 = 4$, $C = 1.5$, $\beta = 5.2$ and $L_E(\equiv K^{\frac{3}{2}}/\epsilon$; K: turbulent kinetic energy, ϵ: mean dissipation rate of turbulent kinetic energy) is the integral length scale of turbulence. The constants c_L and c_η are determined such that the integrals of $E(\kappa)$ and $2\nu\kappa^2 E(\kappa)$ over the whole wavenumber space become K and ϵ, respectively.

2.2 Velocity Fluctuation around a Bluff Body

Fluctuating velocity $\boldsymbol{u}$ around a bluff body is expressed in terms of a stream function $\boldsymbol{\Psi}$ and a velocity potential ϕ, so that

$$\boldsymbol{u} = \boldsymbol{u}_0 + \nabla \times \boldsymbol{\Psi} + \nabla \phi = \boldsymbol{u}_0 + \boldsymbol{u}_d + \boldsymbol{u}_s, \tag{6}$$

where $\boldsymbol{\Psi}$ is chosen to satisfy $\nabla \cdot \boldsymbol{\Psi} = 0$, and $\nabla^2 \phi = 0$. Here, $\boldsymbol{u}_0 C \boldsymbol{u}_d = \nabla \times \boldsymbol{\Psi}$ and $\boldsymbol{u}_s = \nabla \phi$ are called the upstream term, the distortion term, and the source or blocking term, respectively. The upstream turbulence $\boldsymbol{u}_0$ is assumed to be homogeneous and isotropic. The boundary conditions are $\boldsymbol{u}_d \to 0$, $\boldsymbol{u}_s \to 0$ as going away from the body, and $\boldsymbol{u}_d \cdot \boldsymbol{n} = 0$, $\boldsymbol{u}_s \cdot \boldsymbol{n} = -\boldsymbol{u}_0 \cdot \boldsymbol{n}$ on the body surface $\boldsymbol{S}$, where $\boldsymbol{n}$ is the unit normal vector on $\boldsymbol{S}$. To express the fluctuating velocity field $\boldsymbol{u}$ in term of the upstream fluctuating velocity $\boldsymbol{u}_0$, the transfer tensor $\boldsymbol{M}$ is introduced:

$$\hat{\boldsymbol{u}}(x, y, k_z, \omega) = \iint_{-\infty}^{\infty} \boldsymbol{M}(x, y, \boldsymbol{k}) \boldsymbol{S_0}(\boldsymbol{k}, \omega) \mathrm{d}k_x \mathrm{d}k_y \tag{7}$$

$$\boldsymbol{u}(\boldsymbol{x}, t) = \iint_{-\infty}^{\infty} \hat{\boldsymbol{u}}(x, y, k_z, \omega) \exp\left\{i\left(k_z z - \omega t\right)\right\} \mathrm{d}k_z \mathrm{d}\omega, \tag{8}$$

where $\boldsymbol{S_0}(\boldsymbol{k}, \omega)$ is the Fourier transform of $\boldsymbol{u}_0$. The transfer tensor $\boldsymbol{M}$ is modeled according to the eddy sizes relative to the length scale of the body.

For the fluctuating velocity at large scales, the distortion term can be neglected, so that the fluctuating velocity $\boldsymbol{u}$ is modeled only by the upstream term $\boldsymbol{u}_0$ and the source term $\boldsymbol{u}_s$. Regarding the source term, the transfer tensor $\boldsymbol{M}$ is determined by the potential flow theory [7].

For the fluctuating velocity at small scales, the source term expressing irrotational fluctuation became effective only near the surface. Therefore, we ignore the source term in (6) at small scales. On the other hand, the distortion term expressing the rotational motion cannot be neglected even at small scales. In the present model, the change of vorticity $\boldsymbol{\omega}$ is modeled as follows [5].

$$\omega_i(x, y, z, t) = \gamma_{ij}(x, y)\omega_{0j}(x, y - \Delta_y, t - \Delta_T), \tag{9}$$

where ω_{0j} is the j component of the vorticity for the upstream turbulence, $\gamma_{ij}(x, y)$ is the vorticity distortion tensor, Δ_y is the distance in the y-direction (see Fig. 1) that the mean streamline through the observing point $\boldsymbol{x}_A = (x, y, z)$ has been displaced by the body, and Δ_T is the time difference when the

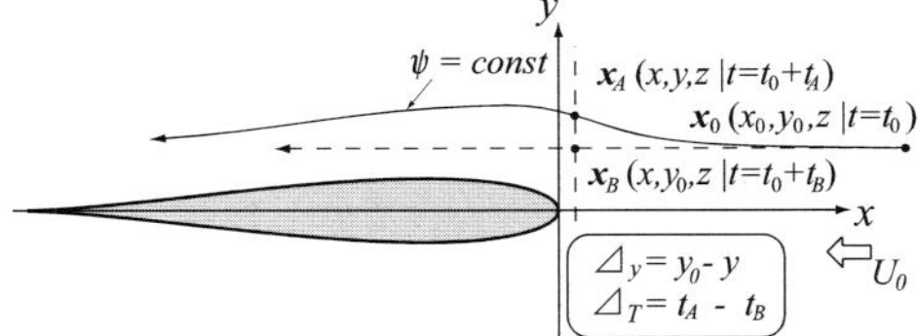

Fig. 1. Definitions of Δ_T and Δ_y

fluid elements should travel along the body surface from the far upstream position $\boldsymbol{x}_0$ to $\boldsymbol{x}_A$, compared with the time when the fluid elements travel to the same x plane as the observing point in the absence of the body.

3 Experimental and Simulation Conditions

The aerofoil used in this study is the Joukowski symmetric aerofoil J012 of the chord length $L = 156$mm. The coordinate system is shown in Fig. 2. Two kinds of grids with mesh sizes $M = 25$ and 100mm are used to generate the grid-turbulence

Table 1. Experimental conditions

M[mm]	R_λ	u'_0/U_0	$\tau_{\max}/L_E$
25	59	0.0175	1.17
100	142	0.0538	0.54

with different integral length scale. The fluctuating velocity field is measured using X-type hot wire probe. The turbulent Reynolds number $R_\lambda = u'\lambda/\nu$ (u': the r.m.s. value of the streamwise velocity fluctuation, λ: Taylor microscale, ν: kinematic viscosity), the relative r.m.s. fluctuating velocity u'_0/U_0, the ratio of the length scale of body (here the maximum aerofoil thickness, $\tau_{\max}$) to the integral length scale of turbulence L_E are tabulated in Table 1.

Measurement are performed along the 4 lines (No.1~No.4) shown in Fig. 3, which are obtained by the Joukowski transform from the radial lines of $\theta = 0, \pi/8, \pi/4$ and $3\pi/8$ around the circle.

With regard to the simulation, the same values of R_λ, u'_0/U_0 and $\tau_{\max}/L_E$ as the experiments are adopted. It was ascertained that the modeled spectra with these parameters agree well with the experimental results of the grid-turbulence. The critical wavenumber κ_c which separates large- and small-scale eddies is chosen to $\kappa_c = \pi/D, D = \tau_{\max}\cos\theta + L\sin\theta$, where D is the length scale of the body which is chosen in consideration of the approaching direction of each line (No.1 $\sim$ No.4) to the aerofoil.

4 Experimental and Simulation Results

Here only the turbulence statistics for the streamwise (x-directional) velocity fluctuation u will be shown because of paper space.

Figure 4 shows the downstream variations of the streamwise r.m.s. velocity u'/u'_0 around the aerofoil (along the lines No.1 $\sim$ No.4). In the figure, r denotes the distance along each line from the aerofoil surface. It is found that on the stagnation line (No.1) u' decreases as approaching to the aerofoil for both cases of $\tau_{\max}/L_E = 0.54$ and 1.17. The decreasing of u' is more notable for $\tau_{\max}/L_E = 0.54$ than $\tau_{\max}/L_E = 1.17$. On the other hand, on the lines (No.2 $\sim$ No.4) off the stagnation line, u' increases near the surface of the aerofoil.

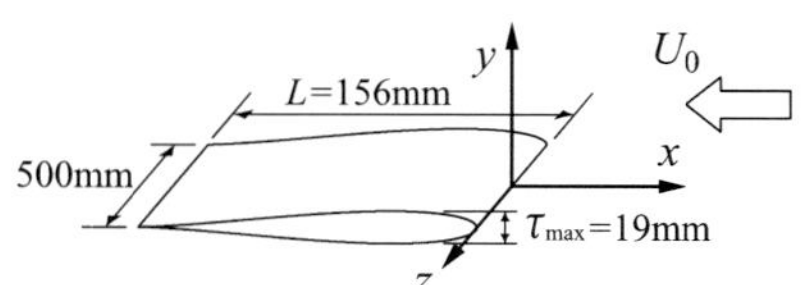

Fig. 2. Coordinate system

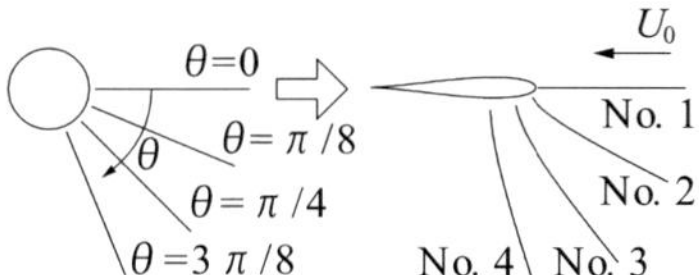

Fig. 3. field of measurement around the aerofoil

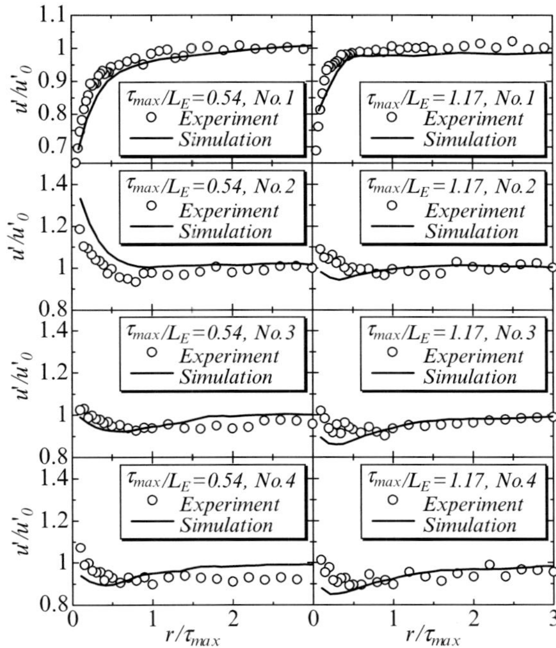

Fig. 4. Variation of the r.m.s. fluctuation velocity u'/u'_0 around the aerofoil

Figure 5 presents the changes of u-spectrum around the aerofoil. The abscissa and the ordinate are non-dimensionalized as $\bar{\kappa}_1 = 2\pi f L_E/U_0$, $E_{11}(\bar{\kappa}_1) = E_1(f)U_0/2\pi u_0'^2 L_E$. It is observed that on the line No.1, the spectrum for large scales decreases as approaching to the aerofoil, while on the line No.2 the distribution of the spectrum for large scales is enhanced. The change of the spectrum is more significant for $\tau_{\max}/L_E = 0.54$ than for $\tau_{\max}/L_E = 1.17$. Here we notice that for the No.2 line of $\tau_{\max}/L_E = 0.54$ the simulation shows the larger values than the experimental results near the surface. This is because of the too strong blocking effect by the simulation model, which is consistent with the results shown in Fig. 4 (see the case of $\tau_{\max}/L_E = 0.54$). On the lines No.3 and No.4, the distribution of the spectrum show the small up and down change according to the variation of u' shown in Fig. 4.

As a whole, the simulation results are in good agreements with the experimental data. However, we still find some deviation between the simulation and experimental results near the surface of the aerofoil, especially for the lines No.2 $\sim$ No.4 in Fig. 4. One of the reasons may be the neglection of the source term for the small-scale turbulence. The modeling of this term is an important future work of this research. Further, although not shown here, it was ascertained that the statistics for the $y-$directional velocity fluctuation v are also consistent with the experimental results.

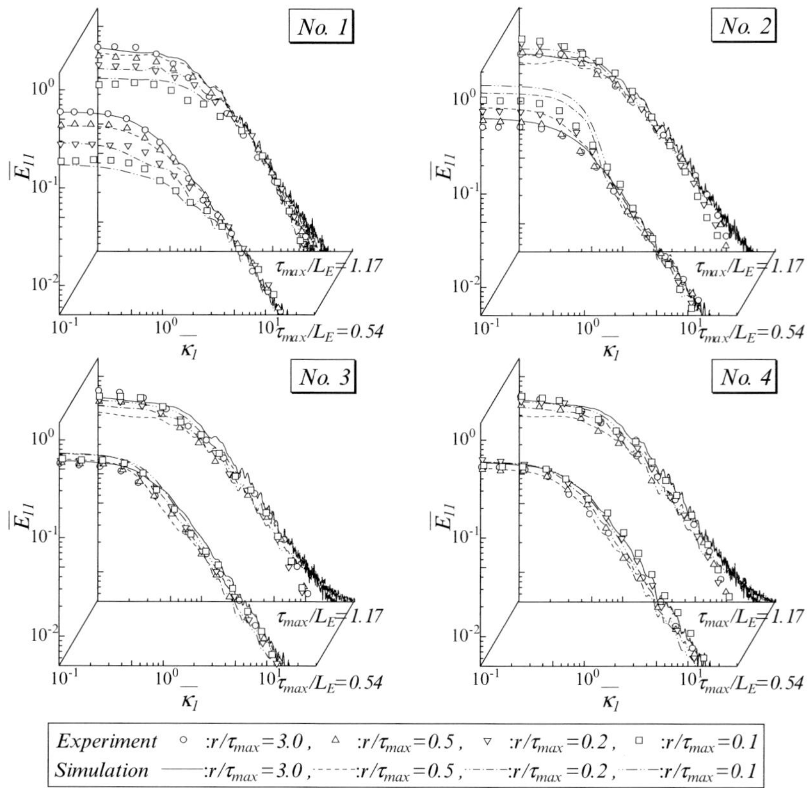

Fig. 5. Variation of the spectrum of streamwise fluctuation velocity u around the aerofoil

5 Conclusions

The turbulent velocity fields around the symmetric aerofoil in two kinds of grid-generated turbulent flows were investigated by the experiments and the simulation using the combined model of random Fourier modes and rapid distortion theory. The results of the simulation are in good agreements with the experimental results, which confirms the validity of the present kinematic simulation method.

References

1. Moin P, Mahesh K, (1998) Annu Rev Fluid Mech 30:539–578
2. Lesieur M, (1997) Turbulence in Fluids, third revised and enlarged edition, Kluwer Acad. Pub., 375–408
3. Fung JCH et al. (1992) J Fluid Mech 236:281–318
4. Sakai Y et al. (1993) Applied Sci Res 51:547–553
5. Hunt JCR (1973) J Fluid Mech 61:625–706
6. Pope SB (2000) Turbulent Flows, Cambridge Univ. Press, 232–234
7. Hunt JCR, Mulhearn PJ (1973) J Fluid Mech 61:245–274

Eddy Viscosity in Magnetohydrodynamic Turbulence

Nobumitsu Yokoi[1], Robert Rubinstein[2] and Akira Yoshizawa[3]

[1] Institute of Industrial Science, University of Tokyo, Komaba, Meguro-ku, Tokyo 153-8505, Japan
`nobyokoi@iis.u-tokyo.ac.jp`
[2] NASA Langley Research Center, `r.rubinstein@larc.nasa.gov`
[3] Emeritus Professor, University of Tokyo, `ay-tsch@mbg.nifty.com`

Abstract. Eddy viscosity in magnetohydrodynamic (MHD) turbulence is investigated with the notion of synthesized timescale composed of the eddy-turnover and the Alfvén times. For the purpose of determining the timescale weight factor, we fully utilize the fact that the decay rate of the turbulent MHD residual energy, the difference between the kinetic and magnetic fluctuation energies, is constituted by the eddy-distortion and the Alfvén effects. Comparison of the spacecraft observations of solar-wind turbulence and the simulation of the turbulence model incorporating the residual-energy equation provides us with the estimate of the timescale weight factor. The spontaneous flow generation observed in the reversed magnetic shear (RS) mode in tokamak plasma is examined by using the numerical simulation of a turbulence model. Flow localization, which could not be reproduced with the usual eddy viscosity, is successfully reproduced with the synthesized viscosity.

Keywords: magnetohydrodynamics (MHD), plasmas, eddy viscosity, Alfven effects, turbulence model

1 Introduction

A uniform magnetic field can alter the properties of turbulence. This is in marked contrast to a uniform velocity, whose effects can be made to vanish by a Galileian transformation. The Alfvén effect as well as the introduction of anisotropy is one of the well known consequences of a strong mean magnetic field in turbulence. [1] Incorporation of the Alfvén effect or its time scale into a turbulence model [2] is of considerable importance in some magnetohydrodynamic (MHD) phenomena in astro/geophysical and fusion plasmas.

The turbulent viscosity is expressed in terms of the spectral density $E(k)$ and the characteristic timescale $\tau(k)$, apart from the constant, as [3]

$$
\nu_{\mathrm{T}} = \int_{k_0}^{\infty} \mathrm{d}k \, \tau(k) E(k),
\tag{1}
$$

279

Y. Kaneda (ed.), *IUTAM Symposium on Computational Physics and New Perspectives in Turbulence*, 279–284.

where k_0 is the characteristic wave number in the energy-containing range. If we assume the Kolmogorov scaling of turbulence: $E(k) = C_K \varepsilon^{2/3} k^{-5/3}$ (ε: turbulent MHD-energy dissipation rate, C_K: model constant) with the eddy-turnover time $\tau_E = K/\varepsilon$ as a timescale, we have the usual eddy viscosity

$$\nu_{TE} \propto K^2/\varepsilon. \qquad (2)$$

Here, use has been made of the spectral expression of the turbulent MHD energy: $K \equiv \langle \mathbf{u}'^2 + \mathbf{b}'^2 \rangle/2 \simeq \int_{k_0}^{\infty} dk E(k)$ (magnetic field expressed in the Alfvén-speed units). On the other hand, if we assume the Iroshnikov–Kraichnan scaling of turbulence: $E(k) = C_{IK}(\varepsilon B)^{1/2} k^{-3/2}$ [$B(=|\mathbf{B}|)$: Alfvén speed of the large-scale magnetic field, C_{IK}: model constant] [4, 5] with the Alfvén timescale $\tau_A = (kB)^{-1}$, the turbulent viscosity is expressed as

$$\nu_{TA} \propto \left(K/B^2\right) K^2/\varepsilon. \qquad (3)$$

We should note that ν_{TA} is expressed in terms of the large-scale quantity B as well as the turbulence quantities, K and ε. We see from Eqs. (2) and (1) that the Alfvénic turbulent viscosity ν_{TA} is reduced by a factor of K/B^2 as compared with the usual eddy viscosity ν_{TE}.

2 Synthesized Timescale

Under the influence of the mean-field inhomogeneity, timescale of turbulence is altered from the usual eddy-turnover time $\tau_E = K/\varepsilon$. Synthesized timescales incorporating the mean rotation and strain effects have been developed in the hydrodynamic turbulence modeling (see [6] and therein). In the presence of the large-scale magnetic field, the Alfvén time $\tau_A = (kB)^{-1}$ associated with the mean field $\mathbf{B}$ may play an important role in turbulence dynamics. So it is reasonable to consider that the timescale of MHD turbulence is expressed in the combination of τ_E and τ_A. [7] For instance, Pouquet et $al.$ composed a triple decay time by taking into account both the eddy-distortion and Alfvén effects in the eddy-damped quasi-normal Markovianized (EDQNM) approximation closure. [2] We consider a synthesized timescale τ_S composed by

$$\tau_S^{-1} = \tau_E^{-1} + \chi \tau_A^{-1}, \qquad (4)$$

where χ is the timescale weight factor. Since the turbulent energy in Alfvénic turbulence is written as $K \sim (\varepsilon B)^{1/2} k_0^{-1/2}$, the Alfvén time is expressed as

$$\tau_A \sim (k_0 B)^{-1} \sim (K/B^2) K/\varepsilon \sim (\mathbf{B}^2/K)^{-1} \tau_E. \qquad (5)$$

This leads to the synthesized timescale [Eq. (4)]:

$$\tau_S = (1 + \chi \mathbf{B}^2/K)^{-1} K/\varepsilon. \qquad (6)$$

The weight factor χ should be determined by comparison of the model simulation with experimental observations or DNS's of MHD turbulence. Note that, for $\mathbf{B}^2 \ll K$, Eq. (6) is reduced to $\tau_S = K/\varepsilon = \tau_E$. On the other hand, for $\mathbf{B}^2 \gg K$, Eq. (6) becomes $\tau_S \simeq \chi^{-1}(K/\mathbf{B}^2)(K/\varepsilon) = \tau_A/\chi$.

3 Turbulent MHD Residual-Energy Equation

The deviation from equipartition between the kinetic and magnetic energies, expressed by the turbulent MHD residual energy $K_{\mathrm{R}}(\equiv \langle \mathbf{u}'^2 - \mathbf{b}'^2 \rangle/2)$, is concerned with the combination of the eddy distortion and the Alfvén effects. It is known that the residual energy plays an important role in dynamo action. [2, 8] In the context of homogeneous MHD turbulence, the residual energy has been studied numerically. [9, 10] Recently the evolution equation of K_{R} has been studied with the aid of the two-scale direct-interaction approximation (TSDIA), a statistical theory for inhomogeneous turbulence. On the basis of the analytical results, a model equation for the turbulent MHD residual ennergy was proposed as [8]

$$\frac{\mathrm{D}K_{\mathrm{R}}}{\mathrm{D}t} = \frac{1}{2}\nu_{\mathrm{R}}\left(\mathsf{S}^2 - \mathsf{M}^2\right) - \varepsilon_{\mathrm{R}} + \nabla \cdot \left(\frac{\nu_{\mathrm{T}}}{\sigma_{\mathrm{R}}}\nabla K_{\mathrm{R}}\right), \tag{7}$$

where S and M are the strain rates of the mean velocity and magnetic field, respectively. Here, ν_{T} is the turbulent viscosity, and ν_{R} is the residual viscosity defined by $\nu_{\mathrm{R}} = \nu_{\mathrm{T}}K_{\mathrm{R}}/K$. The dissipation rate of K_{R}, ε_{R}, is expressed as

$$\varepsilon_{\mathrm{R}} = C_{\varepsilon\mathrm{R}} \left[1 + (C_{r1}/C_{\varepsilon\mathrm{R}})\mathbf{B}^2/K\right] (\varepsilon/K)K_{\mathrm{R}} \tag{8}$$

($C_{\varepsilon\mathrm{R}}$, C_{r1}, σ_{R}: positive model constants). This always works for returrning the MHD turbulence to equipartition. Since the RHS of Eq. (7) is formally linear in K_{R}, the residual energy cannot be generated unless there are some seeds of K_{R}. In other words, if there is a deviation from the equipartition between the kinetic and magnetic energies, the first term in Eq. (7) shows that this deviation can be amplified depending on the mean-field inhomogeneities represented by S^2 and M^2. Note that the presence of mean-velocity strain (S^2), which is indifferent to the mean magnetic field ($\mathbf{B}^2$ and M^2), contributes to the deviation from equipartition. The second or C_{r1}-related term in the square bracket of Eq. (8) shows that the residual energy is suppressed in the presence of the mean magnetic field $\mathbf{B}$. As is the case in the Alfvénic turbulent viscosity expression [Eq. (3)], this is a manifestation of the Alfvén effects leading to the equipartition between the kinetic and magnetic energies in the presence of the mean magnetic field. Equation (8) suggests that the timescale of MHD turbulence is altered by the mean magnetic field $\mathbf{B}$ as $K/\varepsilon \rightarrow \left[1 + (C_{r1}/C_{\varepsilon\mathrm{R}})(\mathbf{B}^2/K)\right]^{-1} K/\varepsilon$. Comparison with Eq. (6) shows that the ratio of constants, $C_{r1}/C_{\varepsilon\mathrm{R}}$, corresponds to the timescale weight factor χ in Eq. (6).

4 Implication from Solar-Wind Turbulence

Solar winds are plasma flows blown away from the coronal base of the Sun. Spacecraft observations of solar winds only provide us with in situ information on velocity and magnetic-field fluctuations in MHD turbulence. Not a few efforts have been made to study MHD turbulence evolution in large-scale solar-wind structures. [11] According to the observations, the solar-wind turbulence shows a high degree of Alfvénicity near the Sun, i.e., high correlation between the velocity and magnetic-field fluctuations and equipartition between the kinetic and magnetic energies. The degree of Alfvénicity decreases as the heliocentric distance increases. [12] One of the unsolved problems of the solar-wind turbulence is the radial behavior of the residual energy. The Alfvén ratio r_A defined by the ratio of the turbulent kinetic to magnetic energies ($r_A = \langle \mathbf{u}'^2 \rangle / \langle \mathbf{b}'^2 \rangle$) is about 0.5 for the heliocentric distance $r > 3$ (AU) without further decline with r. There has been no complete account for this stationary value.

Using a turbulence model incorporating the K_R equation, the radial evolutions of the solar-wind turbulence including the stationary value of $r_A \simeq 0.5$ are successfully reproduced. [13] In the simulation, the model constant associated with the Alfvén-effect decay of K_R, C_{r1}, plays an important role, and is estimated as $C_{r1} = O(10^{-2})$. With $C_{\varepsilon R} = 1$ (which comes from the requirement of the no-magnetic-field limit), we obtain the timescale weight factor $\chi = C_{r1}/C_{\varepsilon R} = O(10^{-2})$. In other words, the solar-wind turbulence observations imply that, for any $\mathbf{B}$, the eddy-distortion and Alfvén times should be synthesized with the timescale weight factor $\chi = O(10^{-2})$. It is worth noting here the universality of the model constants. Namely, once a system of model constants has been optimized in a turbulence model, it should be fixed throughout any applications of the model to various flow problems.

With the synthesized timescale [Eq. (6)], we adopt the MHD turbulence viscosity ν_{TS} as

$$\nu_{TS} = C_\nu \left(1 + \chi \mathbf{B}^2 / K\right)^{-1} K^2 / \varepsilon \quad \text{with} \quad \chi = 0.01, \tag{9}$$

where $C_\nu = 0.09$ as in the usual eddy viscosity model.

5 An Application to Fusion Plasmas

As the first step to the application of synthesized turbulent viscosity ν_{TS}, we may consider an improved confinement in tokamak plasmas, where a strong magnetic field is externally imposed (Fig. 1). It is known that the confinement of high temperature plasmas is much improved in devices with the reversed or negative magnetic shear in the core region (RS mode). [14] A transport barrier was found to be dynamically formed near the region with minimum safety factor $q[= rB^z/(RB^\theta)]$ (Fig. 2) and an associated poloidal plasma rotation was also observed there. [15] From the viewpoint of the turbulent dynamo, the mechanism of poloidal-rotation generation was theoretically investigated and it was shown that the curvature of the electric current coupled with the cross helicity (velocity–magnetic-field correlation) induces a poloidal rotation. [16] This theoretical prediction was also validated by a numerical simulation with the aid of the $K - \varepsilon - W$-type model of MHD turbulence [$W(\equiv \langle \mathbf{u}' \cdot \mathbf{b}' \rangle)$: turbulent cross helicity]. In the simulation, the spontaneous generation of

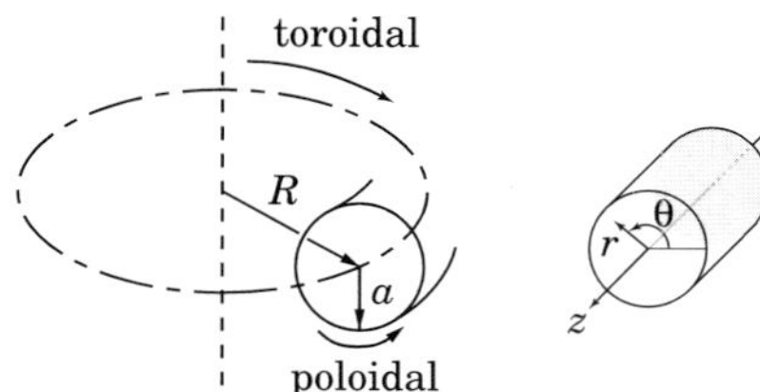

Fig. 1. Tokamak's geometry (R: major radius of tokamak's torus, a: minor radius, r: radial coordinate) and the local cylindrical coordinate system (r, θ, z). The poloidal rotation is expressed by U^θ.

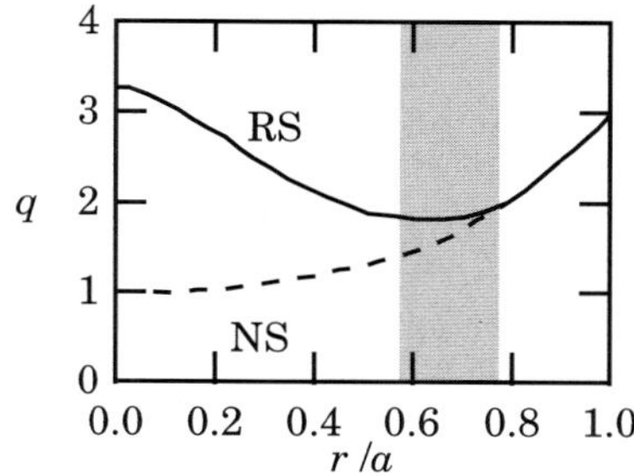

Fig. 2. Radial distributions of the safety factor $q = rB^z/(RB^\theta)$ in the reversed-shear (RS) mode and in the normal-shear (NS) mode. Shaded region corresponds to a transport barrier.

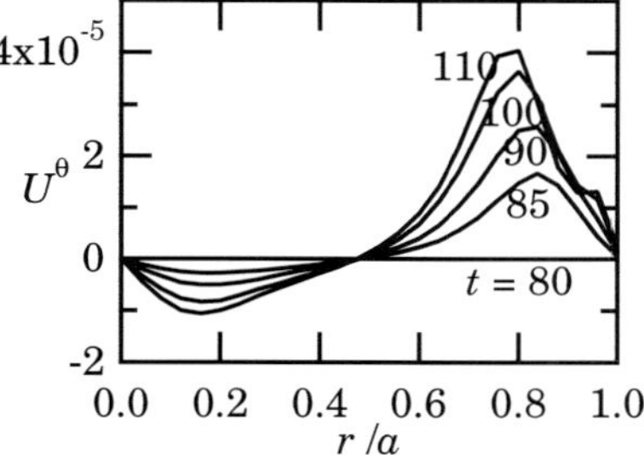

Fig. 3. Radial distribution of the generated poloidal rotation with the eddy viscosity ν_{TE}. The peak of U^θ at $t = 110$ corresponds $O(10)$ km s^{-1} under the present normalization.

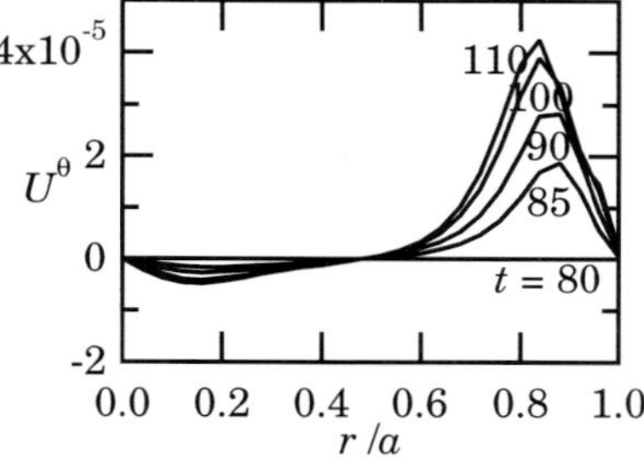

Fig. 4. Radial distribution of the generated poloidal rotation with the synthesized viscosity ν_{TS}.

the poloidal rotation near the region with minimum q was confirmed, but the persistent spatial localization of the observed flows could not be well reproduced. The simulated profile of the poloidal rotation extends to a broad region rather rapidly (Fig. 3).

The typical strength of the toroidal magnetic field in RS mode is a few T, which corresponds to $B^z = O(10^3)$ km s^{-1} in Alfvén-speed units. Since this is much larger than the typical turbulent velocity of a few km s^{-1}, the Alfvénic viscosity is expected to play an important role there. We apply the synthesized turbulent viscosity ν_{TS} [Eq. (9)] to this problem. Figure 4 shows the temporal behavior of the generated rotational velocity in the simulation of the $K - \varepsilon - W$ turbulence model with the synthesized turbulent viscosity ν_{TS}. As compared with the counterpart with the usual eddy viscosity ν_{TE}, the spatial localization of the rotational velocity is remarkable.

6 Conclusions

Using a synthesized timescale composed of the eddy-turnover and the Alfvén times, the eddy viscosity in MHD turbulence was investigated. Utilizing the fact that the residual-energy decay comes from both the eddy-distortion and the Alfvén effects, the timescale weight factor was determined so that the turbulence model incorporating the residual-energy equation might be consistent with the spacecraft observations of the solar-wind turbulence. A turbulence model with this synthesized turbulent viscosity was applied to the RS mode, an improved confinement mode in tokamaks. It was shown that the synthesized turbulent viscosity reproduces the localized flow generation observed in the fusion experiments. This synthesized turbulent viscosity is expected to play an important role in applications of the turbulence model to flow phenomena with a strong mean magnetic field such as the solar internal motions. [17]

References

1. Davidson PA (2001) An Introduction to Magnetohydrodynamics. Cambridge University Press, Cambridge
2. Pouquet A, Frisch U, Léorat J (1976) J Fluid Mech 77:321–354
3. Kraichnan RH (1976) J Atmos Sci 33:1521–1536
4. Iroshnikov PS (1964) Sov Astron 7:566–571
5. Kraichnan RH (1965) Phys Fluids 8:1385–1387
6. Yoshizawa A (2006) Phys Fluids 18:035109 1–17
7. Zhou Y, Matthaeus WH, Dmitruk P (2004) Rev Mod Phys 76:1015–1035
8. Yokoi N (2006) Phys Plasmas 13:062306 1–17
9. Müller W-C, Grappin R (2004) Plasma Phys Contrrol Fusion 46:B91–B96
10. Müller W-C, Grappin R (2005) Phys Rev Lett 95:114502 1–4
11. Tu C-Y, Marsch E (1995) MHD Structures, Waves and Turbulence in the Solar Wind: Observations and Theories. Kluwer Academic, Dordrecht
12. Roberts DA, Goldstein ML, Klein LW (1990) J Geophys Res A 95:4203–4216
13. Yokoi N, Hamba F (2007) An application of the turbulent MHD residual-energy equation to the solar wind. (in preparation)
14. Fujita T, Oikawa T, Suzuki T, et al. (2001) Phys Rev Lett 87:245001 1–4
15. Bell RE, Levinton FM, Batha SH, et al. (1998) Phys Rev Lett 81:1429–1432
16. Yoshizawa A, Yokoi N, Itoh S-I, Itoh K (1999) Phys Plasmas 6:3194–3206
17. Diamond PH, Itoh S-I, Itoh K, Silvers LJ (2006) β-plane MHD turbulence and dissipation in the solar tachocline. In: Hughes, D, Weiss N (eds) Problems of MHD turbulence and dynamo (tentative). Cambridge University Press, Cambridge

Large Eddy Simulation of a Turbulent Boundary Layer Flow over Urban-Like Roughness

Kojiro Nozawa[1] and Tetsuro Tamura[2]

[1] Institute of Technology, Shimizu Corporation, 4-17, Etchujima 3-Chome, Koto-ku, Tokyo 135-8530, Japan
nozawa@shimz.co.jp

[2] Department of Environmental Science and Technology Tokyo Institute of Technology 4259 Nagatsuta, Midori-ku, Yokohama 226-8502, Japan
tamura@depe.titech.ac.jp

Abstract. In this study, Large Eddy Simulation of a boundary layer flow over large-scale roughness was performed targeting the experiments conducted by Cheng and Castro [2]. In order to duplicate the experimental conditions, the quasi-periodic boundary method for rough-wall boundary flows was applied to the inlet boundary conditions. The characteristics of the turbulent boundary layer flows over urban-like roughness (with a roughness area density of 25% and a boundary layer height ratio δ/h around 7) are different from the common homogenous with δ/h larger than 50. The spatial variation of vertical profiles of mean and fluctuation velocities are studied in detail and compared to the experimental data. We focus on the influence on the spatial variations of turbulence structure deduced by large-scale roughness.

Keywords: quasi-periodic boundary condition, random roughness, roughness sublayer, roughness length, zero-plane displacement

1 Introduction

The turbulence characteristics of the boundary layer flows over urban terrains are different from those of well-studied homogenous rough-wall turbulent boundary layer flows. In homogenous rough-wall turbulent boundary layer flows the upper limit of the roughness sublayer is from 2–5h and the inertial sublayer, which correspond with the logarithmic layer, is approximately 15% of boundary layer thickness [1]. While the most of the urban terrain have small δ/h, it could be less than 20 in the case of a huge city. The roughness sublayer may extend to a significant height and the inertial sublayer then becomes squeezed between the roughness sublayer and the outer layer [2]. Cheng and Castro [2] measured the spatially averaged mean velocity of the turbulent boundary layer flows over a urban-type surface ($\delta/h = 7$) in a wind tunnel and identified the upper limit of the inertial sublayer to be 2.3–2.4h, where h

Y. Kaneda (ed.), *IUTAM Symposium on Computational Physics and New Perspectives in Turbulence*, 285–290.

is the cube height. Besides, the turbulence characteristics of turbulent boundary layer flows over relative large-scale roughness has not been well studied compared to those of homogenous rough-wall turbulent boundary layer flows. The objectives of this study are to simulate the flows over large-scale inhomogenous roughness using LES and compare the turbulence characterisitics with those of Cheng and Castro's and to investigate the influence of random roughness on turbulence characteristics of urban-like roughness.

2 Numerical Methods

The Navier-Stokes equations for an incompressible fluid combined with subgrid-scale turbulent viscosity are used for the large-eddy simulation. The dynamic procedure based on the Smagorinsky model is used to identify the model coefficient. The model coefficient is determined through a least-squares minimization procedure [3]. Fourth-order central differencing scheme is used as spatial discretization and second-order time accurate explicit Adams-Bashforth differencing scheme is used for the convective terms and a part of the SGS turbulent diffusion terms. The rest of the diffusion term is treated semi-implicitly by using Crank-Nicolson formulation. The roughness blocks are modeled using the immersed boundary method and twenty grid points are used for a single block in horizontal direction. The Reynolds number based on the free stream velocity and the boundary layer thickness is 9.3×10^4.

2.1 Simulating Spatial Developing Boundary Layer Using LES

The quasi-periodic boundary condition [5] is introduced in streamwise direction to simulate the spatially developing boundary layers. The quasi-periodic boundary condition was proposed by Lund [4] and modified by Nozawa and Tamura [5] to apply to a rough-wall turbulent boundary layer flow. In this method the velocities at the recycle station are rescaled and reintroduced at the inlet and the outflow boundary is set far downstream of the recycle station. The rescaling of the velocity is done by decomposing the velocities into mean and fluctuating parts and applying the appropriate scaling laws to each component separately. Velocity fluctuations in the inner and outer regions are rescaled according to the ratio of friction velocities at the inlet and at the recycle station. The mean velocities are rescaled according to the "law of the wall" in the inner region and a "velocity defect law" in the outer region. The periodic boundary conditions for velocities and pressure are applied in spanwise direction. At the outflow boundary the convective-type boundary condition is applied. The boundary conditions on the top surface of the computational domains are

$$\frac{\partial U}{\partial z} = 0, V = U_0 \frac{d\delta^*}{dx}, \frac{\partial W}{\partial z} = 0, \tag{1}$$

where U_0 is the free stream velocity and δ^* is the displacement thickness of the boundary layer.

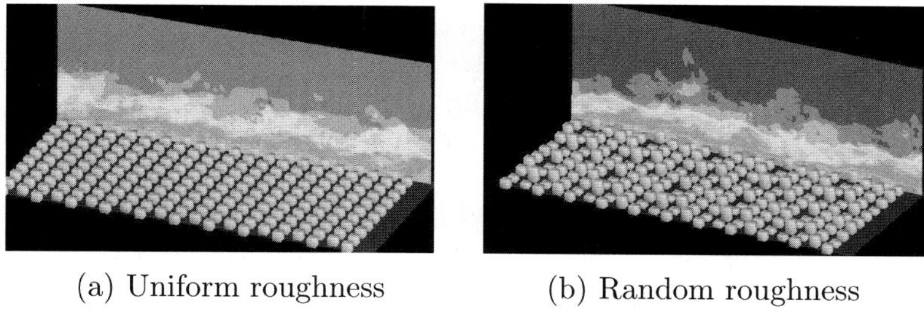

(a) Uniform roughness   (b) Random roughness

Fig. 1. Roughness surface

2.2 Roughness Surfaces

The computational domains for the simulations are $88h \times 20h \times 40h$ in streamwise, spanwise and wall-normal directions, respectively, where h is the height of the cubic roughness elements. The uniform cubic blocks are placed in staggerd pattern following the experiments [2] (Fig. 1). The roughness density defined as the ratio of frontal area to the floor area occupied by a single element is around 25%. In the case of random roughness, the height of the roughness elements is set as random variable, having five different heights chosen from a normal distribution with a mean and a standard deviation of $1.0h$ and $0.37h$ respectively.

3 Results

3.1 Depth of Roughness Sublayer and Inertial Sublayer

Figure 2 shows the vertical profiles of spatially averaged Reynolds stress. Raupach [6] indicated that vertical profiles of single-point velocity moments should collapse to common curves indepedent of wall roughness above the roughness sublayer. The ratios $\sigma_u/u_*, \sigma_v/u_*$ and σ_w/u_* take values of aproximately 2.1, 1.4 and 1.1 respectively at 10% of the boundary layer thickness in the typical wind tunnel test, whose δ/h were 9-20. The vertical profiles of the uniform roughness flow are in good agreement with those of experiments conducted by Cheng and Castro, while the v and w components are overestimating the values of wind tunnel test introduced by Raupach $et~al$ [7]. The stresses are nomalized by u_* deduced from the spatially averaged shear stress in both the roughness sublayer and the inertial sublayer. This u_* would have small values compared with the values of u_* deduced from the constant-stress part of each profiles. The vertical profiles of shear stress at sixteen different locations around a single element are shown in Fig. 3. The quantities are averaged over time and repeating locations in spanwise direction. The upper limit of roughness sublayer(RS) which is defined as the convergence height of vertical profiles of shear stress is estimated $1.8h$ in the uniform roughness flow. Following the method of Cheng and Castro [2], the upper limit of inertial sublayer(IS) was estimated $2.6h$. These values are in good agreement

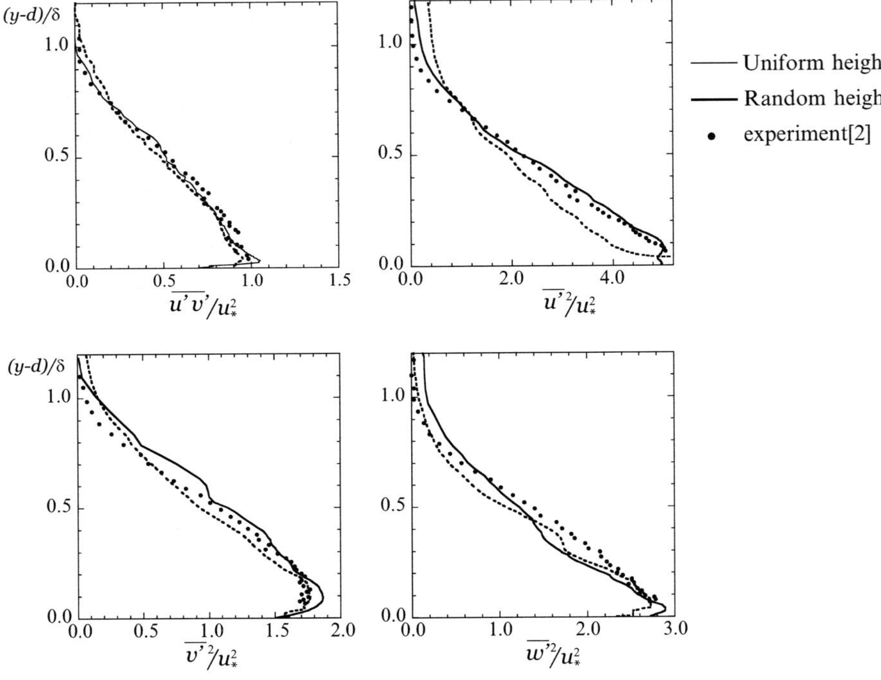

Fig. 2. Spatially averaged Reynolds stress profiles.

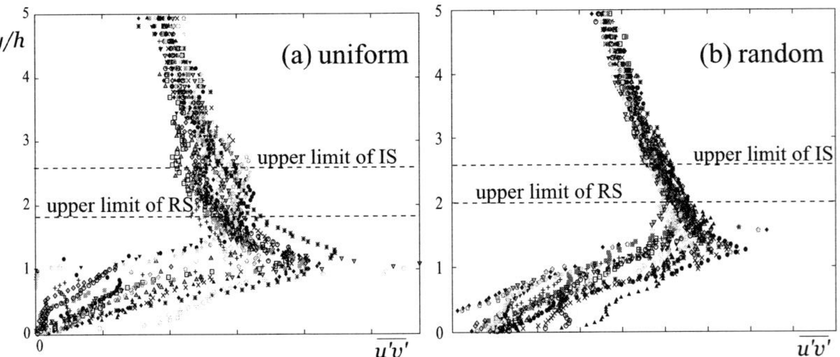

Fig. 3. Vertical profiles of shear stress at 16 difference location around a single element

with the experiment, although these are very small compared to those of the well-studied homogenous rough wall turbulent boundary layer flows. In the case of random roughness the upper limit of IS and RS have values (IS $< 2.6h$ and RS $< 2.0h$) with close to those of uniform roughness flow. We could not find that the random-height roughness has the strong influence to the upper limit of the RS and the IS.

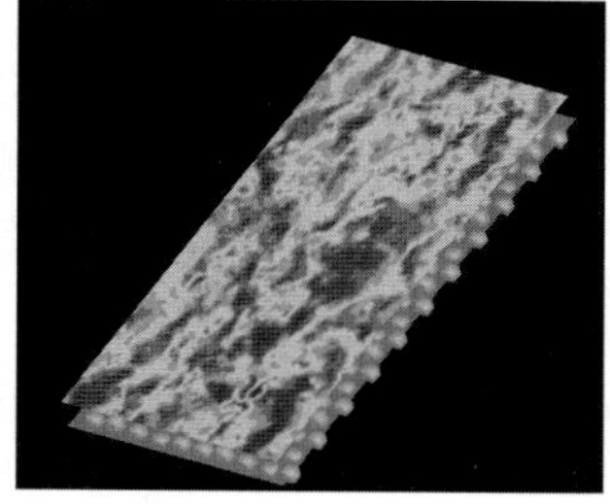

Fig. 4. Instantaneous vortical structure identified using isosurface of second invariant of velocity gradient tensor (Q)

Fig. 5. Large-scale coherent structures above roughness layer ($y = 2.6h$, white: high speed, black: low speed).

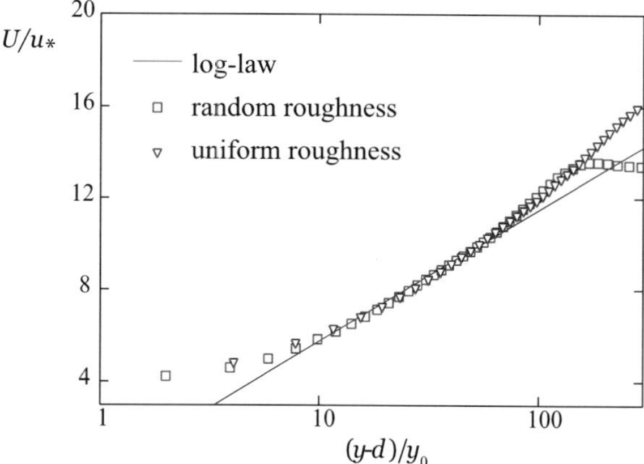

Fig. 6. Mean velocity profiles against log-law

Figure 4 shows the instantaneous vortical structures over random roughness identified using isosurface of second invariant of velocity gradient tensor (Q). Most of the vortical structures are formed below $2h$ while a few vortical structures can be seen above $2h$. This tendency does not change according to the iso-value of the second invariant. These small vortical structures are strongly influenced by the local roughness elements directly. Above the $2h$ line, the large scale structures are dominant in the flows. While high and low speed regions whose length scale in streamwise direction are around $30h$ are dominant above $2h$ (Fig. 5). The change in the dominant structures in the flow around $2h$ is consistent with the value for RS estimated from the vertical profiles of the shear stress.

3.2 Estimation of Roughness Length and Zero-Plane Displacement

The mean velocities are normalized using friction velocity which were deduced by averaging spatially averaged shear stress in the whole surface layer (RS and IS) (Fig. 6). The friction velocity of the random roughness is around 30% larger than that of the uniform roughness. The roughness length and zero-plane displacement are identified fitting the mean velocity profiles to the log-law. The zero-plane displacement of random roughness is only 10% larger than that of uniform roughness, while the roughness length is approximately 2.5 times as large. This results are consistent with the experimental result [2], that roughness length is dependent on the standard deviation of height variability in roughness elements.

4 Summary

The large eddy simulation of spatially developing rough-wall turbulent boundary layer flows were performed to study the effect of urban-like roughness on the turbulence structure. The vertical profiles of the uniform roughness flow were in good agreement with those of the experiment conducted by Cheng and Castro [2]. From the spatial variation of the shear stress vertical profiles the upper limit of the RS was estimated $1.8h$ (uniform) and $2.0h$ (random), the IS was limited to $2.6h$ (both uniform and random). The small scale strong vortices spreaded below $2h$, while the low-speed large scale structures were dominant above $2h$. These could fit with estimation of RS using the vertical profiles of shear stress. The roughness length of random case was almost 2.5 times as large as that of uniform case, while the difference of the zero-plane displacement was limited 10% and this result was consistent with early studies.

References

1. Jiménez J (2004) Annu Rev Fluid Mech. 36:173–196
2. Cheng H, Castro IP (2002) Boundary-Layer Meteorology 104:229–259
3. Lilly DK (1992) Phys Fluids A4 3:633–635
4. Lund TS, Wu X, Squires KD (1998) J Comp Phys 140:233–258
5. Nozawa K, Tamura T (2001) Simulation of rough-wall turbulent boundary layer for LES inflow data Proc Turbulent Shear Flow Phenomena
6. Raupach MR (1981) J Fluid Mech 108:363–382
7. Raupach MR, Antonia RA, Rajagopalan S (1991) Appl Mech Rev 44:1–25

A Spectral Method for Unbounded Domains and Its Application to Wave Equations in Geophysical Fluid Dynamics

Keiichi Ishioka

Division of Earth and Planetary Sciences, Graduate School of Science, Kyoto University, Kitashirakawa Oiwake-cho, Sakyo-ku, Kyoto, 606-8502, Japan
`ishioka@gfd-dennou.org`

Abstract. A spectral method for wave equations on 1D or 2D unbounded domain is proposed. It uses a mapping from a closed domain to an unbounded domain with a standard spectral method for the closed domain. Although the idea to use the mapping is not brand-new, two new ideas are introduced into the spectral method. One is "pseudo-hyper-viscosity", which prevents spurious wavy errors from growing. The other is a variational formulation of the spectral discretization, which leads to enstrophy conservation so that it can avoid numerical instability. Several application examples for GFD wave equations are shown.

Keywords: spectral method, unbounded domain, conformal mapping, geophysical fluid dynamics, wave equation

1 Introduction

In geophysical fluid dynamics (GFD), several kinds of waves such as gravity wave and Rossby wave play important roles. To investigate the radiation of these waves from a fluid motion numerically is a little difficult task because wave reflection at the boundary often makes the result incorrect if the computational domain is bounded. One method to overcome this difficulty is using a non-reflective boundary condition (NRBC). If the property of the waves in the system is suitable for NRBC, this method works well. Waves in GFD, however, are so dispersive that it is difficult to construct an accurate NRBC. Another method is introducing a sponge region near the boundary to damp waves. Although it is applicable for dispersive waves, it is rather difficult to determine the width of the sponge region and the damping coefficient depending on the wave speed.

In this study, another method in which the computational domain is unbounded is explored. This method uses a spectral method on a sphere with a conformal mapping from the infinite plane, which is originally developed by [2]. Making the domain unbounded itself can not prevent the wave reflection, however. We newly introduce a "pseudo-hyper-viscosity", which acts like

291

Y. Kaneda (ed.), *IUTAM Symposium on Computational Physics and New Perspectives in Turbulence*, 291–296.
© 2008 *Springer*.

a sponge far away from the origin and can damp the waves very efficiently. Combining the spectral method with the pseudo-hyper-viscosity, we can conduct accurate numerical experiments of a fluid motion accompanied with wave radiation. Several application examples and non-trivial development of the spectral method itself to treat GFD wave equations are also shown.

2 One-Dimensional Case

Let us consider 1D linear shallow water equation on a rotating frame

$$\frac{\partial \zeta}{\partial t} = -fD \equiv \dot{\zeta}, \quad \frac{\partial D}{\partial t} = -\frac{\partial^2 h}{\partial x^2} + f\zeta \equiv \dot{D}, \quad \frac{\partial h}{\partial t} = -HD \equiv \dot{h}.$$

Here, ζ is the vorticity, D is the divergence, h is the deviation of water depth, H is the mean water depth, and f is the Coriolis parameter. Suppose that we want to solve the equation above numerically in $\mathbb{R}^1$ ($-\infty < x < \infty$) with the initial condition

$$h = \frac{1}{\sqrt{2\pi}\sigma} e^{-x^2/(2\sigma^2)}, \quad \zeta = D = 0,$$

and the boundary condition, $h \to 0$ as $|x| \to \infty$. Time evolution of the exact solution of the problem is shown in Fig. 1a. We adopt a mapping from a circle to $\mathbb{R}^1$ ($\theta \in (-\pi, \pi) \mapsto x \in (-\infty, \infty)$)

$$x = 2R \tan\left(\frac{\theta}{2}\right),$$

where R is the radius of the circle. The dependent variables are expanded as,

$$a(x, t) = \sum_{k=-K}^{K} \hat{a}_k(t) e^{ik\theta}$$

Here, K is the truncation wavenumber, and a represents either h, ζ, or D. The space derivative is calculated on the θ-domain as

$$\frac{\partial^2 h}{\partial x^2} = \frac{1}{8R^2} \left\{ (3 + 4\cos\theta + \cos(2\theta)) \frac{\partial^2 h}{\partial \theta^2} - (2\sin\theta + \sin(2\theta)) \frac{\partial h}{\partial \theta} \right\}.$$

Time derivative of the coefficients are determined by Galerkin method as

$$\frac{d\hat{a}_k}{dt} = \frac{1}{2\pi} \int_0^{2\pi} \dot{a}(x, t) e^{-ik\theta} d\theta.$$

The manipulation does not change basically when the nonlinear advection terms are introduced. The idea of this mapping method is originally introduced by [1]. The mapping itself, however, cannot overcome the wave reflection problem. Spurious waves coming back from far field contaminate the

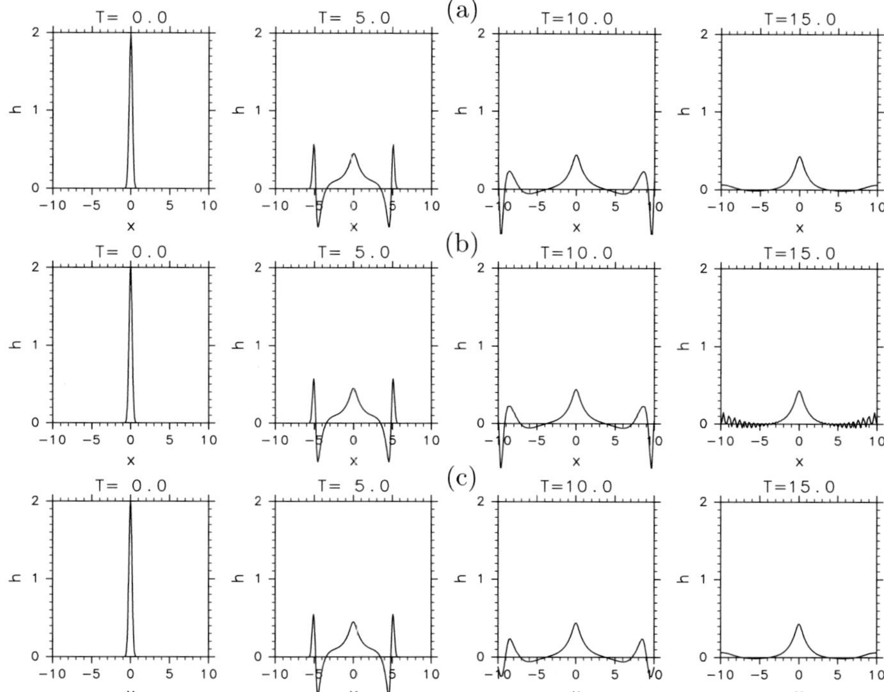

Fig. 1. Time evolution of $h(x,t)$. ($f = 1, H = 1, \sigma = 0.2$). (a): exact solution. (b): numerical solution by the mapping spectral method ($R = 1, K = 319$) without PHV. (c): numerical solution by the mapping spectral method ($R = 1, K = 319$) with PHV ($l = 2, \nu = 10^{-8}$).

numerical solution at last (Fig. 1b). The cause of the spurious waves is explained as follows. Waves going far away from the origin is mapped to waves of shorter and shorter wavelength on the θ-domain. Once the wavelength on the θ-domain falls below the resolution of the expansion, $2\pi/K$, the Gibbs phenomenon occurs and the numerical solution is contaminated by the spurious waves of wavenumber $\sim K$. To overcome the difficulty, we introduce "pseudo-hyper-viscosity"(PHV) term in the right hand side of the equation as,

$$\nu(-1)^{l-1}\left(\frac{\partial^2}{\partial\theta^2}\right)^l a.$$

Here, ν is PHV coefficient and l is the order of PHV. Using the mapping, $\partial^2/\partial\theta^2$ is expressed as

$$\frac{\partial^2}{\partial\theta^2} = R^2\left(1 + \left(\frac{x}{2R}\right)^2\right)^2 \frac{\partial^2}{\partial x^2} + \frac{x}{2}\left(1 + \left(\frac{x}{2R}\right)^2\right)\frac{\partial}{\partial x}.$$

Therefore, the PHV term behaves like a normal hyper-viscosity term near the origin ($|x| \ll R$), but strongly damps waves in far field ($|x| \to \infty$) so that the PHV term behaves like a sponge in far field. Introducing the PHV term, waves going to far field are damped efficiently, so that an accurate solution is obtained near the origin (Fig. 1c). The merit of the PHV is that operating $(\partial^2/\partial\theta^2)^l$ is very easy in coding because it can be done by multiplying the spectral coefficients by $(-k^2)^l$. This 1D spectral method with PHV was already used in [3] to investigate gravity wave radiation from unsteady rotational flow in f-plane shallow water equation.

3 Two-Dimensional Case

Let us consider a linear quasi-geostrophic (QG) vorticity equation on a β-plane

$$\frac{\partial \xi}{\partial t} + \beta \frac{\partial \psi}{\partial x} = 0.$$

Here, ψ is the streamfunction, and ξ is the potential vorticity deviation defined as

$$\xi(x,y,t) = (\triangle - \gamma^2)\psi(x,y,t); \quad \triangle = \frac{\partial^2}{\partial x^2} + \frac{\partial^2}{\partial y^2}.$$

Here, $1/\gamma$ is the Rossby deformation radius. Suppose that we want to solve the equation above numerically in $\mathbb{R}^2$ from the initial condition

$$\xi = \exp\left(-\frac{x^2 + y^2}{2\sigma^2}\right)$$

and the boundary condition, $\psi \to 0$ as $\sqrt{x^2 + y^2} \to \infty$. Time evolution of the exact solution of the problem is shown in Fig. 2a. We adopt a conformal mapping from the surface of a sphere to $\mathbb{R}^2$,

$$x = \rho \cos \lambda, \quad y = \rho \sin \lambda,$$

$$\rho = 2R \tan\left(\frac{\phi}{2} + \frac{\pi}{4}\right) = 2R\sqrt{\frac{1+\mu}{1-\mu}}, \quad \mu = \sin \phi.$$

Here, λ is the longitude, ϕ is the latitude, and R is the radius of the sphere. Using this mapping with spectral method on a sphere is firstly proposed by [2]. However, it is impossible to apply their method naively to the QG equation because of β and γ terms, so that we introduce several non-trivial modifications. First, we construct an approximate solution of the equation by finite spherical harmonics expansion as,

$$\tilde{\psi}(\lambda, \phi, t) = (1 - \mu) \sum_{n=0}^{M} \sum_{m=-n}^{n} \hat{\psi}_n^m(t) Y_n^m(\lambda, \phi).$$

Here, $Y_n^m(\lambda, \phi)$ is spherical harmonics. The factor $(1 - \mu)$ makes $\tilde{\psi} \to 0$ as $\rho = \sqrt{x^2 + y^2} \to \infty$ because $\mu \to 1$ as $\rho \to \infty$. Time derivative of the coefficients $\hat{\psi}_n^m$ are determined by solving the following equation,

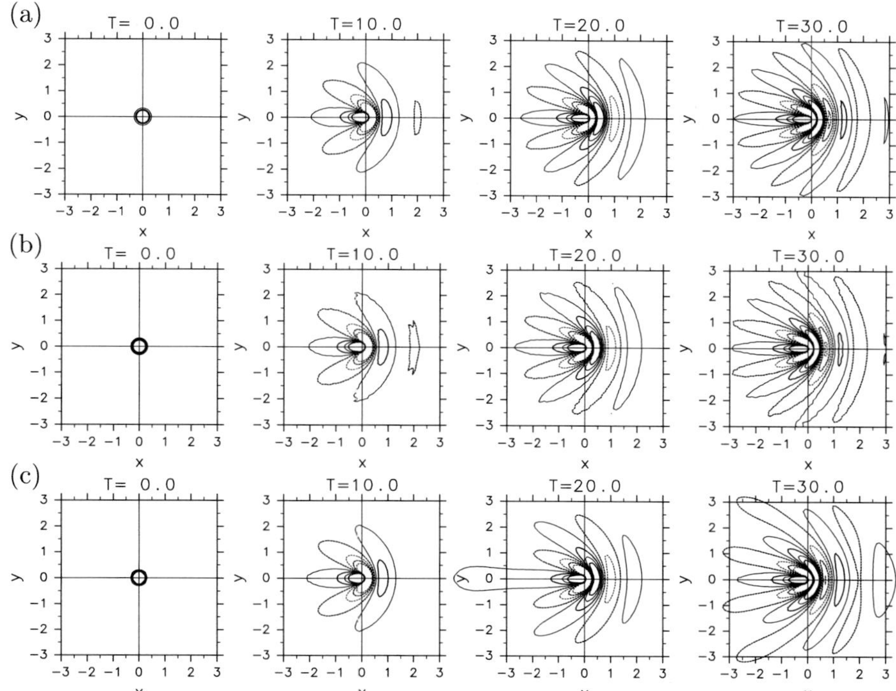

Fig. 2. Time evolution of $\xi(x, y, t)$ field. ($\beta = 1, \gamma = 0, \sigma = 0.1$). (a): exact solution. (b): numerical solution by the mapping spectral method ($R = 2, M = 63$). (c): numerical solution by double Fourier spectral method (domain: $[4\pi \times 4\pi]$).

$$\frac{\partial E}{\partial \dot{\hat{\psi}}_n^m} = 0 \quad (n = 0, 1, \ldots, M; m = -n, -n+1, \ldots, n). \tag{1}$$

Here, $\dot{\hat{\psi}}_n^m = \mathrm{d}\hat{\psi}_n^m/\mathrm{d}t$, and E is the squared residual defined as

$$E = \left\langle \left(\frac{\partial \tilde{\xi}}{\partial t} + \beta \frac{\partial \tilde{\psi}}{\partial x}\right)^2 \right\rangle; \quad \langle A \rangle = \int_{-\infty}^{\infty}\int_{-\infty}^{\infty} A \, dx dy = \int_0^{2\pi}\int_{-1}^{1} A \frac{4R^2}{(1-\mu)^2} d\mu d\lambda,$$

where $\tilde{\xi} = (\triangle - \gamma^2)\tilde{\psi}$. This variational formulation gurantees numerical stability because determining $\dot{\hat{\psi}}_n^m$ by (1) yields the following potential enstrophy conservation automatically,

$$\frac{\mathrm{d}}{\mathrm{d}t}\left\langle \frac{1}{2}(\tilde{\xi})^2 \right\rangle = 0.$$

Using this mapping spectral method, a more accurate numerical solution can be obtained (Fig. 2b) than simple double Fourier spectral method (Fig. 2c) because the mapping spectral method is free from spurious Rossby waves coming back from the boundary. For the example above (Fig. 2b), no numerical viscosity is used because the period of the time integration is so short that the influence of the spurious waves is not serious. When the integration period is long, however, it is necessary to damp waves in far field as in 1D case. In 2D case, PHV term can be introduced as,

$$\nu(-1)^{l-1}\left(\nabla_s^2\right)^l\psi; \quad \nabla_s^2 = \frac{\partial}{\partial\mu}\left\{(1-\mu^2)\frac{\partial}{\partial\mu}\right\} + \frac{1}{1-\mu^2}\frac{\partial^2}{\partial\lambda^2}.$$

Here ∇_s^2 is 2D Laplacian on the sphere. Using the conformal mapping, ∇_s^2 can be expressed as, $\nabla_s^2 = (4R^2/(1-\mu)^2)\triangle$, so that the PHV term can damp waves efficiently in far field where $\mu \to 1$.

One defect of this 2D mapping spectral method is that it needs spherical harmonics transform, which is numerically costly, to deal with nonlinear problems. If the comuputational speed is the most important, one may want to use the 1D mapping introduced in the previous section for both x and y directions to benefit by FFT. Although this method may work, we have not tried it yet because it loses isotropy and the expression of the Laplacian operator on the mapped domain becomes too complicated to solve Poisson problems.

4 Summary

A mapping spectral method for 1D and 2D unbounded domains is proposed. Although the basic mapping idea is not brand-new, two original ideas, including pseudo-hyper-viscosity (PHV), are developed to apply the spectral method to GFD wave equations. It has been shown that the spectral method is promising. In particular, the PHV term seems very effective to damp far going waves and very easy for coding. We are applying this spectral method to nonlinear problems, β-drift of an isolated vortex in 2D QG equation, gravity wave radiation from breakdown of an unstable vortex in shallow water equation, etc, where we have found the spectral method works well. We will report these application examples in our future papers.

The author would like to thank two anonymous reviewers for their helpful and constructive comments on this work.

References

1. Cain AB, Ferziger JH, Reynolds WC (1984) J Comp Phys 56:272–286
2. Matsushima T, Marcus PS (1997) J Comp Phys 137:321–345
3. Sugimoto N, Ishioka K, Yoden S (2005) Theoretical and Applied Mechanics Japan 54:299–304.

A Model for the Far-Field Anisotropic Acoustic Emission of Rotating Turbulence

Fabien Godeferd, Lukas Liechtenstein, Claude Cambon, Julian Scott, and Benjamin Favier

Laboratoire de Mécanique des Fluides – ÉCL, UCBL, INSA, CNRS
36 avenue Guy-de-Collongue, 69134 Écully, France
Fabien.Godeferd@ec-lyon.fr

Abstract. The isotropy of homogeneous turbulence can be broken in flows subject to background rotation. This results in the anisotropic structuration, with large-scale vortices elongated along the rotation axis, and inertial waves with anisotropic mixing properties. Considered as acoustic sources, these anisotropic structures produce noise with characteristics different from the acoustic emission of isotropic turbulence. We propose a computationally efficient model that helps to assess these differences, and to quantify the properties of noise produced by anisotropic turbulence. The simplified configuration is homogeneous turbulence submitted to rotation, for which we evaluate the far-field acoustic emission. A combination of two models is used: Lighthill's acoustic analogy, for the computation of far-field sound emission, and Kinematic Simulation, a stochastic model for homogeneous turbulence, adapted for rotation. The sound spectrum dependence on the rotation rate and the directivity of acoustic intensity are studied and shown to differ from an isotropic emission.

Keywords: acoustic emission, rotating turbulence

1 Introduction

We aim at a better modelling of noise produced by rotating turbulent flows, of interest in aeronautics, *e.g.* production of noise by turboprops wakes. The noise emitting configuration is simple: periodic box of homogeneous turbulence submitted to rotation. The mechanism is the following: the Coriolis force generates inertial waves, and the pressure field adopts an anisotropic state, resulting in sound emission directivity. Note also that Lagrangian dispersion in rotating turbulence is strongly modified by the presence of inertial waves [1, 4], which play a major role when the rotation time is small compared to the eddy turnover time, *i.e.* for fast rotation.

Because of the strong scale separation, computing accurately the pressure in the Eulerian turbulent flow and its acoustic propagation in the mean time is a hard task, all the more if one considers more complex geometrical configurations. To overcome this difficulty, we couple two models which, separately, have been extensively used, one pertaining to the computation of

Y. Kaneda (ed.), *IUTAM Symposium on Computational Physics and New Perspectives in Turbulence*, 297–302.
© 2008 *Springer*.

far-field sound emission, the other being a stochastic model for homogeneous turbulence. First, sound propagation is computed using Lighthill's analogy [2]. In so doing, we assume the incompressibility of the turbulent flow, although compressibility is retained at small scales to permit sound wave propagation. The far-field acoustic pressure is obtained as a by-product of the Eulerian source. Lighthill's integral provides the statistics of the pressure spectrum, if the source, assumed to be localized, is known in terms of two-time velocity correlation statistics. Of course, here we retain the explicit dependence of the velocity and the pressure-probe/source separation vector on the orientation around the axis of rotation, for information on the directivity of the sound source. Second, the velocity field comes from a stochastic model for turbulence, Kinematic Simulation (KS), which provides the Eulerian velocity field, with good two-point and two-time statistics compared to direct numerical simulations (DNS) and to linearized theories, both for isotropic and anisotropic turbulence [3, 4]. We use KS because its computational cost is lower than DNS at equivalent high Reynolds numbers, since capturing the structuration of rotating turbulence at high Reynolds number by DNS is costly [1]. KS provides good statistics for the Eulerian field when averaged over many realisations of the velocity field. Coupled with Lighthill's integral and computed with a standard numerical quadrature scheme, it yields the statistics of the pressure field far from the source. Smoothing is required at the periodic boundaries to avoid spurious sources of sound. This way, we compute the time-dependent pressure signal and the corresponding power spectrum in decibels at arbitrary points away from the source. The dependence of the sound level on the direction of the separation vector characterizes the directivity of rotating turbulence as a source of noise. The parametric analysis is performed in terms of the rotation rate Ω, by checking the dependence of the power of the sound obtained by our model, with asymptotic estimates from simple approximations.

We present briefly the inertial waves solutions in section 2 and Kinematic Simulations in section 3. We recall the basic ideas underlying Lighthill's analogy in section 4, and expose the results in section 5 before proposing some conclusions and perspectives.

2 Inertial Waves in Rotating Turbulence

The Navier-Stokes equations for an incompressible fluid are written in a frame rotating at a rate Ω:

$$\partial_t - Re^{-1}\nabla^2 \boldsymbol{u} = -\nabla(p^* + \boldsymbol{u}^2/2) + \boldsymbol{u} \times \nabla \times \boldsymbol{u} - f\,\boldsymbol{n}_3 \times \boldsymbol{u} \text{ and } \nabla \cdot \boldsymbol{u} = 0.$$

When linearized, these equations admit plane waves solution $u_i(x,t) = \sum_{\epsilon=\pm 1} U_i e^{i(\boldsymbol{k}\cdot\boldsymbol{x}-\epsilon\sigma t)}$ with the anisotropic dispersion relation

$$\sigma = 2\Omega k_3/k = 2\Omega \cos\theta \tag{1}$$

with θ the propagation angle of a wave of wavevector $\boldsymbol{k}$ with respect to the rotation axis, assumed to be $\boldsymbol{x}_3$. The general solution of the linearized equations, for the velocity at time t, given the initial conditions at $t = 0$, may be written in the Craya-Herring frame $\boldsymbol{e}^{(1)}$, $\boldsymbol{e}^{(2)}$, $\boldsymbol{e}^{(3)}$—unit vectors of the spherical coordinates system. This frame is local, *i.e.* attached to $\boldsymbol{k}$, but simple geometrical projections allow to write all quantities in either the fixed or the local frame. Using the Craya-Herring frame provides simpler analytical solutions, from the fact that $\boldsymbol{u}(\boldsymbol{k}) \perp \boldsymbol{k}$ from incompressibility. Denoting the velocity components $u^{(1)}$ and $u^{(2)}$ in the Craya-Herring frame, the solution is

$$\left\{\begin{matrix} u^{(1)} \\ u^{(2)} \end{matrix}\right\}(\boldsymbol{k},t) = \begin{pmatrix} \cos\sigma t & \sin\sigma t \\ -\sin\sigma t & \cos\sigma t \end{pmatrix} \left\{\begin{matrix} u^{(1)} \\ u^{(2)} \end{matrix}\right\}(\boldsymbol{k},t=0) \tag{2}$$

and the general solution field for any kind of initial perturbation is a superposition of many such plane waves. These analytical solutions also formally yield the second order moments of velocity, that is the two-time correlations that are needed for sound computation. Unsteadiness therefore results directly from the presence of inertial waves, at least in this first model.

3 Kinematic Simulation

KS was first used by [3] as a model for the velocity field of isotropic homogeneous turbulence. We have extended the kinematic simulation principle to explicitly include the effect of rotation through the inertial propagating waves, whose amplitudes are taken from (2), such that the KS velocity field is

$$u_i(\boldsymbol{x},t) = \sum_{n=1}^{N_k} e^{i\boldsymbol{k}_n \cdot \boldsymbol{x}} \left(u^{(1)}(\boldsymbol{k}_n,t)e_i^{(1)}(\boldsymbol{k}_n) + u^{(2)}(\boldsymbol{k}_n,t)e_i^{(2)}(\boldsymbol{k}_n) \right)$$

The number of degrees of freedom is chosen large, typically $N_k = 10^5$. Our KS model is very different from the classical one, since we introduce the following two new features. First, the time evolution of $u^{(1)}(\boldsymbol{k}_n,t)$ and $u^{(2)}(\boldsymbol{k}_n,t)$ is analytically known from inertial waves solutions presented above. In previous KS models, unsteadiness came from the introduction of ad'hoc frequencies computed from the kinetic energy spectrum. Second, we avoid using a fixed grid in $\boldsymbol{k}$ space. The energy spectrum $E(k)$ of the velocity field is prescribed analytically. It is numerically reproduced by using random wavevectors k with a probability distribution derived from $E(k)$ (see Fig. 1a). This allows a better control of anisotropy—which was cumbersome in previous KS—since we use uniformly distributed random orientations θ and ϕ for each $\boldsymbol{k}_n$ (Fig. 1b). Initial amplitudes for each initial wave component are the initial velocities $u^{(1)}(\boldsymbol{k}_n,0)$ and $u^{(2)}(\boldsymbol{k}_n,0)$, chosen to be of unit magnitude. The acoustic emission results depend on the large scale behaviour of $E(k)$, but not on its exact shape.

The properties of such "kinematic" velocity field are the following. With respect to DNS, there are no "vertical" coherent structures, due to the absence of dynamic nonlinearity. Hence, there is no nonlinear development of anisotropy in the velocity field (viscous decay is discarded as in forced DNS). On the other hand, spatio-temporal coherence induced by anisotropic linear dynamics of inertial waves is exactly accounted for.

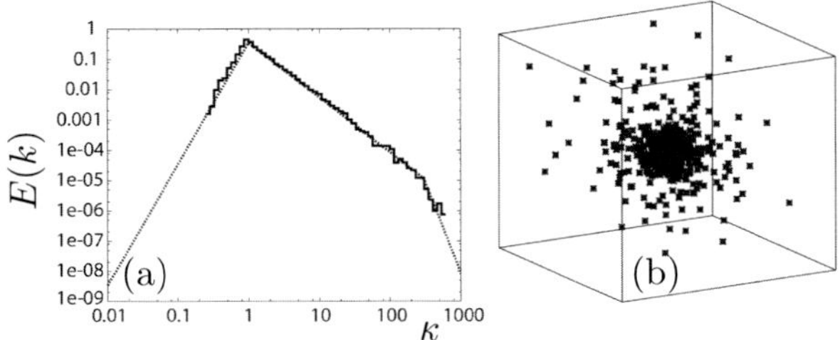

Fig. 1. An illustration of the distribution of wavevectors. (a) The prescribed piecewise linear energy spectrum $E(k)$ is recovered when computing a posteriori the statistics of the KS velocity field; (b) the corresponding random distribution of the wavevectors, more dense at the spectrum peak wavenumber.

4 Lighthill's Analogy

For computing the sound waves propagation from the KS sources, Lighthill's analogy is used [2]. It states the separation of the source from the propagation medium, yielding the density wave equation (c_0 is the sound velocity):

$$\partial^2 \rho / \partial t^2 - c_0^2 \Delta \rho = \partial^2 T_{il} / \partial x_i \partial x_l \tag{3}$$

where Lighthill's tensor contains three contributions: $T_{il} = \rho_0 u_i u_l + (p - \rho_0 c_0^2)\, \delta_{il} - \tau_{il}$. Each corresponds to a given noise generating phenomenon in turbulence: *(a)* convective nonlinear unsteadiness $\rho_0 u_i u_l$, the contribution of fluctuations; *(b)* "entropy" noise; *(c)* viscous stress fluctuations τ_{il}, always neglected. Retaining the dominant, first, term yields $T_{il} = \rho_0 u_i u_l$.

Implicitly, Lighthill's analogy performs a separation of scales, and associates the acoustic production to a volumic unsteady source from external forces, $\frac{\partial^2 T_{il}}{\partial x_i \partial x_l}$. In the absence of obstacles and assuming statistical stationnarity in the far field, the general solution of (3) is obtained using a Green's function, in which appear r/c_0 the retarded time and $r = |\boldsymbol{x}|$, the distance of the observer, with $\boldsymbol{x} = r\boldsymbol{n}$ using the unit vector $\boldsymbol{n}$. From $\partial T_{ij}/\partial x_i = -\dot{T}_{ij} x_i / c_0$, and $\rho = c_0^2 p$ in the far field, one obtains the pressure $p(\boldsymbol{x}, t) = \frac{1}{4\pi c_0^2} \frac{n_i n_j}{r} \int_V \ddot{T}_{ij}(\boldsymbol{y}, t - \frac{r}{c_0} + \frac{\boldsymbol{y}.\boldsymbol{n}}{c_0}) d\boldsymbol{y}$ and Fourier transforming in time, noting ω the frequency

$$\tilde{p}(\boldsymbol{x}, \omega) = \frac{n_i n_j \omega^2}{4\pi c_0^2 r} e^{i\omega \frac{r}{c_0}} \int_V \tilde{T}_{ij}(\boldsymbol{y}, \omega) e^{-i\omega \frac{\boldsymbol{y}.\boldsymbol{n}}{c_0}} d\boldsymbol{y} \tag{4}$$

where one identifies the temporal phase shift in the first exponential.

5 Implementation and Results

The global model for the emission of sound by rotating turbulence is therefore obtained by merging the two models, KS and Lighthill's hypothesis.

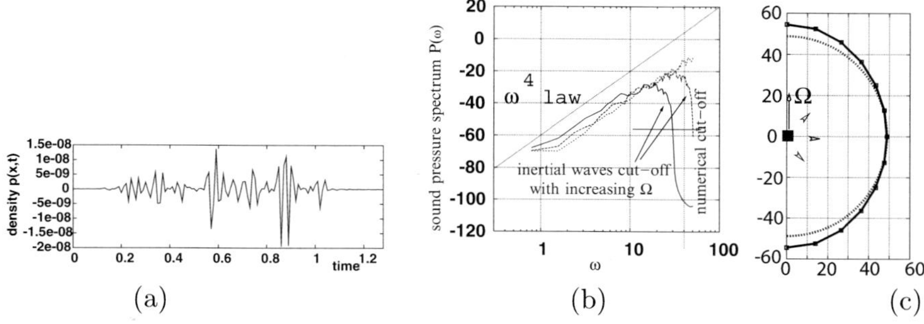

Fig. 2. (a) Pressure signal over the time window; (b) sound pressure spectra at $\Omega = 10\pi, 30\pi, 50\pi$; (c) acoustic intensity variation depending on the angle of propagation, in decibels. (Dashed: isotropic reference.)

The time-dependent KS velocity field uses spectral amplitudes coming from the linear solution (section 2), and random initial phases. Ensemble averages are obtained by varying the initial conditions in a dozen realizations. Two-time correlations are computed numerically to provide Lighthill's tensor T_{ij} at points $\boldsymbol{y}$ in a cartesian grid of a cube of non-dimensional size $L = 1/2$ representing the source, and for a time span $[0, T_f]$. Space and time discontinuities at the space-time frontiers are smoothed by cosine filtering to avoid spurious sound emissiondue to space truncation. After a Fourier transform in time, Lighthill's tensor is used to evaluate the propagating sound pressure at the observer's location, using equation (4). The integral is discretized with a second order quadrature method. A sampling of the pressure signal $p(\boldsymbol{x}, t)$ at $|\boldsymbol{x}| = r \gg L$ is shown on Fig. 2a, with fluctuations consistent with turbulent emission.The results are expected to depend on the ratio r/L since the pressure level decreases away from the source. We compute the pressure spectrum $P(\omega)$ from $< \tilde{p}(\boldsymbol{x}, \omega)\tilde{p}^*(\boldsymbol{x}, \omega) >$. Eq. (1) shows that the maximum attainable frequencies are $\mathrm{Max}(\sigma) = 2\Omega$. The corresponding high-frequency cut-off, and its dependence on Ω is checked on Fig. 2b. Our model is therefore consistent with the physics of inertial waves. Figure 2b also shows the numerical high frequency cut-off, and the ω^4 power law dependence of the pressure spectrum in the low frequency range. The acoustic intensity is $\boldsymbol{I} = p\boldsymbol{u} = p^2/(\rho c_0)\boldsymbol{r}$. It evolves as a power of the Mach number, Ma^β, according to asymptotic simplified arguments. $\beta = 8$ is generally observed in isotropic turbulent noise: $I^{\mathrm{dB}}_{\mathrm{theory}} = \rho_0(L/4\pi r)^2 u'^3 Ma^8$. Our results also exhibit a power law dependence with Ma, although with $\beta = 4$. This difference comes from the particulars of our wave-turbulence field, which is constituted by inertial waves with a specific spectral distribution (Fig. 2a) different from the spectrum of isotropic turbulence. Finally, we measure the directivity of rotating turbulence by computing the acoustic intensity in different directions with respect to the rotation axis, at constant distance r, plotted on Fig. 2c. For comparison, we also plot the isotropic emission at the same intensity as observed in the horizontal direction for rotating turbulence. Figure 2c clearly shows that the acoustic intensity is anisotropic, and is larger in the vertical, *i.e.* along the rotation axis rotation. Although at first glance the departure appears to be limited on this plot in decibels, the emitted sound is 6dB higher in the vertical than in the horizontal direction, a significant difference to the ear.

6 Conclusion and Perspectives

We propose a model for the emission of sound by rotating turbulence in the absence of solid boundaries. The model is designed to be computationally cheaper than a complete coupled simulation of the source velocity field and sound propagation. This is achieved by using Lighthill's analogy, that separates the Eulerian source emitting noise from the far-field noise propagation. Thanks to the introduction of explicit wave solutions in the stochastic KS model, the role of inertial waves is explicitly taken into account. Some anisotropy of the Eulerian pressure field is reproduced, and transmitted to the acoustic field. A noticeable directivity of sound is observed, with levels of acoustic intensity higher by 6dB along the rotation axis than perpendicular to it. The power law dependence of acoustic intensity is also much different than for isotropic turbulence. This simple approach, based on the solutions of the .linearized Navier-Stokes equations, demonstrates that rotation modifies the sound emission of turbulence. We plan to include more complex effects, in order to refine the prediction. First, we will include in KS the spectral information reflecting the presence of intense elongated vortices aligned with the rotation vector in rotating turbulence. Anisotropic spectra computed by DNS or statistical two-point models of rotating homogeneous turbulence will be used. Second, in addition to the deterministic time dependence due to the inertial waves, we will introduce the time dependence due to background turbulent noise as in the initial KS model, by adding frequencies, in the unsteady exponential term, whose spectrum is derived from the kinetic energy spectrum, as proposed by [3]. We shall also compare these results with those from DNS fields taken as acoustic sources, at moderate Reynolds numbers.

References

1. Liechtenstein L, Godeferd F, Cambon C (2005) J Turbul 6:1–18
2. Lighthill J (1952) Proc R Soc Lond A 221:564–587
3. Fung J, Hunt J, Malik N, Perkins R (1992) J Fluid Mech 236:281–318
4. Cambon C, Godeferd F, Nicolleau F, Vassilicos J (2004) J Fluid Mech 499: 231–255

Part IV

Geophysical, Astrophysical and Complex Turbulence

Dynamics of the Small Scales
in Magnetohydrodynamic Turbulence

Annick Pouquet[1], Pablo Mininni[1], David Montgomery[2], and Alexandros Alexakis[1]

[1] NCAR, PO Box 3000, Boulder, CO 80304, USA
`pouquet@ucar.edu, alexakis@ucar.edu, mininni@ucar.edu`
[2] Dept. of Physics and Astronomy, Dartmouth College, Hanover, NH 03755
`david.c.montgomery@dartmouth.edu`

Abstract. Direct numerical simulations of magnetohydrodynamic flows with regular grids of up to 1536^3 points and at a unit magnetic Prandtl number allow for an investigation of the rate at which small scales develop. No forcing term is included and computations for both deterministic and random initial conditions are reported. Parallel current and vorticity sheets form at the same spatial locations, with a strong degree of correlation; these sheets further destabilize and fold or roll-up after an initial exponential growth phase. At high Reynolds numbers, a self-similar evolution of the current and vorticity maxima is found; these maxima grow as a cubic power of time within localized structures. The flow then reaches a finite dissipation rate independent of Reynolds number as was also found in MHD in two space dimensions. This evolution may be linked to the degree of nonlocality of nonlinear interactions in turbulent flows, to the amount of velocity-magnetic field correlation, and to the possibility of occurrence of non-universal behavior in MHD turbulence.

Keywords: turbulence, MHD, nonlocal interactions, dynamo

1 Introduction

Magnetic fields pervade the universe and are often dynamically relevant in flows that are strongly turbulent because Reynolds numbers are high. The origin of magnetic fields likely arises from line stretching overcoming Joule dissipation, through a dynamo process above a critical magnetic Reynolds number that may depend on the magnetic Prandtl number P_M, low in the core of the Earth or in the solar convection zone and high in the interstellar medium. The nonlinear regime of the dynamo is costly to study numerically, and experimental dynamos have not yet reached criticality when embedded in an unconstrained turbulent flow, except at Cadarache [1].

In this context, we tackle a study of MHD turbulence starting with initial conditions centered in the large scale with kinetic and magnetic energy in equipartition. The incompressible MHD equations read, with $\rho_0 = 1$ a constant density and ν and η the kinematic viscosity and magnetic diffusivity:

305

Y. Kaneda (ed.), *IUTAM Symposium on Computational Physics and New Perspectives in Turbulence*, 305–312.

$$\frac{\partial \mathbf{v}}{\partial t} + \mathbf{v} \cdot \nabla \mathbf{v} = -\frac{1}{\rho_0} \nabla \mathcal{P}_0 + \mathbf{b} \cdot \nabla \mathbf{b} + \nu \nabla^2 \mathbf{v}$$

$$\frac{\partial \mathbf{b}}{\partial t} = \nabla \times (\mathbf{v} \times \mathbf{b}) + \eta \nabla^2 \mathbf{b}$$

$$\nabla \cdot \mathbf{v} = 0, \quad \nabla \cdot \mathbf{b} = 0, \tag{1}$$

where $\mathcal{P}_0 = \mathcal{P} + b^2/2$ is the total pressure, $\mathbf{b} = \mathbf{B}/\sqrt{\mu_0 \rho_0}$ is the induction in Alfvénic units and μ_0 the permeability. In the absence of dissipation ($\nu = 0$, $\eta = 0$), the energy $E_t = E_v + E_m = \langle v^2 + b^2 \rangle /2$ and cross helicity $H_c = \langle \mathbf{v} \cdot \mathbf{b} \rangle /2$, or their combinations in terms of the pseudo-energies of the Elsässer variables $E^\pm = < z^{\pm^2} > /2$ are conserved, as well as the magnetic helicity $H_m = \langle \mathbf{A} \cdot \mathbf{b} \rangle$ in 3D, with $\mathbf{b} = \nabla \times \mathbf{A}$, $\mathbf{A}$ being the vector potential.

With the Elsässer variables, one can symmetrize (1). We first define generalized Lagrangian derivatives as:

$$D_\pm/Dt = \partial_t + \mathbf{z}^\pm \cdot \nabla \ ;$$

we then obtain, omitting dissipation and with $z_m^\pm$ being the m_{th} component of the $\mathbf{z}^\pm$ field when taking the curl of (1):

$$\frac{D_\mp \mathbf{z}^\pm}{Dt} = -\nabla \mathcal{P}_0, \tag{2}$$

$$\frac{D_\mp \omega^\pm}{Dt} = \omega^\pm \cdot \nabla \mathbf{z}^\mp + \sum_m \nabla z_m^\pm \times \nabla z_m^\mp. \tag{3}$$

Note that the first term on the r.h.s. of (3) is equal to zero in two space dimensions; the second term is absent in the Navier-Stokes case and may account for extra and faster growth of the generalized vorticities for conducting fluids [2] unless the Elsässer field gradients are parallel.

In order to study the development of structures in MHD turbulence, we solve numerically (1) using a pseudospectral method in a three dimensional box of side 2π with periodic boundary conditions. All computations are de-aliased, using the standard 2/3 rule. At all times, the dissipation wavenumber evaluated using the Kolmogorov spectrum is fully resolved. Initial conditions are either an Orszag-Tang vortex in three dimensions [2], or an ABC flow [3] with superimposed random initial conditions.

2 How Fast Do Structures Form in MHD?

The simulations using a random flow were performed at NCAR on regular grids of up to 1536^3 points (in that latter case, with an initial Reynolds number $R_v \sim 5700$). They allow for the study of the early-time generation of spatial features in MHD turbulence, and its link with non-local interactions and with deviations from self-similarity.

In MHD, early-time behavior leads to a self-similar formation of current and vorticity sheets which later on are destabilized. The initial exponential

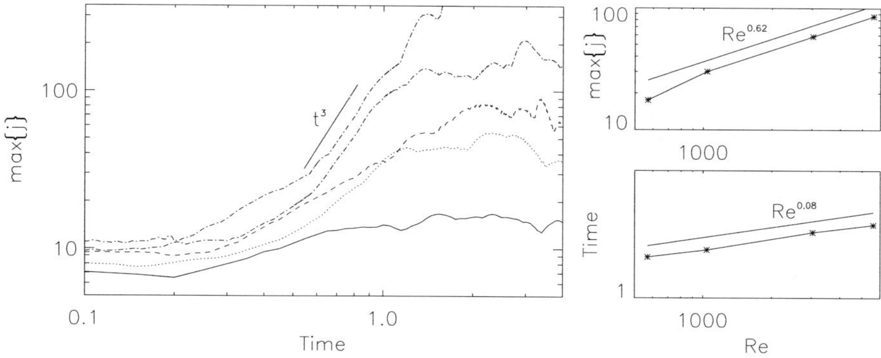

Fig. 1. *Left:* Temporal evolution of current maxima for the random flow at resolutions from 128^3 to 1536^3 grid points. *Right:* Scaling with Reynolds number of the value of first maximum (top) and time of first maximum (bottom) of current extrema in the flow with Orszag-Tang initial conditions.

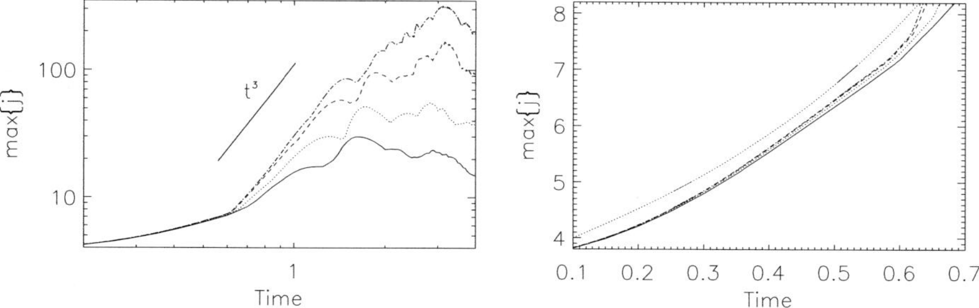

Fig. 2. *Left:* Temporal evolution of current maxima for the OT flow for resolutions from 64^3 to 512^3 grid points, in log-log coordinates. Note the t^3 behavior after the first exponential phase. *Right:* Scaling of current maximum with time with a fit to the MHD equivalent [5] to the Beale-Kato-Majda theory of singular behavior [4]. Solid, dotted, dashed, and dot-dashed lines correspond to four increasing Reynolds numbers, and the long dot is the fit.

growth of current and vorticity maxima is followed by a self-similar algebraic growth in the nonlinear regime, with $j_{max} \sim t^3$ as shown in Fig. 1 (left) for the random flow and Fig. 2 (left) for the OT vortex. Note that a fit of the type $max \sim [t_* - t]^{-1}$ as proposed in [4] for fluids and in [5] for MHD – in the context of singular behavior in the non-dissipative cases – does not seem to work as well (see Fig. 2, right) as the self-similar solution shown earlier. The jagged temporal behavior of j_{max} for early times in the random case corresponds to jumping from one structure to the next; we verified that in the t^3 regime, the evolution of j_{max} followed one single structure.

The time of saturation of the early self-similar growth of j_{max} scales weakly with Reynolds number, namely $\sim R_v^{0.08}$ (see Fig. 1, right, for the OT runs); similar slow scaling obtains for the global maximum of total enstrophy (not shown). The actual value of the first maximum of current, on the other hand, scales as $R_v^{0.62}$ (see Fig. 1, right). We also note that the dissipation of total energy for the OT vortex becomes quasi-constant as the Taylor Reynolds number R_λ grows (up to 880, with $P_M = 1$); similar results obtain for random initial conditions, as was previously reported in two space dimensions [6, 7].

It is not clear what are the dissipative structures that appear in MHD thereafter for the three-dimensional case. At moderate resolutions (up to 512^3) vortex tubes (observed in 3D hydrodynamic simulations) cannot be identified, and seem to be replaced by vorticity and current sheets , and possibly ribbons [2], apparently aligned with the large-scale magnetic field [8] but otherwise randomly distributed in the flow. This behavior is observed with random initial conditions and on the three-dimensional Orszag-Tang (OT) vortex. However, in the 1536^3 simulation with random initial conditions, multiple roll-up of current sheets occur, with the rolls sometimes occurring in opposite directions in the same cluster complex, as in the structures shown in [8]. As a result, tubular structures are observed at high Reynolds numbers, with the tube formed as a result of folding and rolling, aligned with the direction of the local magnetic field. Examining the temporal evolution of such structures confirms that the evolution is similar to the one observed in a Kelvin-Helmholtz (KH) instability when the magnetic field lines are parallel to the current and vortex sheets. Note that the KH instability [9] and cylindrical structures parallel to the ambient magnetic field have both been observed in the magnetosphere of the Earth [10], in particular with the Cluster multi-satellite configuration.

In the vicinity of these current and vorticity sheets, the normalized velocity-magnetic field correlations are strong, as can be expected in Alfvén vortices [11], although they are weak globally; furthermore, all terms appearing in the equation for the temporal evolution of the generalized vorticities (3) contribute in comparable amounts to their dynamical evolution.

3 How Non-Local Is Energy Transfer in MHD?

In fluid turbulence, the generation of structures that appear in the flow (vortex tubes) and the structures themselves are associated with non-local interactions, with a correlation between large-scale shear and strong small-scale gradients [12]. MHD turbulence is found to be more non-local than its hydrodynamic counterpart [12]-[13] in forced flows (see [14] for the decay case). As soon as one exits the initial linear phase of exponential growth of field gradients, nonlinear transfer involves a wide range of scales with equal contributions from all smaller wavenumbers (with the exception, however, of an excess of transfer from the energy containing scale), a non-locality first modeled by Iroshnikov-Kraichnan [15] in an isotropic framework. Note that current sheets can be formed by compressing two large-scale flux tubes against each other, a process that naturally generates small scales non-locally in Fourier space. Non-locality of non-linear interactions in Fourier space indicates that the large scales have an imprint on the small-scale dynamics, consistent with

Parker's theory which shows that MHD develops sharp gradients. The greater degree of non-locality in MHD (compared to the fluid case) can be attributed to the enhanced dynamics of strong localized structures in MHD, which is also found to be more intermittent that neutral fluids [16, 17]. Such nonlocality was already invoked when studying dynamos at high magnetic Prandtl number since in that case the magnetic field may grow at scales in the viscous sub-range and hence it is the large-scale velocity that must amplify the small-scale magnetic field [19, 18].

4 Non-Locality and Non-Universality in MHD Turbulence

It has been known for some time that there may be a breakdown of complete universality in MHD turbulence. When two independent forcing functions in the velocity and induction equations are introduced, the Renormalization Group approach in its dynamical formulation indicates for the non-helical case [20] that there are two non-trivial fixed points, whereas only one appears in the pure fluid case. They correspond to (i) a state dominated by the velocity, in which the magnetic field behaves as a passive vector with transport coefficients renormalized by the small scales of the velocity field; and (ii) a state dominated by the magnetic field above a critical dimension $d_c \sim 2.8$, with transport coefficients renormalized by the magnetic small scales and with a renormalization of the Lorentz force vertex. Universality is explicitly broken in the sense that overlap of regimes can take place (note that the helical case leads to runaway linked with the inverse cascade of magnetic helicity.

A thorough numerical search of parameter space in two space dimensions [21] showed that there was also a range of parameters for which equipartition $E_v = E_m$ obtains; note that when $\mathbf{v} = \mathbf{b}$, nonlinear terms in the MHD equations cancel exactly (and thus the RG analysis becomes irrelevant); furthermore, this solution implies strong correlations between the velocity and the magnetic field. This numerical study considered the parameter space made up of the ratio of the two invariants H_c and H_m to the energy invariant E_t, an indication that the role of invariants in the dynamics of turbulent flow is indeed essential.

Are the energy spectra different or the same in these three regimes? Using the Eddy Damped Quasi Normal Markovian (EDQNM) closure, it was shown [22] that when the amount of correlation between the velocity and the magnetic field is varied, the spectral indices of the $E^\pm$ energies of the Elsässer variables, namely $m^\pm$, differ with $m^+ + m^- = 3$ corresponding to the Iroshnikov-Kraichnan (IK) spectrum ($m^\pm = 3/2$ in the uncorrelated case [15]). Numerical simulations in two dimensions confirmed this result [7], but this point may not have been checked numerically in 3D. A similar (but not identical) relationship is found analytically for weak MHD turbulence [23], confirming the validity of the approach. Note that the EDQNM spectra are compatible with the IK phenomenology by including in the decorrelation time of third-order moments the Alfvén time, whereas a Kolmogorov spectrum would obtain when only the eddy turn-over time is considered. The Lagrangian renormalized approximation also gives spectra compatible with the IK model [24], a result confirmed by numerical simulations in which again both the velocity and magnetic fields are forced with no correlation between the two driving mechanisms; the results also show that the residual spectrum $E_r(k) = |E_v(k) - E_m(k)| \sim k^{-2}$ as already conjectured in [22]. Similarly, in a magnetically dominated regime obtained by imposing a strong uniform field B_0 and forcing both $\mathbf{v}$ and $\mathbf{b}$, the IK regime obtains for the perpendicular (to B_0) spectrum [25, 26], whereas, for a decaying flow, $E_t(k) \sim k^{-5/3}$ and $E_r(k) \sim k^{-7/3}$, the latter resulting from an interplay of dynamo and Alfvén effects [26].

The exact laws found in MHD turbulence [27]-[28] provide constraints on the dynamics of the flow. In the simplest case in dimension d, they read:

$$\langle\, \delta z_L^-(\mathbf{r})\, [\delta z_i^+(\mathbf{r})^2]\, \rangle = -\frac{4}{d}\, \epsilon^+\, r \tag{4}$$

with $\delta\mathbf{w}(\mathbf{r}) = \mathbf{w}(\mathbf{x}+\mathbf{r}) - \mathbf{w}(\mathbf{x})$ the structure function of the $\mathbf{w}$ field, w_L its longitudinal component and ϵ^+ its energy transfer rate. The fact that triple correlators – involving either both $\mathbf{v}$ and $\mathbf{b}$, or both $\mathbf{z}^\pm$ – scale linearly with distance does not imply that the variables themselves scale as $r^{1/3}$ since correlations between the velocity and the magnetic fields or between $\mathbf{z}^\pm$ (implying a role for the residual energy) may alter this scaling [27], as found in [16]. A further remark [29] is that, if velocity-magnetic field correlations are important in the sense that these two fields are strongly aligned, one can (and should) measure this correlation as a function of scale. Indeed, in the above relationship, a dimensionless quantity, e.g. an angle, can be introduced that has a non-trivial scaling and thus the relationship $\sim r^{1/3}$ which could appear as the simplest solution to (4) is broken; such a scaling, with $\theta \sim r^{1/4}$ has been observed numerically in the presence of forcing in the momentum equation and a strong uniform magnetic field, where θ is the anble between $\mathbf{v}$ and $\mathbf{b}$ ([29] and references therein).

Nonlinear transfer of energy to small scales, as stated in the preceding section, is non-local in the forced (dynamo) regime [13]-[14] but more local in decaying flows [14]; this enhanced non-locality, at the level of shell-to-shell energy transfer, is due to the stretching term in the induction equation and the Lorentz force in the momentum equation, not to the advection terms. As a result, it seems apparent that different forcings (e.g. mechanical, electromotive, or both) can change the relevance of non-local vs. local contributions in the total energy transfer and flux. Moreover, the differences observed in the energy transfer of free-decaying and forced MHD turbulence also suggest a breakdown of complete universality, a topic that requires further study.

Anisotropies are not considered in the above remarks although it is certainly present in MHD flows. Note that the $k_\perp \sim k_\parallel^{2/3}$ relationship – where $\perp$ and $\parallel$ refer to a uniform field $\mathbf{B_0}$ – proposed in [30] and observed e.g. in [25, 31] is not necessarily a consequence of a Kolmogorov scaling [32]; yet, alignment between fields play an essential role in the dynamics (see [29] and references therein, also [33]); it is also known that Beltrami flows (parallel velocity-vorticity) and Alfvénic solutions (parallel velocity-induction) are found in the small scales (note that Alfvén vortices [10, 11] follow this rule).

A further question concerns intermittency: the computations obtaining a 3/2 energy spectrum show little departure from this spectral index and hence, likely, little intermittency. These simulations also have an imposed external magnetic field, that gives a preferred transfer of energy to wavenumbers perpendicular to the imposed field. Are the smaller intermittency corrections linked with the similar finding that in strongly rotating flows, intermittency is greatly reduced?

5 Conclusion and Perspectives

Models of turbulent flows should be able to take into consideration the nonlocality of nonlinear energy transfer to small scales, as for example in Rapid Distortion Theory, or with Lagrangian averaged models (its evaluation for MHD in three dimensions can be found in [34]), as well as the correlation (and more generally, the alignment) between the various dynamical fields which may lead to non-universal behavior. These non-local interactions are also responsible for the spectral shape of the magnetic energy spectrum in the early phase of the kinematic growth of magnetic fields, a feature observed at low magnetic Prandtl numbers P_M [35] down to $P_M \sim 0.005$.

The studies presented here must be pursued at higher Reynolds numbers in order to be able to reach clear conclusions. Indeed, singular behavior is difficult to detect (see [36] for the MHD case); similarly, the origin of the Kelvin-Helmoltz roll-up of current sheets and of the growth of current maxima as t^3 in the nonlinear phase remains to be investigated. In the future, spectral element adaptive codes [37] could also be used to study MHD flows in order to reach higher Reynolds numbers.

References

1. Monchaux R et al. (2007) Phys Rev Lett 98:044502
2. Politano H, Pouquet A, Sulem PL (1995) Phys Plasmas 2:2931–2939
3. Arnol'd V, (1986) C R Acad Sci Paris 261:17–20; Galloway D, Frisch U, (1986) Geophys Astrophys Fluid Dyn 36:53–83
4. Beale J, Kato T, Majda A, (1984) Comm Math Phys 94:61–66
5. Calflisch R, Klapper I, Steele G, (1997) Comm Math Phys 184:443–455
6. Biskamp D, Welter H, (1989) Phys Fluids B 1:1964–1979
7. Politano H, Pouquet A, Sulem PL (1989) Physics Fluids B 1:2330–2339
8. Mininni P, Montgomery DC, Pouquet A (2006) Phys Rev Lett 97:244503
9. Hasegawa H et al. (2004) Nature 430:755–758; Sundkvist D et al. (2005) Nature 436:825–829; Nykyri K et al. (2006) Annales Geophysicae 24:1–25
10. Alexandrova O et al. (2004) J Geophys Res 109:A05207

11. Petviashvili V, Pokhotelov O (1992) Solitary waves in plasmas and in the atmosphere, Gordon and Breach (New York)
12. Alexakis A, Mininni PD, Pouquet A (2005) Phys Rev Lett 95:264503; Alexakis A, Mininni PD, Pouquet A (2006) Phys Rev E 74:056320
13. Alexakis A, Mininni PD, Pouquet A (2005) Phys Rev E 72:046301; Alexakis A, Mininni PD, Pouquet A (2005) Phys Rev E 72:046302
14. Debliquy O, Verma M, Carati D (2005) Phys Plasmas 12:042309; D Carrati et al. (2006) J Turbul 7:1–12
15. Iroshnikov P (1963) Sov Astron 7:566–571; Kraichnan RH (1965) Phys Fluids 8:1385–1387
16. Politano H, Pouquet A, Carbone V (1998) EuroPhys Lett 43:516–521
17. Bruno R , Carbone V (2005) Living Reviews in Solar Physics 2:4
18. Schekochihin A et al. (2004) ApJ 612:276–307
19. Zel'Dovich Ya et al (1984) J Fluid Mech 144:1–11
20. Fournier J D, Sulem P L, Pouquet A (1982) J Phys A 15:1393–1420
21. Ting A, Matthaeus W, Montgomery DC (1986) Phys Fluids 29:3261–3274
22. Grappin R, Pouquet A, Léorat J (1983) Astron Astrophys 126:51–56
23. Galtier S et al. (2000) J Plasma Phys 63:447–488
24. Yoshida K, Arimitsu T (2006) Inertial subrange structures of isotropic incompressible MHD turbulence in the Lagrangian renormalized approximation, preprint, University of Tsukuba (Tsukuba)
25. Maron J, Goldreich P (2001) Astrophys J 554:1175–1196
26. Müller WC, Grappin R (2005) Phys Rev Lett 95:114502
27. Politano H, Pouquet A (1998) Geophys Res Lett 25:273–276
28. Politano H, Gomez T, Pouquet A (2003) Phys Rev E 68:026315
29. Boldyrev S, Mason J and Cattaneo F (2006) Dynamic alignment and exact scaling laws in MHD turbulence, arXiv:astro-ph/0605233
30. Goldreich P, Sridhar S (1995) ApJ 438:763–777
31. Cho J, Lazarian A, Vichniac E (2001) Astrophys J 564:291–301
32. Galtier S, Pouquet A, Mangeney A (2005) Phys Plasmas 12:092310
33. Gibbon JD, Holm DD (2006) Lagrangian analysis of alignment dynamics for isentropic compressible MHD, preprint, Imperial College (London)
34. Mininni PD, Montgomery DC, Pouquet A (2005) Phys Rev E 71:046304
35. Ponty Y et al. (2005) Phys Rev Lett 94:164502
36. Kerr R, Brandenburg A, (1999) Phys Rev Lett 83:1155–1158; Grauer R, Marliani C (2000) Phys Rev Lett 84:4850–4853
37. Rosenberg D et al. (2006) J Comp Phys 215:59–80

Recent Experimental and Computational Studies Related to the Fluid Dynamics of Clouds

R. Narasimha[1] and G.S. Bhat[2]

[1] Jawaharlal Nehru Centre for Advanced Scientific Research, Bangalore 560 064,
    `roddam@caos.iisc.ernet.in`
[2] Centre for Atmospheric and Oceanic Sciences, Indian Institute of Science,
    Bangalore 560 012, `bhat@caos.iisc.ernet.in`

Abstract. A fresh attack has been made in the last 15 years on the long-standing problem of the counter-intuitive nature of entrainment in clouds. Experimental and computational studies show that turbulent jets and plumes subjected to off-source heating, simulating latent heat release on condensation of water vapour in a cloud, illuminate many aspects of the fluid dynamics of clouds. The paper presents a review and a critical reanalysis of published results.

Keywords: cloud-like flows, jets/plumes with off source heating, entrainment coefficient, coherent structures, numerical solution of Boussinesq equations

1 Introduction

Cumulus and cumulonimbus clouds (clouds henceforth) are a special kind of natural free shear flow, in the class of thermals or plumes [1]. Their distinguishing features include, most importantly, the additional generation of buoyancy away from the source, following the release of latent heat when water vapour condenses to liquid water above cloud base. Although cloud-plumes develop in a stratified environment, may involve all three phases of water and generally develop in cross-wind, we concentrate here chiefly on the effect of additional buoyancy, which may even result in *acceleration* of the plume fluid instead of the usual decay with streamwise distance ([2], p 166).

A fascinating account of early ideas on entrainment and mixing processes in clouds is given by Simpson [3]. The chief problem is highlighted by the finding of Paluch [4], based on aircraft measurements in tall Colorado cumuli, that observed cloud thermodynamic properties can be accounted for only if air from cloud base ascends to cloud top without laterally mixing with ambient air. This goes counter to expectations from classical entrainment ideas [5, 6]. Several experimental studies have been made employing chemical reactions for producing buoyancy increase [7, 8], but these are difficult to interpret for

Y. Kaneda (ed.), *IUTAM Symposium on Computational Physics and New Perspectives in Turbulence*, 313–320.
© 2008 *Springer*.

clouds as the buoyancy enhancement in such experiments is a *result* of mixing, whereas the question at hand is the effect of buoyancy enhancement *on* mixing.

In the last ten to fifteen years, several experimental [9-16] and computational [17, 18] studies relevant to entrainment in shear flows subjected to off-source buoyancy have been made. It is our objective here to review these studies, reanalyse their results and assess the present position.

2 Experimental Studies

These studies began with the development of a new technique for volumetric addition of heat over a specified region in a shear flow created in a water tank [9]. The technique involves ohmic heating of water, rendered electrically conductive by adding a small quantity of hydrochloric acid, restoring the original density by adding acetone; correspondingly the ambient fluid is made non-conductive by deionizing it. Typically 5 to 6 electrode meshes made of thin Pt wire 90 μm dia. and 10 mm apart, and netted on a supporting frame, are placed horizontally across the flow at desired heights. The voltage applied across the electrodes has a high frequency (20 kHz), which serves to prevent electrolysis of water and release of gas bubbles. Figure 1 shows the experimental set up at the Indian Institute of Science (IISc).

Figure 2 compares a visualization of jet flows subjected to such volumetric heating in a nearly discontinuous ambient stratification [9] with a picture of clouds in nature (downloaded from the internet). (The lab pictures are taken at different stages in the temporal evolution of flow at different heating levels.) The resemblance between nature and laboratory simulation is striking, and suggests that the technique may be capable of capturing some major features of the fluid dynamics of clouds.

Beginning with the experiments first carried out at IISc [9–11], groups at Florida State University (FSU), Tallahassee [14] and Delaware [15, 16] have also made studies with set ups virtually identical to the one shown in Fig. 1.

Experiments at IISc were carried out on round jets and plumes, while those at FSU and Delaware involved round jets only. Although the plume is a closer approximation to a cloud [19], the consequences of volumetric heating on jets and plumes are broadly similar [12, 13], so we focus here on the jet as all three groups have studied it.

The experimental tank is big enough to ignore the effects of confinement [11, = BN below; 20]. Referring to Fig. 1, heat is volumetrically added in the region $z_b < z < z_t$ with z_b chosen to be in the fully developed state. To simulate the effect of the heating experienced in clouds, BN introduced the non-dimensional heat release number $G = (\alpha g/\rho C_p)(z_b^3/d^3)(Q/U_o^3)$, where U_o is the efflux velocity through the orifice of diameter d at $z = 0$, α is the coefficient of thermal expansion, g acceleration due to gravity and Q is total heat injected. G is a measure of the ratio of buoyancy to inertial forces, like a bulk Richardson number Ri. The experiments carried out at IISc, going up

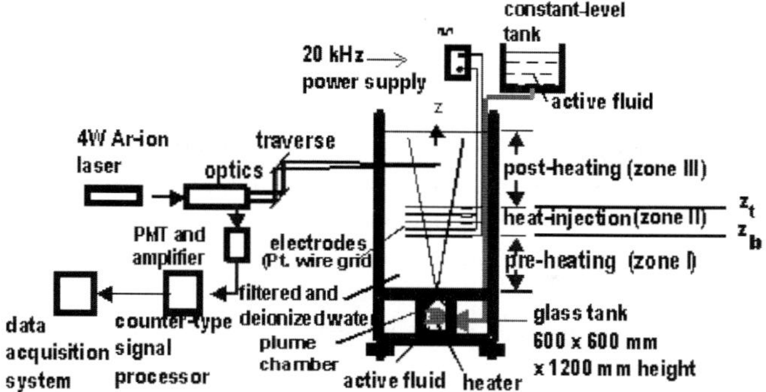

Fig. 1. The apparatus for simulation of cloud like flows in the laboratory [13].

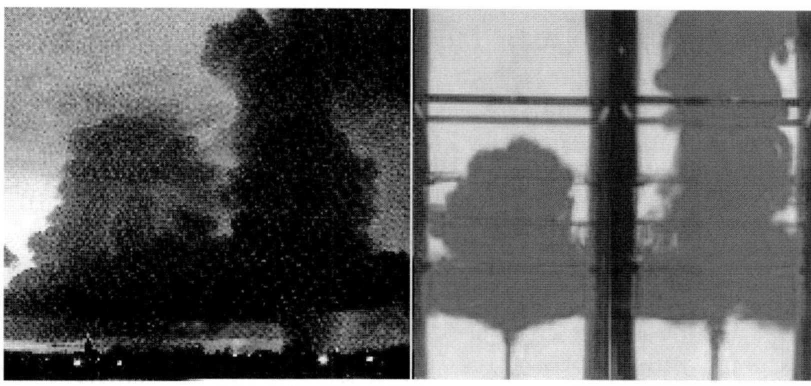

Fig. 2. (a) Natural clouds. (b) Image from a dye flow visualization of a jet subjected to off-source volumetric heating, neutral with respect to ambient in the lower, denser layer below where the jet spreads out horizontally.

to $G = 4.7$ (Ri $\simeq 0.64$), covered the range of Ri characteristic of clouds [2, 13, 21].

Laser fluorescence visualization and laser doppler velocimetry were used at IISc, whereas particle image velocimetry (PIV) was used at FSU and Delaware. Micro-encapsulated thermochromic liquid crystals of 40 μm dia., illuminated by a laser sheet, were used at Delaware for temperature measurements [16].

Both instantaneous and time-averaged flow visualization sections in these experiments show major structural changes in the jet upon off-source heating. In particular, the large structures characteristic of free shear flows are progressively disrupted in the heat injection zone (= HIZ below) [10; 12, Fig. 2]. As G increases, both velocity and concentration (scalar) profile widths (b_u, b_s respectively) decrease, after a slight initial increase near z_b. Volumetric heating changes the ratio $\beta \equiv b_s/b_u$ from about 1.25 [22] to near unity or even lower [11]. Mixing with ambient fluid is confined to the outer part of the flow which insulates an unmixed core, totally unlike in an ordinary jet ([11]; [12], Figure 9). Entrainment and mixing clearly change drastically on heating.

Measuring the mass flux directly between the electrode grids in the heating zone presents problems. Near the edges of the jet the velocity is small and fluctuating, and so suffers from scatter, especially in the short 20 min. runs available in the apparatus used. The concentration width of the jet, b_s, is dramatically reduced by the heating, as is clear from Fig. 2 (by about 50% of the unheated value at the top of the heating zone at $G = 4.7$ [11]). If the velocity profiles are similar, the mass flux is proportional to $U_c b_u^2 = U_c b_s^2/\beta^2$. In order to show that assuming a constant β is unjustified, BN made a hypothetical estimate of the mass flux putting $\beta = \text{const.}$, and using experimental data for b_s and U_c (Fig. 3a). However, using their own experimental evidence on the reduction of β with heat injection [11, Fig. 11], they developed a variable β integral model whose predictions of U_c were shown to be in excellent agreement with experiment. They did this because values of U_c were far more accurate than those of mass flux could be. The mass flux values estimated from the variable β model are also shown in Fig. 3a. It is seen that they are generally in excess of the mass flux in an unheated jet, especially in the lower HIZ, in general agreement with data reported earlier from the same laboratory [10]. Ignoring all this Agrawal & Prasad [15; = AP below] make the extraordinary and unjustified assertion that BN report a decrease in mass flux with heating; nothing could be farther from the truth, as we have just shown.

BN pointed to a decrease in the entrainment coefficient α_E, in the upper HIZ and beyond. A direct measurement of the entrainment using PIV has been reported in [14]. (Incidentally the variation of α_E with z seen in [14] is mainly due to the relatively short averaging time used, which was not enough to remove the contribution of individual large vortical structures.) The average value of α_E for the axial range covered is 0.057 for the unheated case (close to Turner's value of 0.056 [1]), and only 0.017 for the heated case (i.e. 70% lower). These results are compared in Fig. 3b with estimates we obtain here from the experiments of BN and AP. A drastic reduction in α_E in the upper HIZ and beyond is once again something that all experiments agree on.

The only real difference between the IISc and Delaware experiments concerns the normalized turbulent intensity (say u'/U). In the IISc experiments u_c'/U_c, the center-line value of this parameter, decreases with increasing G [10, 11], whereas AP report significant increase with heating (about 50% higher than in unheated jets in the lower HIZ; see Fig. 3c). Also, the flow slows down drastically in the Delaware experiments immediately above z_b, while this was not observed in IISc. A plausible explanation for these differences is the much higher conductivity in the Delaware experiments, due to the higher concentration of acid in the active fluid (15 ml/l, against 2 ml/l at IISc). The

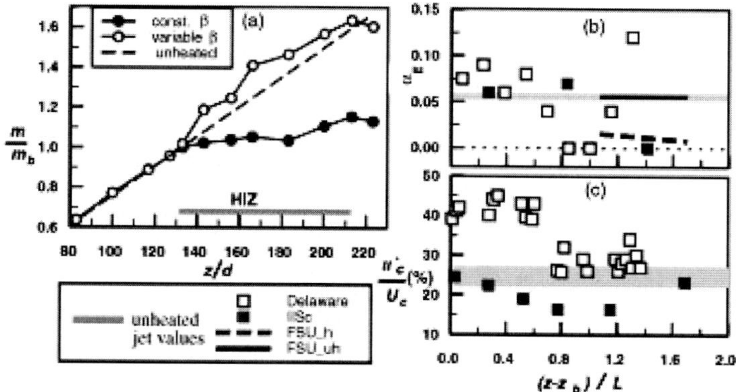

Fig. 3. (a) Estimates of mass flux (i) assuming β = constant and (ii) adopting the variable β model of [11]. (b) Axial variation of entrainment coefficient α_E in heated and unheated jets. L is the length of the heat injection zone. Solid and dashed lines represent mean trends in FSU data. (c) Axial variation of normalized streamwise velocity fluctuations.

resulting high current density near the electrodes creates narrow, 2D plumes which generate more intense fine scale turbulence and the high value of u'_c/U_c observed even at $z \simeq z_b$ by AP.

3 Computer Simulations

Almost immediately following the first IISc experiments, a computational study of the flow was reported by Basu & Narasimha [17]. This was a temporal simulation of the Boussinesq equations which, allowing for heat injection, take the form (apart from the incompressibility condition div $\boldsymbol{u} = 0$),

$$(\partial \boldsymbol{u}/\partial t) + (\boldsymbol{u}.\nabla)\boldsymbol{u} = -(1/\rho)\nabla p + \nu\nabla^2\boldsymbol{u} - \boldsymbol{g}\alpha T, \tag{1}$$

$$(\partial T/\partial t) + (\boldsymbol{u}.\nabla)T = \kappa\nabla^2 T + J/\rho C_p \tag{2}$$

in standard notation; J is rate of heat addition per unit volume, T is change in temperature above ambient. The equations were solved in a box with periodic boundary conditions, with heat injection over a time interval $t_f - t_i$ corresponding to the HIZ $z_b < z < z_t$ in the experiments. The solution method uses a spectral Fourier-Galerkin scheme [23]. Simulations were performed for both 64^3 and 128^3 grids. The heat J was injected with a radial distribution similar to that of the velocity at $t = t_i$, when self-similarity in the velocity profile had been achieved.

Although a precise comparison of these simulations with experiment is clearly not possible, there is remarkable agreement between them on the qualitative effect of heating on the flow. Thus, heating accelerates the flow and narrows the jet; absolute values of turbulence intensity increase but not as

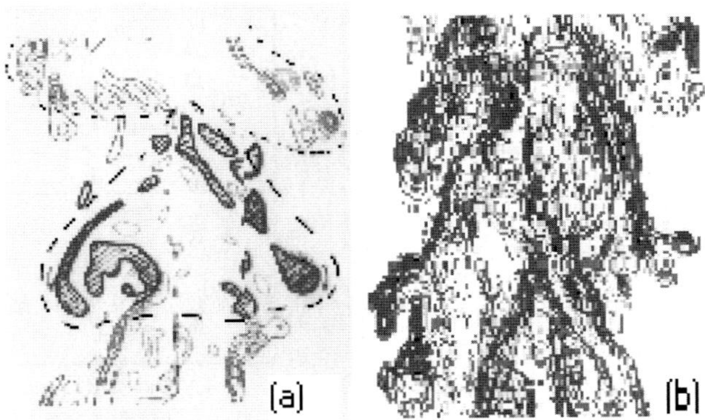

Fig. 4. Azimuthal vorticity in axial section of jet. (a) Unheated jet: note the arrow-head shaped coherent structure within the boundary marked by the dashed curve. (b) Heated jet, showing much higher vorticity and disruption of the coherent structure.

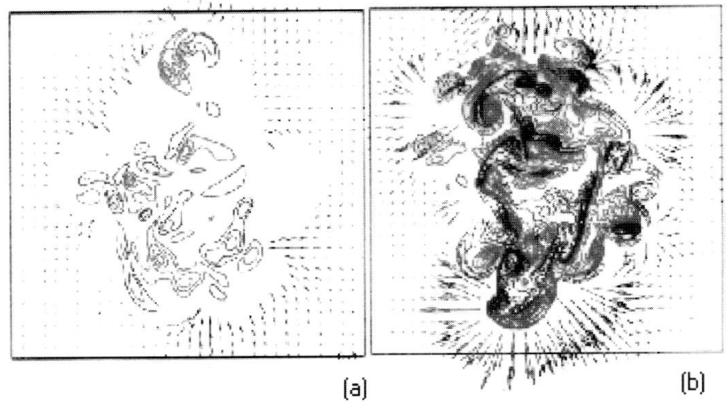

Fig. 5. Streamwise vorticity with entraining velocity field in typical diametral section. (a) unheated jet; (b) heated jet, showing strong expulsive motions at the edges. After [17].

rapidly as the mean velocities, so normalized turbulence intensities are lower, as observed in [11]. The coherent structures in the jet are considerably distorted and even disrupted (Fig. 4a,b).

Further, the simulations showed that heating leads to dramatic increases in the vorticity and its gradients in the flow, and to striking differences in the entraining flow field (Fig. 5a,b). Due in part to vortex stretching induced by the acceleration of the flow due to buoyancy, the streamwise vortices become intense, and appear to be responsible for strong expulsive motions in the immediate ambient neighbourhood of the core flow, especially in the plane of the base of what would have been a coherent structure in the jet. However no definite statements could be made about entrainment itself, as the net mass flux across opposite sides of the box always remains zero because of the periodic boundary conditions.

Agrawal, Boersma & Prasad [18] have recently reported a spatial simulation of a jet with buoyancy-induced acceleration. Here they introduce an additional equation to the system (1,2), governing the scalar (acid) concentration C,

$$(\partial C/\partial t) + (\boldsymbol{u}.\nabla)C = (1/\mathrm{ScRe})\nabla^2 C, \tag{3}$$

where Sc is the Schmidt number. Furthermore they assume that the heating term J in (3) is proportional to C. The coupled system is solved in a spherical polar coordinate system with outflow boundary conditions proposed by Akselvoll & Moin [24], and traction-free conditions on the lateral boundary.

Unfortunately the parameters taken for these simulations make it virtually impossible to compare their results with experiment. In the first place their Richardson number, at 12, is about 40 times higher than in their experiments (where it was around 0.3). Furthermore they set $\mathrm{Pr} = 1$, $\mathrm{Sc} = 1$, whereas in the experiments $\mathrm{Pr} = 6$, $\mathrm{Sc} = 40$. As their heating is coupled to the scalar C, the vastly different diffusivities in the experiments and the computations make the heating to which the jet is subjected very different in character from that in the experiments. Finally the Reynolds number, at 1000, was lower than in the experiments (1600 to 3200 in IISc, results from $\mathrm{Re} = 1450$ to 3000 in Delaware; 1600 in the simulations of Basu and Narasimha [17]). For all these reasons the conclusions drawn from this set of DNS results cannot be usefully compared with experiments.

4 Conclusions

A reanalysis of published experimental results on jets subjected to off-source volumetric heating shows that (i) in the beginning of the heating zone momentum and scalar widths show a slight increase, followed by (ii) slower, negligible or even negative growth in width; (iii) total mass flux may increase above its value in the unheated flow, but (iv) there is a reduction in entrainment coefficient in the upper heat injection zone and beyond. The only disagreement concerns the turbulence intensity; this is traced here to the precise way in which heat is injected into the flow. Temporal simulations of solutions of the Boussinesq equations with off-source heating reproduce these observations qualitatively, including the increase in absolute intensity and decrease in normalized intensity of streamwise turbulent velocity fluctuations found in our experiments. The simulations further show dramatic increases in vorticity due to the operation of the baroclinic torque. A proper spatio-temporal computer simulation of the flow still needs to be carried out.

RN is grateful to DRDO for support of his work.

References

1. Turner JS (1973) Buoyancy Effects in Fluids, Cambridge Uni Press
2. Ludlam FH (1980) Clouds and Storms, Penn State Uni Press, University Park PA
3. Simpson J (1983) In: DK Lilly, T Gal-Chen (eds) Mesoscale Meteorology D Reidel, Netherlands, pp 355–445
4. Paluch LR (1979) J Atm Sci 36:2467–2478
5. Morton BR, Taylor GI, Turner JS (1956) Proc Roy Soc A 234:1–23
6. Turner JS (1986) J Fluid Mech 173:431–471
7. Turner JS (1963) Q J Royal Met Soc 89:62–74
8. Johari H, Breidenthal RE (1987) Proc 3rd Intl Symp Stratified Flows, Caltech, 3–5 Feb
9. Bhat GS, Narasimha R, Arakri VH (1989) J Fluid Mech 7:99–102
10. Elavarasan R, Bhat GS, Prabhu A, Narasimha R (1995) Fluid Dyn Res 6:189–202
11. Bhat GS, Narasimha R (1996) J Fluid Mech 325:303–330
12. Venkatakrishnan L, Bhat GS, Prabhu A, Narasimha R (1998) Curr Sci 74:597–606
13. Venkatakrishnan L, Bhat GS, Narasimha R (1999) J Geo Res 104:14271–14281
14. Venkatakrishnan L, Elavarasan R, Bhat GS, Krothapalli A, Laurenco L (2003) Curr Sci 85:778–785
15. Agrawal A, Prasad AK (2004) J Fluid Mech 511:95–123
16. Agrawal A, Sreenivas KR, Prasad AK (2004) Int J Heat Mass Transfer 47:1433–1444
17. Basu A J, Narasimha R (1999) J Fluid Mech 385:199–228
18. Agrawal A, Boersma BJ, Prasad AK (2004) Flow Turbulence and Combustion 73:277–305
19. Emanuel KA (1994) Atmospheric Convection, Oxford Uni Press, New York
20. Hussein JH, Capp SP, George HK (1994) J Fluid Mech 258:31–75
21. Le Mone MA, Zipser E (1980) J Atm Sci 37:2444–2457
22. List EJ, Imberger J (1973) J Hydr Div ASCE 99:1461–1474
23. Orszag SA (1971) Stud Appl Maths 50:293–327
24. Akselvoll K, Moin P (1996) J Fluid Mech 315:387–411

Strongly Anisotropic Turbulence Using Statistical Theory: Still a Computationally Demanding Problem

Claude Cambon

Laboratoire de Mécanique des Fluides et d' Acoustique, École Centrale de Lyon, Université Claude Bernard Lyon, INSA, CNRS, F-69134 Écully Cedex, France
`claude.cambon@ec-lyon.fr`

Abstract. Strongly anisotropic turbulence is described by a minimal set of angle-dependent spectra, including in general both directional anisotropy — which includes dimensionality — and polarization anisotropy. The first kind of anisotropy is addressed in rotating turbulence, with various statistical approaches including wave-turbulence theory, and DNS. In addition to the long-term relevance of resonant triads, the role of anisotropic phase mixing is essential for explaining various, more or less transient, trends from the simplest properties of linear solutions for second-order statistics to the more complex dynamics of triads. A new approach to the asymmetry in terms of cyclonic and anticyclonic vertical vorticity distribution in rotating turbulence is discussed, with encouraging preliminary results. The case of stably stratified turbulence is touched upon at the end, in order to identify some analogies, as the relevance of gravity waves phase mixing for vertical — more generally poloidal — velocity, and differences with the rotating case. The main difference is that the horizontal layering can be described by an 'anti-2D' toroidal nonlinear energy drain, which is essentially disconnected from rapid gravity waves dynamics. In all cases, however, statistical theory, with strong anisotropy including dimensionality, appears as a useful complement to DNS/LES, and can offer alternative explanations with respect to oversimplified stability analysis.

Keywords: anisotropy, rotation, stratification, waves, phase-mixing

1 Introduction

To the question: what remains to be done in turbulence? responses often include further investigations of intermittency, anisotropy, and inhomogeneity. The emphasis is placed here on anisotropy, considering that important classes of flow, relevant e.g. to geophysical applications, are strongly anisotropic. On the other hand, inhomogenity is often not crucial far from boundaries, and anisotropy affects low-order statistical moments, the ones which deal with energy spectra and interscale energy transfer (cascade), without need for a precise knowledge of higher order (up to 3) statistical moments, which are

321

Y. Kaneda (ed.), *IUTAM Symposium on Computational Physics and New Perspectives in Turbulence*, 321–330.

possibly affected by intermittency. Anisotropy is crucial in many applications because it reflects *dimensionality*, and allows us to describe possible transition from 3D to 2D structure, as investigated in rotating turbulence (see Sect. 3). A very different transition, from 3D to 1D structure, can be described in stably stratified turbulence using similar tools (see Sect. 5).

Classical 'triadic' closures, including EDQNM and wave-turbulence theory, were shown to be relevant for calculating angle-dependent spectra, and gave a useful complementary *statistical* approach with respect to DNS, and even with respect to qualitative arguments from simplified stability analysis. On the other hand, solving numerically such closure equations in the relevant anisotropic context, remains a computationally demanding problem: the cost and the complexity greatly exceeds that of the 3D isotropic case, in which solving the Lin equation for the energy spectrum, closed by EDQNM for instance, is inexpensive.

These different aspects are discussed and partly revisited in this article, which is organised as follows. The anisotropic description is introduced in Sect. 2, with applications focussing on the axisymmetric case. The case of pure rotation is addressed in Sect. 3, with a review of recent results. A general review of anisotropic phase mixing is offered in Sect. 4, illustrating the subtle interplay between linear and nonlinear effects in rotating turbulence; promising applications to third-order vorticity statistics are touched upon. The case of stably stratified turbulence is briefly addressed in Sect. 5, which is devoted to conclusions.

2 Anisotropic Description for Second Order Statistics in Homogeneous Turbulence

General Case

The complete anisotropic description is mainly applied to the second order spectral tensor $\hat{R}_{ij}(\mathbf{k})$, which is obtained by Fourier transformation of the two-point velocity correlations $R_{ij}(\mathbf{r}) = \langle u_i(\mathbf{x})u_j(\mathbf{x}+\mathbf{r})\rangle$. The four-fold decomposition [1, 2] (revisited in [3]):

$$\hat{R}_{ij} = \underbrace{\frac{E(k)}{4\pi k^2}P_{ij}}_{\text{isotropic}} + \underbrace{\underbrace{\mathcal{E}(\mathbf{k})P_{ij}}_{\text{directional}} + \underbrace{\Re\left(Z(\mathbf{k})N_iN_j\right)}_{\text{polarization}}}_{\text{anisotropy}} + \underbrace{i\epsilon_{ijm}\frac{k_m}{k^2}\mathcal{H}(\mathbf{k})}_{\text{helicity}} \qquad (1)$$

accounts for all the properties coming from hermitian symmetry, incompressibility, and homogeneity, for arbitrary anisotropy. This equation involves the fully 3D (scalar) energy $U(\mathbf{k}) = E(k)/(4\pi k^2) + \mathcal{E}(\mathbf{k}) = (1/2)\hat{R}_{ii}(\mathbf{k})$, and helicity $\mathcal{H}(\mathbf{k})$ spectra, as well as the complex pseudo-scalar $Z(\mathbf{k})$ which characterizes polarization anisotropy. Tensorial bases, as the well-known projection operator P_{ij} and the 'helical' mode $\mathbf{N}$ [1, 4], only depend on the orientation of the wavevector $\mathbf{k}$, involving only the three eigenvectors $(\mathbf{k}/k, \mathbf{N}, \mathbf{N}^*)$

of the orthonormal rotation matrix around $\mathbf{k}$. In the isotropic case, only the spherically averaged energy spectrum $E(k)$, which depends on the modulus $k =| \mathbf{k} |$, is involved, and (1) reduces to its first term. Of course, 3D Fourier space is only a mathematical convenience; looking at a pure scalar correlation, a decomposition of $U = (1/2)\hat{R}_{ii}(\mathbf{k})$ in terms of scalar spherical harmonics in $\mathbf{k}$-space has an exact counterpart (same degree, term-to-term relationship in [5], see also [6]) for $(1/2)R_{ii}(\mathbf{r})$. The advantage of Fourier space is to drastically reduce the number of tensorial terms (see [7] for an analysis in physical space), *a priori* taking into account the divergence-free condition by means of algebraic relationship (e.g. Craya/Herring decomposition [8]). Given the high degree of anisotropy we are faced with, low-degree expansions in terms of angular harmonics, which are only valid for weak anisotropy, are avoided here. The first non-trivial coefficients (second degree) for instance generates the dimensionality structure tensor [9] (see [10], appendix, for details).

Axisymetric Turbulence around n

Using a polar-spherical system of coordinates for $\mathbf{k}$ with the fixed polar axis (fixed unit vector) $\mathbf{n}$, the terms U ($U = E/(4\pi k^2) + \mathcal{E}$ in (1)) and Z depend only on the set $(k, \cos\theta)$ or equivalently on $(k_{\parallel}, k_{\perp})$, with

$$\cos\theta = \mathbf{k} \cdot \mathbf{n}/k, \quad k_{\parallel} = \mathbf{k} \cdot \mathbf{n}, \quad k_{\perp} =| \mathbf{k} \times \mathbf{n} | . \tag{2}$$

It is important to notice that 2D turbulence is an extreme case of axisymmetric turbulence, with

$$U(\mathbf{k}, t) = \frac{E(k_{\perp}, t)}{2\pi k_{\perp}}\delta(k_{\parallel}), \tag{3}$$

and $Z = 0$ or $Z = -e$, depending on 2D-3C or 2D-2C cases [11]. The presence of the 'Dirac delta function' indicates how incomplete is a parameterisation of anisotropy in terms of the first non-trivial spherical harmonics (e.g. the k-spectrum of the dimensionality structure tensor [9]) if one want to describe a detailed transition from 3D to 2D structure.

3 Angle-Dependent Spectra for Rotating Turbulence

Looking at Fig. 1, construction of a spectrum $E(k)$ is shown in the purely isotropic case at zero laminar viscosity (top left, from [12]) and in rapidly rotating turbulence at vanishing Rossby number (top right, from [13]). Even if similar Quasi-Normal closure theories are used in both cases, the second is much more computationally demanding, so that computation of the thermalized tail is impractical and a finite Reynolds number is used instead when the inertial range extents until the maximum wavenumber. In comparison with the isotropic case, the k^{-3} inertial range slope for $E(k)$ is notable; an even more prominent difference is the strong dependence of the spectra $4\pi k^2 U(k, \cos\theta)$ on the anglular variable $\cos\theta = k_{\parallel}/k$ (bottom left, from [13]). Recall that

324   C. Cambon

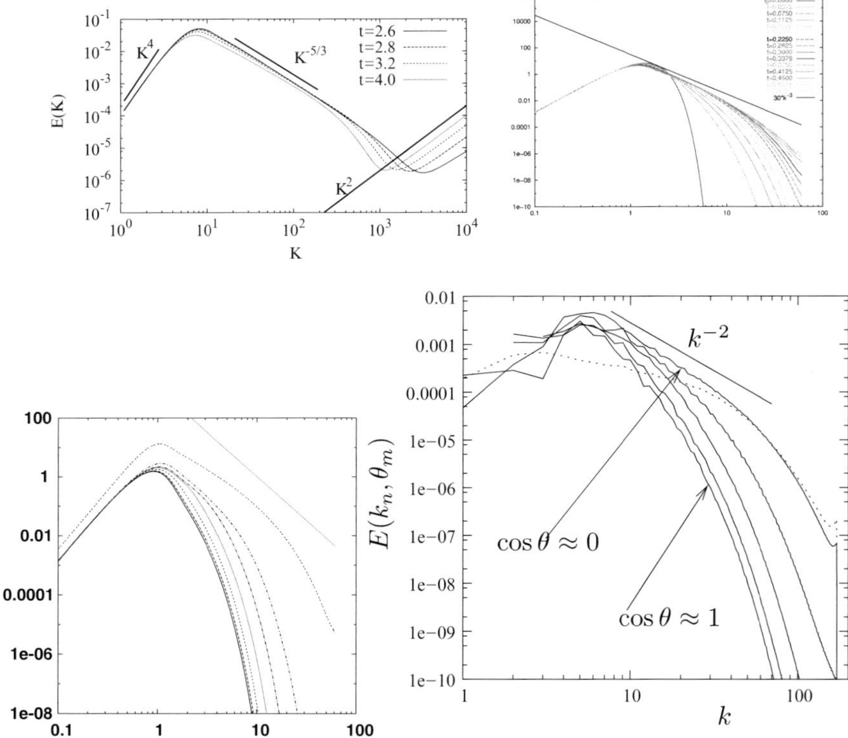

Fig. 1. Top: spherically averaged energy spectrum, isotropic EDQNM (left), AQNM for rapid rotation (right). Bottom: angle-dependent spectra, AQNM (left), DNS (right)

EDQNM [14] is one of the simplest 'triadic' closure for the spectral transfer term in the following Lin equation

$$\frac{\partial E}{\partial t} + 2\nu k^2 E(k,t) = T(k,t). \tag{4}$$

In rotating turbulence, similar equations are exactly found for the complete set U ($U = E/(4\pi k^2) + \mathcal{E}$), Z and $\mathcal{H}$ in (1), for instance

$$\frac{\partial U}{\partial t} + 2\nu k^2 U(k, \cos\theta, t) = T^{(e)}(k, \cos\theta, t), \tag{5}$$

but the isotropy is broken by the angle-dependent transfer term $T^{(e)}$, via a resonance operator

$$\exp[\mathrm{i}(\sigma_k \pm \sigma_p \pm \sigma_q)(t - t')], \tag{6}$$

which is discussed in more detail in Sect. 4. The numerical problem, even restricted to solving (5) with AQNM [13] on the surfaces of resonant triads — given by the zero value of the phase in (6) — was very cumbersome to handle.

The resonant surfaces were imbedded in a volume of about 300^3 points; in contrast with analytical approaches using similar wave-turbulence equations [15], no approximation, such as $k_\parallel/k$ being small, was made and all the orientations were numerically addressed with the same care.The strong directional anisotropy constructed by the AQNM model, in the asymptotic limit, is also qualitatively recovered in 512^3 DNS (Fig. 1, bottom right [16]). The most surprizing feature is that the anisotropy, linked to the prevalence of 2D properties in agreement with (3), increases with the wavenumber, similarly in AQNM and DNS.

4 Linear or Nonlinear Mechanisms? Anisotropic Phase-Mixing Applied to Various Statistics: Pure Rotation

Basic Linear Operator

Solutions of the linear part of Navier-Stokes equations in a rotating frame are found in 3D Fourier space, in a local and algebraic form [18, 1], as

$$\hat{u}_i(\mathbf{k}, t) = G_{ij}(\mathbf{k}/k, \sigma_k t)\hat{u}_j(\mathbf{k}, 0) \tag{7}$$

with

$$G_{ij} = \Re[N_i N_j^* e^{i\sigma_k t}], \quad \sigma_k = f\frac{k_\parallel}{k}. \tag{8}$$

The helical mode $\mathbf{N}$ is the same as in (1), and σ_k is the unsigned dispersion law of inertial waves, f being the system vorticity, to be replaced by the local Coriolis parameter in the geophysical context. Of course the G_{ij} operator is highly anisotropic, changing isotropic statistical symmetry into axisymmetry with no mirror symmetry if the initial velocity field is isotropic; but its impact depends on the statistical quantity under consideration.

Single-Time Double Correlations

Any single-time second-order correlation tensor can be generated from the $U, Z, \mathcal{H}$ set, which, according to linear solutions (7) of the initial-value problem reduces to [1]

$$U(\mathbf{k}, t) = U(\mathbf{k}, 0); \quad \mathcal{H}(\mathbf{k}, t) = \mathcal{H}(\mathbf{k}, 0); \quad Z(\mathbf{k}, t) = Z(\mathbf{k}, 0)\exp(2ift\cos\theta). \tag{9}$$

One recovers that isotropy *at this simple level of description* can only be broken by nonlinear terms, via the transfer terms on the right-hand side of e.g. (5). On the one hand, this poor relevance of the linear solution reflects the fact that phase information is lost in homogeneous turbulence considering only second-order statistics [19] because the time dependence cancels out on multiplying $e^{\pm i\sigma t}$ by its complex conjugate. On the other hand, the linear solution keeps some relevance if turbulence is started with anisotropic structure, including polarization anisotropy. In the latter case, a sudden change of anisotropy can be predicted for any single-point second-order correlation, resulting from strict conservation of directional anisotropy and rapid damping of polarization anisotropy. This damping is one of the simplest effects of the phase-mixing, so that

$$\int_0^1 C(k,x)\exp(2\mathrm{i}ftx)dx \to 0 \quad \text{if} \quad t \to \infty, \quad x = \cos\theta, \tag{10}$$

provided $C(k,x)$, which is directly linked to the initial value of Z, has no singularity. For instance, starting from a Reynolds stress tensor of the 'cigar' type, possibly created by an axisymmetric contraction, rapid rotation yields a pancake-type Reynolds stress tensor [1, 2].

Two-Time Double Correlations

Isotropy breaking is directly obtained for second order statistics in the linear limit if *two-time* correlations are addressed, since time dependency in terms like $\mathrm{e}^{\pm i\sigma t}$ and $\mathrm{e}^{\pm i\sigma t'}$ cannot cancel out by product. As a useful and simple application, a general method was introduced by Kaneda [20] for calculating single-particle dispersion, by means of two-time linear solutions and a simplified Corrsin hypothesis. The reader is referred to [21] and [16] for further analytical calculations in rotating and rotating/stratified cases, with comparisons to KS and DNS (these developments were emphasized in my talk).

Triple Correlations

The ambiguity of what is linear and what is nonlinear (see also [19]) appears at this level of description. On the one hand, the three-fold product of Green's function directly reflects purely linear dynamics at the level of triple correlations: the resonance factor in (6) modulates any transfer term in the r-h-s of equations such as (5). On the other hand, the presence of these r-h-s terms results from quadratic nonlinearity in the basic Navier-Stokes equations. As the more specific result, anisotropic EDQNM(2-3), asymptotic AQNM (also in agreement with the qualitative prediction by [4]) and DNS results support the sketch of an angular energy drain towards horizontal wavevectors (see Fig. 2-(c)) with the rise of more coherent vortices about the vertical direction (see Fig. 2-(d) and Fig. 3-(right) from DNS).

A New Approach Towards Asymmetry of Cyclonic and Anticyclonic Vertical Vorticity

In order to show that the interplay of linear solutions and of nonlinear spectral closures is not only useful for calculating those triple correlations which are

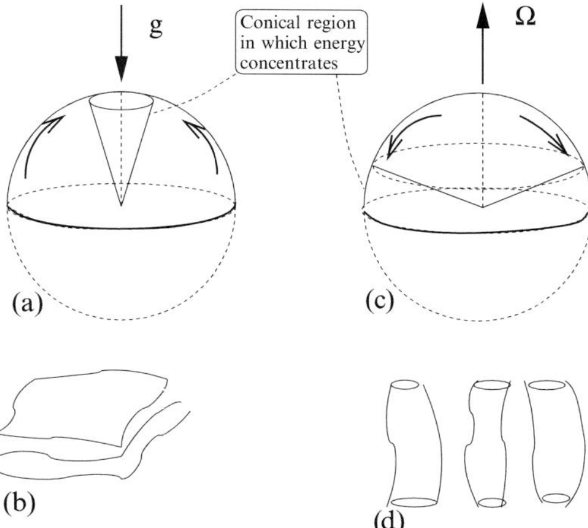

Fig. 2. Sketch of anisotropic structure induced by nonlinearity: stratified (left), rotating (right), spectral (top) and physical (bottom) space, from [25]

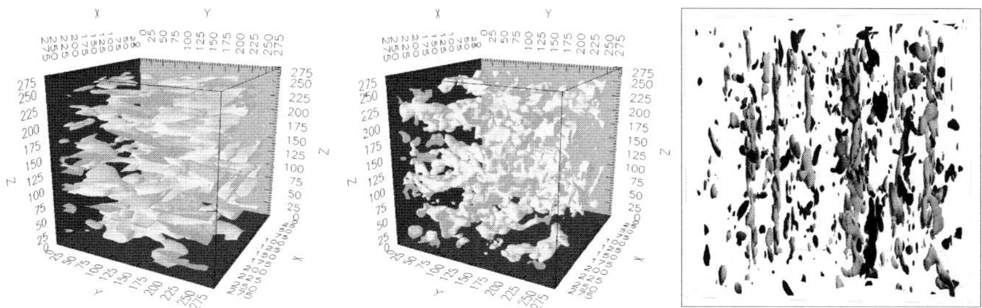

Fig. 3. DNS results, from [17]. Stratified turbulence, decomposing the velocity field into toroidal (left) and poloidal (midle) velocity components. Rotating turbulence (right) with cyclonic eddies in grey.

involved on the nonlinear energy transfer terms, another type of triple correlation is now addressed. The asymmetry in terms of cyclonic and anticyclonic vertical vorticity can be found in several physical or numerical experiments, as well as in observations in nature (see Fig. 3-(right) here). For instance, [22] plotted the skewness of vertical vorticity to quantify this effect, showing the rise of a very significant positive value if the Rossby number is not too small. Explanations based on instability theory (e.g. centrifugal instability) are not convincing if real turbulence with a 'democratic crowd' (Yukio Kaneda's talk) of vortices is considered, and the relevance of under-resolved DNS with hyperviscosity (therefore LES implicitely) is questionable from a quantitative viewpoint. A recent experimental study [23] of decaying rotating turbulence showed the relevance of the linear time-scale to collect different cases with the same scaling: a clear power law $t^{.7-.75}$ is exhibited for the rise of the vorticity skewness in Fig. 4-(left), and the final rapid collapse is attributed to the

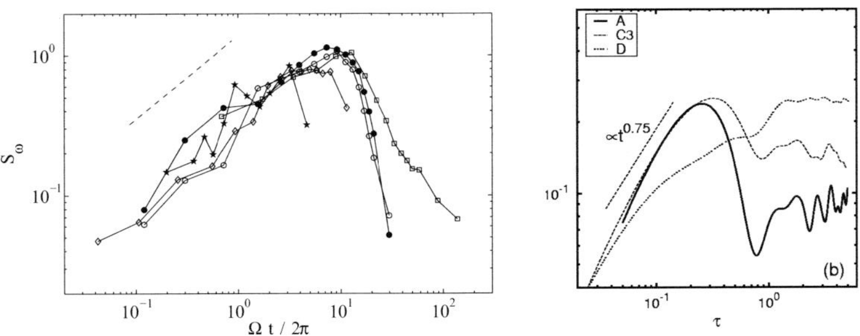

Fig. 4. Vorticity skewness: experiment (left) [23] and DNS [24] (right)

rise of nonhomegeneous mechanismes, such as Ekmann pumping. Preliminary DNS [24] plotted in terms of the linear time-scale look similar, but the final collapse can be interpreted as a final stage of linear 'triadic' phase-mixing, in the absence of strong enough nonlinearity in decaying turbulence at moderate initial Reynolds number. Classically, RDT (or more conveniently here, linear theory) is used to predict the evolution of second order statistics, but it is possible to predict correlations of any order, and we will address triple correlations below. Formulation in Fourier space allows an accurate description of phase-mixing, and one can readily pass from velocity to vorticity. Triple vorticity correlations in Fourier space display a third order spectral tensor as

$$< \hat{\omega}_i(\mathbf{q}, t)\hat{\omega}_j(\mathbf{k}, t)\hat{\omega}_m(\mathbf{p}, t) >= \Phi_{ijm}(\mathbf{k}, \mathbf{p}, t)\delta(\mathbf{k} + \mathbf{p} + \mathbf{q}),$$

and equation (4) — the same for vorticity — yields

$$\Phi_{ijm}(\mathbf{k}, \mathbf{p}, t) = G_{ir}(-\mathbf{k} - \mathbf{p}, t)G_{js}(\mathbf{k}, t)G_{mn}(\mathbf{p}, t)\Phi_{rsn}(\mathbf{k}, \mathbf{p}, 0) \qquad (11)$$

Going back to a single-point correlation, or $< \omega_i\omega_j\omega_m >$, we recover an integral formulation via $< \omega_i\omega_j\omega_m > (t) = \int_{R^6} \Phi_{ijm}(\mathbf{k}, \mathbf{p}, t)d^3\mathbf{k}d^3\mathbf{p}$, so that for instance

$$< \omega_3^3 > (t) = \int_{R^6} G_{3r}(-\mathbf{k} - \mathbf{p}, t)G_{3s}(\mathbf{k}, t)G_{3n}(\mathbf{p}, t)\Phi_{rsn}(\mathbf{k}, \mathbf{p}, 0)d^3\mathbf{k}d^3\mathbf{p}. \quad (12)$$

Finally, it is necessary to know the contribution from initial triple correlations for all triads[1] in order to solve the latter equation, but the problem is much better documented than in physical space: robust spectral theories such as EDQNM provide a systematic way to express initial, isotropic, Φ_{ijn} in terms of the initial scalar energy spectrum $E(k)$. More generally, more complex EDQNM(2-3) versions can be used to solve the fully nonlinear problem, not only for generating isotropic initial data in (12). As a common point of all,

---

[1] or equivalently, in physical space, for any triple correlation at three points, information which cannot be provided by the third-order structure function alone.

linear as well as nonlinear, formulations, the product of three Green's functions amounts to the phase term (6) which controls the phase-mixing, its zero value corresponding to exact resonance condition for triads.

5 Conclusions and Perspectives: Wave-Turbulence Versus 'Toroidal' Strong Turbulence

Rapid rotation is essentially 'weak' inertial wave-turbulence; strongly stably-stratified turbulence illustrates the interplay of 'strong' turbulence with wave-turbulence, since the toroidal part of the flow (essentially the same as the Riley's vortex mode [26] in this case), whose spectral energy is $U - Z$ using (1), is not affected by gravity waves at low Froude number. Only the poloidal velocity component — which includes the entire vertical component but also a 'divergent' part of the horizontal field — is directly affected. Implications for one-particle diffusion were discussed in my talk, following [20, 21, 16] and considering combined effects of rotation and stable stratification. Looking at anisotropic DNS and EDQNM2 [25] results, which compare very well [27], the sketch for the horizontal layering (Fig. 2-(left)) was confirmed: partial transition from 3D isotropic to 1D VSHF mode (Vertically Sheared Horizontal Flow in physical space: the term was coined by [28]) is essentially due to an angular energy drain towards vertical wave-vectors. In addition, this *nonlinear energy transfer* effect mainly affects the toroidal (or vertical vortex) mode. As shown on Fig. 3-(left), the contribution of the toroidal mode to the anisotropic layering is completely dominant. In this sense, the behaviour of strongly stratified flow *is not quasi-2D but anti-2D* (comparing sketchs on the left and on the right in Fig. 2). On the one hand, our approach is consistent with the one by Billant & Chomaz [29] and by Lindborg [30], who stressed that the dynamics of stably stratified turbulence had nothing to do with 2D dynamics. On the other hand, our angle of attack is essentially statistical, allowing anisotropic structuring, but without need for pre-existing [29] or forced [30] large coherent vertical vortices, as needed for triggering a zig-zag instability.

In short, anisotropic phase mixing is essential for describing and understanding transient effects in turbulence affected by dispersive waves. If the entire flow is dominated by waves, as in the rapidly rotating case, resonant wave interactions govern the long-term nonlinear dynamics. In the different case of strongly stratified fluid, the relevant nonlinearity only involves the toroidal part of the flow at moderate times, nonlinear wave-turbulence dynamics remaining a secondary delayed mechanism. Wave-turbulence theory, from linear to weakly nonlinear, is essential in rotating turbulence, and remains useful in various aspects (e.g. Lagrangian diffusion) of stratified turbulence, but the horizontal layering can be explained by 'strong' toroidal cascade, excluding wave effects. Statistical theory remains a valuable alternative to DNS, but detailed description of strong anisotropy yields new challenges, and offers new opportunities for numerical computations. Strong anisotropy at small scales is another non-conventional aspect, which must be accounted for in LES.

References

1. Cambon C, Jacquin L (1989) J Fluid Mech 202:295–317
2. Cambon C, Jacquin L, Lubrano JL (1992) Phys Fluids A 4:812–824
3. Cambon C, Rubinstein R (2006) Phys Fluids 18:085106
4. Waleffe F (1993) Phys Fluids A 5:677–685
5. Cambon C, Teissèdre C (1985) C R Acad Sci Paris 301, Série II, 2:65–70
6. Sreenivasan KR, Narasimha R (1978) J Fluid Mech 84:497–516
7. Arad I, L'vov VS, Procaccia I (1999) Phys Rev E 59:6753–6765
8. Herring JR (1974) Phys Fluids 17:859–872
9. Kassinos S, Reynolds WC, Rogers M (2000) J Fluid Mech 428:213–248
10. Salhi A, Cambon C (2006) J Appl Mech 73:449–460
11. Cambon C, Mansour NN , Godeferd FS (1997) J Fluid Mech 337:303–332
12. Bos WJT, Bertoglio JP (2006) Phys Fluids 18:071701
13. Bellet F, Godeferd FS, Scott JF, Cambon C (2006) J Fluid Mech 562:83–121
14. Orszag SA (1970) J Fluid Mech 41:363–386
15. Galtier S (2003) Phys Rev E 68:1–4
16. Liechtenstein L, Godeferd FS, Cambon C (2005) J Turbul 6:1–18
17. Liechtenstein L, unpublished.
18. Greenspan HP (1968) The theory of rotating fluids. Cambridge University Press
19. Davidson PA, Stapelhurst PJ, Dalziel SB (2006) J Fluid Mech 557:135–144
20. Kaneda Y (2000) J Phys Soc Japan 69:3847–3852
21. Cambon C, Godeferd FS, Nicolleau F, Vassilicos JC (2004) J Fluid Mech 499:231–255
22. Bartello P, Métais O, Lesieur M (1994) J Fluid Mech 273:1–29
23. Morize C, Moisy F, Rabaud M (2005) Phys Fluids 17, 9:095105
24. van Bokhoven LJA, Cambon C, Liechtenstein L, Godeferd FS, Clercx HJH (2006) In: Proc. Euromech 477 meeting, Twente, The Netherland, to appear
25. Godeferd FS, Cambon C (1994) Phys Fluids 6:2084–2100
26. Riley JJ, Metcalfe RW, Weissman MA (1981) DNS of homogeneous turbulence in density-stratified fluids. In: Proc. of AIP Conference on Nonlinear Properties of Internal Waves, B. J. West (Ed.). New York, American Institute of Physics
27. Godeferd FS, Staquet C (2003) J Fluid Mech 486:115–159
28. Smith LM, Waleffe F (2002) J Fluid Mech 451:145–168
29. Billant P, Chomaz JM (2001) Phys Fluids 13:1645–1651
30. Lindborg E (2006) J Fluid Mech 550:207–242

Vortical Interactions with Interfacial Shear Layers

Julian C.R. Hunt[1], Ian Eames[1] and Jerry Westerweel[2]

[1] University College London, Gower Street, London WC1E 7JE, UK
[2] J.H. Burgers Centre, Delft University of Technology, 2628 CA Delft,
The Netherlands

Abstract. Physical models based on linearised calculations are developed to provide insight into some of the complex processes which occur adjacent to shearing interfaces. Specifically the case of vortices interacting with strong and weak shearing interfaces are developed. These provide a starting point to interpret previous detailed experimental and theoretical studies.

Keywords: turbulence, shear layer, entrainment, jets

1 Introduction

We examine aspects of eddy motion that occur at the interfacial shear layers that exist between regions of turbulent and non-turbulent motion, such as the outer edge of jets and wakes. The inhomogeneous turbulence in these interfacial regions play a critical role in many engineering and natural flows (*e.g.* [1]). With the aid of Particle Imaging Velocimetry (PIV) systems and large computers running at mega- and tera-flops speed, the flow and scalar fields in these layers can be measured and computed in sufficient detail, to test various hypotheses and theoretical models ([2], [3], [4]). In some previous models it was assumed that the dynamics of the interface was determined by the 'nibbling' action of small scale turbulent eddies (*e.g.* [5]) while others were based on the 'engulfing' motions of large scale eddies and the 'elastic' dynamics of turbulence distorted by these eddies [6].

The interaction between internal turbulence and interfacial shear layers has been previously studied theoretically using linearised RDT calculations where the interface is approximately flat (*e.g.* [7]) or weakly deformed (*e.g.* [8]). To delve deeper into the physical processes, we examine the effect of impinging vortices on the distortion of shear layers. Here we consider the effect of vortices interacting with relatively strong interfaces which leads to the strengthening of the shear layer and vortices distorting a weak interfacial layer – both processes are observed in detailed numerical and experimental studies.

331

Y. Kaneda (ed.), *IUTAM Symposium on Computational Physics and New Perspectives in Turbulence*, 331–338.

2 Dynamics of Shear Interface Layers

At an interface located at $x_3 = 0$ bounding flows with mean shear (*e.g.* jets, wakes), the mean vorticity $\Omega_2 = \partial U_1/\partial x_2$ has the form $\langle \Omega_2 \rangle = \delta(-x_3)\Delta U_1 + H(-x_3)\langle dU_1/dx_3 \rangle_c$. The local turbulence continually enhanced by the Kelvin-Helmholtz instability of the shear layer leads to the interface moving outwards at the boundary $E_b = \langle dx_3/dt \rangle(x_3 = 0)$ [2]. Although previous studies have suggested $\Delta U_1 = 0$ (*e.g.* [9]), where $u_0 \sim \sqrt{u_3^2}$, numerical simulations, experiments and theory indicate that at high Reynolds numbers, a vortex sheet exists where $\Delta U_1 \neq 0$. The vortex sheet acts as a barrier to velocity fluctuations relative to the sheet (ie $u_3 = 0$ at $x_3 = 0$) provided $u_0 < \Delta U_1$ through the shear sheltering mechanism [8]. When the eddies $x_3 < 0$ move with the local mean velocity they produce a weak pressure fluctuation at the interface, but their straining motion distorts the vorticity in the shear layer so that it induces an impulse that opposes that of the impinging vortex.

2.1 Vortical Interaction with a Strong Interface

We consider an idealisation of the typical flow at the edge of a turbulent shear layer where there is a relatively strong discontinuity in the conditionally averaged velocity field at the interface, *i.e.* $u_0 \ll \Delta U_1$. In the interior turbulent region ($x_3 < 0$) near the interface, the mean external velocity profile is $U_1 = \Delta U_1^{(-)}(x_3)H(-x_3) - x_3 H(-x_3)\langle dU_1/dx_3 \rangle_c$, and the mean shear $\Omega_2^{(-)} = \partial U_1^{(-)}/\partial x_3$, is small compared to the strain rate of the turbulent eddies and $\Omega_2^{(-)} \to 0$ for $|x_3|/L_x \gg 1$. Pressure fluctuations $O(u_0^2)$ affect the interface displacing it by $\sim (u_0/\Delta U_1)L_x$ so to first order the turbulence is blocked by the shear layer, *i.e.* $u_3^{(-)} = 0$ at $x_3 = 0$.

The aim here is to calculate the mean vorticity $\Omega_2^{(-)}$ below the interface as it is perturbed during the period that a fluid element spends within an eddy near the interface. Fluid lines parallel to the x_2-direction are stretched which produces changes to the mean vorticity field below the interface. The local value of the spanwise component of vorticity $\Omega_2^{(-)}$ is smaller than the vorticity fluctuations $\omega^{(-)}$ of the impinging turbulence. The flow generated by a largescale disturbance moving towards the interface is blocked kinematically initially generating an approximately local linear straining flow,

$$\mathbf{u}^{(-)} = ((\lambda_3 - 1)\Sigma^{(-)}x_1, \Sigma^{(-)}x_2, -\lambda_3\Sigma^{(-)}x_3), \tag{1}$$

where $\Sigma^{(-)} > 0$ and λ_3 depends on the nature of the eddy (see Fig. 1). For short time, the linearised vorticity equation for $\Omega_2^{(-)}$ is

$$\frac{d\Omega_2^{(-)}}{dt} \left(\equiv -\lambda_3 \Sigma^{(-)} x_3 \frac{d\Omega_2^{(-)}}{dx_3} \right) = \frac{\partial u_2^{(-)}}{\partial x_2}\Omega_2^{(-)} = \Sigma^{(-)}\Omega_2^{(-)}. \tag{2}$$

The results can be expressed in a Lagrangian time coordinate t or in terms of x_3. Since for an impinging eddy $\Omega_2^{(-)} = \Omega_{20}^{(-)}$ when $|x_3| \sim L_x$ and $t = 0$, from (2),

$$\Omega_2^{(-)}(t)/\Omega_{20}^{(-)} = \exp(\Sigma^{(-)}t), \qquad \text{or} \qquad \Omega_2^{(-)}(x_3)/\Omega_{20}^{(-)} = (L_x/x_3)^{1/\lambda_3} . \tag{3}$$

This shows how the distortion of the mean vorticity profile depends on the form of the eddies in the external flow below the interface, as characterised by the parameter λ_3. As the stretching by the impinging eddies increases towards the interface, the viscous diffusion of vorticity increases until it limits any further growth at the edge of the viscous internal layer at the interface where $|x_3| \sim l_v = \sqrt{\nu/\Sigma^{(-)}}$ at the sheared interface. Defining the eddy Reynolds number $Re_v = u_0 L_x/\nu$, where $u_0 \sim \Sigma^{(-)} L_x$, the maximum value of the perturbed external vorticity is $\Omega_{2max}^{(-)}/\Omega_{20}^{(-)} \sim (L_x/l_v)^{\frac{1}{2}} \sim Re_v^{\frac{1}{4}}$, for axisymmetric straining flows/vortices (where $\lambda_3 = 2$), while $\Omega_{2max}^{(-)}/\Omega_{20}^{(-)} \sim L_x/l_v \sim Re_v^{\frac{1}{2}}$, for planar straining/longitudinal vortices (where $\lambda_3 = 1$) (see Fig. 1).

In the next phase of their 'life-cycle', the fluid elements typically move towards the separation/stagnation regions (where $x_1 = x_{sep} \sim L_x$, see Fig. 1) where the streamlines converge along the surface before moving downwards into the turbulent region, $i.e.$

$$\mathbf{u}^{(-)} = (-\lambda_1 \Sigma^{(-)}(x_1 - x_{sep}), -\Sigma^{(-)}x_2, (1 + \lambda_1)\Sigma^{(-)}x_3), \tag{4}$$

where $\lambda_1 \sim 1$ and $\Sigma^{(-)} > 0$. Perot & Moin [10] computed the streamlines in the typically axisymmetric 'anti-splat' processes in their direct numerical

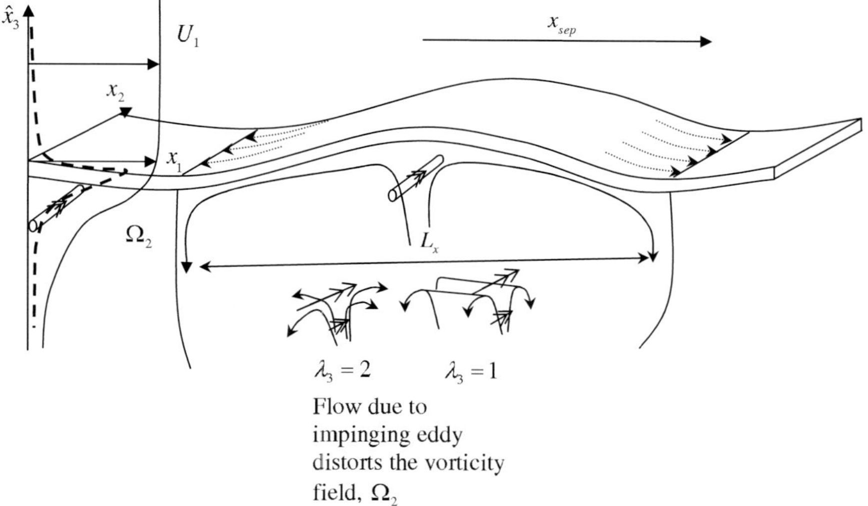

Fig. 1. Schematic showing how the distortion of vorticity by impacting planar ($\lambda_3 = 1$) and axisymmetric ($\lambda_3 = 2$) eddies on the shearing interface (located at $x_3 = 0$) serves to strengthen and maintain the interface.

334   J.C.R. Hunt et al.

simulations of turbulence in this kind of flow for the case of a shear-free layer. Along the surface where $x_3 \sim l_v$,

$$\frac{\mathrm{d}\Omega_2^{(-)}}{\mathrm{d}t} \left(= -\lambda_1 \Sigma^{(-)}(x_1 - x_{sep})\frac{\mathrm{d}\Omega_2^{(-)}}{\mathrm{d}x_1} \right) = -\Omega_2^{(-)}\Sigma^{(-)}, \tag{5}$$

so that

$$\Omega_2^{(-)}/\Omega_{2max}^{(-)} = ((x_1 - x_{sep})/L_x)^{1/\lambda_1}. \tag{6}$$

Thus although $\Omega_2^{(-)}$ tends to where the flow separates the average value of $\Omega_2^{(-)}$ along the surface is much greater than the mean value of $\Omega_2^{(-)}$, i.e. $\langle \Omega_{20} \rangle$, in the external turbulent region, i.e.

$$\langle \Omega_2^{(-)} \rangle(\tilde{x}_3 \sim -l_v) \sim \frac{1}{2}\langle \Omega_{20}^{(-)} \rangle/(l_v/L_x)^{1/\lambda_3}. \tag{7}$$

For highly elongated eddies, $(\lambda_3 \sim 1)$ the variation of mean vorticity is given by

$$\langle \Omega_2^{(-)} \rangle \sim \Omega_{20} Re_v^{\frac{1}{2}}|x_3/L_x|^{-1} \tag{8}$$

giving rise to an approximately logarithmic profile adjacent to the interface. While for axisymmetric vortices $(\lambda_3 \sim 2)$,

$$\langle \Omega_2^{(-)} \rangle \sim \Omega_{20} Re_v^{\frac{1}{4}}|x_3/L_x|^{-2} \tag{9}$$

a much more localised (but weaker) perturbation is created. This mechanism of inhomogeneous straining of large scale turbulence leads to a finite amplification of the mean vorticity in a layer of finite thickness.

2.2 Vortical Interaction with a Weak Interface

Observations, measurements and detailed numerical simulations of the edges of turbulent shear flows ([2], [3], [4]) show that periodically the interface forms cusps that point inwards towards the turbulent region ($x_3 < 0$). In the opposite direction smooth bulges tend to form. The local processes associated with these inward cusps are significant for the overall flow because they affect the transport of material (and scalars) across the shear layer and contribute significantly to the dissipation of energy in the turbulent region. We consider an idealised local analysis to study how the vortex sheet changes as it is entrained by the large eddy motion (see Fig. 2a). During these events the straining flow generated by the vortices or turbulence is stronger than the ambient flow, i.e. $u_0 \sim \Sigma^{(-)}L_x \gg \Delta U_1$ and the velocity of the large scale eddies u_0 in the turbulent region are large compared to the velocity jump ΔU_1 across the interface.

Initially $(t = 0)$ the sheared interface near $x_3 = x_3(0)$ has thickness δ_0. In our model calculation, the initial cross-stream vorticity has a gaussian form,

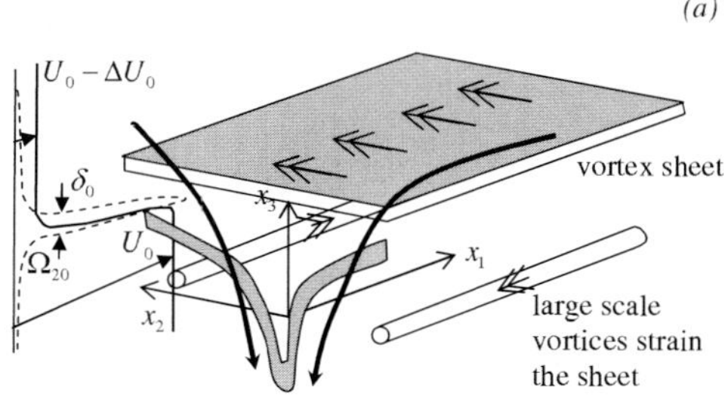

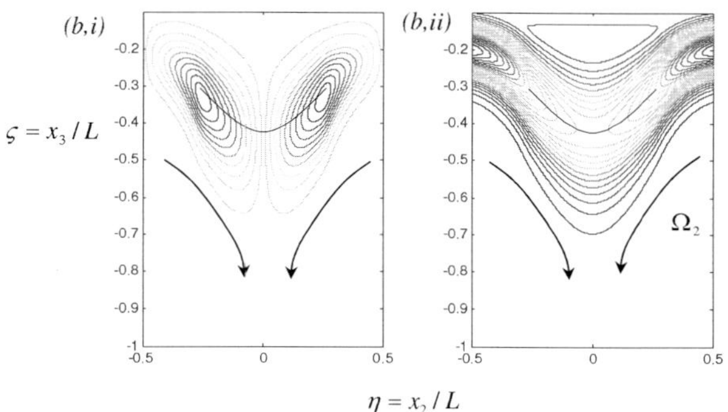

Fig. 2. (a) Schematic showing the engulfment of a sharp interface with a weak vortex sheet by a non-uniform straining flow generated by strong vortices. (b) Numerical solutions for contours of (i) Ω_3, (ii) Ω_2 to the linearised vorticity equation (13) are presented which illustrate the straining and the diffusion of vorticity which limits the strength of Ω_3. Here $\delta_0/L = 0.1$, $\beta = 2$.

$$\Omega_2(t = 0) = \Omega_{20} \exp\left(-x_3^2/2\delta_0^2\right), \; \Omega_{20} = \Delta U_1/\delta_0\sqrt{2\pi}, \tag{10}$$

where ΔU_1 is the jump across the interface, which is expressed in terms of local dimensionless coordinates, corresponding to the velocity profile being an error function. The essential feature of the distortion and entrainment of the interface is a non-uniform straining flow generated by vortical pairs; in this approximate model, we consider a non-uniform straining incompressible flow described by

$$\mathbf{u} = \left(0, -\Sigma^{(-)}x_2\left(1 - \beta x_2^2/L^2\right), \Sigma^{(-)}(x_3 - L)\left(1 - 3\beta x_2^2/L^2\right)\right), \tag{11}$$

where β is a constant and a measure of the non-uniformity of the straining field. The intensity of turbulence is $u_0 \sim \Sigma^{(-)}L$. The distortion of a material sheet, representing the initial position of the vortex sheet, initially located at $x_3 = x_{3I}(t = 0)(< L)$ by the flow field (11), is expressed in non-dimensional coordinates $\eta = x_2/L$, $\zeta = (x_3 - L)/L$, and time as $\tau = \Sigma^{(-)}t$.

336   J.C.R. Hunt et al.

We are interested in the entrained vortex sheet in the centre of the straining flow where $\eta \ll 1$. Here a point initially at (η_0, ζ_0) is advected to

$$\eta(\tau) \approx \eta_0 \exp(-\Sigma^{(-)}\tau), \qquad \zeta(\tau) \approx \zeta_0 \exp\left(\Sigma^{(-)}\tau + \frac{3}{2}\Sigma^{(-)}\beta\eta_0^2(e^{-2\tau} - 1)\right).$$

$$(12)$$

Thus the shape of the interface as it deforms is a parabola whose curvature at the position of maximum penetration $(\eta_0 = 0)$ $\mathrm{d}^2\zeta/\mathrm{d}\eta^2|_{\eta=0} = 3\beta\Sigma^{(-)}(e^{2\tau}-1)$, increases rapidly. The vorticity field in the deformed interface $\boldsymbol{\Omega}$ satisfies the linear vorticity equation

$$\frac{\partial \boldsymbol{\Omega}}{\partial t} + (\boldsymbol{u} \cdot \boldsymbol{\nabla})\boldsymbol{\Omega} = (\boldsymbol{\Omega} \cdot \boldsymbol{\nabla})\boldsymbol{u} + \nu \left(\frac{\partial^2}{\partial x_2^2} + \frac{\partial^2}{\partial x_3^2}\right)\boldsymbol{\Omega}. \tag{13}$$

The vortical dynamics are initially dominated by inviscid stretching/compression and later by viscous effects. For time $\tau \ll 1$, since the local Reynolds number is large, *i.e.* $Re = \Delta U_0 \delta_0/\nu \gg 1$, inviscid dynamics prevail with

$$\frac{\mathrm{d}\Omega_2}{\mathrm{d}\tau} = -\Omega_2,$$

or

$$\Omega_2(\eta, \zeta, \tau) = \Omega_{20} \exp\left(-\frac{(\zeta\exp(-\Sigma^{(-)}\tau - \Sigma^{(-)}\frac{3}{2}\beta\eta^2(1 - e^{2\tau})) - \zeta_I)^2}{2\delta_0^2} - \tau\right),$$

$$(14)$$

so that the maximum spanwise vorticity in the vortex sheet decreases as

$$\Omega_{2max}/\Omega_{20} = \zeta_I/\zeta, \qquad \zeta_I = (x_{3I} - L)/L, \tag{15}$$

due to the vortex lines being compressed (see Fig. 2(b,i)). At the same time, the vortex lines of the interface are being rotated generating a normal or x_3-component of vorticity. Along the centreline of the deforming sheet, where $\eta = 0$, the interface is pinched, and combining (11) and (13) shows that, when viscous diffusion is small,

$$\frac{\mathrm{d}\Omega_3}{\mathrm{d}\tau} \approx \Omega_3 - 6\beta\eta\zeta\Omega_2. \tag{16}$$

Transforming η, ζ into Lagrangian coordinates, the approximate solution to (16), $\Omega_3(\eta \ll 1) \sim -6\beta\eta\zeta\Omega_2(\eta, \zeta, \tau)\sinh\tau$, where Ω_2 is given by (14). Thus the normal component of vorticity is rapidly amplified in the deformed parabolic interface with positive and negative signs for Ω_3. The maximum value of Ω_3 quickly becomes comparable to the undisturbed spanwise vorticity $\beta\Omega_{20}$, while Ω_2 decreases rapidly (see Fig. 2(b,ii))

In the asymptotic stage of the development and close to the centreline, where $3\beta(e^{2\tau} - 1)/2L \sim 1/\delta_0$ or $\tau \sim \log(L/\delta_0\beta)$, the interface is engulfed and the two branches of the interface come together. Enhanced diffusion serves to decrease Ω_3. There is ultimately a balance for Ω_3 between advection in the η-direction/ζ-direction, stretching in the ζ-direction and viscous diffusion in the η-direction so that

$$\zeta \frac{\partial \Omega_3}{\partial \zeta} - \eta \frac{\partial \Omega_3}{\partial \eta} = \Omega_3 + \frac{\nu}{\Sigma^{(-)}} \frac{\partial^2 \Omega_3}{\partial \eta^2}. \tag{17}$$

The similarity solution to (17), is

$$\frac{\Omega_3}{\Omega_{3max}} = -\frac{\hat{\eta}}{(\pi/2)^{1/2}} \exp\left(-\frac{1}{2}\hat{\eta}^2\right), \tag{18}$$

where $\hat{\eta} = \eta/\delta(t)$, $\delta(t)^2 = \nu/\Sigma^{(-)} + (\delta_1^2 - \nu/\Sigma^{(-)})(\zeta_1/\zeta)^2$ and $\Omega_{3max}\delta(t)\zeta = const$, where $\delta = \delta_1$ when $\zeta = \zeta_1$ (see [11], [12]). Thus the asymptotic thickness of the layer $\delta \to \sqrt{\nu/\Sigma^{(-)}}$ due to the balance between straining and diffusion (see [11], [12], [13]). The effect of cross-stream diffusion of vorticity is to reduce Ω_{3max} as ζ increases. For an initially thick interface ($\delta_0 \gg \sqrt{\nu/\Sigma^{(-)}}$), the interface thickness is rapidly reduced by straining but the decrease in the maximum vorticity is negligible. For an initially thin interface ($\delta_0 \ll \sqrt{\nu/\Sigma^{(-)}}$) straining increases the interface thickness and the maximum vorticity decreases rapidly. Thus the peak vorticity in the entrained interface $|\Omega_{3max}|$ is comparable to that at the interface.

The amplification of the vorticity component Ω_3 induces a jet parallel to the streamwise direction that lies between the distorted interfaces. For an initially thick interface, $\Omega_{3max} \sim \pm\beta\Omega_{20}$, the volume flux of the jet is approximately constant in time and of order - $|\Omega_{3max}|\delta_0 L \sim -\Delta U_1 L$ (for the flow in Fig. 2). As the average location of the interface $\langle x_{3I}\rangle$ moves outwards as a result of the bounding entrainment velocity E_b, it follows that there is an integral contribution to the average Reynolds stress produced by the engulfing motion, which is of the same order as that produced by turbulence, namely $-E_b\Delta U_1$ (see [14], §3). Since the mean velocity in the turbulent region is sheared, it follows that Ω_3 is also distorted in the x_1-direction so that a streamwise component Ω_1 of the interface vorticity is generated, which affects the entrainment velocity E_b.

3 Conclusions

Vortices impinging on a strong free shear layer distort the ambient vorticity field with the cross-stream component of vorticity increasing due to the kinematic blocking effect of the interface. Longitudinal vortices create a streamwise flow perturbation which increases as the inverse of distance from the interface or locally a logarithmic flow profile. This provides an alternative mechanistic explaination for the occurance of a logarithmic profiles near rigid walls suddenly introduced into turbulent flows. Compact vortices create a weaker, and more localised, flow perturbation decaying as the inverse of distance from the interface. The entrainment of a weak shear layer by longitudinal vortices generates a cusped vortex sheet. As a consequence of the rotation of the vortical elements, a streamwise jet is created and an associated Reynolds stress.

These linearised calculations provide physical insight into some of the complex processes which are occuring and as consistent with the generally observations. The next step is a more detailed comparison against computational models.

References

1. Davey M, Hunt JCR (2006) Critical layers in geophysical flows Mathematics Today
2. Westerweel J, Pedersen JM, Fukushima C, Hunt JCR (2005) Phys Rev Lett 95(17):174501
3. Bisset DK, Hunt JCR, Rogers MM (2002) J Fluid Mech 451:383–410
4. Mathew J, Basu AJ (2002) Phys Fluids 14:2065–2072
5. Corrsin S, Kistler A (1955) Tech Rep Nasa 1244, Washington DC
6. Townsend AA (1966) J Fluid Mech 26:689–715
7. Fernando HJS, Hunt, JCR (1997) J Fluid Mech 347:197–234
8. Hunt JCR, Durbin PA (1999) Fluid Dyn Res 24:375–404
9. Reynolds WC (1972) J Fluid Mech 54:481–488
10. Perot B, Moin P (1995) J Fluid Mech 295:199–227
11. Bazant MZ, Moffatt HK (2005) J Fluid Mech 541:55–64
12. Hunt JCR, Eames I (2002) J Fluid Mech 457:111–132
13. Hunt JCR, Eames I, Westerweel JC (2006) J Fluid Mech 554:499–519
14. Turner JS (1986) J Fluid Mech 173:431–471

Magnetostrophic Turbulence
and the Geodynamo

H. Keith Moffatt

Department of Applied Mathematics and Theoretical Physics, University
of Cambridge, Wilberforce Road, Cambridge CB3 0WA, UK
`h.k.moffatt@damtp.cam.ac.uk`

Abstract. The flow generated by a random buoyancy field in a rotating medium
permeated by a dynamo-generated magnetic field is considered, under the assumptions that the Rossby number and the magnetic Reynolds number (based on the
scale of the buoyancy fluctuations) are both small. This permits linearisation of
the governing evolution equations. Provided 'up-down' symmetry is broken, a mean
helicity and an associated α-effect are generated. These are calculated in terms of
the spectrum function of the buoyancy field. Expressions are also obtained for the
buoyancy flux and the Reynolds stresses (kinetic and magnetic), and an outline dynamo scenario is proposed. The nature of this type of magnetostrophic turbulence
is briefly discussed.

Keywords: magnetostrophic, geodynamo, helicity, α-effect, mean-field
theory

1 Introduction

Dynamo theory is concerned with the generation of magnetic fields by fluid
motion. When the system considered is sufficiently large, the inductive action
of flow across any weak magnetic field induces currents which can, for suitable
flow configurations, lead to amplification of the magnetic field. The flow is
then unstable to the growth of this field, which grows exponentially until the
associated Lorentz force reacts back upon the flow, leading to field saturation.

It is well known that the helicity of the fluid motion (i.e. correlation of
velocity and vorticity) is highly conducive to dynamo action. Helicity is a
pseudo-scalar quantity, and is generally present only if the fluid is rotating
and if the source of energy for the flow breaks the symmetry with respect
to the rotation vector, for example if the motion consists of buoyant packets
of fluid rising through an otherwise quiescent medium. The precise condition
will be clarified in §3 below.

It has been customary in turbulent dynamo theory to start with a kinematic approach in which the statistics of the velocity field are assumed known,
and to focus attention on the evolution equation for the magnetic field $\mathbf{B}(\mathbf{x}, t)$

339

Y. Kaneda (ed.), *IUTAM Symposium on Computational Physics and New Perspectives in
Turbulence*, 339–346.
© 2008 *Springer*.

(i.e. the induction equation). The techniques of mean-field electrodynamics are then available to determine a simplified equation for the slow evolution of the large-scale field (averaged over scales characteristic of the turbulence) in terms of an α-effect and a turbulent diffusivity (both tensor in character). It has generally been found difficult to extend this approach to the nonlinear regime in which the back-reaction of the field on the flow is considered; this of course requires parallel analysis of the Navier-Stokes equation, including both Coriolis and Lorentz forces and incorporating explicitly the forces, whether of convective origin or otherwise, that drive the fluid motion.

I propose here an approach, relevant to the particular case of the geodynamo, which overcomes at least some of these difficulties. This takes as its starting point the idealised model of [1], [2], in which the ultimate sources of energy for the geodynamo are assumed to be of thermal and gravitational origin: slow cooling of the Earth leads to slow solidification of the liquid metal core onto the solid inner core. This solidification process takes place in a 'mushy zone', whose depth is of the order of 1km, and within which lighter elements (sulphur, oxygen,...) are 'rejected', the resulting density of the inner core being about 5% greater than the mean density of the liquid core. This results in the continuous creation of a buoyant layer which intermittently erupts from the mushy zone, driving what is primarily a state of compositional convection in the liquid core. The upward flux of buoyancy (equivalently the downward flux of mass) can be estimated on the assumption that the inner core has been growing at a roughly uniform rate over the lifetime of the Earth. This, coupled with the reasonable assumption of geostrophic balance between buoyancy and Coriolis forces, leads to estimates $V \sim 0.2\text{mm/s}$ and $\theta = \Delta\rho/\rho \sim 3 \times 10^{-9}$ for the typical upward velocity V and buoyancy θ of upwardly mobile buoyant elements.

It was assumed in [2] for want of better that these buoyant elements remain coherent and roughly spherical throughout their rise towards the core-mantle boundary. Subsequent computations [3] revealed however that this assumption is untenable: when a (dynamo-generated) toroidal magnetic field is present, a spherical blob is apparently subject to a 'slicing' instability as a result of the combined effect of Coriolis and Lorentz forces. Typical localised motions may also be expected to be strongly anisotropic (see for example [4]). It therefore becomes necessary to abandon any assumption concerning the shape of rising elements, this being determined by the full complex dynamics of the convective process.

The approach adopted here (somewhat in the spirit of kinematic dynamo theory, but now including essential features of the dynamics) is to suppose that the *statistics* of the buoyancy field $\theta(\mathbf{x}, t)$ are prescribed, and that (following [2]) the scale L of the convective turbulence lies in the range $V/\Omega \ll L \ll \eta/V$ where Ω is the angular velocity of the Earth, and η is the resistivity (or magnetic diffusivity) of the fluid medium, i.e. that the Rossby number $Ro = V/\Omega L$ and the magnetic Reynolds number $R_m = VL/\eta$ based on the 'blob scales' L and V are both small. This restricts L to the range between 10m and 100km, not unreasonable for a buoyancy distribution that is supposed to originate from a layer of thickness $\sim 1\,\mathrm{km}$. These assumptions allow linearisation of both the induction equation and the equation of motion; the sole remaining nonlinearity is the advective term in the advection-diffusion equation for buoyancy θ; but since we assume that the statistics of θ are prescribed, this difficulty can be postponed, if not avoided altogether.

2 The Magnetostrophic Equations

We suppose that the magnetic field $\mathbf{B}(\mathbf{x}, t)$ in the core of the Earth consists of a mean part $\mathbf{B_0}$, which results from dynamo action and can be considered as locally uniform and steady, and a perturbation field $\mathbf{b}(\mathbf{x}, t)$ induced by the flow $\mathbf{u}(\mathbf{x}, t)$ across $\mathbf{B_0}$. Under the above assumptions that the Rossby number and magnetic Reynolds number are both small, the governing evolution equations may be linearised in the form

$$2\,\mathbf{\Omega} \wedge \mathbf{u} = -\nabla P + (\mu_0 \rho)^{-1} \mathbf{B_0} \cdot \nabla \mathbf{b} - \theta \mathbf{g}\,, \tag{1}$$

$$\partial \mathbf{b}/\partial t = \mathbf{B_0} \cdot \nabla \mathbf{u} + \eta \nabla^2 \mathbf{b}\,, \tag{2}$$

$$\nabla \cdot \mathbf{u} = \nabla \cdot \mathbf{b} = 0\,, \tag{3}$$

where $\mathbf{\Omega}$ is the angular velocity of the Earth, $\mathbf{g}$ is the local gravitational acceleration, ρ is the mean density of the fluid, and ρP is the sum of fluid and magnetic pressure. These are the 'magnetostrophic equations' used in [2], except that here, recognising that magnetic field diffusion in a changing environment is not instantaneous, we retain the local time derivative $\partial \mathbf{b}/\partial t$. With this term included, the equations describe magnetostrophic waves [7] driven by the buoyancy term $-\theta \mathbf{g}$. The approach is thus within the spirit of Braginski's *MAC*-wave scenario [5].

It is perhaps mildly inconsistent to retain the term $\partial \mathbf{b}/\partial t$ in (2) while dropping the nonlinear term $\{\mathbf{u} \cdot \nabla \mathbf{b} - \mathbf{b} \cdot \nabla \mathbf{u}\}$ which may be expected to be of similar order of magnitude; this is just a matter of practicality, and we expect that, if the analysis were pursued to higher order in R_m, this nonlinear term would have a similar qualitative effect to that of the 'local' time-derivative $\partial \mathbf{b}/\partial t$, although involving higher-order statistics of the θ-field.

Neglecting transients, as is appropriate under statistically steady conditions, the equations (1)–(3) evidently establish a linear relationship between $\mathbf{u}$ and θ (and also between $\mathbf{b}$ and θ); it follows that mean quadratic quantities such as the helicity $\mathcal{H} = <\mathbf{u} \cdot \omega >$ (where $\omega = \nabla \wedge \mathbf{u}$) and the Reynolds stress $\tau_{ij} = <u_i u_j>$ are quadratically related to θ, and should therefore emerge as weighted integrals of the spectrum function of θ. The detailed forms of such integrals are obtained in the following sections.

3 Expressions for Mean Helicity and α-Effect

We shall suppose that the θ-field is statistically stationary and locally homogeneous (though certainly not isotropic). It then admits Fourier decomposition in the form

$$\theta(\mathbf{x}, t) = \int \hat{\theta}(\mathbf{k}, \omega) e^{i(\mathbf{k}\cdot\mathbf{x} - \omega t)} \, d\mathbf{k} \, d\omega \,. \tag{4}$$

The spectrum function $\Gamma(\mathbf{k}, \omega)$ is related to the Fourier transform $\hat{\theta}(\mathbf{k}, \omega)$ by

$$< \hat{\theta}^*(\mathbf{k}, \omega)\hat{\theta}(\mathbf{k}', \omega') > = \Gamma(\mathbf{k}, \omega)\delta(\mathbf{k} - \mathbf{k}')\delta(\omega - \omega'). \tag{5}$$

$\Gamma(\mathbf{k}, \omega)$ is real, and, by virtue of the reality of the field θ, satisfies the condition $\Gamma(-\mathbf{k}, -\omega) = \Gamma(\mathbf{k}, \omega)$. Note however that, in general,

$$\Gamma(\mathbf{k}, \omega) \neq \Gamma(\mathbf{k}, -\omega) \,. \tag{6}$$

This inequality corresponds to a breaking of 'up-down' symmetry; thus for example, if the θ-field consists of a distribution of non-overlapping buoyant blobs that rise without change of shape with constant velocity $\mathbf{V}$, with a compensating downward flow of the ambient fluid around the blobs, then

$$\Gamma(\mathbf{k}, \omega) = \Gamma_0(\mathbf{k})\delta(\omega - \mathbf{k} \cdot \mathbf{V}) \,, \tag{7}$$

where $\Gamma_0(\mathbf{k})$ is the spectrum function of θ in the frame of the blobs; this expression is evidently not equal to $\Gamma(\mathbf{k}, -\omega)$. This breaking of up-down symmetry will turn out to be important in what follows.

Equations (1)–(3) can be Fourier transformed, giving

$$2\mathbf{\Omega} \wedge \hat{\mathbf{u}} = -ik\hat{P} + i(\mathbf{B_0} \cdot \mathbf{k})\hat{\mathbf{b}} - \hat{\theta}\,\mathbf{g} \,, \tag{8}$$

$$\hat{\mathbf{b}} = i(\eta k^2 - i\omega)^{-1}(\mathbf{B_0} \cdot \mathbf{k})\hat{\mathbf{u}} \,, \tag{9}$$

$$\mathbf{k} \cdot \hat{\mathbf{u}} = \mathbf{k} \cdot \hat{\mathbf{b}} = 0 \,. \tag{10}$$

(Here, we use Alfven units for $\mathbf{B_0}$ and $\mathbf{b}$, so that the prefactor $(\mu_0\rho)^{-1}$ in (1) disappears.) We may easily solve these equations for $\hat{\mathbf{u}}$ in the form $\hat{\mathbf{u}} = [\mathbf{A}(\mathbf{k}, \omega)/D(\mathbf{k}, \omega)]\hat{\theta}$ where

$$\mathbf{A} = -(\mathbf{k} \cdot \mathbf{B_0})^2(\eta k^2 - i\omega)^{-1}\mathbf{k} \wedge (\mathbf{k} \wedge \mathbf{g}) - 2(\mathbf{k} \cdot \mathbf{\Omega})\mathbf{k} \wedge \mathbf{g} \,, \tag{11}$$

$$D = 4(\mathbf{k} \cdot \mathbf{\Omega})^2 + (\mathbf{k} \cdot \mathbf{B_0})^4(\eta k^2 - i\omega)^{-2}k^2 \,. \tag{12}$$

Here, we may note that $D = 0$ is the dispersion relation for magnetostrophic waves damped by magnetic diffusivity; if the forcing by the buoyancy distribution includes contributions from regions of $(\mathbf{k}, \omega)$-space for which $D = 0$ when $\eta = 0$, then a resonant response is to be expected, controlled by magnetic diffusivity. This control is strong in the low-R_m regime. Note further that $\mathbf{A}$ is complex; writing $\mathbf{A} = \mathbf{P} + i\mathbf{Q}$, we have

$$\mathbf{A}^* \wedge \mathbf{A} = 2i\mathbf{P} \wedge \mathbf{Q} = 4i(\mathbf{k} \cdot \mathbf{\Omega})\omega(\mathbf{k} \cdot \mathbf{B_0})^2(\eta^2 k^4 + \omega^2)^{-1}(\mathbf{k} \wedge \mathbf{g})^2\mathbf{k} \,. \tag{13}$$

We may now construct the mean helicity $\mathcal{H}$ in the form

$$\mathcal{H} = <\mathbf{u}\cdot\omega> = -\iint \frac{i\mathbf{k}\cdot(\mathbf{A}^*\wedge\mathbf{A})}{|D|^2}\Gamma(\mathbf{k},\omega)\,d\mathbf{k}\,d\omega\,, \qquad (14)$$

or, using (13),

$$\mathcal{H} = 4\iint \frac{(\mathbf{k}\cdot\mathbf{\Omega})\omega}{|D|^2}\frac{(\mathbf{k}\cdot\mathbf{B_0})^2}{\eta^2 k^4 + \omega^2}k^2(\mathbf{k}\wedge\mathbf{g})^2\Gamma(\mathbf{k},\omega)\,d\mathbf{k}\,d\omega\,. \qquad (15)$$

It is here that the breaking of up-down symmetry is important; for if $\Gamma(\mathbf{k},-\omega) = \Gamma(\mathbf{k},\omega)$ for all $(\mathbf{k},\omega)$, then the integrand in (15) is an odd function of ω, and so the integral (over all ω) vanishes. Thus to get a non-vanishing helicity, we need $\Gamma(\mathbf{k},-\omega) \neq \mathbf{\Gamma}(\mathbf{k},\omega)$, i.e. up-down symmetry must be broken, as is the case if the convection is characterised by rising blobs, the compensating downward motion being topologically connected; incidentally, this is just the sort of convection that gives rise to downward topological pumping of horizontal magnetic flux, a process that is recognised and well-understood in stellar contexts [6].

The presence of helicity immediately implies an α-effect, i.e. an electromotive force $\mathcal{E}$ linearly related to the large-scale (mean) magnetic field $\mathbf{B_0}$: $\mathcal{E}_i = \alpha_{ij}B_{0j}$, where the pseudo-tensor α_{ij} depends on $\mathbf{g}$, $\mathbf{\Omega}$ and $\mathbf{B_0}$, as well as on η and the statistical properties of the θ-field. Thus α_{ij} is certainly highly anisotropic; however, we can still easily evaluate the trace $\alpha = (1/3)\alpha_{ii}$, which is given, for low R_m, by

$$\alpha = -\tfrac{1}{3}\eta\iint \frac{k^2\mathcal{H}(\mathbf{k},\omega)}{\omega^2 + \eta^2 k^4}\,d\mathbf{k}\,d\omega\,, \qquad (16)$$

([7], chapter 6) where we now have

$$\mathcal{H}(\mathbf{k},\omega) = \frac{4(\mathbf{k}\cdot\mathbf{\Omega})\omega}{|D|^2}\frac{(\mathbf{k}\cdot\mathbf{B_0})^2}{\eta^2 k^4 + \omega^2}k^2(\mathbf{k}\wedge\mathbf{g})^2\Gamma(\mathbf{k},\omega)\,. \qquad (17)$$

Here we may note the phenomenon of 'α-quenching', i.e. the reduction of the α-effect with increasing B_0: since $|D|^2 \sim B_0^8$, (16) gives $\alpha \sim B_0^{-6}$, provided there are no complications associated with resonances (cf [7], pp 254-5). If however resonant magnetostrophic waves are excited, then the quenching effect may be weaker; see [8], where a quenching effect $\alpha \sim B_0^{-3/2}$ was obtained by integrating over the sub-regions of $(\mathbf{k},\omega)$-space where resonance occurred.

4 Buoyancy Flux and Reynolds Stresses

The buoyancy flux $F_z = <\mathbf{u}\theta>_z$, where the average is here over horizontal planes, is also a weighted integral of the spectrum function $\Gamma(\mathbf{k},\omega)$, and is given by

344   H.K. Moffatt

$$F_z = \iint \frac{(\mathbf{k} \cdot \mathbf{B_0})^2 \eta k^2 [\mathbf{k} \wedge (\mathbf{k} \wedge \mathbf{g})]_z \Gamma(\mathbf{k}, \omega)}{4(\mathbf{k} \cdot \mathbf{\Omega})^2 (\eta^2 k^4 + \omega^2) + (\mathbf{k} \cdot \mathbf{B_0})^4 (\eta^2 k^4 - \omega^2) k^2} \, d\mathbf{k} \, d\omega \,. \tag{18}$$

This flux is in effect provided by a downward mass flux which can be estimated from the slow rate of growth of the inner core (on the assumption that the convection is indeed predominantly compositional in character). This therefore places an implicit constraint on the spectrum function $\Gamma(\mathbf{k}, \omega)$ at each horizontal level z.

The Reynolds stress tensor $\tau_{ij} = <u_i u_j>$ is likewise a weighted integral of $\Gamma(\mathbf{k}, \omega)$,

$$\tau_{ij} = \iint \frac{A_i^* A_j}{|D|^2} \Gamma(\mathbf{k}, \omega) \, d\mathbf{k} \, d\omega \,. \tag{19}$$

Similarly, the magnetic counterpart of the Reynolds stress tensor is

$$\tau_{ij}^M = <b_i b_j> = \iint \frac{(\mathbf{k} \cdot \mathbf{B_0})^2}{\eta^2 k^4 + \omega^2} \frac{A_i^* A_j}{|D|^2} \Gamma(\mathbf{k}, \omega) \, d\mathbf{k} \, d\omega \,. \tag{20}$$

These stresses drive an axisymmetric mean flow $\mathbf{U}$ via an equation of the form

$$2(\mathbf{\Omega} \wedge \mathbf{U})_i = -\frac{\partial}{\partial x_i}(\tau_{ij} - \tau_{ij}^M) + \dots \,. \tag{21}$$

This mean flow is predominantly differential rotation, associated with the tendency of convecting elements to conserve angular momentum as they rise; it may also however include mean meridional circulation associated with the convection process. This mean flow provides a strain field which is locally uniform on the scale of the turbulence, and whose effect could in principle be included in (1) by an iterative procedure. This mean flow is axisymmetric, and it is important to recognise that, by virtue of Cowling's theorem, it is incapable on its own of maintaining a magnetic field with the same axis of symmetry. The helicity of the turbulence is, in this scenario, essential for dynamo action, the mean velocity field playing a secondary role.

5 Summary and Comments on the Nature of Magnetostrophic Turbulence

The picture that we have developed is thus as follows. Buoyancy flux F_z associated with compositional convection establishes a stationary random distribution of buoyancy θ, whose spectrum function $\Gamma(\mathbf{k}, \omega)$ exhibits a breaking of up-down symmetry. This θ-field drives an associated velocity field with non-zero helicity, leading to an α-effect, and a dynamo process. The mean field $\mathbf{B_0}$ that is generated ultimately leads to α-quenching and saturation of the dynamo process. A mean velocity is generated by a combination of dynamic and magnetic Reynolds stresses, and this has a secondary influence on the dynamo process.

This is of course just an outline scenario: the longer term aim must be to obtain a self-consistent set of equations for the axisymmetric mean fields

$\mathbf{B_0}$ and $\mathbf{U}$ (with sole input F_z), capable of explaining the evolution of the geomagnetic field over geological time.

The nature of the turbulence considered here is very different from conventional magnetohydrodynamic turbulence, since, as we have seen, all nonlinearities in the equation of motion and the induction equation are here negligible. There is however one remaining nonlinearity which cannot be neglected: this is the advective term in the advection-diffusion equation for θ

$$\frac{\partial \theta}{\partial t} + \mathbf{u}(\theta) \cdot \nabla \theta = S + \kappa \nabla^2 \theta \,. \tag{22}$$

Here, as we have seen, $\mathbf{u}(\theta)$ is linearly related to θ, so we have a quadratic nonlinearity in the equation. This is by no means a 'passive' but rather an 'active' scalar equation, in which all the action is provided by the scalar field itself. The diffusivity κ for chemical inhomogeneity in the core is extremely small, and the Péclet number $Pe = VL/\kappa$ correspondingly large (at least 10^8 in the core). The nonlinearity is therefore dominant. We have included also a source term S which must be interpreted as a source of buoyancy originating in the mushy zone, but in effect continuously regenerating the statistically stationary buoyancy distribution throughout the core.

Now, in the spirit of [9], we may construct an equation for 'thetergy' $<\theta^2>$, which is stationary by assumption:

$$\frac{1}{2}\frac{\mathrm{d}}{\mathrm{d}t} <\theta^2> \; = \; \chi - \kappa <(\nabla\theta)^2> \; = \; 0 \,, \tag{23}$$

where $\chi \; = <S\theta>$ is the rate of injection of thetergy at the scale L. This thetergy cascades to small scales, at which it is ultimately destroyed by molecular diffusion. This cascade is controlled both by the rate of cascade χ and the velocity scale V that relates $\mathbf{u}(\theta)$ to θ; here, $V \sim g/\Omega$, and $\hat{\mathbf{u}} \sim V\hat{\theta}$ at all k. Hence $<\mathbf{u}^2>$ satisfies an equation like (23) but with χ replaced by $\epsilon = V^2\chi$. Dimensional argument then gives the familiar Kolmogorov spectrum and correspondingly

$$\Gamma(k) \sim (\chi/V)^{2/3} k^{-5/3} \tag{24}$$

in an 'inertial' range of wave-numbers $L^{-1} \ll k \ll k_c$, where the conduction cut-off k_c is determined by the diffusive process, i.e.

$$k_c \sim (\kappa^3/V^2\chi)^{1/4} \,. \tag{25}$$

Quite apart from the geomagnetic context considered above, there is an interesting general class of problems here that calls for investigation by direct numerical simulation. Equation (22) with $\mathbf{u}$ an arbitrarily prescribed linear solenoidal functional of θ is in some respects similar to the Burgers equation, but it has a clearer physical basis, and the velocity field $\mathbf{u}(\theta)$ can here be chosen to be fully three-dimensional. The quadratic nonlinearity is similar to that in the Navier-Stokes equation, but we are dealing here with a simpler problem of scalar field evolution, and complications associated with pressure are absent. These features alone suggest that the problem deserves numerical investigation; perhaps someone at this meeting may be induced to take up this challenge!

References

1. Moffatt HK (1989) In: Lielpeteris J, Moreau R (eds) Liquid metal magnetohydrodynamics. Kluwer:403–412
2. Moffatt HK, Loper D (1994) Geophys J Int 117:394–402
3. St Pierre MG (1996) Geophys Astrophys Fluid Dyn 83:293–296
4. Siso-Nadal F, Davidson PA (2004) Phys Fluids 16:1242–1254
5. Braginski SI (1967) Geomag Aeron 4:572–583
6. Drobyshevsky EM, Yuferev VS (1974) J Fluid Mech 65:41–43
7. Moffatt HK (1978) Magnetic Field Generation in Electrically Conducting Fluids. Cambridge University Press (see also <www.moffatt.tc>)
8. Moffatt HK (1972) J Fluid Mech 53:385–399
9. Batchelor GK (1959) J Fluid Mech 5:113–133

MHD Turbulence: Nonlocal, Anisotropic, Nonuniversal?

Alexander A. Schekochihin[1,2,3], Steven C. Cowley[1,4], and Tarek A. Yousef[3]

[1] Blackett Laboratory, Imperial College, London SW7 2BW, UK
[2] King's College, Cambridge CB2 1ST, UK
[3] DAMTP, University of Cambridge, Cambridge CB3 0WA, UK
[4] Department of Physics, UCLA, Los Angeles, California 90095-1547, USA

Abstract. Kolmogorov's theory and philosophy of turbulence are based on a number of assumptions that have become standard notions with which one approaches turbulence in many, including non-hydrodynamic, systems. However, it turns out that in MHD turbulence, locality of interactions in scale space, isotropy of small scales or even universality cannot be taken for granted and, in fact, can be shown to fail. This note focuses on these unconventional aspects of MHD turbulence and on the related phenomenon of small-scale dynamo using a combination of simple physical reasoning and numerical evidence.

Keywords: MHD turbulence, dynamo, anisotropy, nonlocality, nonuniversality

MHD turbulence is the turbulence of a highly conducting fluid in which chaotic velocity and magnetic-field fluctuations exist and interact in a broad range of spatial and temporal scales. It is the prevailing form of turbulence that is presumed to occur and in many cases is, indeed, observed in astrophysical plasmas: examples are solar wind, interstellar medium, accretion discs, galaxy clusters etc. (see, e.g., a recent review [22] and references therein). It is important to remember that the MHD approximation is strictly only appropriate for low-frequency fluctuations at collisional scales, although for strongly magnetised plasmas, it can under certain assumptions be extended to collisionless scales as well [27]. In this short note, we shall not venture beyond the incompressible MHD equations, which read

$$\partial_t \mathbf{u} + \mathbf{u} \cdot \nabla \mathbf{u} = -\nabla p + \mathbf{B} \cdot \nabla \mathbf{B} + \nu \nabla^2 \mathbf{u} + \mathbf{f}, \tag{1}$$

$$\partial_t \mathbf{B} + \mathbf{u} \cdot \nabla \mathbf{B} = \mathbf{B} \cdot \nabla \mathbf{u} + \eta \nabla^2 \mathbf{B}, \tag{2}$$

where $\mathbf{u}$ is velocity, $\mathbf{B}$ magnetic field, p the total pressure determined by the incompressibility condition $\nabla \cdot \mathbf{u} = 0$, ν viscosity, η magnetic diffusivity, and $\mathbf{f}$ a body force (we use units in which p is scaled by ρ and $\mathbf{B}$ by $\sqrt{4\pi\rho}$, where $\rho = \mathrm{const}$ is the density of the medium). The body force models some

347

Y. Kaneda (ed.), *IUTAM Symposium on Computational Physics and New Perspectives in Turbulence*, 347–354.

system-specific energy input at the outer scale ℓ_0. As usual, we assume that the properties of turbulence at scales much smaller than ℓ_0 are not sensitive to the particular choice of $\mathbf{f}$ and only depend on the average injected power per unit volume $\epsilon = \langle \mathbf{u} \cdot \mathbf{f} \rangle$, where $\langle \ldots \rangle$ means volume averaging.

The first theories of MHD turbulence date back to mid-20 century [2, 29, 8, 10] and a vast amount of literature has since been generated. However, our current understanding of this type of turbulence has yet to reach the level comparable to that which exists in the area of neutral-fluid turbulence. Not surprisingly, most theories of MHD turbulence have, in one form or another, attempted to generalise the spirit and the method of Kolmogorov's 1941 dimensional theory (K41) to the MHD case. This course of action involves making assumptions that, while certainly not automatically valid, might appear natural. Roughly speaking, here are the notions that have to various degrees dominated the study of MHD turbulence:

- there is only one **universal** kind of MHD turbulence;
- velocity and magnetic fluctuations are in **scale-by-scale equipartition**;
- interactions are **local** in scale space;
- there is **isotropy** at sufficiently small scales;
- numerical simulations confirm all of the above, while astronomical observational evidence is "consistent" with it.

In this short note, we use simple physical reasoning and numerical evidence to argue that all of these statements are in general untrue.

The **non-universality** of MHD turbulence has to do with the presence or absence of a strong mean (large-scale) field $\mathbf{B}_0$. When B_0 is zero or dynamically insignificant, the magnetic field mostly consists of randomly tangled fluctuations, $\delta B/B_0 \gg 1$. Galaxy-cluster plasmas [21] and some parts of the interstellar medium [3, 7, 27] appear to be in this regime. In the opposite asymptotic situation when $\mathbf{B}_0$ is imposed externally and is such that its energy is larger than the energy of the turbulence motions, $B_0 \gg u_{\mathrm{rms}}$, the physics is clearly different: only small perturbations of the mean field are possible, so $\delta B/B_0 \sim u_{\mathrm{rms}}/B_0 \ll 1$. These perturbations take the form of Alfvén and slow MHD waves, with the dispersion relation $\omega = \pm k_\parallel v_A$, where $v_A = B_0$ is the Alfvén speed. The MHD turbulence in this regime, which is realised most clearly in the solar wind [18, 27], is a turbulence of these waves, or, more precisely, wave packets that interact nonlinearly and give rise to an energy cascade. The assumption of **locality** of interactions is probably acceptable here and the velocity and magnetic field should be at least approximately in a **scale-by-scale equipartition** because $\delta \mathbf{B} \sim \delta \mathbf{u}$ in an Alfvén wave. However, the isotropy assumption, originally considered plausible [8, 10] on the same philosophical grounds as in K41, turns out to be incorrect, as evidenced both by numerical simulations [11, 6] and solar-wind measurements [18]. Instead, the Alfvénic turbulence seems to be correctly described if one assumes that the Alfvén-wave-propagation time and the nonlinear interaction time are balanced at each scale, $k_\parallel v_A \sim k_\perp u_\perp$, i.e., interactions are strong — an assumption due to Goldreich & Sridhar [16]. This leads to a Kolmogorov spectrum in field-perpendicular wave numbers, $k_\perp^{-5/3}$, but also to a nontrivial relation between the parallel and perpendicular scales: $k_\parallel \sim k_\perp^{2/3}$ (see [22] for a short review). Thus, the turbulent cascade has an intrinsic **anisotropy** that increases at small scales — a unique situation in turbulence theory.

It is often thought that this is, in fact, the only type of MHD turbulence and that in systems where no mean field is present, large-scale magnetic fluctuations will dominate energetically and play the role of the mean field to the small-scale Alfvénic fluctuations — an assumption originally proposed by Kraichnan [10]. This may indeed be what happens in *decaying* MHD-turbulence simulations, which seem to be dominated by large-scale force-free magnetic structures [4, 13]. The situation that emerges in the numerical simulations of the *forced* case [12, 26, 17] is very different: while the velocity spectra suggest a cascade of a usual kind, the magnetic field is dominated by long folded flux sheets (or ribbons) with direction reversals at the diffusive scale (see Fig. 2b) — spectrally, this *folded structure* manifests itself as a large energy excess at small scales (see Fig. 3a, inset).

The key to making sense of these results is to realise that in MHD turbulence without a mean field, magnetic field has to be generated self-consistently by the action of the velocity field [the induction equation (2) is not forced — a modeling choice mandated by the physical constraint of magnetic flux conservation]. As long as the (random) forcing in (1) is spatially homogeneous and nonhelical, no mean (large-scale) field is produced by the turbulent velocity and the only generation process is the *small-scale dynamo* — stretching of the magnetic field by the random velocity gradients [2, 9]. This is a fundamentally **nonlocal** type of interaction (cf. [14]): indeed, even a linear-in-space (i.e., infinite-scale) velocity field will, in 3D, lead to exponential growth of magnetic energy from any weak initial field and the magnetic structures created in the process are small-scale in the sense that they are characterised by direction reversals at the diffusive scale [15, 31, 19, 22] (see inset in Fig. 1a).

In what follows, it will be convenient to concentrate on the asymptotic limit of large magnetic Prandtl number, $\mathrm{Pm} = \nu/\eta \gg 1$, i.e., $\mathrm{Rm} \gg \mathrm{Re} \gg 1$, where $\mathrm{Re} = u_{\mathrm{rms}}\ell_0/\nu$ and $\mathrm{Rm} = u_{\mathrm{rms}}\ell_0/\eta$ are the Reynolds and magnetic Reynolds numbers, respectively. The favourite numerical case of $\mathrm{Pm} = 1$, while nonasymptotic and, therefore, obscure in many ways, belongs to the same universality class (the limit $\mathrm{Pm} \ll 1$, which occurs in liquid-metal applications and in the stellar convective zones, is fundamentally different [28] and will not be considered here). When $\mathrm{Pm} \gg 1$, the random but spatially smooth motions at the viscous scale $\ell_\nu \sim \mathrm{Re}^{-3/4}\ell_0$ are the ones that contribute dominantly to the random stretching of an initially weak magnetic field because they have the largest turnover rate, $\gamma \sim (\epsilon/\nu)^{1/2} \sim (u_{\mathrm{rms}}/\ell_0)\mathrm{Re}^{1/2}$, in Kolmogorov turbulence. As a result, the magnetic field grows exponentially at the rate γ and is organised into folded structures with direction reversals at the diffusive scale $\ell_\eta \sim (\eta/\gamma)^{1/2} \sim \mathrm{Pm}^{-1/2}\ell_\nu \ll \ell_\nu$. While the reversal scale is small,

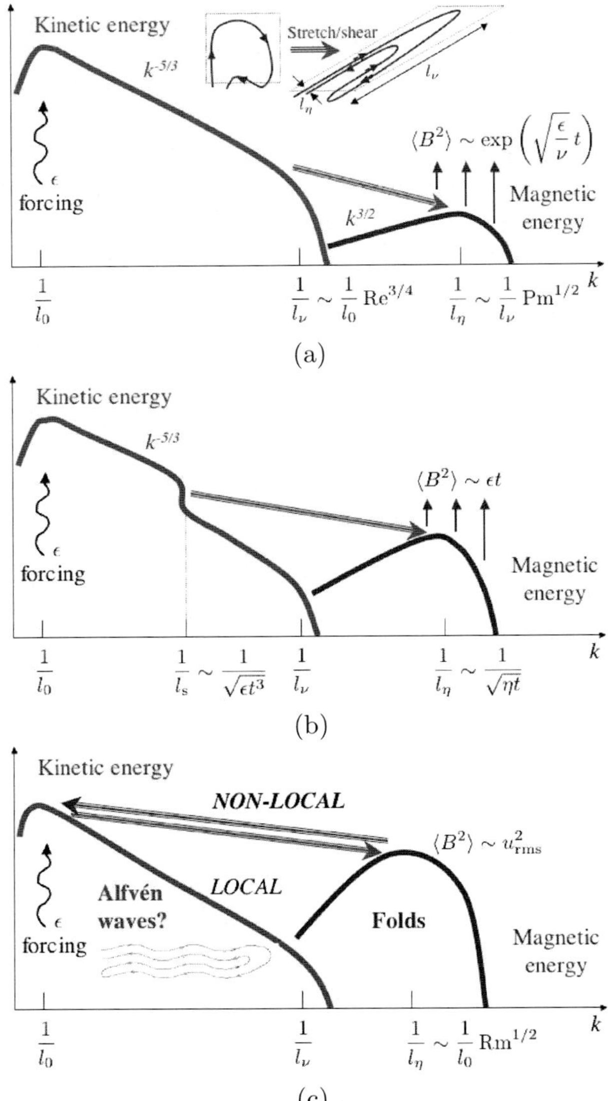

Fig. 1. Schematic illustration of the three stages of the evolution of MHD turbulence with $B_0 = 0$ and $\mathrm{Rm} \gg \mathrm{Re} \gg 1$: (a) kinematic (weak-field) stage: small-scale dynamo, exponential growth, magnetic field is stretched by the viscous-scale motions (the amplification mechanism and the emerging folded structure with direction reversals are illustrated by the inset); (b) intermediate stage: secular growth, field is stretched by the inertial-range motions at scale $\ell_s(t)$; (c) saturated state: balance between stretching by the outer-scale motions and back reaction from folds, possibly a cascade of Alfvénic perturbations of the folded structure (illustrated by the inset).

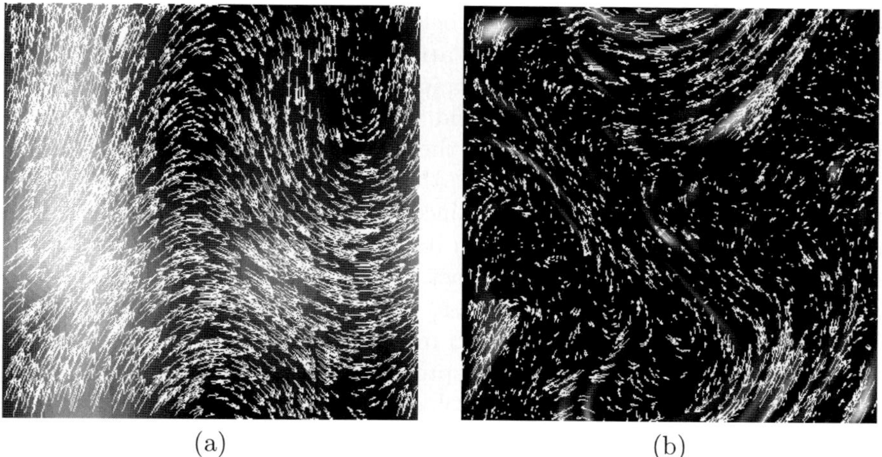

(a)   (b)

Fig. 2. Example of saturated state of MHD turbulence ($B_0 = 0$) dominated by **nonlocal** interactions. These are cross sections of (a) $|\mathbf{u}|$ and (b) $|\mathbf{B}|$ for a simulation with Pm = 1250, Re ~ 1 (run S5 of [26], 256^3). Velocity is random in time due to random forcing, but spatially smooth (single-scale), while magnetic field has folded structure. Arrows indicate the in-plane direction of the fields.

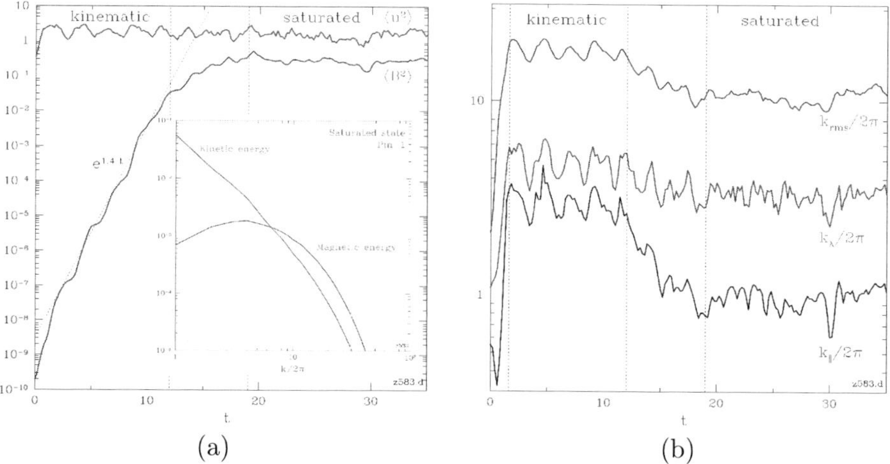

(a)   (b)

Fig. 3. Evolution of MHD turbulence ($B_0 = 0$) in a simulation with Pm = 1, Re $\simeq 400$ (Re$_\lambda \simeq 155$ during the weak-field stage, Re$_\lambda \simeq 116$ in the saturated state; this is run A of [26], 256^3): (a) kinetic and magnetic energies vs. time (inset shows kinetic- and magnetic-energy spectra in the saturated state — red and blue lines, respectively); (b) characteristic wave numbers $k_{\mathrm{rms}} = \left(\langle|\nabla\mathbf{B}|^2\rangle/\langle B^2\rangle\right)^{1/2}$ (measure of inverse reversal scale, blue line), $k_\lambda = \left(\langle|\nabla\mathbf{u}|^2\rangle/\langle u^2\rangle\right)^{1/2}$ (Taylor-microscale wave number, red line), $k_\parallel = \left(\langle|\mathbf{B}\cdot\nabla\mathbf{B}|^2\rangle/\langle B^4\rangle\right)^{1/2}$ (measure of inverse fold length, black line). Note the manifest presence of the intermediate nonlinear stage characterised by secular field growth and elongation of folds.

the field remains straight at scales below the scale of the velocity field, so $\ell_\parallel \sim \ell_\nu \gg \ell_\perp \sim \ell_\eta$ — a case of **local anisotropy** (the orientation of the folds is random, so the turbulence is isotropic globally, in a strict statistical sense).

This geometric property of the field, which can be analytically and numerically demonstrated by considering the statistics of the field-line curvature [23, 26], has a direct implication for the physics of nonlinear back reaction that the field exerts on the fluid: since the Lorentz tension force in (1) is quadratic in **B** and depends only on its variation along itself, it is oblivious to the direction reversals and can act on the scale of the velocity field — another **nonlocal** interaction process, which ends up saturating the small-scale dynamo. In order for saturation to be achieved, the stretching action of the turbulent motions should be suppressed in some way that is not quantitatively fully understood yet but probably amounts to some form of partial statistical **anisotropisation** of the velocity gradients, making them locally more two-dimensional and unable to act as small-scale dynamo [25, 26] — the precise diagnosing of this effect is a subject of ongoing investigations. Figure 2 presents such a saturated state in a numerical simulation with $\mathrm{Rm} \gg \mathrm{Re} \sim 1$ (the velocity field is random in time due to random forcing but is effectively single-scale): the folded structure and a large-scale spatially smooth velocity field coexist, while a balance of the two nonlocal interaction processes — small-scale dynamo and the field's back reaction — maintains a statistically stationary field level $\langle B^2 \rangle \sim u_{\mathrm{rms}}^2$.

In the more realistic case of $\mathrm{Re} \gg 1$, the saturation is preceded by an intermediate nonlinear stage [24, 26, 22] (see Fig. 1b). The back reaction from the field on the viscous-scale motions should become important when the energy of the field approaches the energy of these motions: $\langle B^2 \rangle \sim \delta u_{\ell_\nu}^2 \sim u_{\mathrm{rms}}^2 \mathrm{Re}^{-1/2}$ (then $\mathbf{B} \cdot \nabla \mathbf{B} \sim \langle B^2 \rangle / \ell_\parallel \sim \delta u_{\ell_\nu}^2 / \ell_\nu \sim \mathbf{u} \cdot \nabla \mathbf{u}$). Once this happens, the ability of the viscous-scale motions to amplify the field must be suppressed. During the period the follows, the back reaction gradually suppresses also the inertial-range motions — these have slower turnover rates than the viscous-scale ones, so amplify the field less quickly, but are more energetic and, therefore, require larger magnetic energy to be suppressed. If at any given time t during this period, $\ell_{\mathrm{s}}(t) \in (\ell_\nu, \ell_0)$ is the scale in the inertial range such that the motions associated with it have energy comparable to the energy of the magnetic fields, $\delta u_{\ell_{\mathrm{s}}(t)}^2 \sim \langle B^2 \rangle(t)$, we may argue that the magnetic field is amplified exponentially at the instantaneous rate $\gamma(t) \sim \delta u_{\ell_{\mathrm{s}}(t)} / \ell_{\mathrm{s}}(t)$:

$$\frac{d}{dt}\langle B^2 \rangle \sim \gamma(t)\langle B^2 \rangle(t) \sim \frac{\delta u_{\ell_{\mathrm{s}}}^3}{\ell_{\mathrm{s}}} \sim \epsilon = \mathrm{const}. \tag{3}$$

Thus, the magnetic energy grows secularly: $\langle B^2 \rangle \sim \epsilon t$ (cf. [29]). Assuming that the turbulence is still Kolmogorov at scales above ℓ_{s}, we have $\delta u_{\ell_{\mathrm{s}}} \sim (\epsilon \ell_{\mathrm{s}})^{1/3}$, whence the scale $\ell_{\mathrm{s}}(t) \sim \epsilon^{1/2} t^{3/2}$ — grows until it is comparable to the outer scale ℓ_0 after time $t \sim \ell_0^{2/3} \epsilon^{-1/3} \sim \ell_0 / u_{\mathrm{rms}}$ (the turnover time at the outer scale). At this point, $\langle B^2 \rangle \sim u_{\mathrm{rms}}^2$ and the saturated state is reached. By the logic of the above, at any given time t, the field amplification process is a small-scale dynamo done by the motions of scale $\ell_s(t)$, so the length of the folds increases during the intermediate stage, $\ell_\parallel(t) \sim \ell_s(t)$, and is comparable to the outer scale ℓ_0 in saturation. The diffusive (field-reversal) scale also increases, but slower: $\ell_\eta(t) \sim (\eta/\gamma(t))^{1/2} \sim (\eta t)^{1/2}$ until it is $\ell_\eta \sim \mathrm{Rm}^{-1/2}\ell_0$ in saturation. Figure 3 shows quantitatively how the evolution described above appears in a (nonasymptotic) numerical simulation with $\mathrm{Pm} = 1$ (see [26, 30] for further numerical tests and discussion).

Thus, the fully developed state of forced MHD turbulence in the absence of a strong externally imposed mean field is the saturated state of the small-scale dynamo, in which statistically stationary level of magnetic energy $\langle B^2 \rangle \sim u_{\mathrm{rms}}^2$ is achieved via a highly **nonlocal** interaction (stretching balanced by back reaction) between the outer-scale velocity field and the small-scale direction-reversing folded magnetic fields (see Fig. 1c). A detailed numerical investigation of energy transfer in the wave-number space [1, 14] lends further support to this view (see also A Pouquet's contribution in these Proceedings). Another way to corroborate it is to examine the scale-by-scale budgets of MHD turbulence in terms of the exact laws analogous to the 4/5 and 4/3 laws of turbulence [5, 20] — this is done in [30], where it becomes evident that the energy cascade is short-circuited at the outer scale with most or, at least, a large fraction of the kinetic energy injected by the forcing diverted directly into maintaining the small-scale folded structure against magnetic diffusion. Whether there is also a significant amount of **local** energy transfer is still unclear. It is possible that a local cascade does exist and consists of Alfvénic perturbations of the folded structure (see inset in Fig. 1c) — these are mathematically possible [24] and physically allowed because they do not stretch the field, but remain a conjecture as it is unclear whether they can support a viable turbulent cascade and also whether there is a practical way of detecting them at currently accessible numerical resolutions.

Another unsolved problem is the precise shape of the magnetic-energy spectrum in the saturated state (see Fig. 3a, inset). A simplistic model of the folds as uniform bundles of stripy, direction-reversing straight fields gives a k^{-1} spectrum: this follows from the fact that for such a field an average field increment across some distance ℓ ($\ell_\eta \ll \ell \ll \ell_0$) is constant: $\delta B_\ell \sim B_{\mathrm{rms}} \sim \mathrm{const}$ [30]. While it seems clear that the spectrum cannot be steeper than k^{-1} if it is energetically dominated by folds, it may be shallower and possibly have a small positive exponent [25, 26] — similar to its form in the small-scale-dynamo (weak-field) growth stage when the scaling is $k^{+3/2}$ [9].

As genuinely high-resolution numerical simulations are now becoming possible, a full quantitative picture of MHD turbulence is starting to emerge. Much, however, remains to be understood — indeed, a true quantification and verification of the ideas proposed above will require much more than the 256^3 runs quoted here [26]; even the 1024^3 record achieved in [17] appears to be insufficient. However, progress is inevitable in the near future. Our motivation in writing this short note has been the view that this progress will be greatly accelerated if numerical and analytical studies are done in full understanding that none of the standard received notions that one brings into the field from the conventional turbulence theory — such as the expectation of locality, isotropy etc. — is automatically guaranteed to be justified and that many of them can, in fact, be shown to fail by very simple physical arguments.

This paper has benefited from discussions with P Mininni, A Tsinober, F Rincon and K Yoshida. AAS was supported by a PPARC Advanced Fellowship. He also thanks the Nagoya conference organisers for travel support. TAY was supported by a UKAFF Fellowship and the Newton Trust.

References

1. Alexakis A, Mininni PD, Pouquet A (2005) Phys Rev E 72:046301
2. Batchelor GK (1950) Proc Roy Soc London A 201:405–416
3. Beck R (2006) in: Boulanger F, Miville-Deschenes MA (eds) Polarisation 2005. EAS Pub Ser, in press (e-print astro-ph/0603531)
4. Biskamp D, Müller W-C (2000) Phys Plasmas 7:4889–4900
5. Chandrasekhar S (1951) Proc Roy Soc London Ser A 204:435–449
6. Cho J, Lazarian A, Vishniac ET (2002) Astrophys J 564:291–301
7. Haverkorn M, Gaensler BM et al (2005) Astrophys J 637:L33–L35
8. Iroshnikov RS (1963) Astron Zh 40:742–750
9. Kazantsev AP (1967) Zh Eksp Teor Fiz 53:1806–1813
10. Kraichnan RH (1965) Phys Fluids 8:1385–1387
11. Maron J, Goldreich P (2001) Astrophys J 554:1175–1196
12. Meneguzzi M, Frisch U, Pouquet A (1981) Phys Rev Lett 47:1060–1064
13. Mininni PD (2006) private communication
14. Mininni P, Alexakis A, Pouquet A (2005) Phys Rev E 72:046302
15. Moffatt HK, Saffman PG (1964) Phys Fluids 7:155–155
16. Goldreich P, Sridhar S (1995) Astrophys J 438:763–775
17. Haugen NEL, Brandenburg A, Dobler W (2004) Phys Rev E 70:016308
18. Horbury TS, Forman MA, Oughton S (2005) Plasma Phys Control Fusion 47:B703–B717
19. Ott E (1998) Phys Plasmas 5:1636–1646
20. Politano H, Pouquet A (1998) Geophys Res Lett 25:273–276
21. Schekochihin AA, Cowley SC (2006) Phys Plasmas 13:056501
22. Schekochihin AA, Cowley SC (2007) in: Molokov S, Moreau R, Moffatt HK (eds) Magnetohydrodynamics: Historical Evolution and Trends. Springer, Berlin (e-print astro-ph/0507686)
23. Schekochihin AA, Cowley SC et al (2002) Phys Rev E 65:016305
24. Schekochihin AA, Cowley SC et al (2002) New J Phys 4:84
25. Schekochihin AA, Cowley SC et al (2004) Phys Rev Lett 92:084504
26. Schekochihin AA, Cowley SC et al (2004) Astrophys J 612:276–307
27. Schekochihin AA, Cowley SC, Dorland W (2007) Plasma Phys Control Fusion, in press (e-print astro-ph/0610810)
28. Schekochihin AA, Haugen NEL et al (2005) Astrophys J 625:L115–L118
29. Schlüter A, Biermann L (1950) Z Naturforsch 5a:237–251
30. Yousef TA, Rincon F, Schekochihin AA (2006) J Fluid Mech, in press (e-print astro-ph/0611692)
31. Zeldovich YaB, Ruzmaikin AA et al (1984) J Fluid Mech 144:1–11

Structure Formation in Stratified Turbulence

Jackson R. Herring[1] and Yoshifumi Kimura[2]

[1] National Center for Atmospheric Research, Boulder, Colorado, U.S.A.
   herring@ucar.edu
[2] Graduate School of Mathematics, Nagoya University, Furo–cho, Chikusa–ku,
   Nagoya 464–8602, Japan
   kimura@math.nagoya-u.ac.jp

Abstract. We discuss the effects of stable stratification on homogeneous turbulence
Our tools are direct numerical simulation (DNS) and rudimentary ideas from the
statistical theory of turbulence. We focus on the structures that emerge from stratification and their effects the flow's statistics. Stratification imposes an anisotropy
by inhibiting vertical motion and associated migration of particles in the direction
of stratification. Early DNS suggests that the evolution of decaying flows may remain sensitive to initial conditions throughout the lifetime of the flow. Hence a lack
of universality usually associated with isotropic flows. The structures of decaying
stratified flow may be described as "scattered pancakes", and they frequently form
doublets. These structures imply a diminution of energy cascade to small scales, and
we employ some simple ideas from the statistical theory in an attempt to estimate
this diminution and its effect on the decay rate. Randomly forced flows develop
banded structures perpendicular to the direction of stratification, rather than pancakes, and we present DNS data on the inertial range associated with such flows.
Finally we describe the degree of non-Gaussianity of such flows. As measured by the
flow's acceleration, we find that the "stretched exponential" distribution function of
isotropic flows is replaced by a near Gaussian form, with a much broader distribution in the transverse direction than that along the direction of stratification. This
means that particle dispersion in the transverse direction is stronger than the later.
Using DNS, we estimate the velocity Lagrangian autocorrelation functions.

Keywords: scattered pancakes, stably stratified turbulence

1 Introduction

Here we study stratified turbulence in the simplest context of the Boussinesq
approximation to homogeneous flows. The non-dimensional equations of motion are:

$$\{\partial_t - \nu\nabla^2\}\mathbf{u} = -\mathbf{u}\cdot\nabla\mathbf{u} - \nabla p + \hat{\mathbf{g}}\theta - 2\boldsymbol{\Omega}\times\mathbf{u} \tag{1}$$

$$\{\partial_t - \kappa\nabla^2\}\theta = -N^2 w - \mathbf{u}\cdot\nabla\theta \tag{2}$$

$$\nabla\cdot\mathbf{u} = 0 \tag{3}$$

Y. Kaneda (ed.), *IUTAM Symposium on Computational Physics and New Perspectives in
Turbulence*, 355–360.

We note the frequency of linear part of (1), and (2):

$$\omega = \sqrt{N^2 \sin^2(\vartheta) + 4\Omega^2 \cos^2(\vartheta)} \tag{4}$$

We are primarily here interested in the case $\Omega = 0$, and large N.

2 Does Stratified Turbulence Evolve Toward Universality?

In isotropic turbulence, it is usually found that the eventual state of the small-scales is independent of initial conditions. For stratified flow, such is suspect. To illustrate this point, consider two initial conditions of (1)–(4): the flow is initially isotropic; and (2) the horizontal motion is initially the dominant component. Results from [1] (Fig. 1) shows the "vortical" energy, $\Phi_1 \equiv\ <|\phi_1(t)|^2>$ and the "toroidal" component, $\Phi_2 \equiv<|\phi_2(t)|^2>$ comprising the total energy. Here we write the equations of motion for $(\mathbf{u},\ \theta)$ in terms of $\phi_1(\mathbf{k}, t),\ \phi_2(\mathbf{k}, t)$ and $\theta(\mathbf{k}, t)$:

$$\mathbf{u} = \mathbf{e_1}(\mathbf{k})\phi_1(\mathbf{k}) + \mathbf{e_2}(\mathbf{k})\phi_2(\mathbf{k})$$
$$\mathbf{e_1}(\mathbf{k}) = \mathbf{k} \times \mathbf{g}/|\mathbf{k} \times \mathbf{g}|$$
$$\mathbf{e_2}(\mathbf{l}) = \mathbf{k} \times \mathbf{e_1}(\mathbf{k})/|\mathbf{k} \times \mathbf{e_1}(\mathbf{k})|$$

In terms of $\boldsymbol{\Psi} = \{\phi_1,\ \phi_2,\ \theta\}$, equations of motion are:

$$\{\frac{\partial}{\partial t} + \mathcal{L}\}\boldsymbol{\Psi} = \mathcal{NL}$$

where

$$\mathcal{L} = \begin{pmatrix} 0 & -2\Omega\cos\vartheta & 0 \\ 2\Omega\cos\vartheta & 0 & -N^2\sin\vartheta \\ 0 & \sin\vartheta & 0 \end{pmatrix}$$

Here, $\cos\vartheta = k_z/k$, and $\mathcal{NL}$ stands for the nonlinear terms.

Notice the dramatic difference between the flow that begins isotropically as in the left panel of Fig. 1, and that which begins dominantly two dimensional (Fig. 1, right panel). In the former case, the poloidal component bears a near constant–but small–fraction of energy during the latter portion of the evolutionary track. In the latter case, the poloidal component makes a brief appearance and then all but disappears. It could be that the latter case is describable by the theoretical framework of [2]; the former not.

3 Properties of Decaying Stratified Turbulence

In stratified turbulence, the familiar bent, vortex tubes of isotropic turbulence are replaced by "scattered pancakes", which frequently exist in pairs as shown in Fig. 2. Associated with these structures is a reduction in the cascade of energy to small scales. This implies a reduction in the rate of decay of total energy. Thus, the total energy decays $\sim t^{-1}$ as depicted in the lower right-hand panel of Fig. 2, instead of $t^{-10/7}$.

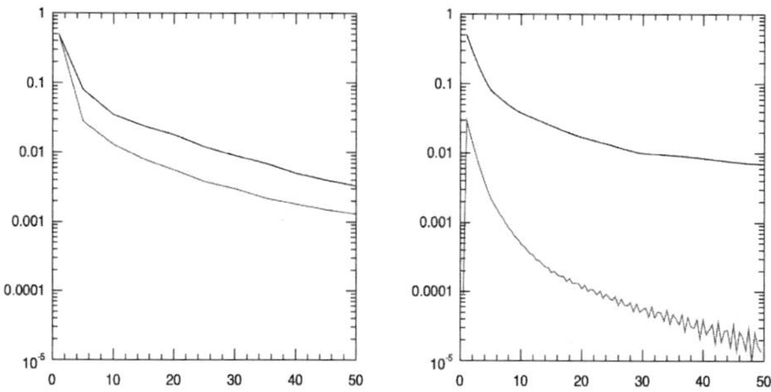

Fig. 1. Decay of $\Phi_1(t)$ (top) and $\Phi_2(t)$ (bottom) for the case in which : Left; $\Phi_2(0) = \Phi_1(0)$, Right; $\Phi_2(0) = 0$.

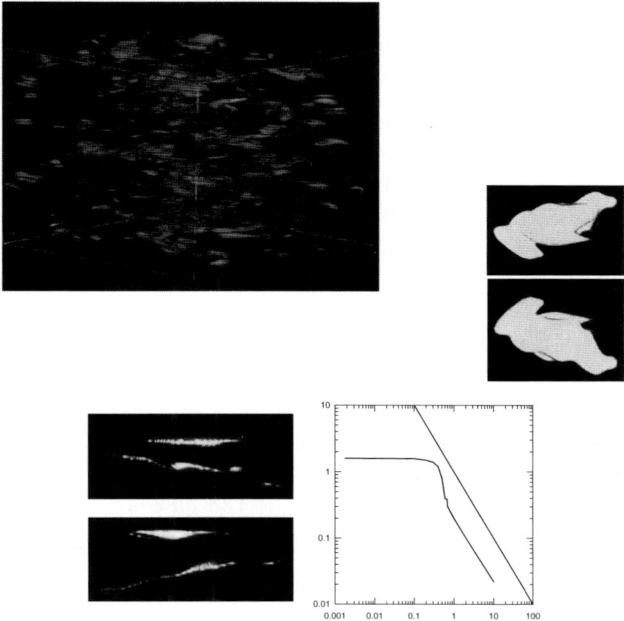

Fig. 2. Left top: enstrophy distribution after stratification and substantial evolution; bottom left; side view of the enstrophy doublet shown in the upper right corner: right top; top view of same doublet: bottom right; decay of total energy and comparison with t^{-1} (straight line). Here, $N = 10$ (see Eqn. (1), & (2)).

4 The Spectra of Stratified Turbulence

Horizontally forced, stationary spectra for $\Phi_1(k_\perp)$, $\Phi_2(k_\perp)$, $\Theta(k_\perp)$ are shown in Fig. 3. We note a near isotropizing at high $k_\perp$ between Φ_1 and Φ_2, with an excess of Φ_1 at large scales. The Θ spectrum nearly parallels Φ_2.

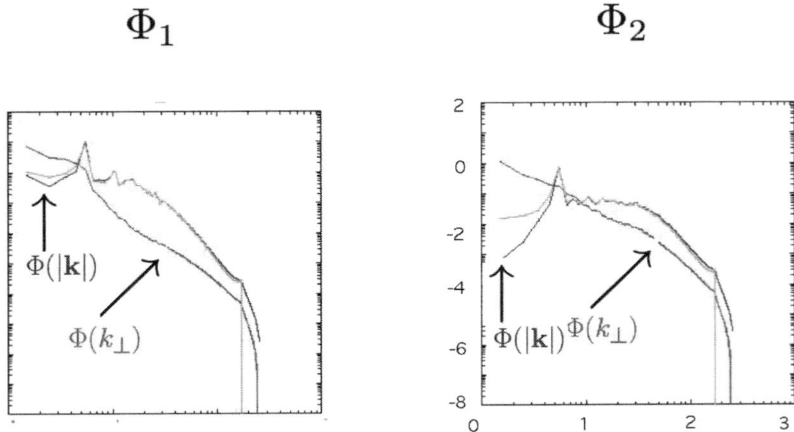

$$\Phi_1 \qquad \Phi_2$$

Fig. 3. Logarithms of Spectra Φ_1 (left) and Φ_2 (right) as functions of the logarithms of: $k_\perp$, $|\mathbf{k}|$, and k_z. $\Phi(k_z)$ is unlabeled. Here $k_{force} = 5$.
Ordinate scale the is central column.

The spectrum, $\Phi_2(k_\perp)$ is somewhat shallower than k^{-2}.[1], and is perhaps closer to $k^{-5/3}$. Such has also been found by [4], [5], and others at this conference. These are numerical results. There is also theoretical evidence for -5/3. Thus, Sukoriansky & Galperin [6] find that -5/3 emerges from their RNG theory.[2] Also to be noted in Fig. 3 (right) is a possible high wave number k_z^{-3} range for Φ_2.

5 Non-Gaussianity of Stratified Flows

Isotropic turbulence is known to be strongly non-Gaussian. Does such survive for strong stratification? We expect stratification acts to lower the effective Reynolds number, since it increases the linearity of Eq'n. (1)–(2). We have noted elsewhere [3] that such seems to be the case for stratified turbulence. For example, the Eulerian acceleration for stratified flow is changed from "stretched exponential" to quite Gaussian shapes, as shown in Fig. 4 (right). Note that the horizontal acceleration exceeding that in the vertical. Thus an enhanced horizontal diffusion and a much reduced vertical diffusion of particles [3].

Perhaps of more interest are velocity Lagrangian auto–correlations and their differences in the vertical and horizontal. These are shown in the left panels of Fig. 4. As is well known, [7] the time integral of these give the

---

[1] We earlier [3] suggested $\sim\sqrt{N\epsilon}k^{-2}$ for this spectrum, and hence $E(t) \sim t^{-5/7}$ for decaying flow.

[2] The structures of forced stratified turbulence are bands of flow whose widths are determined by the condition that their irregular boundaries dissipate the input energy.

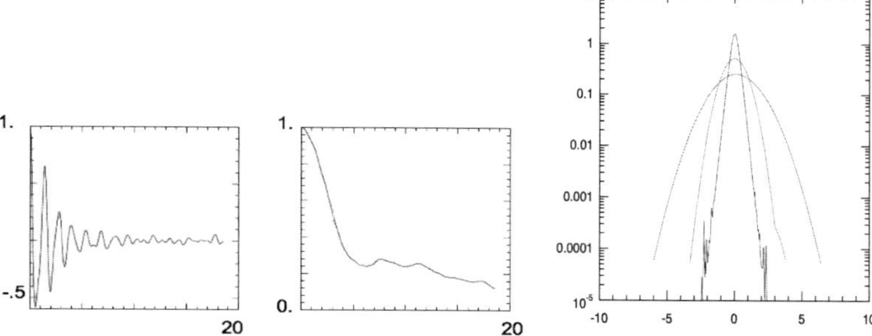

Fig. 4. Left and middle panels: Lagrangian auto correlation functions, $< u_i(t) \cdot u_i(t+\tau) > / < u_i(t)^2 >$ as a function of τ. Left panel; i=3, Middle; average of i=(1,2). Here $N = 10$. Right panel: PDF of $\mathbf{a} \equiv \partial\mathbf{u}/\partial t$. Gaussian curves depict a_z and $a_\perp$ for stratified flow with broadest curve $a_\perp$. The near exponential shown here is $\mathbf{a}$ for the unstratified case.

(rms) particle displacement in the two directions. Here we see that the rapid (oscillatory) accelerations in the vertical tend to cancel, with little net effect, while that in the horizontal acts coherently to effect a much larger horizontal displacement. The horizontal displacement is actually larger than that of an equivalent isotropic flow. This was predicted in [8].

6 Summary

Stratified turbulence is strongly anisotropic; its evolution depending on initial condition. Its non–Gaussian aspects (Eulerian acceleration) are reduced as compared to isotropic flow, along with the cascade of energy to small scales, suggesting that a theory as proposed in [9] may be more successful here than in isotropic turbulence. Simulations at 512^3 are insufficient to discriminate between $k^{-5/3}$ (proposed in [4],[5], and [6]) and k^{-2}, (as in [3],[10] and [11]).

Acknowledgement. This research was facilitated by funds from the Japan-US NSF Cooperative Agreement No. ATM-9209181 SPO7 ATM-94 17693.

References

1. Métais O, Herring JR (1989) J Fluid Mech 202:117–148
2. Majda AJ, Grote MJ (1997) Phys Fluids 10:2932–2940
3. Kimura Y, Herring JR (1996) J Fluid Mech 328:253–269
4. Lindborg E, (2005) Simulations of the energy cascade in a strongly stratified fluid. Geophysical Turbulence Workshop on Coherent Structures in Atmosphere and Oceans, at NCAR, 11–14 July, 2005
5. Riley JR, de Bruyn Kops SM (2003) Phys Fluids 15:2047–2059
6. Sukoriansky S, Galperin B (2006) Nonlinear Processes in Geophysics 13:9–22
7. Taylor GI (1921) Proc Lon Math Soc 20:10–25

8. Mezić I, Wiggins S (1994) Phys Fluids 6:2227–2229
9. Godeferd FS, Cambon C (1994) Phys Fluids 6 (6):2084–2100
10. Herring JR, Kimura Y, James R, Clyne J and Davidson PA (2006) In: Cannon J, Shivamoggi B (eds) Mathematical and Physical Theory of Turbulence. Chapman & Hall/CRC, Taylor & Francis Group, Boca Raton London New York
11. Caillol P, Zeitlin V (1998) In: Gyr A, Kinzelbach, W, and Tsinober A (eds) Proceedings of the II Monte Verita Colloquium On Fundamental Problematic Issues in Turbulence, Birkhäuser Verlag, Basel

Reduced Models for Rotating and Stratified Flows

Leslie Smith[1] and Mark Remmel[2]

[1] Departments of Mathematics and Engineering Physics, University of Wisconsin, Madison, WI, 53706 USA, `lsmith@math.wisc.edu`

[2] Department of Mathematics, University of Wisconsin, Madison, WI, 53706 USA, `remmel@math.wisc.edu`

Abstract. It has been shown numerically that near-resonant three-wave interactions are responsible for the self-generation of large-scale coherent structures from small-scale, random forcing in some dispersive-wave systems describing geophysical flows. However, a reduced model of near-resonant triad interactions is not a partial differential equation (PDE) and is thus computationally inefficient. A new approach is introduced to derive PDE reduced models by extracting different classes of wave-mode and vortical-mode interactions, which may contain a subset of near-resonances. The Rotating Shallow Water (RSW) equations are studied as proof of concept.

Keywords: dispersive-wave turbulence, eigenmode decomposition

1 Introduction

The dynamics of rotating, stratified fluids include the complex interaction of waves and turbulence, and are the basis for understanding the earth's oceans and atmosphere. The wide range of spatial scales in geophysical flows, from thousands of kilometers to meters, is one reason why they are so rich in behavior, so costly to compute and so difficult to understand. In certain scale regimes, intermediate-scale motions self-organize to generate larger-scale structures such as hurricanes, while in other regimes, energy is transferred from large-scale winds and tides to small-scale turbulent fluctuations. On large scales the flow becomes quasi two-dimensional (2D), with motions mainly parallel to the surface of the earth, whereas at small scales it is often fully three-dimensional (3D) and turbulent.

Several dynamical systems used to describe atmospheric and oceanic flows have a common mathematical structure, *e.g.* the Boussinesq equations, the Rotating Shallow Water (RSW) equations, and the β-plane model. These systems include a skew-symmetric linear operator leading to dispersive waves, as well as quadratic nonlinearities leading to turbulence at high flow rates or small values of the viscosity. The present work exploits this common framework to

361

Y. Kaneda (ed.), *IUTAM Symposium on Computational Physics and New Perspectives in Turbulence*, 361–366.

better understand the role of dispersive waves for the transfer of energy from turbulent intermediate-scale motions to coherent large-scale structures such as jets, layers and vortices.

2 Governing Equations

The Boussinesq equations for vertically stratified flow rotating about the vertical $\hat{\mathbf{z}}$-axis, in non-dimensional form, are given by [1]

$$\frac{D\mathbf{u}}{Dt} + \frac{\hat{\mathbf{z}}\times\mathbf{u}}{Ro} + \Gamma\rho'\hat{\mathbf{z}} = -\overline{P}\nabla\phi + \frac{\nabla^2\mathbf{u}}{Re}, \quad \frac{D\rho'}{Dt} - \frac{w}{(Fr\ \Gamma)} = \frac{\nabla^2\rho'}{(Re\ Pr)}, \quad (1)$$

where the velocity $\mathbf{u}(x,y,z,t) = u\hat{\mathbf{x}} + v\hat{\mathbf{y}} + w\hat{\mathbf{z}}$ satisfies $\nabla\cdot\mathbf{u} = 0$, $\rho = \rho_o - bz + \rho'$ is the density, $Ro = U/(fL)$ is the Rossby number, $\Omega = f/2$ is the (constant) frame rotation rate, $Fr = U/(NL)$ is the Froude number, $N = (gb/\rho_o)^{1/2}$ is the (constant) buoyancy frequency, g is the acceleration of gravity, $\Gamma = BgL/U^2$, $B\rho_o$ is a density scale, $\overline{P}$ is a non-dimensional pressure, Re is the Reynolds number, Pr is the Prandtl number and U, L are characteristic velocity and length scales.

The RSW equations can be derived from (1) assuming (i) $D/L << 1$ where $D\ (L)$ is the vertical (horizontal) scale, (ii) the hydrostatic approximation, and (iii) the density is constant and normalized to $\rho_o = 1$ [1]. With $\mathbf{u}(x,y,t) = u\hat{\mathbf{x}} + v\hat{\mathbf{y}}$, the RSW equations are given by

$$\frac{D\mathbf{u}}{Dt} + \frac{\hat{\mathbf{z}}\times\mathbf{u}}{Ro} = -\frac{\Theta\nabla h}{Fr^2}, \quad \Theta\frac{Dh}{Dt} + \left(1 + \Theta h - \frac{h_b}{H_o}\right)\nabla\cdot\mathbf{u} = \mathbf{u}\cdot\nabla\left(\frac{h_b}{H_o}\right). \quad (2)$$

Here the total depth of the fluid $H(x,y,t)$ is given by $H = H_o + h - h_b$ where H_o is the mean depth, $h(x,y,t)$ is the top fluid surface and $h_b(x,y)$ is the bottom surface topography. The Rossby number is $U/(Lf)$, the Froude number is now $Fr = U/(gH_o)^{1/2}$, $\Theta = N_o/H_o$ and N_o is a characteristic height variation.

Also related to (1)-(2) is the β-plane model: a local, planar approximation to 2D motion on the surface of a sphere, which accounts for mid-latitude variation β of the normal component of the Coriolis parameter [2]. The equation for the vertical vorticity $\zeta(x,y) = v_x(x,y) - u_y(x,y)$ is

$$\partial_t\zeta + J(\nabla^{-2}\zeta,\zeta) + (Rh)^{-1}\partial_x\nabla^{-2}\zeta = Re^{-1}\nabla^2\zeta \quad (3)$$

where $J(g,h) = g_x h_y - g_y h_x$ is the Jacobian, and $Rh = U/(\beta L^2)$ is the Rhines number.

In the inviscid, linear limits, and subject to periodic boundary conditions, equation (1), (2) and(3) separately admit wave solutions:

$$\mathbf{v} \propto \exp\left[i\left(\mathbf{k}\cdot\mathbf{x} - \sigma(\mathbf{k})\frac{t}{R_*}\right)\right] + \text{c.c.}, \quad (4)$$

where c.c. denotes the complex conjugate, and the state variable $\mathbf{v}$ is $[\mathbf{u}, \rho']$ for (1), $[\mathbf{u}, h]$ for (2) and ζ for (3). The non-dimensional R_* is $R_* = Rh$ on the β-plane (3), whereas for Boussinesq flow (1) and RSW (2), R_* is conveniently chosen as $R_* = Ro$ in rotation-dominated flow and $R_* = Fr$ in stratification dominated flow. In all cases (1)-(3), resonant triad interactions with $\sigma(\mathbf{k}) + \sigma(\mathbf{p}) + \sigma(\mathbf{q}) = 0$ dominate on $O(1)$ time scales. One can show analytically that resonant triad interactions cannot transfer energy from fast waves to slow modes with zero frequency [1], [3]. However, numerical experiments clearly show the generation of large-scale slow modes from fast waves. Examples of slow modes are zonal flows on the β-plane, vertically sheared horizontal flows in 3D stratification dominated flows, and vortical columns in 3D rotation dominated flows. Thus it is necessary to look beyond exact triad resonances to understand the generation of jets, layers and vortices from fast waves. It is natural to investigate *near*-resonant triad interactions with $\sigma(\mathbf{k}) + \sigma(\mathbf{p}) + \sigma(\mathbf{q}) = O(R_*)$ since these interactions become important on longer time scales $T = O(1/R_*)$ for small R_* (see, *e.g.*, [4]).

3 Reduced Models of Near-Resonant Triads

Numerical calculations of reduced models including only near-resonant triad interactions with $|\sigma(\mathbf{k}) + \sigma(\mathbf{p}) + \sigma(\mathbf{q})| < R_*$ have been compared to simulations of full systems in [5, 6]. In such reduced models, the physical-space form of the nonlinear term is not known, and thus spectral calculations are required to study flow in a periodic domain. The greater cost of spectral calculations as compared with pseudo-spectral calculations means that the comparisons are restricted to low resolutions and moderate Ro, Fr and Rh [5, 6].

First consider (1) with $\rho = \rho_o$ (the non-stratified, rotating case) in a periodic domain. Figure 1 (left) compares same-time, instantaneous energy spectra for a full simulation and the reduced model of near resonances. Both runs have $Ro = 0.085$, and energy is injected by a 3D white-noise force with Gaussian spectrum and peak forcing wavenumber $k_f = 10$. The run of near resonances reproduces the energy spectrum of the full simulation surprisingly well, considering that the near-resonant triads are only about 12% of the total number of triad interactions. The large-scale spectra scale steeper than $k^{-5/3}$ indicating that the transfer of energy from small to large scales is not an inverse cascade among 2D modes only. In both the full simulation and the reduced model calculation, energy injected into 3D modes at intermediate scales $k \approx 10$ is partially transferred to larger 2D slow modes with $k_z = 0$ corresponding to vortical columns. Figure 1 (right) shows the strong asymmetry between cyclones (with positive vorticity) and anticyclones (with negative vorticity), and the asymmetry in favor of cyclones is at least as strong in the run of near resonances as in the full simulation. Here PDFs are sampled over points $Q = u_x v_y - u_y v_x > 0.25 \max Q$. In contrast, the reduced model including the complement set of non-resonant triad interactions does not show significant energy transfer from intermediate forced scales to larger scales.

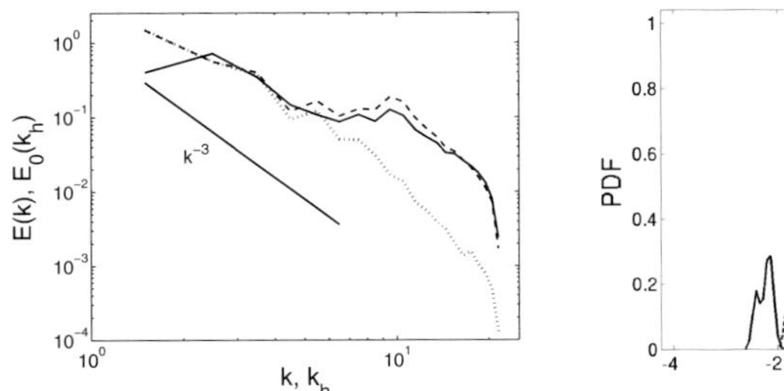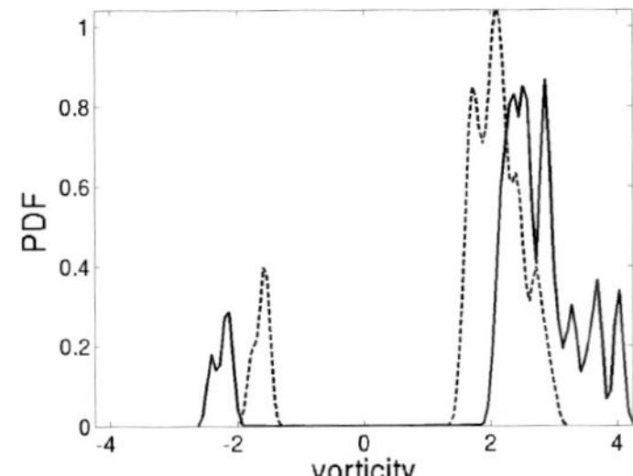

Fig. 1. Left: $E(k)$ (solid) for a full simulation of 3D rotation; $E(k)$ (dashed) and $E_0(k_h) = E_0(k_h; k_z = 0)$ (dotted) for the run of near resonances. In both, $Ro = 0.085$. A k^{-3} line is also shown for reference. Right: PDF of vertical vorticity in the $\hat{z}$-averaged velocity fields: near resonances (solid) and the full simulation (dashed).

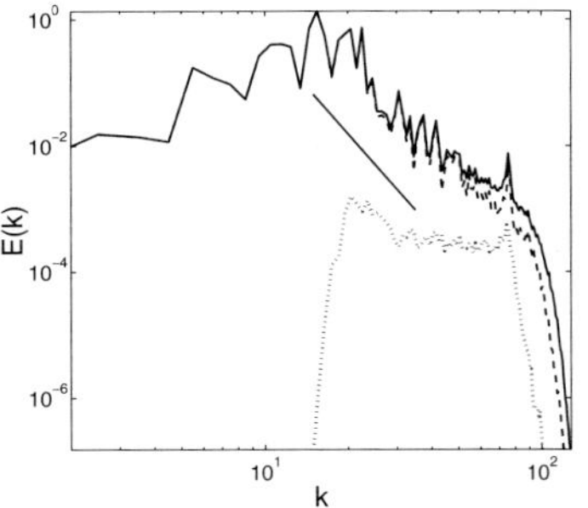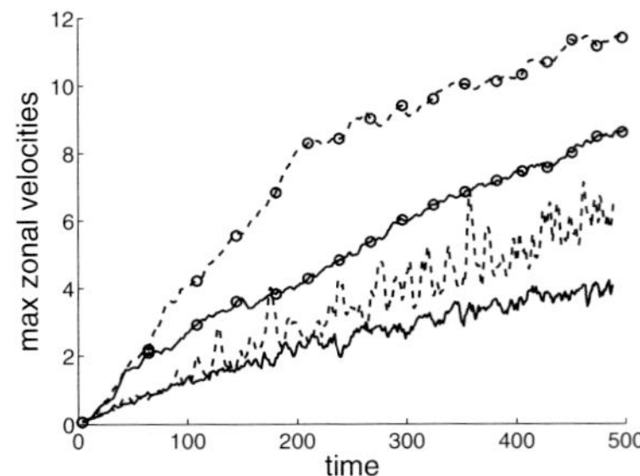

Fig. 2. Left: Energy spectra $E(k)$ (solid), $E_0(k_y)$ (dashed) and $E_0(k_x)$ (dotted) for the reduced model of near resonances only. The line is k^{-5}. Right: Maximum eastward (solid) and westward (dashed) velocities: full simulation (no symbols); near resonances (circles). In both runs $Rh = 0.5$.

Next consider flow on the β-plane (3). Figure 2 (left) shows single-time energy spectra for the reduced model of near resonances in a periodic domain with resolution 384^2. In this run with $Rh = 0.5$, near-resonant triads inter-actions are about 33% of the total number of triad interactions. Energy is injected by a white-noise force with Gaussian spectrum peaked at wavenum-ber $k_f = 80$. One can see that $E(k) \approx E_0(k_y)$ at large scales, showing that the large-scale flow is predominantly zonal. Here $E(k)$ is the 2D energy spectrum; $E_0(k_y)$ ($E_0(k_x)$) is the zonal (meridional) spectrum, with energy in a small sector $\pi/12$ about $k_x = 0$ ($k_y = 0$). As in the case of 3D rotation, instanta-neous spectra for the run of near resonances is qualitatively and quantitatively similar to same-time spectra (not shown) for the full β-plane simulation [6]. As can be seen in Fig. 2 (right), the time developing β-plane flow (without linear drag) has a clear asymmetry in favor of stronger westward jets. That asymmetry is enhanced in the simulation of near resonances only. The reduced model of non-resonant triads does not produce strongly zonal flows.

4 PDE Reduced Models

For each of the dispersive-wave systems (1)-(3), flow in the nonlinear regime can be expressed as a superposition of linear waves:

$$\mathbf{v}(\mathbf{x}, t) = \sum_{\mathbf{k}} \sum_{m} b_m(\mathbf{k}, t) \Psi^m(\mathbf{k}) \exp\left[i\left(\mathbf{k}\cdot\mathbf{x} - \sigma_m(\mathbf{k})\frac{t}{R_*}\right)\right] \tag{5}$$

where m is the number of orthogonal linear eigenfunctions $\Psi(\mathbf{k})$. Such an eigenmode decomposition has been used in statistical theories of wave turbulence and to derive models for two-point correlations $e.g.$ [7, 8]. To illustrate the new approach to derive PDE reduced models, we consider the RSW equations (3), which have a non-wave (vortical) mode Ψ^0 and two gravity wave modes Ψ^+ and Ψ^-. The 2D quasi-geostrophic (QG) equations result from interactions of vortical modes Ψ^0 only. Although the full RSW equations (3) develop anti-cyclones in a particular parameter range, 2D QG flow is symmetric with respect to cyclones and anti-cyclones. To improve upon the 2D QG model by including more physics, we derived a model denoted PPG including interactions among (i) vortical modes Ψ^0 only, and (ii) two vortical modes Ψ^0 and one gravity-wave mode $\Psi^\pm$. For $Fr = 0.25$ and $Ro = 0.4$, Fig. 3 shows a comparison of the PPG model to the full RSW and 2D QG. In all cases, the flow decays from random initial conditions. As observed in [9], the full RSW flow develops a negative skewness of vorticity corresponding to a predominance of anti-cyclones. Although the 2D QG reduced model shows near-zero skewness, the skewness of the PPG model is just slightly less negative than that of the full RSW equations. The PPG reduced model seems to include the interactions primarily responsible for the development of anti-cyclones in a range of Fr and Ro, and adding more interactions leads to very little change in the character of the flow.

It should be noted that the PPG model contains exact resonances between three vortical modes Ψ^0, but that no more exact or near resonances are included with interactions between two vortical modes Ψ^0 and one wave mode $\Psi^\pm$. Thus near-resonant triad interactions appear not to play an essential role

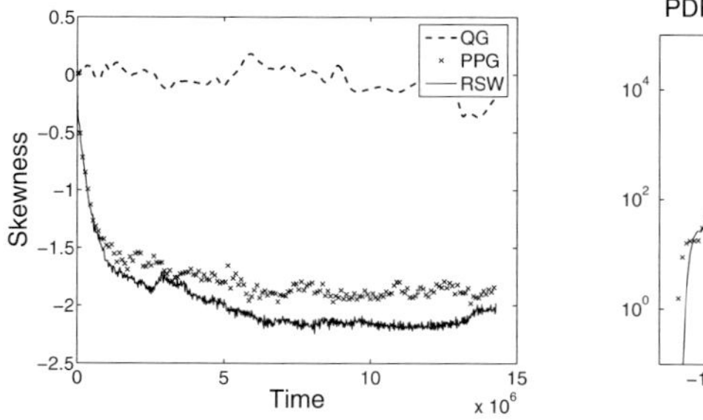
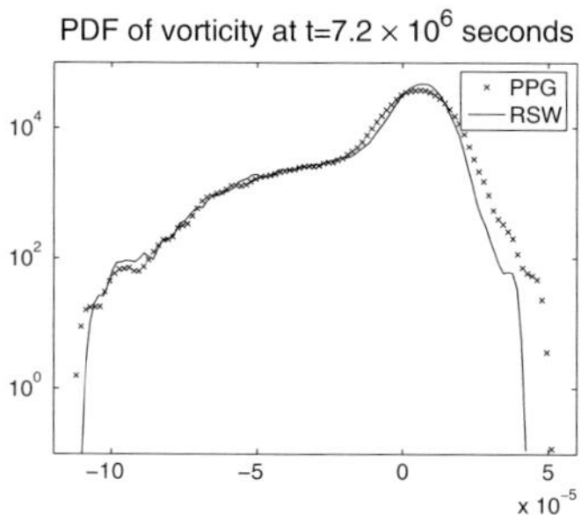

Fig. 3. Left: Skewness of vorticity vs. time; Right: PDF of vorticity.

in structure formation during SWE decay, at least for $0.1 \leq Fr \leq 0.3$ and $0.25 \leq Ro \leq 1$. It is interesting to study SWE with random forcing at small scales in order to determine if near-resonant triads play an important role in large-scale structure formation, as they do in 3D rotation and β-plane flows.

5 Summary

Based on the decomposition (5), one can derive PDE reduced models consisting of different combinations of three-mode interactions between linear eignemodes Ψ^m. For systems without a vortical mode Ψ^0 (as in 3D rotation and β-plane flow), one can define Ψ^0 to be the slow wave mode with $\sigma(\mathbf{k}) = 0$ in order to isolate interactions explicitly including slow modes. The approach is general and can be used to investigate physics in many dispersive-wave systems including (1)-(3). Specifically, one may determine which interactions are primarily responsible for the formation of coherent structures in both decaying and forced flows. In some cases, the new PDE models will naturally contain a subset of near-resonant triad interactions, shown to be important for structure formation in 3D rotation and β-plane flow forced randomly at small scales. [5, 6]. The present results indicate that near-resonant triads are not important for structure formation in RSW decay with $0.1 \leq Fr \leq 0.3$ and $0.25 \leq Ro \leq 1$; a future investigation will determine if near resonances in RSW are important for large-scale structure formation from small-scale random forcing.

References

1. Majda A (2000) Introduction to PDEs and Waves for the Atmosphere and Ocean CIMS New York University New York, AMS Providence RI
2. Pedlosky J (1986) Geophysical Fluid Dynamics Springer-Verlag New York
3. Longuet-Higgins M, Gill A (1967) Proc Roy Soc A 299:120–140
4. Newell A (1969) J Fluid Mech 35:255–271
5. Smith L, Lee Y (2005) J Fluid Mech 535:111–142
6. Lee Y, Smith L (2006) J Fluid Mech in press
7. Cambon C, Jacquin L (1989) J Fluid Mech 202:295–317
8. Galtier S (2003) Phys Rev E 68:015301–1
9. Polvani L, McWilliams J, Spall M, Ford R (1994) Chaos 4:177–186

Three-Dimensional Stability of Vortices in a Stratified Fluid

Pantxika Otheguy, Axel Deloncle, Paul Billant, and Jean-Marc Chomaz

Laboratoire d'hydrodynamique (LadHyX), CNRS - École Polytechnique, 91128 Palaiseau cedex, France
`chomaz@ladhyx.polytechnique.fr`

Abstract. The three-dimensional linear stabilities of vertically uniform shear flows and vortex configurations (dipole, couple, von Karman street and double symmetric row) are investigated through experiments, theoretical and numerical analysis when the fluid is stratified. For strong stratification, all the vortex configurations are unstable to the zigzag instability associated to vertically sheared horizontal translations that develop spontaneously. The most unstable wavelength decreases with the strength of the stratification, whereas the maximum growthrate is independent of the stratification and solely proportional to the strain felt by the vortex core. Experiments and direct numerical simulation show that the zigzag instability eventually decorrelates the flow on the vertical. The zigzag instability is therefore a generic instability that constrains turbulent energy cascade in stratified fluid and contributes to structure oceanic and atmospheric flows.

Keywords: stratified and rotating flows, zigzag instability, von Karman street, dipole, turbulence cascade

1 Introduction

The atmosphere and the ocean are characterized by a stable stratification that limits vertical motions and makes the flow mainly horizontal. Riley, Metcalfe & Weissman (1981) have shown that, if both the horizontal and the vertical scales of the flow are large compared to the buoyancy length scale, the leading order dynamics is then two-dimensional. Lilly (1983) has proposed that the kinetic energy spectra observed in the atmosphere at mesoscale are a manifestation of this two-dimensional dynamics with a transfer of energy from small ($\sim$1 km) to large ($\sim$500 km) scales.

Recently, Lindborg (1999) has questioned Lilly's conjecture by deducing from high order statistical moments that the energy cascade is in the opposite direction : from large to small scales. Billant & Chomaz (2000, 2001) proposed that the dynamic is not two-dimensional because the vertical scale selected by the flow is the local buoyancy length scale $L_B = U/N$ (where U is the horizontal velocity scale and N the Brunt-Väisälä frequency), invalidating the hypothesis of Riley, Metcalfe & Weissman (1981). They have also shown,

367

Y. Kaneda (ed.), *IUTAM Symposium on Computational Physics and New Perspectives in Turbulence*, 367–372.
© 2008 *Springer*.

in the specific case of a counter-rotating vortex pair, that the vertical scale selection is due to an instability which they named zigzag instability. In the present paper, we extend their work and show that zigzag instability is generic and affects other classical vortex configurations (pair of co-rotating vortices, von Karman street and double symmetric row) in a stratified fluid.

Of course, for real geophysical flows the forcing is not uniform on the vertical and the actual importance of the zigzag instability in accounting for the observed scaling laws is not known since it is definitely not the ultimate explanation of the layering in strongly stratified flows. In particular, it has been shown that the layering can be obtained in the absence of pre-existing large 2D structures, and in the absence of 2D horizontal forcing. For example by an angle-dependent energy drain which affects the toroidal part of the velocity field, and tends to transfer spectral energy from horizontal to vertical wavevectors (Godeferd & Cambon,1994). This selective damping mechanism, which related to the formation of Verticaly Sheared Horizontal Flow as described by Smith & Waleffe (2002), has recently been confirmed by the direct numerical simulation of Godeferd & Staquet (2003) and Liechtenstein, Godeferd & Cambon (2005) and is certainly crucial in the turbulence dynamics.

2 Stability of a Vortex Pair

For the vortex pair, the vertical vorticity field of the eigenmode (Fig. 1) shows that the zigzag instability translates the two vortices bringing the vortices closer or farther, alternatively along the vertical. The growth rate of the zigzag

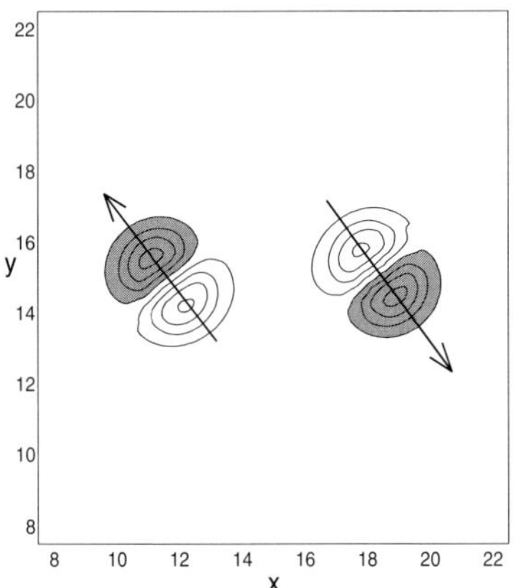

Fig. 1. Zigzag instability of two co-rotating vortices in a strongly stratified fluid. Vertical vorticity of the eigenmode of the zigzag instability for $F_h = 1$, $Re = 1000$, $a/b = 0.15$ and $k_z = 1.5$. The arrows indicate the direction of corresponding translation.

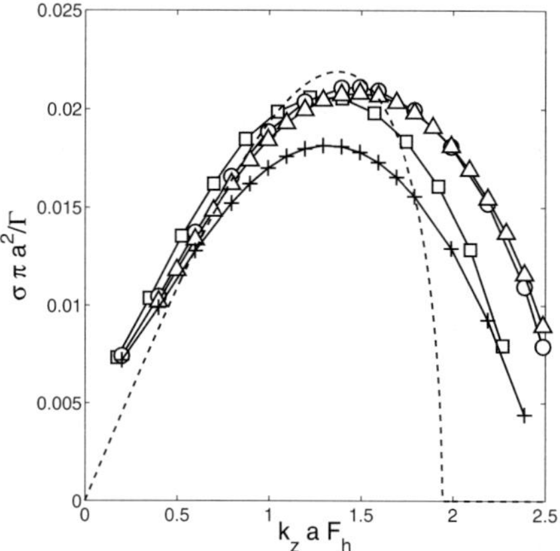

Fig. 2. Zigzag instability of two co-rotating vortices in a strongly stratified fluid. Growth rate as a function of $k_z F_h$ for $Re = 15000$, $F_h = 1.7$ (⊔), $F_h = 1$ (○), $F_h = 0.5$ (△) and $F_h = 0.2$ (+). The dotted line represents the asymptotic growth rate.

instability is a function of the rescaled wavenumber $k_z F_h$ (Fig. 2). The most unstable wavelength decreases therefore as the stratification increases. The nonlinear evolution of this zigzag instability induces the formation of thin horizontal layers with a thickness inversely proportional to the Brunt-Väisälä frequency.

The existence of the zigzag instability on the co-rotating vortex pair has been confirmed experimentally (Fig. 3). The experiments have been performed in a 100cm wide, 100cm long and 70cm deep glass tank filled with a linear stratified salt solution. Two co-rotating columnar vortices are created by quickly rotating two flaps with an apparatus similar to the one used by Meunier & Leweke (2001). Initially, the two vortices rotate one around the other and are straight along the vertical. At time t_1, the zigzag instability distorts symmetrically the two vortices. The distance between the vortices varies sinusoidally along the vertical generating layers where merging is accelerated or delayed resulting in a complex entanglement of the vortices (time t_2).

The effect of a planetary rotation has been also investigated. The maximum growth rate of the zigzag instability is approximately independent of the Rossby number ($Ro = \Gamma/\pi a^2 f$, where f is the Coriolis parameter). In contrast, the most unstable wavenumber k_{zm} varies continuously and scales as the Rossby number for small Ro : $k_{zm} \propto Ro/F_h$ i.e. $k_{zm} \propto N/f$. This means that, for small F_h and Ro, zigzag instability is nothing but the tall-column instability discovered in the quasi-geostrophic model by Dritschel, de la Torre Juárez & Ambaum (1999).

$t_1 = 81s\ t_2 = 135s$

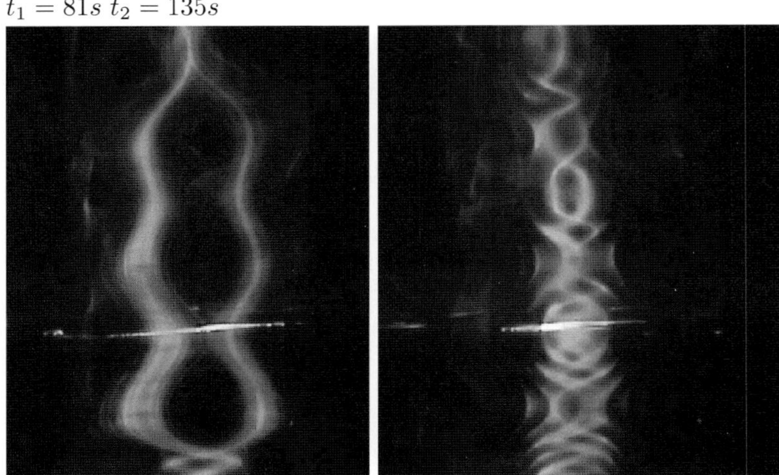

Fig. 3. Side view visualizations of the Zigzag instability of two co-rotating vortices in a strongly stratified fluid at different times. The flow is visualized by UV light and fluoresceine dye.

3 Generalisation of the Zigzag Instability to Vortex Arrays

We further demonstrate that both the von Karman street and the double symmetric row of columnar vertical vortices are also unstable to the zigzag instability.

By means of an asymptotic theory in the limit of long-vertical wavelength and well-separated vortices, the most unstable wavelength is found to be proportional to bF_h, where b is the separation distance between the vortices and F_h the horizontal Froude number ($F_h = \Gamma/\pi a^2 N$ with Γ the circulation of the vortices, a their core radius. The maximum growthrate is independent of the stratification and uniquely proportional to the strain $S = \Gamma/2\pi b^2$. A direct linear stability analysis performed numerically, confirms the theoretical predictions (Fig. 4).

The non-linear evolution of vortex couples, vortex dipoles, vortex arrays as well as random vortex distributions computed by direct numerical simulation shows that the zigzag instability ultimately *slices* the flow into horizontal layers. Figure 5, presents a vortex dipole so deformed by the zigzag instability that it is about to break into independent layers creating a vertical shear so intense, at the time it is tear in two, that Kelvin-Helmholtz instability can occur (example not shown here). The zigzag instability is therefore a generic mechanism that might explain the observations of layered structures in experiments and numerical simulations of stratified turbulence, in particular when forcing is homogeneous on the vertical.

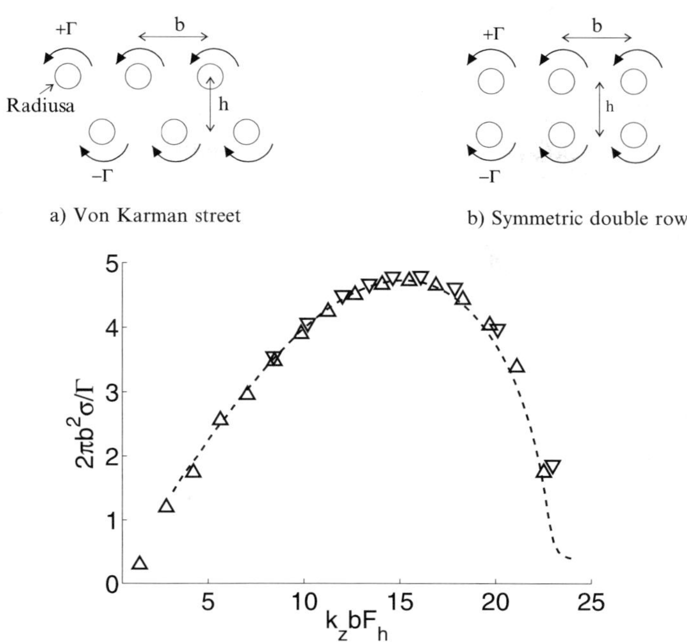

Fig. 4. (a, b) Vortex arrays configurations. (c) Zigzag instability of a Von Karman vortex street with $\kappa = h/b = 0.281$. Plots of the non-dimensional growthrate $2\pi b^2 \sigma/\Gamma$ as a function of the rescaled vertical wavenumber $k_z bF_h$. $\triangle$ and $\triangledown$ symbols show the growth rates given by a numerical simulation for two different values of F_h: $F_h = 0.2$ and $F_h = 0.4$ respectively. The dash-dotted line represents the inviscid prediction obtained by an asymptotic analysis.

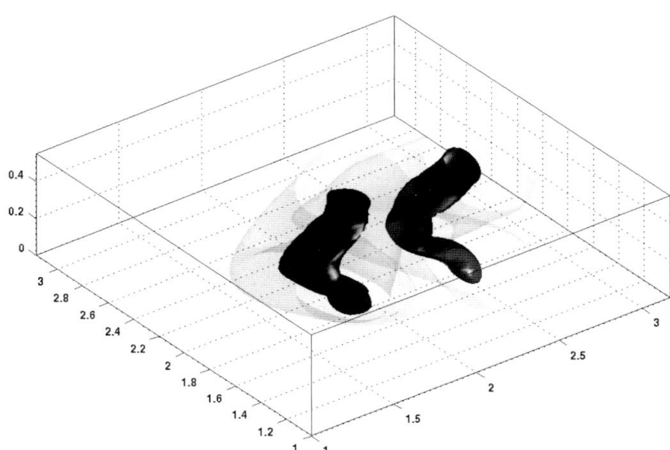

Fig. 5. Nonlinear simulation of counter-rotating vortices (dipole) subject to the zigzag instability.

References

1. Riley JJ, Metcalfe RW, Weissman MA (1981) Proc AIP Conf Nonlinear properties of internal waves (ed West BJ), 79–112
2. Lilly DK (1983) J Atmos Sci 40:749–761
3. Lindborg E (1999) J Fluid Mech 388:259–288
4. Billant P, Chomaz J-M (2000) J Fluid Mech 419:29–63
5. Billant P, Chomaz J-M (2001) Phys Fluids 13:1645–1651
6. Godeferd FS, Cambon C (1994) Phys Fluids 6:2084–2100
7. Smith LM, Waleffe F (2002) J Fluid Mech 451:145–168
8. Godeferd FS, Staquet C (2003) J Fluid Mech 486:115–159
9. Liechtenstein L, Godeferd FS, Cambon C (2005) J Turbul 6:115–159
10. Meunier P, Leweke T (2001) Phys Fluids 13:2747–2750
11. Dritschel DG, de la Torre Juárez M (1996) J Fluid Mech 328:129–160

Simulation of Strongly Stratified Fluids

G. Brethouwer[1], P. Billant[2] and E. Lindborg[1]

[1] Department of Mechanics, KTH, SE-100 44 Stockholm, Sweden
   geert@mech.kth.se
[2] LadHyX, Ecole Polytechnique, F-91128 Palaiseau Cedex, France

Abstract. Stably and strongly stratified turbulent flows have been studied by employing scaling analysis of the governing equations along the lines of [1], [2] and [3]. The scaling analysis suggests the existence of two different dynamical states. The parameter determining the state is $\mathcal{R} = ReF_h^2$, where Re and F_h are the Reynolds number and horizontal Froude number, respectively. If $\mathcal{R} \gg 1$, viscous forces are negligible and the turbulence is strongly anisotropic but three-dimensional and causes a forward energy cascade. The vertical length scale l_v scales as $l_v \sim U/N$ (U is a horizontal velocity scale and N is the Brunt-Väisälä frequency). If $\mathcal{R} \ll 1$, horizontal inertial forces are balanced by vertical viscous shearing and $l_v \sim l_h Re^{-1/2}$ (l_h is a horizontal length scale). The scaling analysis has been confirmed by direct numerical simulations of homogeneous stratified turbulence. Spectra have been studied as well.

Keywords: stratification, DNS, turbulence, geophysical flows

1 Introduction

Measurements in the middle atmosphere show commonly horizontal $k_h^{-5/3}$-spectra, where k_h is the horizontal wave number. Several hypothesis exist that try to explain these measured spectra. Two-dimensional turbulence with an inverse energy cascade has been suggested as well as a forward energy cascade produced by gravity waves. Recently, strongly anisotropic turbulence with a forward energy cascade was suggested by [3]. This hypothesis was confirmed by simulations which also produced the horizontal $k_h^{-5/3}$-spectrum [3]. Such a spectrum was also seen in direct numerical simulations (DNS) by [4]. In other simulations and laboratory experiments, however, the horizontal $k_h^{-5/3}$-spectrum and a strong forward energy cascade were not observed, e.g. [5], [6], [7]. These discrepancies must be due to the different conditions under which these investigations have been carried out. Presumably F_h and Re are the key parameters in this respect but the studies do not fully clarify how the dynamics of stratified flows is related to these parameters. The objective of our study is to address this issue.

Y. Kaneda (ed.), *IUTAM Symposium on Computational Physics and New Perspectives in Turbulence*, 373–378.

2 Scaling Analysis

To understand the dynamics of stably stratified fluids, the governing equations under the Boussinesq approximation are analysed using scaling arguments. Details of such an analysis can be found in [1], [2], [3] and [8]. Here we present only some of the main results.

We introduce a horizontal and vertical length scale l_h and l_v respectively, a horizontal velocity scale U and the nondimensional parameters $Re = Ul_h/\nu$, $Sc = \nu/\kappa$ (ν and κ are the viscosity and diffusivity respectively), $F_h = U/Nl_h$ and $\alpha = l_v/l_h$. Assuming $Re \gg 1$ and $F_h \to 0$, and following [1], we obtain the following nondimensional equations

$$\frac{\partial \boldsymbol{u}_h}{\partial t} + \boldsymbol{u}_h \cdot \nabla_h \boldsymbol{u}_h + \frac{F_h^2}{\alpha^2} u_z \frac{\partial \boldsymbol{u}_h}{\partial z} = -\nabla_h p + \frac{1}{Re\,\alpha^2} \frac{\partial^2 \boldsymbol{u}_h}{\partial z^2} \tag{1}$$

$$0 = -\frac{\partial p}{\partial z} - \rho \tag{2}$$

$$\nabla_h \cdot \boldsymbol{u}_h + \frac{F_h^2}{\alpha^2} \frac{\partial u_z}{\partial z} = 0 \tag{3}$$

$$\frac{\partial \rho}{\partial t} + \boldsymbol{u}_h \cdot \nabla_h \rho + \frac{F_h^2}{\alpha^2} u_z \frac{\partial \rho}{\partial z} = u_z + \frac{1}{ReSc\,\alpha^2} \frac{\partial^2 \rho}{\partial z^2} . \tag{4}$$

Here $\boldsymbol{u}_h$ and u_z are the horizontal and vertical velocity respectively, z is the vertical coordinate, ρ is the density perturbation and ∇_h is the horizontal gradient, all in nondimensional form. We have kept $\mathcal{O}(F_h^2/\alpha^2)$ and $\mathcal{O}(Re\alpha^2)$ terms because the value of α is unknown yet. Instead, we adopt the idea that l_v is a free parameter determined by the dynamics of the flow and not determined by other conditions. Assuming $Re \gg 1$ and $F_h \ll 1$, we see that two different regimes are possible depending on the ratio of the vertical advection term and the viscous term in (1), which is given by the parameter $\mathcal{R} = ReF_h^2$.

$\mathcal{R} \gg 1$: the strongly stratified turbulence regime

If $\mathcal{R} \gg 1$, the viscous and diffusive terms in (1) and (4) can be neglected and the equations are self-similar with respect to the scaled vertical coordinate $z'N/U$. Hence, [1] suggested the scaling $l_v \sim U/N$. The two advection terms in (1) are then of the same order. In [3] the scaling $l_v \sim U/N$ was adopted and this led to the hypothesis of strongly anisotropic but three-dimensional turbulence with a forward energy cascade, which we will call strongly stratified turbulence. It is assumed that $l_h \sim U^3/\varepsilon$, where ε is the kinetic energy dissipation. The parameter $\mathcal{R}$ can then be estimated as

$$\mathcal{R} = \frac{\varepsilon}{\nu N^2} . \tag{5}$$

We can argue that if $\mathcal{R} \gg 1$ (see [8]) an inertial range can exist. Using scaling arguments, [3] suggested that in this inertial range the horizontal kinetic and potential energy spectra have a $k_h^{-5/3}$ behaviour.

$\mathcal{R} \ll 1$: the viscosity affected stratified flow regime

If $\mathcal{R} \ll 1$, the vertical advection term in (1) is much smaller than the viscous term and can be neglected. Assuming that l_v is a free parameter selected by the flow itself, [2] have argued that it is determined by a balance between the horizontal advection term and the vertical diffusion term in (1). This leads to

$$l_v \sim l_h Re^{-1/2}. \tag{6}$$

The vertical length scale is thus independent of the stratification strength.

3 Direct Numerical Simulations

DNS of homogeneous turbulence with a uniform stratification is carried out. A pseudospectral method with periodic boundary conditions is employed to solve the Boussinesq equations. Energy is injected at the large scales by means of a forcing. The forcing is restricted to the horizontal velocity components and to horizontal modes and does therefore not introduce a vertical length scale. DNS are carried out at different Re and Fr and the parameters are chosen so that both regimes, $\mathcal{R} > 1$ and $\mathcal{R} < 1$, are covered by the DNS. Furthermore, $Sc = 0.70$. The numerical and physical parameters used in the DNS are listed in Table 1. The physical parameters are extracted from the DNS after the flow reached statistical stationarity. For further details and a more extensive set of DNS we refer to [8].

Table 1. Overview of the numerical and physical parameters used in the simulations. L_h and L_v are the horizontal and vertical dimension of the box respectively, and N_h, N_v are the number of nodes in the horizontal and vertical direction, respectively. l_v is computed in the same way as in [3] and $\mathcal{R}$ according to (5).

run	Re ($\times 10^3$)	F_h ($\times 10^{-2}$)	$\mathcal{R}$	L_h/L_v	$N_h \times N_v$	l_v/l_h
A0.06	11	0.23	0.058	4	256×64	0.020
A0.2	7.3	0.53	0.21	4	256×64	0.025
A0.8	5.0	1.2	0.75	4	256×64	0.030
A1.8	7.5	1.5	1.75	4	256×80	0.026
A2.8	5.5	2.3	2.84	4	256×64	0.040
A9.3	5.3	4.2	9.3	2.9	256×96	0.049
B0.1	21	0.23	0.11	6	512×96	0.014
B0.4	20	0.45	0.40	6	512×96	0.014
B1.1	17	0.81	1.09	6	512×96	0.017
B3.0	12	1.5	2.97	5	512×128	0.025
B9.3	13	2.7	9.3	4	512×144	0.032

4 Results

After the transition period the flow approaches a statistically stationary state in all simulations whereby the kinetic and potential energy stay approximately constant and the dissipation balances on average the power input by the forcing. The statistics and visualisations are extracted after the simulations reached such a steady state.

Figure 1 presents snapshots of the fluctuating density field on a vertical plane extracted from a simulation with $\mathcal{R} < 1$ and $\mathcal{R} > 1$. When $\mathcal{R} < 1$, the density field is smooth with relatively flat large-scale structures, which points to the significant influence of viscosity on the dynamics. Only a few disturbances are visible in the snapshot. On the other hand, when $\mathcal{R} > 1$ large-scale structures are visible, but small-scale disturbances and turbulent-like motions are abundantly present in the flow. These are presumably caused by the sharp vertical gradients between the large structures. Note that the layer formation in the present simulations are entirely the result of dynamical processes in the stratified fluid because the forcing is purely horizontal.

In Fig. 2, l_v scaled by $l_h Re^{-1/2}$ and scaled by U/N is displayed. Here l_v is computed as in [3]. Note that Fig. 2 includes the results of many DNS with a fairly wide range of F_h and Re. The vertical length scale extracted from the

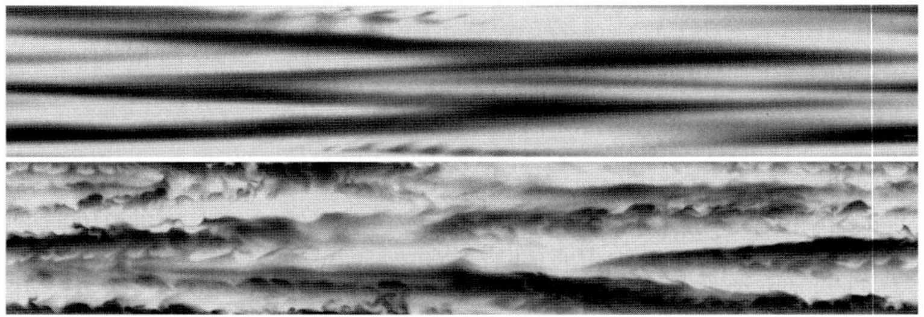

Fig. 1. Snapshots of the density fluctuations in a vertical plane. $\mathcal{R} < 1$ (run B0.4) in the top figure and $\mathcal{R} > 1$ (run B9.3) in the bottom figure.

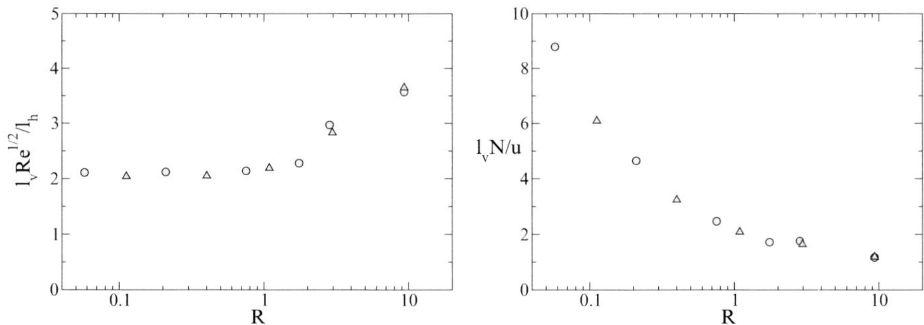

Fig. 2. The scaled vertical length scale (left figure) $l_v Re^{1/2}/l_h$ and (right figure) $l_v N/U$ as a function of $\mathcal{R}$. Circles, runs A; triangles, runs B.

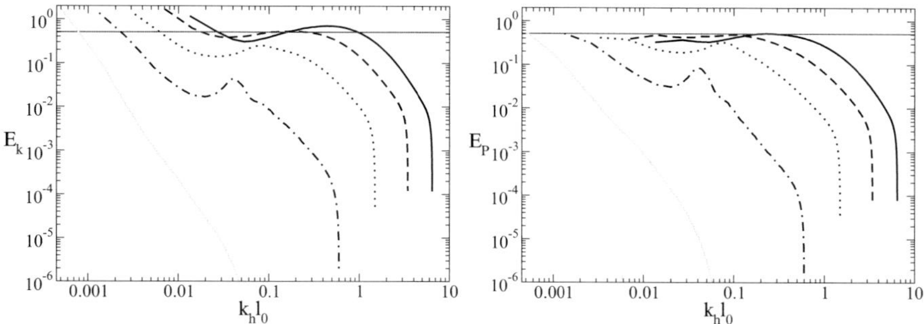

Fig. 3. The compensated horizontal one-dimensional kinetic energy spectra $E_K(k_h)k_h^{5/3}/\varepsilon^{2/3}$ (left figure) and potential energy spectra $E_P(k_h)k_h^{5/3}\varepsilon^{1/3}/\varepsilon_P$ (right figure), where ε_P is the potential energy dissipation.. The horizontal wavenumber k_h is scaled by the Ozmidov length scale l_O. (———), B9.3; (− − −), B3.0; ($\cdots\cdots$), B1.1; (− · · − · · −), B0.4; (———, gray line), B0.1.

simulations shows the scaling $l_v \sim l_h Re^{-1/2}$ when $\mathcal{R} < 1$ and supports the scaling $l_v \sim U/N$ when $\mathcal{R} > 1$. Moreover, Fig. 2 shows that the transition between the two regimes takes place around $\mathcal{R} = 1$. These results are consistent with the hypotheses presented here before.

In Fig. 3, compensated horizontal one-dimensional kinetic and potential energy spectra are shown. The constant value 0.51 for the compensated spectra in the inertial range obtained by [3] using simulations with hyperviscosity is represented by the straight line in Fig. 3. We do not see a clear inertial range in our simulations because $\mathcal{R}$ is presumably not large enough in the DNS to have an inertial range virtually free of viscous effects. Nevertheless, the spectra extracted from the runs B3.0 and B9.3 show a wave number range quite close the straight line, suggesting an approach to the $k_h^{-5/3}$-power-law behaviour. Spectra extracted from the simulations with $\mathcal{R} < 1$ have a steep slope and there is obviously no wave number range that resembles a $k_h^{-5/3}$-range.

To conclude, Fig. 4 presents a diagram with the different regimes that are found in stably stratified fluids depending on the value of Re and F_h as suggested by this study. The strongly stratified turbulence regime is bounded by the thresholds $\mathcal{R} > 1$ and $F_h < 0.02$ (see [8]). When $\mathcal{R} < 1$, we find the viscosity affected stratified flow regime with large, stable and quasi-two-dimensional layers with strong vertical viscous shearing. Weakly stratified turbulence which does not obey the scaling $l_v \sim U/N$ is found for $0.02 > F_h > 0$. For geophysical flows, the strongly stratified turbulence regime is arguably of much interest. However, this regime is hard to achieve in laboratory experiments and numerical simulations because both a low F_h and a high Re are required. In particular, at moderately high Re, the F_h range where strongly stratified turbulence is found is limited. The diagram shows the conditions under which the DNS have been performed. Some of the present simulations have just met the conditions $\mathcal{R} > 1$ and $F_h < 0.02$ but unfortunately, with

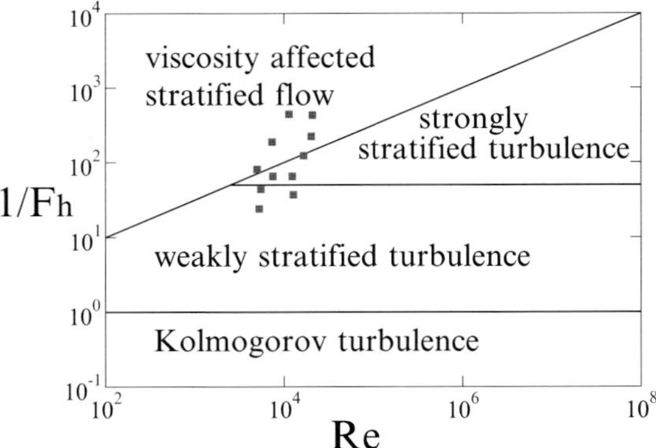

Fig. 4. Regimes in stably stratified fluids. The conditions under which the DNS are carried out are represented by the symbols.

the present computer capacities it is not yet possible to simulate stratified turbulence with $F_h \ll 1$ and at the same time $\mathcal{R} \gg 1$ so that there is a clear inertial subrange with negligible viscous influences.

References

1. Billant P, Chomaz JM (2001) Phys Fluids 13:1645–1651
2. Godoy-Diana R, Chomaz JM, Billant P (2004) J Fluid Mech 504:229–238
3. Lindborg E (2006) J Fluid Mech 550:207–242
4. Riley JJ, deBruynKops SM (2003) Phys Fluids 15:2047–2059
5. Smith LM, Waleffe, F (2002) J Fluid Mech 451:145–168
6. Waite ML, Bartello P (2004) J Fluid Mech 517:281–308
7. Praud O, Fincham AM, Sommeria J (2005) J Fluid Mech 522:1–33
8. Brethouwer G, Billant P, Lindborg E, Chomaz JM (2006) J Fluid Mech, submitted

Numerical Simulation of Quantum Fluid Turbulence

Kyo Yoshida and Toshihico Arimitsu

Graduate School of Pure and Applied Sciences, University of Tsukuba, 1-1-1
Tennoudai, Tsukuba, Ibaraki 305-8571, Japan
yoshida@sakura.cc.tsukuba.ac.jp

Abstract. Dynamics of quantum fluids such as low-temperature superfluids and
Bose-Einstein condensates are described by the Gross-Pitaevskii (GP) equation under a certain approximation. Numerical simulations of quasi-isotropic turbulence
obeying the GP equation with forcing and dissipation are performed. The domain
of the simulations is a cube with periodic boundary conditions and the numbers of
grid points are up to 512^3. It is found that the interaction energy spectrum $E^{\text{int}}(k)$
obeys the scaling $\propto k^{-3/2}$. In contrast to the preceding numerical studies, the energy
spectrum $E^{\text{wi}}(k)$ associated with the incompressible part of the weighted velocity
field does not obey the Kolmogorov scaling $\propto k^{-5/3}$. Detailed analyses of the results
from the numerical simulations are found in the paper by the same authors [Yoshida
K, Arimitsu T (2006) J Low Temp Phys 145:219–230].

Keywords: turbulence, quantum fluid, Gross-Pitaievski equation, energy
spectrum

1 Introduction

Dynamics of low-temperature superfluids and Bose-Einstein condensates are
described by the Gross-Pitaevskii (GP) equation [1, 2] (also called the nonlinear Schrödinger equation),

$$i\hbar\frac{\partial\psi}{\partial t} = -\left(\frac{\hbar^2}{2m}\nabla^2 + \mu\right)\psi + g|\psi|^2\psi, \tag{1}$$

under a certain approximation. Here, $\psi := \langle\hat{\psi}\rangle$ is the order parameter, the
vacuum expectation value of the boson field $\hat{\psi}$, μ is the chemical potential
and g is the coupling constant. The chemical potential μ and $n := |\psi|^2$ are
related by $\mu = g\bar{n}$ where $\bar{\cdot}$ denotes the spatial average. By using Madelung's
transformation $\psi = \sqrt{\rho/m}\exp(i\varphi)$ with $\mathbf{v} := (\hbar/m)\nabla\varphi$, one can transform
(1) into the equations of motion for fluid with density ρ and velocity $\mathbf{v}$. Here,
we call the fluid *quantum fluid*.

A remarkable feature of the quantum fluid is the existence of quantized
vortex lines. Since $\mathbf{v}$ is a potential flow, the vorticity $\omega := \nabla \times \mathbf{v}$ is $\mathbf{0}$ wherever

Y. Kaneda (ed.), *IUTAM Symposium on Computational Physics and New Perspectives in
Turbulence*, 379–384.

$\mathbf{v}$ is defined, i.e. $\rho \neq 0$. The vorticity is concentrated in the lines where $\rho = 0$ and the circulations around such lines, *i.e.* vortex lines, are quantized due to the the uniqueness of the phase φ up to modulus of 2π. On the other hand, circulations can take arbitrary values in classical fluid which obeys the Navier-Stokes equations.

Turbulence of quantum fluid may be caused by tangles of the quantized vortex lines. There are some evidence from recent experiments [3, 4] of turbulence in superfluid phase of liquid ^{4}He that the energy spectrum in quantum fluid turbulence obeys the same Kolmogorov scaling $\propto k^{-5/3}$ as in classical fluid turbulence. The Kolmogorov energy spectrum is also observed in numerical simulations of the GP equation [5, 6]. These experimental and numerical results suggest that there is an energy cascade process not only in classical fluid turbulence but also in quantum fluid turbulence.

In order to investigate the statistical structures and energy cascade process in quantum fluid turbulence, we performed numerical simulations of quasi-isotropic turbulence obeying GP equation with forcing and dissipation. In this paper, we will show some basic results from the simulation. Detailed analysis of the results is given in Ref. [7].

2 Numerical Simulation

2.1 Set Up

We employ following normalizations, $\tilde{\mathbf{x}} := \mathbf{x}/\ell, \tilde{t} := (g\bar{n}/\hbar)t, \tilde{\psi} := (1/\sqrt{\bar{n}})\psi$, to obtain a normalized GP equation,

$$i\frac{\partial \tilde{\psi}}{\partial \tilde{t}} = -\tilde{\xi}^2 \tilde{\nabla}^2 \tilde{\psi} - \tilde{\psi} + |\tilde{\psi}|^2 \tilde{\psi}, \tag{2}$$

with $\tilde{\xi} := \xi/\ell$ where $\xi := \hbar/\sqrt{2mg\bar{n}}$ is the healing length, and ℓ is an arbitrary unit length scale. Since we will deal with the normalized variables, we will omit the tilde $\tilde{\ }$ in the following. The density field and the velocity field of the quantum fluid are given by $\rho := |\psi|^2$ and $\mathbf{v} := 2\xi^2 \nabla\varphi$, respectively, in the present normalization. The density fluctuation is given by $\delta\rho := \rho - \bar{\rho}$, where $\bar{\rho} = 1$. We apply periodic boundary conditions with periods 2π in each of three directions in the Cartesian coordinates. This boundary condition enable us to work with the Fourier space representation of (2), *i.e.*,

$$\frac{\partial}{\partial t}\psi_{\mathbf{k}} = -i\xi^2 k^2 \psi_{\mathbf{k}} + i\psi_{\mathbf{k}} - i \sum_{\mathbf{k}+\mathbf{p}-\mathbf{q}-\mathbf{r}=0} \psi_{\mathbf{p}}^* \psi_{\mathbf{q}} \psi_{\mathbf{r}}$$
$$+ D_{\mathbf{k}} + F_{\mathbf{k}}, \tag{3}$$

where $f_{\mathbf{k}}$ is the Fourier transform of an arbitrary function $f(\mathbf{x})$. In (3), we added a dissipation term $D_{\mathbf{k}}$ and a forcing term $F_{\mathbf{k}}$ to the original equation (2).

To the authors' knowledge, the mechanism of dissipation in quantum fluid is not well understood. It should have its origin in the interaction between ψ

and the fluctuation, $\delta\hat{\psi} := \hat{\psi} - \psi$, which is neglected in (2). In the present study, we will not go into the details of the dissipation mechanism and will assume a simple model, $i.e.$, a Laplacian type model,

$$D_{\mathbf{k}} = -\nu k^2 \psi_{\mathbf{k}}. \tag{4}$$

The dissipation term $D_{\mathbf{k}}$ mainly acts in a high wavenumber range due to its k^2 dependence. We expect that some of the statistical structures in the lower wavenumber range, $i.e.$ the inertial subrange, are insensitive to the detail of the model for the dissipation. Note that this expectation is an assumption at the present stage since, to the authors' knowledge, localness or nonlocalness of the interaction in the wavevector space for the GP turbulence has not been investigated in detail yet.

The forcing term $F_{\mathbf{k}}$ is introduce to compensate the density loss due to the dissipation. The density is pumped by the forcing which amplifies low wavenumber modes, $i.e.$,

$$F_{\mathbf{k}} = \begin{cases} \alpha\psi_{\mathbf{k}} & (k < k_f) \\ 0 & (k \geq k_f) \end{cases}, \tag{5}$$

where α is determined at every time step so as to keep $\bar{\rho}$ almost unity.

We performed the numerical simulations of (3) with (4) and (5). The convolution sum was computed by utilizing a Fast Fourier transform and a phase shift dealiasing. The time integration was computed by a 4th-order Runge-Kutta method. We performed three simulations with 128^3, 256^3, and 512^3 grid points which will be denoted by RUN128, RUN256, and RUN512, respectively. In the present study, the smallest scale of interest is the healing length ξ. In order that the length scales of the order of ξ are resolved in the simulations, we set ξ to satisfy $\xi \sim 3k_{\mathrm{max}}^{-1}$, where k_{max} is the maximum resolved wavenumber. ($\xi = 0.05, 0.025$ and 0.0125 for RUN128, RUN256 and RUN512, respectively.) The dissipation coefficient is set to $\nu = \xi^2$, within the present normalization, in order that the dissipation mainly acts at length scales in the order of or smaller than ξ. We set $\bar{\rho} = 1$, $\Delta t = 0.01$ and $k_f = 2.5$ for all the runs.

2.2 Results

The kinetic and interaction energy density per unit mass, E^{kin} and E^{int}, respectively, are given by

$$E^{\mathrm{kin}} := \frac{1}{V} \int d\mathbf{x}\xi^2|\nabla\psi|^2, \qquad E^{\mathrm{int}} := \frac{1}{2V} \int d\mathbf{x}(\delta\rho)^2, \tag{6}$$

where $V = (2\pi)^3$ is the volume of the periodic boundary box domain. Following Ref. [5], we introduce a weighted velocity field $\mathbf{w} := (\sqrt{2}\xi)^{-1}\sqrt{\rho}\mathbf{v}$, which

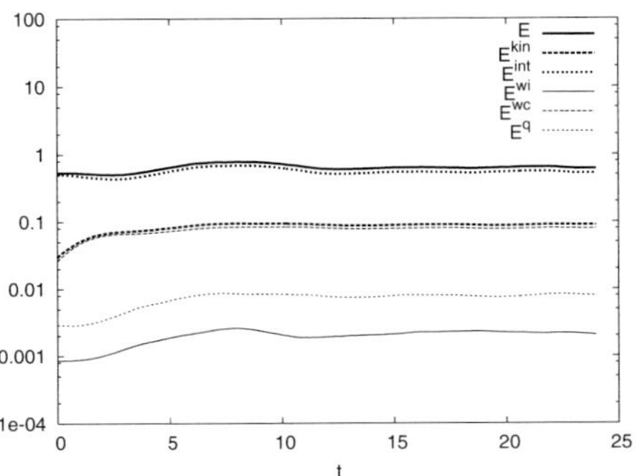

Fig. 1. The time evolution of the energies E, E^{int}, E^{kin}, E^{wi}, E^{wc}, and E^{q} in RUN512. (Fig. 1 of Ref. [7], reproduced with permission.)

enables us to divide E^{kin} into three components, *i.e.*, $E^{\mathrm{kin}} = E^{\mathrm{wi}} + E^{\mathrm{wc}} + E^{\mathrm{q}}$ with

$$E^{\mathrm{wi}} := \frac{1}{2V} \int d\mathbf{x}|\mathbf{w}^{\mathrm{i}}|^2, \quad E^{\mathrm{wc}} := \frac{1}{2V} \int d\mathbf{x}|\mathbf{w}^{\mathrm{c}}|^2, \quad E^{\mathrm{q}} := \frac{1}{V} \int d\mathbf{x}\xi^2|\nabla\sqrt{\rho}|^2, \quad (7)$$

where $\mathbf{w}^{\mathrm{i}}$ and $\mathbf{w}^{\mathrm{c}}$ are, respectively, incompressible and compressible parts of $\mathbf{w}$. Figure 1 shows the time evolution of the energies (6) and (7) together with the total energy $E := E^{\mathrm{kin}} + E^{\mathrm{int}}$. One can see from the figure that E^{int} is dominant, *i.e.*, $E^{\mathrm{int}} = 0.85E$, and the rest of energy, *i.e.*, $E^{\mathrm{kin}} = 0.15E$, is shared among E^{wi}, E^{wc} and E^{q}. The energy E^{wc} exceeds E^{wi} by the factor of about 40.

Energy spectra associated with the energies (6) and (7) are given by

$$E^{\mathrm{kin}}(k) := \sum_{k'=k} \xi^2 k'^2 |\psi_{\mathbf{k}'}|^2, \qquad E^{\mathrm{int}}(k) := \sum_{k'=k} |(\delta\rho)_{\mathbf{k}'}|^2, \qquad (8)$$

$$E^{\mathrm{wi}}(k) := \frac{1}{2}\sum_{k'=k} |\mathbf{w}^{\mathrm{i}}_{\mathbf{k}'}|^2, \qquad E^{\mathrm{wc}}(k) := \frac{1}{2}\sum_{k'=k} |\mathbf{w}^{\mathrm{c}}_{\mathbf{k}'}|^2, \qquad (9)$$

$$E^{\mathrm{q}}(k) := \sum_{k'=k} \xi^2 k'^2 |(\sqrt{\rho})_{\mathbf{k}'}|^2, \qquad (10)$$

where $\sum_{k'=k}$ denotes the summation with respect to $\mathbf{k}'$ over the shell $k-1/2 < |\mathbf{k}'| \le k + 1/2$. Note that, since ρ and $\mathbf{w}$ are nonlinear functions of ψ, the spectra $E^{\mathrm{int}}(k)$, $E^{\mathrm{wi}}(k)$, $E^{\mathrm{wc}}(k)$, and $E^{\mathrm{q}}(k)$ include contribution of $\psi_{\mathbf{k}'}$ outside the shell.

The energy spectra (8)–(10) in RUN512 at $t = 24$ are shown in Fig. 2. It is found that $E^{\mathrm{int}}(k)$ and $E^{\mathrm{kin}}(k)$ obey scaling laws $E^{\mathrm{int}}(k) \propto k^{-3/2}$ and $E^{\mathrm{kin}}(k) \propto k^{4/3}$, respectively, in a wavenumber range $k_f \sim< k \sim< \xi^{-1}$. The energy spectrum $E^{\mathrm{wi}}(k)$ in quantum fluid turbulence is the analogue of the energy spectrum in incompressible classical fluid turbulence. In preceding numerical simulation of GP equation by Kobayashi and Tsubota [6] (hereafter

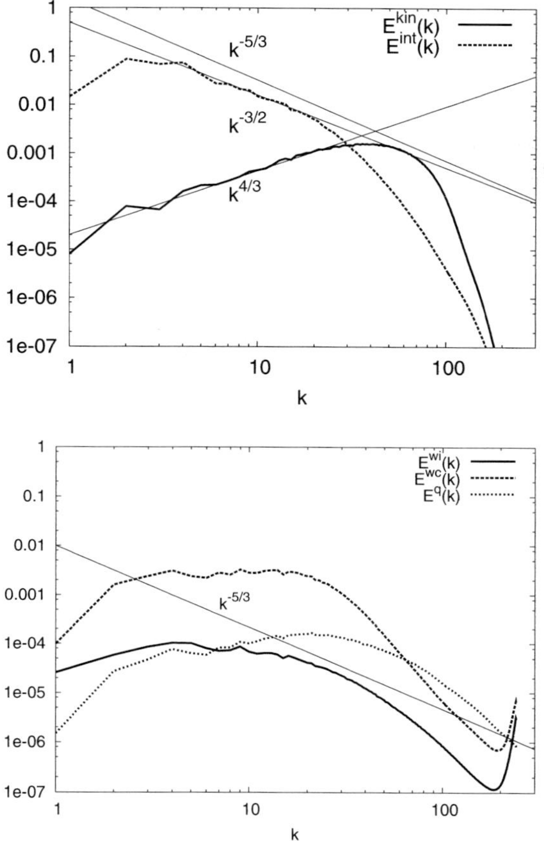

Fig. 2. Energy Spectra in RUN512 at $t = 24$. $E^{\mathrm{kin}}(k)$ and $E^{\mathrm{int}}(k)$ in the top figure, and $E^{\mathrm{wi}}(k)$, $E^{\mathrm{wc}}(k)$, and $E^{\mathrm{q}}(k)$ in the bottom figure. (Fig. 2 of Ref. [7], reproduced with permission. Data from RUN256 in the original figure is removed in the reproduction.)

KT), the Kolmogorov scaling $E^{\mathrm{wi}}(k) \propto k^{-5/3}$ is observed. However, in the present study, the scaling of $E^{\mathrm{wi}}(k)$ is not clearly observed.

3 Discussion

The scaling $E^{\mathrm{int}}(k) \propto k^{-3/2}$ found in the present numerical simulation is consistent with the weak turbulence analysis by Dyachenko *et al.* [8]. However, from further data analysis of the present numerical simulations, it turns out that the basic assumptions in the weak turbulence analysis, *i.e.*, the fluctuation of ψ from its mean is small and ψ is approximated by an ensemble of waves of the linearized equation of (1), are not satisfied. See Ref. [7] for the details of the analysis. Thus, the present results of the numerical simulations are calling for another explanation for the scaling $E^{\mathrm{int}}(k) \propto k^{-3/2}$. A possible scenario for the explanation is discussed in Ref. [7].

As for the scaling $E^{\mathrm{kin}}(k) \propto k^{4/3}$, we have not found its physical interpretation yet.

In the present numerical simulations, E^{wi} is much smaller than E^{wc} and the scaling of $E^{\mathrm{wi}}(k)$ is not clearly observed. However, in the preceding numerical simulations by KT, $E^{\mathrm{wi}} > E^{\mathrm{wc}}$ and the Kolmogorov scaling $E^{\mathrm{wi}}(k) \propto k^{-5/3}$ are observed. The origin of the discrepancy may be the differences of forcing term and dissipation term between the simulations in KT and those in the present study. Since the forcing term dominates low wavenumber modes in which energy is concentrated, we may expect that the difference in the forcing is more responsible than that in the dissipation. In contrast to the forcing of the form (5), the forcing in the numerical simulations by KT modifies not only the amplitude but also the phase of ψ. Presumably, the forcing of the form (5) supplies energy to the system with a ratio $E^{\mathrm{wi}}/E^{\mathrm{wc}} \ll 1$ and the supplied E^{wc} is not efficiently transferred to E^{wi}. When energy is not sufficiently supplied to the the field $\mathbf{w}^{\mathrm{i}}$ through the forcing or the transfer, turbulence would be moderately developed with respect to $\mathbf{w}^{\mathrm{i}}$, $i.e.$, the scaling law of $E^{\mathrm{wi}}(k)$ would not be observed as in the present simulation.

Further analysis of the dependence of the energy spectra on the forcing and the dissipation is left for future studies.

Acknowledgment

The DNSs were performed on a Fujitsu HPC2500 system at the Information Technology Center, Nagoya University. This research is partially supported by Grant-in-Aid for Young Scientists (B) 17740246 from Ministry of Education, Culture, Sports, Science, and Technology of Japan, and by Grant-in-Aid for Scientific Research (B) 17340117 from the Japan Society for the Promotion of Science.

References

1. Pitaevskii LP (1961) Soviet Phys JETP 13: 451–454
2. Gross EP (1963) J Math Phys 4:195–207
3. Maurer J, Tabeling P (1998) Europhys Lett 43:29–34
4. Stalp SR, Skrbek L, Donnelly RJ (1999) Phys Rev Lett 82:4831–4834
5. Nore C, Abid M, Brachet, ME (1997) Phys Rev Lett 78:3896–3899
6. Kobayashi M, Tsubota M (2005) J Phys Soc Jpn 74:3248–3258
7. Yoshida K, Arimitsu T (2006) J Low Temp Phys 145:219–230
8. Dyachenko S, Newell AC, Pushkarev A, Zakharov VE (1992) Physica D 57:96–160

Turbulent Drag Reduction by Wall Deformation Synchronized with Flow Acceleration

Ryo Matsumura[1], Shuhei Koyama[2], and Yoshimichi Hagiwara[1]

[1] Department of Mechanical and System Engineering, Graduate School of Science and Technology, Kyoto Institute of Technology, Matsugasaki, Kyoto city, 606-8585, Japan
yoshi@kit.ac.jp

[2] Mitsubishi Electric Co., Wakayama city, 640-8319, Japan

Abstract. Direct numerical simulation has been conducted for turbulent pulsating flow over a deforming wall. The wall simulates the skin of dolphins. The wall deformation is synchronized with an increase in the bulk mean velocity because the deformation of skin is observed only when the dolphins swim fast. The computational results show that the wall deformation contributes to drag reduction by decreasing both the viscous shear stress and the Reynolds shear stress. The decrease in the viscous shear stress is due to a decrease in the Q2 event in the downhill region. The decrease in the Reynolds shear stress is caused by the transport of some hairpin vortices away from the downhill and the decrease in the Q2 event.

Keywords: DNS, turbulent flow, deforming wall, drag reduction

1 Introduction

The reduction of turbulent friction drag in swimming dolphins has been widely discussed in marine biology and fluid dynamics. One of the probable mechanisms for the drag reduction is skin folds [1]. However, a high-speed propagation of skin folds referred to in [1] is suspicious because of the papillae and muscle structures. Large-scale skin folds referred to in [1] are also suspicious because they may increase friction drag and pressure drag.

The present authors have tried to elucidate drag reduction by small-scale skin folds with the fixed nodes on the skin. In the previous study [2], we carried out DNS for turbulent flow over a wall with deformation whose amplitude was proportional to the mean wall shear stress in a specific region. Although we predicted turbulence modifications due to the wall deformation, we did not discuss the drag reduction.

Since we observe the skin folds only when dolphins swim swiftly [3], we have considered that the amplitude of skin folds increases with the swimming speed of dolphins. In the present study, we carry out DNS for pulsating flow over a wall with deformation whose amplitude increases with flow velocity.

385

Y. Kaneda (ed.), *IUTAM Symposium on Computational Physics and New Perspectives in Turbulence*, 385–390.

2 Computational Method

We considered turbulent horizontal flow in a domain between an upper shear-free lid and a lower non-slip deforming wall. The x-, y- and z-axes were positioned in the streamwise, vertical and transverse directions, respectively. The ξ- and η-axes were positioned along the wall and the wall-normal direction, respectively. The domain was converted to the computational domain, which was a rectangular box of $2\pi h \times h \times \pi h$, by using an unsteady generalized curvilinear coordinate system to fit the deformed wall at any given moment [2]. The Reynolds number was $Re = hu_{\tau 0}/\nu = 200$ where ν is the kinematic viscosity and $u_{\tau 0}$ is the friction velocity at the initial state.

A staggered grid system was adopted. The grid spacing was identical both in the ξ and the z directions. It increased from the lower wall in the η direction based on a hyperbolic tangent. The grid resolution was $\Delta x^+ = 9.81$, $\Delta y^+ = 0.25$ - 4.00 and $\Delta z^+ = 4.90$. The time increment was $\Delta t^+ = 0.02$.

The computational schemes were the same as those in our previous study [2]. The periodic boundary condition was applied for velocity and pressure in the ξ and z directions.

The database of fully developed non-pulsating flow in the case of flat lower wall was adopted as the initial velocity field. We confirmed that the mean velocity and turbulence intensities for this database are in agreement with those for non-pulsating turbulent flow in an open channel obtained by Yamamoto et al. [4] (figure omitted).

The pulsating flow was obtained by imposing a sinusoidal fluctuation with time on the mean pressure gradient. The non-dimensional period of pulsation was $T_p^+ = 1200$. The bulk mean velocity was increased and decreased at most approximately 22% by the flow pulsation.

The two-dimensional sinusoidal deformation of the wall started when the bulk mean velocity was at its minimum. The locations of nodes for the deformation were fixed. The region between the first and second nodes and that between the third and fourth nodes ascended through the acceleration period of pulsating flow (See Fig. 1). The other regions descended through the period. The defomed wall returned to its flat state through the deceleration period. Thus, the deformation is not the same as the traveling wave, which is

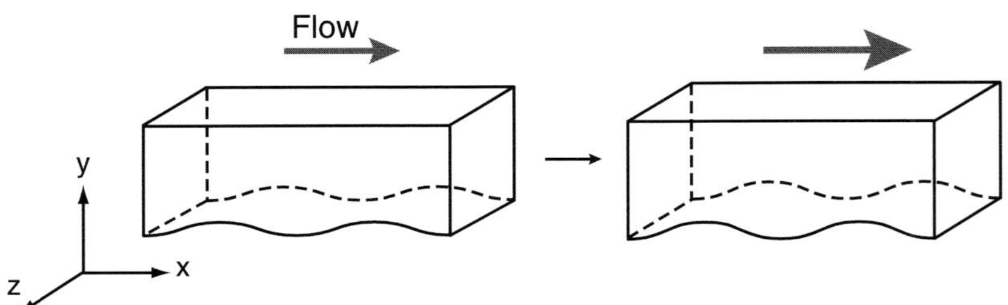

Fig. 1. Time change of flow domain.

effective for the drag reduction [5]. The ratio of maximum amplitude for the deformation ($= 8\nu/u_{\tau 0}$) to the wavelength λ was 0.0127.

3 Results and Discussion

3.1 Mean Velocity Profile, Turbulence Intensities and Turbulent Kinetic Energy

We deal with the result in the case of pulsating flow over a flat wall (hereafter called case 1) and that in the case of pulsating flow over the deforming wall (hereafter called case 2).

The mean velocity in case 2 was slightly lower in the buffer region than that in case 1 (figure omitted).

Figure 2 indicates the turbulence intensity profiles only for the acceleration periods. All the intensities in case 2 were lower than those in case 1.

Figure 3 exhibits the budget of turbulent kinetic energy (TKE) only for the acceleration periods. The gain of TKE by the viscous diffusion D_K in case 2 is much higher than that in case 1 in the vicinity of the wall. This is due to the expansion of the linear sublayer in the valley region. Furthermore, the absolute values of the dissipation rate of TKE, ϵ_K, in case 2 are also much higher than those in case 1 in the near-wall region.

3.2 Shear Stresses

Figure 4 depicts the profiles of shear stress which were calculated from the averages over space and time only for the acceleration periods. Since the mean velocity and its gradient in the near-wall region are high due to the flow acceleration, the viscous stress and thus the total shear stress in case 1 are higher than those in the case of non-pulsating flow [4] drawn with a broken line in the figure. These stresses in case 2 are 1.5% lower than those in case1. This shows that the friction drag was reduced by the wall deformation. The

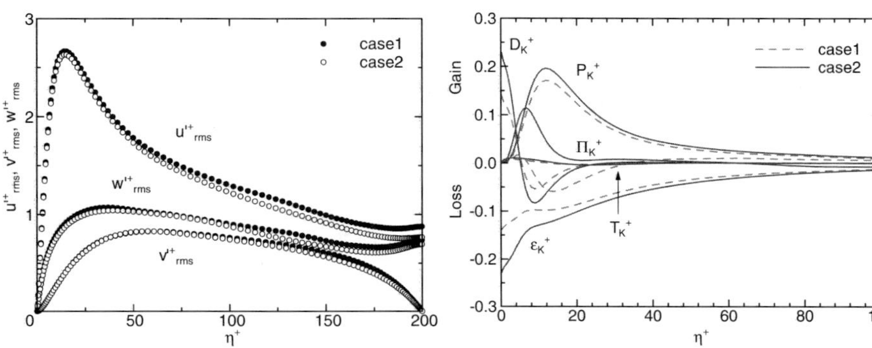

Fig. 2. Turbulence intensities

Fig. 3. Budget of turbulent kinetic energy

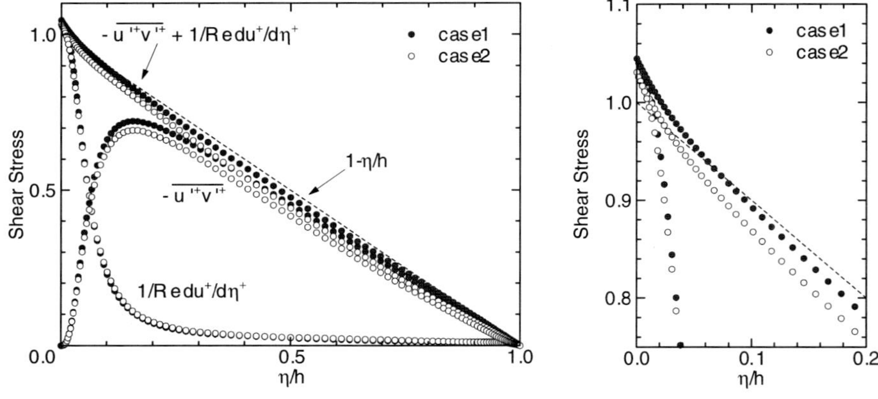

Fig. 4. Shear stresses

Reynolds shear stress was also reduced by the wall deformation. Note that the increase in the pressure drag due to the wall deformation is much smaller than the decrease in the friction drag.

The modifications of turbulence intensities and the Reynolds shear stress in the present study are opposite to those obtained by De Angelis et al. [6]. This is because they dealt with a rigid wavy wall whose ratio of amplitude to λ was 0.025 and thus they had the spread of average streamline spacing in the valley of the wall.

3.3 Wall Shear Stress Profile

Hereafter, statistics are focused on the instant when the wall is deformed most. The region between the valley and the downstream node is referred to as "uphill 1", that between the node and the hilltop, "uphill 2", that between the hilltop and the downstream node, "downhill 2", and that between the node and the valley, "downhill 1".

Figure 5 shows the streamwise profile for the averaged value of local wall shear stress in the transverse direction. The wall configuration is also exhibited in the lower part of this figure. The minimum is located at $x/\lambda = 0.17$ (x is measured from the upstream hilltop.). That is, the minimum wall shear stress is located in the downhill 2. This value is different from $x/\lambda = 0.45$ obtained by De Angelis et al. [6]. On the other hand, the maximum occurs at $x/\lambda = 0.99$. That is, the maximum wall shear stress occurs at the hilltop. This value is higher than $x/\lambda = 0.86$ obtained by De Angelis et al. There is no other maximum or minimum corresponding to the other hilltop.

3.4 Modification of Coherent Structure

Figure 6 indicates the cross-correlation coefficient for the streamwise fluctuating velocity in the transverse direction, R_{uu}, for the plane of $\eta^+ = 10$ in

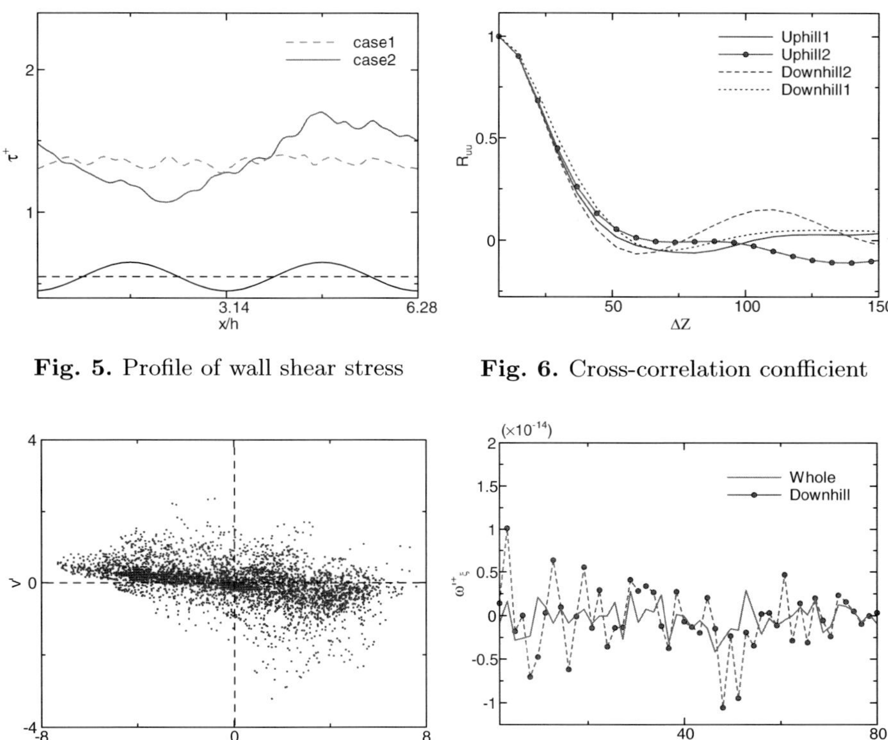

Fig. 5. Profile of wall shear stress   **Fig. 6.** Cross-correlation conffi* cient

Fig. 7. Quadrant analysis for the fluctu- **Fig. 8.** Streamwise fluctuating vorticity
ating velocities($\eta^+ = 10$).

case 2. Kim et al. [7] estimated the average spacing of the low-speed streaks in
the transverse direction as twice the transverse length, ΔZ, between the positions where R_{uu} takes its maximum and minimum. The transverse length was
$\Delta Z_0^+ \approx 75$ in case 1 (figure omitted). It is found from this figure that ΔZ^+ for
downhill 1 and uphill 1 are in near agreement with ΔZ_0^+. On the other hand,
ΔZ for uphill 2 is much higher than ΔZ_0^+. This result and the maximum of
wall shear stress in Fig. 5 suggest that the sweep becomes predominant in
uphill 2. In contrast, ΔZ for downhill 2 is much lower than ΔZ_0^+.

Figure 7 demonstrates the quadrant analysis for the fluctuating velocities
on $\eta^+ = 10$ in the downhill 2. The outward fluctuating velocity($v' > 0$) is
not noticeable in the first and second quadrants. Thus, the Q2 event is not
predominant there. This leads to low-speed fluid remaining adjacent to the
wall. This contributes to the decrease in the wall shear stress in the downhill 2.
Also, this decrease in the Q2 event contributes to the decrease in the Reynolds
shear stress.

Figure 8 exhibits the profile of streamwise fluctuating vorticity, ω_ξ', in
the η direction. The solid line shows the profile for the whole region, while

the broken line shows the profile for the downhill 2. High absolute values of vorticity are seen not only the near-wall region but also around $\eta^+ = 50$ in the downhill 2. This indicates some hairpin vortices were transported to the edge of the buffer region. This leads to a decrease in the Reynolds shear stress.

4 Conclusions

The direct numerical simulation was carried out for turbulent flow over the wall whose deformation was synchronized with flow pulsation. The main conclusions are as follows:

1. The wall shear stress and the viscous shear stress were decreased by the deforming wall. This is because the Q2 event is not predominant in the downhill region. The increase in the pressure drag due to the deformation was negligible.

2. The Reynolds shear stress and thus turbulence intensities were attenuated by the deforming wall. This is because some hairpin vortices were transported to the edge of the buffer region in the downhill and because of the attenuation of the Q2 event.

The authors acknowledge Mr. R. Matsubara at Echizen-Matsushima aquarium for his comments on dolphins.

References

1. Fish FE (2006) Bioinsp Biomim 1:R17-R25
2. Nagamine H, Yamahata K, Hagiwara Y, Matsubara R (2004) J Turbul 5, article no. 018:1-25
3. Zhang H, Hagiwara Y, Nakamura S, Matsubara R, Saito T (2006) Proc 3rd Int Symp on Aero Aqua Biomechanisms (CD-ROM), paper no.S43:1-8
4. Yamamoto Y, Kunugi T, Serizawa A (2001) J Turbul 2, article no. 010:1-16
5. Min T, Kang MS, Jason LS, Kim J (2006) J Fluid Mech 558:309-318
6. De Angelis V, Lombardi P, Banerjee S (1997) Phys Fluids 9:2429-2442
7. Kim J, Moin P, Moser R (1987) J Fluid Mech 177:133-166

On the Large Scale Evolution of Rotating Turbulence

Philip Staplehurst,[1] Peter Davidson[1] and Stuart Dalziel.[2]

[1] Department of Engineering, University of Cambridge, Trumpington Street, Cambridge, CB2 1PZ, UK

[2] Department of Applied Mathematics and Theoretical Physics, University of Cambridge, Wilberforce Rd, Cambridge, CB3 0WA, UK

Abstract. For a number of years the fundamental processes behind rotating turbulence have been debated. Whilst the Coriolis force appears as a linear term in the Navier-Stokes equation, many believe that the columnar eddies, which are so evident in experiments and simulations, are created by non-linear mechanisms. However, new findings have recently re-established the importance of linear processes, suggesting that, under typical laboratory situations, linear inertial waves play an important role in the formation of columnar eddies. These findings are both analytical and experimental.

Analytical work, conducted in the limit $Ro \to 0$, shows that an initially compact eddy evolves into two separate elongated structures, which propagate along the rotation axis. This change in the morphology of the eddy is achieved through inertial wave propagation, a prediction that has now been confirmed experimentally.

Our laboratory experiment consisted of a single grid oscillation in the bottom of a rotating tank of water. This created a cloud of turbulence from which elongated columnar structures emerged. These propagate linearly with time into the quiescent region above. The distance travelled by these structures was then tracked over a sequence of images, and the speed of propagation is found to be proportional to both the speed of rotation and the bar size of the grid, which is consistent with structure formation due to linear wave propagation.

Keywords: turbulence, rotating, inertial, waves, linear

1 Introduction

In this paper we are concerned with freely decaying, rotating turbulence. The development of the flow is characterised by the Rossby number, Ro, defined as $u/2\Omega l$, where u is the fluctuating velocity, Ω is the bulk rotation of the fluid and l is the large-scale turbulent lengthscale. We consider an initially turbulent field dominated by inertia ($Ro > 1$) in solid body rotation. As the turbulence then decays the Rossby number falls and the influence of the Coriolis force becomes more significant. Previous research [4], [5] has shown that once $Ro < 1$ the flow is dominated by columnar eddies, elongated in

Y. Kaneda (ed.), *IUTAM Symposium on Computational Physics and New Perspectives in Turbulence*, 391–396.

the direction of bulk rotation. The stability of these anisotropic structures is determined by their component of vorticity parallel to $\boldsymbol{\Omega}$; whilst anti-cyclonic structures are observed to rapidly destabilise, cyclonic structures tend to be stable and persist for far longer than the typical large eddy turnover time [1]. Consequently, the flow evolves towards a two-dimensional, symmetry breaking state.

In recent years, explanations for this phenomenon have relied on the existence of inertial waves. At any given time, the flow can be regarded as a collection of inertial waves, and as the turbulence develops it is the transfer of energy by these waves that drives the flow towards a two-dimensional state. The group velocity of these waves is set by the initial distribution of wavevectors, $\mathbf{k}$, according to

$$\mathbf{c}_g = \pm 2\mathbf{k} \times (\boldsymbol{\Omega} \times \mathbf{k})/|\mathbf{k}|^3. \tag{1}$$

Since the magnitude of the group velocity is largest parallel to $\boldsymbol{\Omega}$, one may tentatively expect the distribution of energy, from a random collection of inertial waves, to become elongated along the direction of rotation. In turn, this linear wave propagation argument gives a possible explanation for the evolution towards the two-dimensional state of rotating turbulence. However, a number of studies suggest that it is the non-linear transfer of energy between different inertial waves, rather than the linear propagation of individual waves, that produces the columnar structures. This non-linear hypothesis is based on near-resonant triadic interactions systematically shifting energy towards wavevectors in a plane perpendicular to the rotation axis. Nonetheless, we show here that columnar structure formation can be the result simply of linear wave propagation. This is established with theoretical work conducted in the limit of $Ro \to 0$, which is subsequently confirmed by experiment.

2 General Concepts in the Linear Limit

To highlight the importance of linear dynamics, let us consider a compact blob of vorticity, with arbitrary complexity, situated near the origin. For $t > 0$, energy will radiate away from the localised vorticity field in accordance with the group velocity and although, in general, energy will radiate away in all directions, we will show that the distribution of energy will (nearly) always be biased towards the direction of rotation. Our starting point is the governing equation for the angular momentum density, in the linearised, inviscid limit,

$$\partial(\mathbf{x} \times \mathbf{u})/\partial t = 2\mathbf{x} \times (\mathbf{u} \times \boldsymbol{\Omega}) + \nabla \times (p\mathbf{x}/\rho), \tag{2}$$

whose axial component is given by,

$$\partial(\mathbf{x} \times \mathbf{u})_z/\partial t = -\nabla \cdot \left[(\mathbf{x}^2 - z^2)\boldsymbol{\Omega}\mathbf{u}\right] + \left[\nabla \times (p\mathbf{x}/\rho)\right]_z. \tag{3}$$

Now let us integrate this expression over a cylindrical volume, V_R, of infinite length and radius R, centred on an axis parallel to $\boldsymbol{\Omega}$. By choosing this particular volume, both terms on the right-hand side of (3) integrate to zero;

the first term has no net contribution due to continuity and the second term must also be zero because pressure can not exert a net torque on the cylinder. Consequently, the axial component of angular momentum, H_z, within the cylindrical volume is conserved and linear waves cannot support a horizontal flux of axial angular momentum. Furthermore, if the cylindrical volume contains all of the vorticity, $\boldsymbol{\omega}$, at $t = 0$, H_z is equal to the angular impulse of the fluid:

$$H_z = \int_{V_R} (\mathbf{x} \times \mathbf{u})_z dV = \frac{1}{3} \int_{V_\infty} (\mathbf{x} \times (\mathbf{x} \times \boldsymbol{\omega}))_z \, dV. \tag{4}$$

where V_∞ is the entire volume. This confinement of axial angular momentum places crucial restrictions on how the energy within the initial blob of vorticity can be redistributed by inertial waves. Within the cylinder the characteristic velocity, $|\mathbf{u}|$, can fall no faster than $|\mathbf{u}| \sim t^{-1}$, whilst outside the cylinder inertial waves spread energy in all directions and so the characteristic velocity falls at a faster rate of $|\mathbf{u}| \sim t^{-3/2}$. Hence, through linear dynamics alone, inertial wave propagation provides a preferential channelling of energy parallel to the rotation axis. For a more detailed discussion, see [2].

3 Experiments on Rotating Turbulence

Having established the potential importance of linear dynamics, we now determine whether these concepts are relevant to the columnar structures which form in laboratory rotating turbulence. Previous work in [5] suggests that the Coriolis force begins to have a significant influence on the development of the flow once the Rossby number falls to the order of unity. At this point the non-linear inertial force and the Coriolis force are comparable in magnitude, and under these conditions it is not clear whether the flow can develop in a mannar predicted by Section 2.

To investigate this matter we have conducted a variant of the experiments of Dickinson and Long [3] who created a cloud of turbulence in the top of a tank of water by continually oscillating a grid. During the transition to a statistically steady state it was observed that the rate of turbulent diffusion, into the quiescent region below, depended on the strength of rotation. When the rotation was weak the turbulent cloud spread at a rate of $L_Z \sim t^{1/2}$, but when rotation was dominant, it spread at a faster rate of $L_Z \sim t$. As the turbulence was continually forced, it is unclear whether the linear diffusion rate was due to the evolution of the rotating turbulence or the direct emission of inertial waves from the oscillating surface.

We have repeated the experiment, except we have initiated the turbulence by a single oscillation of the grid in the bottom of the tank. The tank was 45cm square and 60cm deep, filled to a depth of 50cm (Fig. 1). To remove any mean circulation within the tank a square tube was placed on top of the grid, and both the tube and the grid were oscillated to create a cloud of turbulence. To prevent inertial waves being emitted from waves on the free surface a perspex tray was placed 5cm below the surface of the water. Two

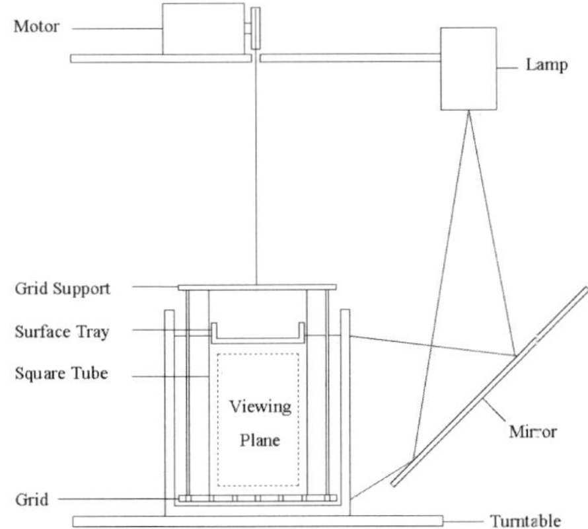

Fig. 1. Schematic of the experimental set-up. All apparatus shown rotates at a speed Ω.

grids were used, with mesh sizes, M, of 5cm and 8cm, 64% porosity, and a bar width, $b = M/5$. For each grid, experiments were conducted with rotation rates of $\Omega = 1$ and 2 rad/s. The single grid oscillation was based on a cycle of 0.6Hz, with an amplitude of 8cm.

To view the turbulence, approximately 2ml of Pearlescence was mixed into the water, until it was uniformly distributed. Pearlescence consists of light reflecting platelets, approximately 30μm in diameter and 6μm thick. In a region of shear the platelets continually rotate but at a non-constant speed, so that they spend most of their time closely aligned with the local shear direction. The platelets rotate due to their finite thickness; viscous forces act on the surfaces of the platelets, creating a torque and causing the platelets to flip over. As the majority of platelets will be aligned with the shear these regions can be identified by the relatively strong light that is reflected off them; in regions of weak shear the platelets are randomly orientated and any incoming light is scattered in all directions. For more information see [6].

The Pearlescence was illuminated by a 1kW halogen lamp forming a 1cm thick, vertical light-sheet, located in the middle of the tank. The reflected light, from the Pearlescence, was then recorded using a digital camera. Both the lamp and camera were mounted on top of the turntable, and 45° mirrors were used to produce and record the images respectively (only the lamp is shown in figure 1 for simplicity).

Figure 2 shows a typical sequence of images from a single experiment at $2\Omega t = 17.5$, 23.6 and 35. From these images we can clearly see elongated structures that have emerged from the decaying turbulent cloud and propagate into the region above. To ensure that these structures have not been emitted directly from the bars of the grid, each experiment was conducted fifty times. The images in Fig. 2 were then produced by taking single images from one experiment and subtracting the ensemble average images of all fifty experiments. As the structures are markedly visible, they must have been produced

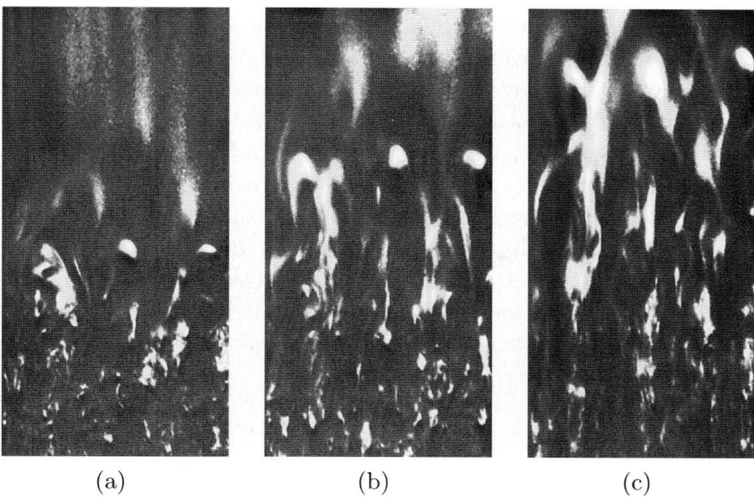

(a)   (b)   (c)

Fig. 2. Images of the flow taken at different times after the single grid oscillation. For this particular experiment $M = 8$cm and $\Omega = 2$ rad/s. (a) $2\Omega t = 17.5$, (b) $2\Omega t = 26.3$, (c) $2\Omega t = 35$. Image dimensions are 40cm x 20cm.

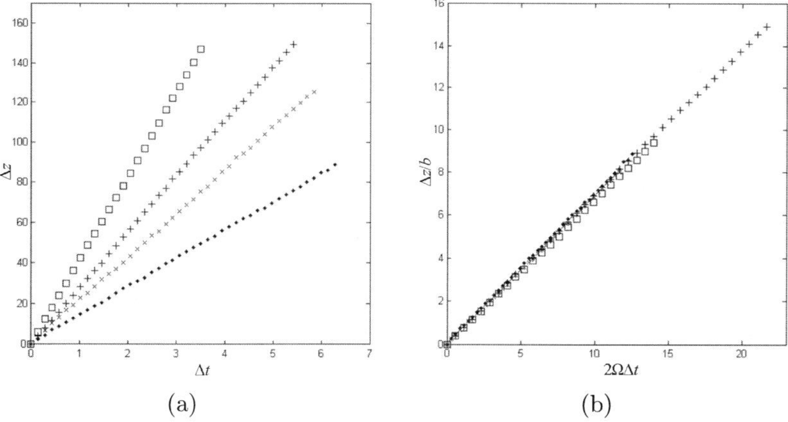

(a)   (b)

Fig. 3. The distance travelled by the leading edge of typical columnar structures as a function of time. The graph on the right has Δz and t normalised by bar size, b, and Ω^{-1}, respectively. $\square$ ($M = 8$cm, $\Omega = 2$rad/s), $\times$ ($M = 8$cm, $\Omega = 1$rad/s), $+$ ($M = 5$cm, $\Omega = 2$rad/s), $\bullet$ ($M = 5$cm, $\Omega = 1$rad/s).

by the non-repeatable initiation of the turbulence rather than the systematic grid oscillation

The evolution of these columnar structures was determined for both rotation rates and bar sizes. By identifying a constant value of light intensity at the tip of these structures, their location was determined to an accuracy of better than ± 2mm. Figure 3a shows the distance travelled by the leading edge of the columnar structures, Δz, as a function of time, with each set of data being an ensemble average of ten nominally identical structures. In all cases the structures propagate linearly with time. Furthermore, in order to compare them with the linear theory presented in §2, Fig. 3b shows the same data but with distance normalised by bar size and time normalised by rotation rate. According to (1), if the structures have been produced by linear dynamics, such a normalisation should collapse the data from all of the experiments onto a straight line, and indeed it does. In turn, it can be concluded that these elongated structures are created by the constructive interference of inertial waves. At later times, the entire tank is filled with these elongated structures and the flow is reminiscent of homogeneous rotating turbulence [2].

4 Conclusions

Linear dynamics are able to evolve an inhomogeneous cloud of turbulence into a two-dimensional state dominated by columnar structures. The propagation rate of these structures is constant and proportional to both rotation rate and bar size, which is in accordance with linear theory.

References

1. Bartello P, Metais O, Lesieur M (1994) J Fluid Mech 237:1–29
2. Davidson P, Staplehurst P, Dalziel S (2006) J Fluid Mech 557:135–144
3. Dickinson S, Long R (1983) J Fluid Mech 126:315–333
4. Hopfinger E, Browand F, Gagne Y (1982) J Fluid Mech 125:505–534
5. Jacquin L, Leuchter O, Cambon C, Mathieu J (1990) J Fluid Mech 220:1–52
6. Savas O (1985) J Fluid Mech 152:235–248

Effects of the Slow Modes in the Differential Diffusion in Stratified Sheared Turbulence

Hideshi Hanazaki

Department of Mechanical Engineering and Science, Kyoto University, Kyoto
606-8501, Japan
hanazaki@mech.kyoto-u.ac.jp

Abstract. Differential diffusion of a passive scalar and an active scalar/density in
stratified shear flow is considered when both the density and the passive scalar have
a mean vertical gradient. The vertical diffusion coefficient of the passive scalar
$K_c = -\overline{cu_3}(t)/(d\bar{c}/dx_3)$ (c: passive scalar fluctuation, $\bar{c}(x_3)$: mean passive scalar
distribution) has been obtained by the rapid distortion theory (RDT), and the
comparison with the density diffusion coefficient $K_\rho = \overline{\rho u_3}/(d\bar{\rho}/dx_3)$ (ρ: density
fluctuation, $\bar{\rho}(x_3)$: mean density distribution) shows that the difference between K_c
and K_ρ comes from the initial density fluctuations. Namely, passive scalar flux has a
'slow mode' which oscillates at half frequency of the density flux, whose magnitude
is proportional to the initial potential energy due to density fluctuations.

Keywords: differential diffusion, stratified flow, passive scalar, turbulence,
slow mode

1 Introduction

Differential diffusion of a passive scalar and an active scalar/density in stratified shear flow has recently become an important problem in the ocean circulation modeling [1], since the difference in the turbulent diffusivity coefficient
in the numerical model leads to the large difference in the mean circulation
velocity, mean temperature and the mean salinity distribution of the global
ocean.

In ocean models, equal value of turbulent diffusivity coefficients has been
traditionally used for salt (K_S) and temperature (K_T), on the assumption that
the Reynolds number is sufficiently high. However, some differences could be
expected since the molecular diffusivities are very different between temperature and salt. Indeed, recent observations [2], laboratory experiments [3] and
direct numerical simulations [4], [5] show that the turbulent diffusivities are
affected by the molecular diffusivities, with the ratio as low as $K_S/K_T \sim 0.1$.

In this study, diffusion of passive scalar in stratified turbulence is considered when both the density and the passive scalar have a mean vertical
gradient. Extending the solution obtained by the rapid distortion theory for

397

Y. Kaneda (ed.), *IUTAM Symposium on Computational Physics and New Perspectives in
Turbulence*, 397–401.

stratified sheared turbulence [2], we have obtained the turbulent diffusion coefficient of the passive scalar K_c, and considered the possibility of its difference from the density diffusion coefficient K_ρ.

For inviscid flow, the results for K_c and K_ρ show dependence on initial conditions which is similar to the unsheared stratified flow [6]. Namely, passive scalar flux $\overline{cu_3}$ has a 'slow mode' oscillating at the half frequency of the density flux $\overline{\rho u_3}$ only if there is an initial potential energy due to density fluctuations ($PE_0 \neq 0$). We call this mode the 'inviscid slow mode'.

We will show how this initial condition leads to the difference in the turbulent fluxes of passive and active scalars, which are often assumed to be the same in many geophysical and engineering applications.

2 Solutions by RDT Equations and the Scalar Fluxes

The governing equations are the Boussinesq equations for the momentum, advection equations for the density and passive-scalar, and the condition of incompressibility. These equations can be written as

$$\frac{\partial \boldsymbol{u}}{\partial t} + (\boldsymbol{u} \cdot \nabla)\boldsymbol{u} = -\frac{1}{\rho_0}\nabla p - g\hat{x}_3 \frac{\rho}{\rho_0} + \nu\nabla^2 \boldsymbol{u}, \tag{1}$$

$$\frac{\partial \rho}{\partial t} + (\boldsymbol{u} \cdot \nabla)\rho + u_3 \frac{d\overline{\rho}}{dx_3} = \kappa\nabla^2\rho, \tag{2}$$

$$\frac{\partial c}{\partial t} + (\boldsymbol{u} \cdot \nabla)c + u_3 \frac{d\overline{c}}{dx_3} = D\nabla^2 c, \tag{3}$$

$$\nabla \cdot \boldsymbol{u} = 0, \tag{4}$$

where $\boldsymbol{u} = (u_1, u_2, u_3)$ is the velocity, $\rho_0 (= \overline{\rho}(x_3 = 0))$ is the mean density, ρ and c are the density and passive scalar perturbation from $\overline{\rho}(x_3)$ and $\overline{c}(x_3)$ respectively, g is the acceleration due to gravity, and $\hat{x}_3$ is the unit vector in the vertical direction. We substitute the spectral decompositions into the governing equations (1)-(4) and obtain equations for the spectral components $\hat{u}_i$, $\hat{\rho}$ and $\hat{c}$, where $\hat{\rho}$ is the Fourier component of $(g/\rho_0)\rho$, and three dimensional spectra such as $\Phi_{\rho 3}$ and the flux $\overline{\rho u_3}$ are defined using this $\hat{\rho}$.

If we neglect the nonlinear terms, we obtain the equations in the frame work of the rapid distortion theory (RDT).

$$\left(\frac{d}{dt} + \nu k^2\right)\hat{u}_i = S\hat{u}_3\left(\frac{2k_i k_1}{k^2} - \delta_{i1}\right) + \left(\frac{k_i k_3}{k^2} - \delta_{i3}\right)\hat{\rho}, \tag{5}$$

$$\left(\frac{d}{dt} + \kappa k^2\right)\hat{\rho} = N^2\hat{u}_3, \tag{6}$$

$$\left(\frac{d}{dt} + Dk^2\right)\hat{c} = -\gamma\hat{u}_3, \tag{7}$$

where $N^2 = -(g/\rho_0)(d\bar\rho/dx_3)$ is the Brunt-Väisälä frequency and $\gamma = d\bar{c}/dx_3$ is the vertical gradient of the mean passive scalar, both of which are the constants, and $S = dU(x_3)/dx_3$ is the mean shear.

The applicability condition of this linear theory to the stratified shear flow with a passive scalar would be a combination of the applicability conditon for the stratified shear flow and the stratified flow with a passive scalar diffusion. Both gives the condition $Fr(l) = u(l)/Nl \ll 1$ (if $Ri = N^2/S^2 \le O(1)$ and $u_3/u_1 = O(1)$) [6],[7]. The condition of $u_3/u_1 = O(1)$ is usually satisfied in DNS[8] and experiments[9]. If $Fr(l) \ll 1$ is satisfied at the integral scale ($l = L_0$), namely if $Fr(L_0) \ll 1$, RDT would be applicable to the energy-containing scales with larger scale than L_0, since $l/u(l)$, i.e., the time scale of an eddy with size l, is usually an increasing function of l. Then, the energies and fluxes would be well approximated by RDT. We can conclude that $Fr(L_0) \sim u_{rms}/NL_0 \ll 1$ (if $Ri = O(1)$) is the applicability condition of RDT to the fluxes and energies in stratified shear flow with passive scalar diffusion.

The solutions of RDT equations (5)–(7) give the spectra as

$$\hat\rho(t) = A\,P_\nu(z) + B\,Q_\nu(z), \quad \text{(Legendre functions } P_\nu(z),\, Q_\nu(z)), \tag{8}$$

$$\hat{u}_3(t) = -\frac{iS\cos\phi_0}{N^2}\left(A\,P'_\nu(z) + B\,Q'_\nu(z)\right), \tag{9}$$

$$\hat{c}(t) = \hat{c}(0) - \frac{\gamma}{N^2}\left(A\,P_\nu(z) + B\,Q_\nu(z) - \hat\rho(0)\right), \tag{10}$$

where

$$\nu = \frac{1}{2}\left[-1 \pm \left(1 - 4\frac{Ri}{\cos^2\phi_0}\right)^{1/2}\right], \tag{11}$$

$$A = \frac{1}{\sin^2\theta_0}Q'_\nu(z_0)\hat\rho_0 + \frac{N^2}{iS\cos\phi_0\sin^2\theta_0}Q_\nu(z_0)\hat{u}_{30}, \tag{12}$$

$$B = -\frac{1}{\sin^2\theta_0}P'_\nu(z_0)\hat\rho_0 - \frac{N^2}{iS\cos\phi_0\sin^2\theta_0}P_\nu(z_0)\hat{u}_{30}, \tag{13}$$

$$z = i(\cot\theta_0 - St\cos\phi_0), \tag{14}$$

and z_0 is the initial value of z and prime ($'$) denotes the derivative ($\partial/\partial z$).

Then, the vertical passive scalar flux and the vertical density flux can be obtained by integration in the whole spectral space as

$$\overline{cu_3}(t) = \frac{1}{2}\overline{\hat{c}^*\hat{u}_3 + \hat{c}\hat{u}_3^*}$$

$$= -\frac{S}{4\pi N^2}\overline{c\rho}(0)\int d\theta_0 d\phi_0 \frac{\cos\phi_0}{\sin\theta_0}\operatorname{Re}\left[i(P'_\nu(z)Q'_\nu(z_0) - P'_\nu(z_0)Q'_\nu(z))\right]$$

$$-\frac{S\gamma}{2\pi N^2}PE_0\int d\theta_0 d\phi_0 \frac{\cos\phi_0}{\sin^3\theta_0}\operatorname{Re}[i(P_\nu(z)Q'_\nu(z_0) - P'_\nu(z_0)Q_\nu(z) - \sin^2\theta_0)$$

$$\times(P'_\nu(z)Q'_\nu(z_0) - P'_\nu(z_0)Q'_\nu(z))^*]$$

$$- \frac{\gamma}{4\pi S} KE_0 \int d\theta_0 d\phi_0 \frac{1}{\sin\theta_0 \cos\phi_0} \mathrm{Re}[\mathrm{i}(P_\nu(z)Q_\nu(z_0) - P_\nu(z_0)Q_\nu(z))$$
$$\times (P'_\nu(z)Q_\nu(z_0) - P_\nu(z_0)Q'_\nu(z))^*]. \tag{15}$$

$$\overline{\rho u_3}(t) = \frac{S}{2\pi} PE_0 \int d\theta_0 d\phi_0 \frac{\cos\phi_0}{\sin^3\theta_0} \mathrm{Re}[\mathrm{i}(P_\nu(z)Q'_\nu(z_0) - P'_\nu(z_0)Q_\nu(z))$$
$$\times (P'_\nu(z)Q'_\nu(z_0) - P'_\nu(z_0)Q'_\nu(z))^*]$$
$$+ \frac{N^2}{4\pi S} KE_0 \int d\theta_0 d\phi_0 \frac{1}{\sin\theta_0 \cos\phi_0} \mathrm{Re}[\mathrm{i}(P_\nu(z)Q_\nu(z_0) - P_\nu(z_0)Q_\nu(z))$$
$$\times (P'_\nu(z)Q_\nu(z_0) - P_\nu(z_0)Q'_\nu(z))^*]. \tag{16}$$

Comparison of (15) and (16) shows that if $PE_0 \neq 0$ or $\overline{c\rho}(0) \neq 0$, passive scalar flux K_c has additional underlined terms which are not contained in K_ρ. Those terms would oscillate at half frequency of the other terms since they have the factor of PQ and not of $(PQ)^2$.

Figure 1 shows the difference between the correlation coefficients of density ($R_{\rho3} = \overline{\rho u_3}/\rho' u'_3$) and passive scalar ($R_{c3} = \overline{cu_3}/c'u'_3$) at the Richardson number of $Ri = N^2/S^2 = 1$. This figure illustrates the 'period doubling' in the passive scalar flux due to the slow mode. If we compare the results with the unsheared flow[6], we note that the effect of initial density fluctuations (PE_0) to the period doubling is similar in sheared and unsheared flow.

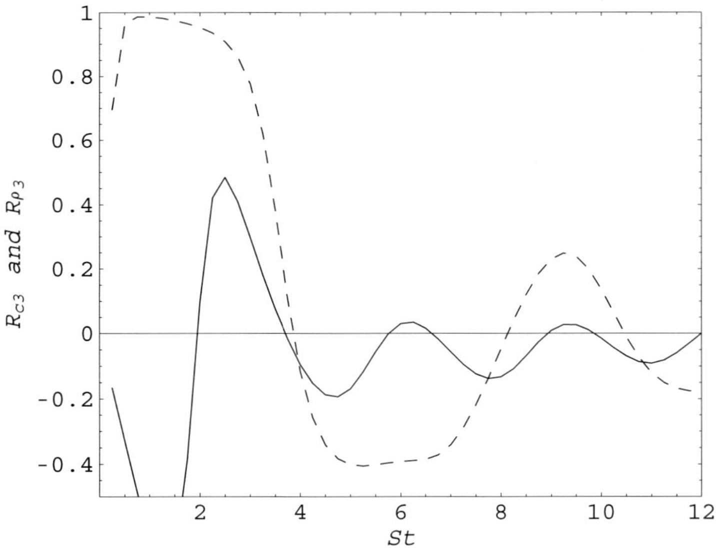

Fig. 1. Time development of the correlation coefficients of density ($R_{\rho3}$: solid line) and passive scalar (R_{c3}: dashed line) when the initial condition is the almost 'fossil' turbulence, i.e. $PE_0/KE_0 = 10^4 (\gg 1)$ (PE_0: initial potential energy of turbulence due to density fluctuations, KE_0: initial kinetic energy). Time is scaled by the mean vertical shear S.

Then, the effect of initial conditions are more essential than shear in making difference between K_ρ and K_c. Although the results only for a very large $PE_0/KE_0(= 10^4)$ is given here, clear period doubling could be clearly observed at much lower values of $PE_0/KE_0(\sim 0.5)$. Indeed, in unsheared flow, only the slow mode exists when $PE_0/KE_0 = 0.5$.

3 Summary

Passive scalar diffusions in turbulent stratified shear flow is considered when there is a mean vertical passive scalar gradient. Solutions of the linearized governing equations, i.e. RDT equations, show that the passive scalar flux becomes different from the density flux when there are initial density fluctuations (turbulent potential energy), and the main difference is in the period doubling. Since the slowly oscillating mode becomes significant even at small values of $PE_0/KE_0 \sim 0.1$ which are usually observed in laboratory experiments, the period doubling would appear in very general conditions. Although the solutions given here are limited to the inviscid fluid, extentions to the case of $Pr = Sc = 1$ is easy, and we can also expect similar effects in the more general cases of $Pr \neq 1$ and $Sc \neq 1$.

From the oceanic point of view, time-averaged value of K_c and K_ρ would be important. Since both K_c and K_ρ decay with time, initial difference due to the different oscillation period leads to $K_c/K_\rho \neq 1$ also in a long-time average, suggesting the importnace of $K_c/K_\rho \neq 1$ in the real ocean.

References

1. Gargett AE (2003) Progress in Oceanography 56:559–570
2. Nash JD, Moum JN (2002) J Phys Oceanogr 32:2312–2333
3. Jackson PR, Rehmann CR, Sáenz JA, Hanazaki H (2005) Geophys Res Lett 32:L10601
4. Gargett AE, Merryfield WJ, Holloway G (2003) J Phys Oceanogr 33:1758–1782
5. Hanazaki H, Konishi K (2007) Differential diffusion in double-diffusive stratified turbulence. European Turbulence Conference 11
6. Hanazaki H (2003) Phys Fluids 15:841–848
7. Hanazaki H, Hunt JCR (2004) J Fluid Mech 507:1–42
8. Métais O, Herring JR (1989) J Fluid Mech 202:117–148
9. Lienhard JH, Van Atta CW (1990) J Fluid Mech 210:57–112

Numerical Study of 3-D Free Convection under Rotation: Mean Wind and Bolgiano–Obukhov Scaling

Takeshi Matsumoto, Takuya Mizukami, and Sadayoshi Toh

Division of Physics and Astronomy, Graduate School of Science, Kyoto University, Kitashirakawa Oiwakecho Sakyoku Kyoto 606-8502, Japan
`takeshi@kyoryu.scphys.kyoto-u.ac.jp`

Abstract. 3D thermal convection turbulence in a triply periodic domain with zero mean-temperature gradient under rotation is numerically studied. The system is forced by a large scale temperature forcing. The rotation axis is perpendicular to the direction of the gravity since the rotation is a model of the mean wind observed in recent experiments. With rotation the Bolgiano–Obukhov spectra ($k^{-11/5}$ for the kinetic energy and $k^{-7/5}$ for the temperature variance) are observed for a range of Ω, whereas the Kolmogorov and Obukhov–Corrsin spectra ($k^{-5/3}$ for both) are obtained without rotation. Flow structures in the presence of rotation are also discussed.

Keywords: convection turbulence, rotating turbulence, Bolgiano–Obukhov scalings, mean wind

1 System Studied

We perform a direct numerical simulation (DNS) of three-dimensional incompressible Boussinesq equations in a rotating frame with constant angular velocity $\boldsymbol{\Omega}$:

$$\partial_t \mathbf{u} + (\mathbf{u} \cdot \nabla)\mathbf{u} = -\nabla p/\rho_0 + \alpha \mathbf{g} T + 2\boldsymbol{\Omega} \times \mathbf{u} + (-1)^{h+1} \nu \nabla^{2h} \mathbf{u} + D, \quad (1)$$

$$\nabla \cdot \mathbf{u} = 0, \quad (2)$$

$$\partial_t T + (\mathbf{u} \cdot \nabla)T = (-1)^{h+1} \kappa \nabla^{2h} T + f. \quad (3)$$

Here $\mathbf{u}, p, \rho_0, \alpha, \mathbf{g}, T, \nu, \kappa$ are respectively, the velocity, pressure, (constant) density, thermal expansion coefficient, gravitational acceleration, temperature, hyper kinematic viscosity and hyper thermal diffusivity. The order of hyper viscosity/diffusivity is $h = 8$ and Prandtl number unity $\nu = \kappa$ case is considered here. The terms D in (2) and f in (3) are respectively the large scale friction acting on the velocity and the large scale forcing acting on the temperature [1]. The boundary condition of the system is periodic in all three directions. Notice that no mean temperature gradient is present (neutrally

403

Y. Kaneda (ed.), *IUTAM Symposium on Computational Physics and New Perspectives in Turbulence*, 403–407.
© 2008 *Springer*.

stable stratification). Instead, the system is driven by the large scale forcing f (convection in a periodic domain without mean temperature gradient is henceforth called free convection). We here consider the geometrical setting that the rotation axis is perpendicular to the gravity: $\Omega \perp \mathbf{g}$. Specifically we take $\Omega = (0, 0, \Omega)$ and $\mathbf{g} = (g \sin \Omega t,\ g \cos \Omega t,\ 0)$ in the rotating frame. A fully dealiased spectral method with 4th order Runge-Kutta scheme is employed.

We focus on (i) spectral properties, in particular whether the Bolgiano–Obukhov scaling (BO scaling) is observed or not and (ii) spontaneous flow organizations emerged in this system. Our motivation for this rotating free convection comes from the recent experimental discoveries, which shall be explained below.

2 Motivation

Recent experimental studies of high Rayleigh number convection turbulence have revealed existence of a mean flow circulating in the convection container [2]. It was also shown that this mean flow (so-called mean wind) changes its direction in a rather short time and this reversal transition occurs not in a periodic way. Later the direct measurement of the instantaneous velocity field inside the convection container was made without disrupting the flow [3]: not only characteristics of the mean wind were investigated thoroughly [4], but also the kinetic energy spectrum $E(k)$ of the small-scale flow was obtained without using Taylor's frozen flow hypothesis. Solid evidence is obtained that the velocity in the small scale follows the BO scaling $E(k) \propto k^{-11/5}$ [3].

However the BO scaling is hardly observed in DNS of 3-D convection turbulence in a periodic domain (without rotation) [5]. Here, based on these solid experimental findings, it is conceivable that the BO scaling in the small scale and the mean wind in the large scale are linked in some important manner. We study this point via DNS of (1)–(3). The reason why we study the free convection system under the rotation is that the periodic boundary condition and zero mean temperature gradient can be thought as a model of the central region (region far from the boundary layers) of the convection container; the large scale forcing f in (2) reflects injection of large-scale thermal plumes in the region; we further assume that the mean wind is felt in the central region as solid body rotation with the rotation axis perpendicular to the direction of the gravity (the issue of the reversal of the mean wind is not addressed here). In other words, the system we simulate numerically is a mean-wind-effect-incorporated model for the velocity and the temperature in the small scales in the central region of high Rayleigh number convection turbulence.

3 Results

In 2D, a free convection DNS (without rotation) produced the BO scaling [7]. However in 3D, our free convection DNS without rotation does not produce the BO scaling but the Kolmogorov and Obukhov–Corrsin scaling (Fig. 1(a))

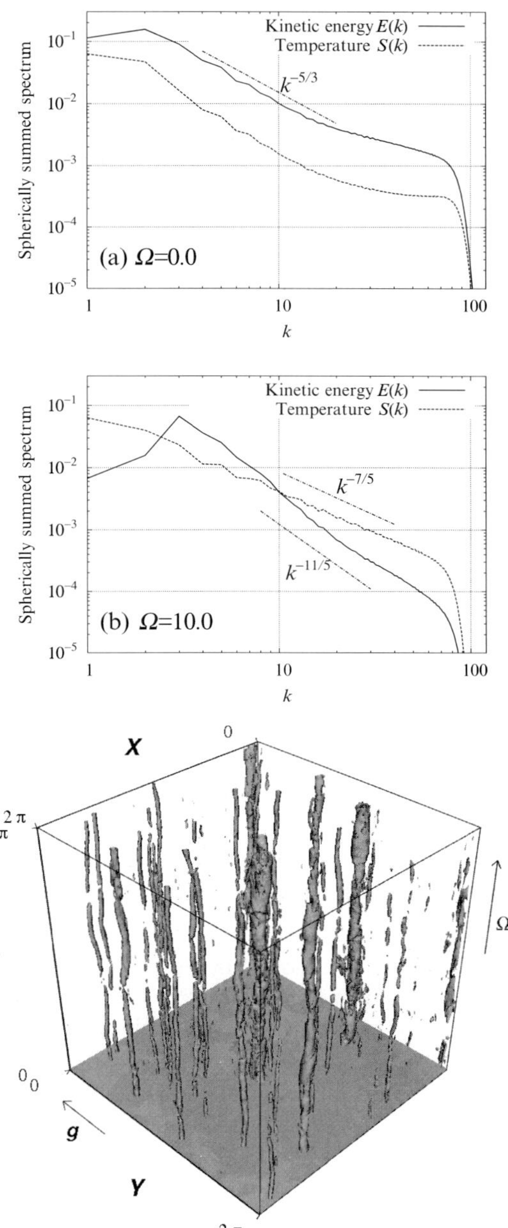

Fig. 1. (a) Energy and temperature variance spectra for $\Omega = 0.0$ (resolution 256^3). Bottleneck effect is pronounced in the large wavenumber region. (b) For $\Omega = 10.0$ (256^3). (c) Vorticity modulus $|\omega|$ isosurface for $\Omega = 10.0$, which is basically the same as that of the large positive vorticity component parallel to the rotation axis (256^3). In the $z = 0$ plane, the temperature field is shown. Plume structure is seen. Both the non-rotating (a) and the rotating (b) cases have the same initial condition and the same parameters except for Ω.

even though the turbulence is driven by the large-scale temperature forcing. Indeed, without rotation, the energy cascade dominates over the temperature variance cascade. In such a case the BO scaling does not hold according to the Bolgiano–Obukhov phenomenology. Whereas, in the presence of strong rotation, we observe the spectra consistent with the BO scaling (Fig. 1(b)), suggesting that in 3D the mean wind or some kind of large-scale effect is crucial for the BO scaling in small scales. In this case, the temperature variance cascade overwhelms that of the energy. In terms of the Rossby number based on the temperature input and forcing, $Ro_f = [\epsilon_\theta k_f^2 (\alpha g)^2 / \Omega]^{1/5}$ (ϵ_θ: temperature variance dissipation rate, k_f: forcing wavenumber ($= 2$ here)), the BO scaling is observed around $Ro_f \simeq 0.1$.

Another parameter for the BO scaling is the Bolgiano–Obukhov wavenumber $k_{BO} = [(\alpha g)^6 \epsilon_\theta^3 / \epsilon^5]^{1/4}$, where ϵ is the energy dissipation rate. In the wavenumber range $k < k_{BO}$, we can expect the BO scaling. We observe that, as we increase Ω, the Bolgiano–Obukhov wavenumber increases. In the non-rotating case shown in Fig. 1(a), $k_{BO} \simeq 1$. While in the case shown in Fig. 1(b), $k_{BO} \simeq 60$, which indeed gives a range for the BO scaling.

The suppression of the energy cascade down to small scales is due to "two dimensionalization" of the flow caused by the rotation (see the physical space structure shown in Fig. 1(c)). Under strong rotation, we observe that the nonlinear transfer of the energy changes its direction, namely going up to large scales. This can be a signature of the "two dimensionalization". In the physical space, as we can see from Fig.1 (c), strong asymmetry emerges in the distribution of ω_z, the vorticity component parallel to the rotation axis (z-axis). We speculate that such asymmetric organization formed in the central region of the convection may give positive feedback on sustainment of the mean wind. Detailed study about Rossby number dependence and anisotropy of the flow will be reported elsewhere.

Acknowledgement. This study is supported by the Grant-in-Aid for the 21st Century COE "Center for Diversity and Universality in Physics" from the Japanese Ministry of Education. The computation is done on SX8 at Yukawa Institute for Theoretical Physics in Kyoto University.

References

1. The friction D employed here is given in the Fourier space by $-d_0 \hat{\mathbf{u}}(\mathbf{k})/|\mathbf{k}|^2$ which is effective only in $0 < |\mathbf{k}| \leq 3$ (d_0 is a constant). This friction is needed to avoid accumulation of the energy in large scales, which is caused by rotation. The forcing f is realized by keeping the temperature Fourier coefficients in $0 < |\mathbf{k}| \leq 2$ constant.
2. Niemela JJ, Skrbek L, Sreenivasan KR, Donnelly RJ (2001) J Fluid Mech 449:169–178
3. Mashiko T, Tsuji Y, Mizuno T, Sano M (2004) Phys Rev E 69:036306

4. Tsuji Y, Mizuno T, Mashiko T, Sano M (2005) Phys Rev Lett 94:034501
5. Borue B, Orszag SA (1997) J Sci Comp 12:305–351 (for a case of unstable stratification in a periodic domain without rotation.)
6. Siggia ED (1994) Ann Rev Fluid Mech 26:137–168
7. Toh S, Suzuki E (1994) Phys Rev Lett 73:1501–1504

The Horizontal Energy Spectra and Cascade Processes in Rotating Stratified Turbulence

Yuji Kitamura[1] and Yoshihisa Matsuda[2]

[1] Division of Earth and Planetary Sciences, Graduate School of Science,
Kyoto University, Sakyo-ku, Kyoto 606-8502, Japan
`kitamura@kugi.kyoto-u.ac.jp`

[2] Department of Astronomy and Earth Science, Tokyo Gakugei University,
Koganei-city, Tokyo 184-8501, Japan
`ymatsuda@u-gakugei.ac.jp`

Abstract. Numerical experiments on rotating stratified turbulence are conducted with a high-resolution model to investigate the horizontal energy spectrum seen in the troposphere. Both the k_H^{-3} and $k_H^{-5/3}$ regimes observed in the atmosphere are reproduced in our experiments by downscale cascade from the synoptic scale at which forcing is given. The transition scale between the two regimes becomes smaller with the Rossby deformation radius NH/f. The energy associated with divergent motion is predominant in the $k_H^{-5/3}$ region and the triads composed of one vortical and two divergent modes mainly contribute to energy transfer to the higher wavenumbers.

Keywords: rotating stratified turbulence, spectral power law, energy cascade, numerical experiment

1 Introduction

It is well known that the horizontal energy spectrum derived from the observations in the troposphere follows the k_H^{-3} power law in the synoptic scales and the $k_H^{-5/3}$ in the mesoscales and the transition between these spectral ranges exists at about 500 km [7]. While a formation mechanism of the k_H^{-3} spectrum has been regarded as enstrophy cascade appearing in quasi-geostrophic turbulence [2], the interpretation of the $k_H^{-5/3}$ spectrum is still controversial. [5] proposed that upscale energy cascade is allowed in stratified turbulence as in 2D turbulence. On the other hand, [8] interpreted the mesoscale spectrum in terms of downscale cascade of inertial gravity waves. Energy cascade in stratified turbulence have been investigated numerically by several authors, e.g. [1], [3]. [6] found from numerical experiments that a $-5/3$ spectral slope can be obtained from downscale cascade produced by forcing at large scale in rotating stratified turbulence. While most of numerical studies are focused on only the $k_H^{-5/3}$ spectral range, [4] showed that both -3 and $-5/3$ slopes can

409

Y. Kaneda (ed.), *IUTAM Symposium on Computational Physics and New Perspectives in Turbulence*, 409–414.

be obtained by downscale cascade from the synoptic scale in rotating stratified turbulence. They found that the divergent motion plays an essential role
in the formation of the $k_H^{-5/3}$ spectrum even if the external forcing excites
only geostrophic modes. In the present study, we focus on dependence of the
energy spectrum on stratification and energy transfer to divergent motion,
which are not treated in [4].

2 Numerical Procedure

We adopt a three-dimensional Boussinesq model on an f-plane with the horizontal domain size of 6000 km $\times$ 6000 km and vertical one of 10 km. The
boundary conditions are assumed to be cyclic in the horizontal directions and
rigid at the top and bottom. In the present study, we focus on the terrestrial
parameter range and perform experiments with $2\pi/N = 10$ or 20 minutes (N:
the Brunt-Väisälä frequency) of the buoyant oscillation period and $f = \Omega$ or
2Ω (f: the Coriolis parameter, Ω: the terrestrial angular velocity). Integrations
for 30 days are carried out with a low resolution model with $600 \times 600 \times 40$
grid points for a spin-up process and subsequent integrations for more five
days are done with a high resolution model with $2000 \times 2000 \times 80$ grid points.
The results for the last two days in the high resolution model are analyzed.
[9] pointed out that the vertical grid spacing should be sufficiently smaller
than the overturning vertical scale U/N. We confirm a posteriori that this
condition is satisfied in all our simulations.

In the present study, we focus on energy cascade process from a synoptic
scale. The forcing distribution is confined in lower wavenumbers and the horizontal scale corresponding to the spectral peak is the domain size, as assumed
in [4]. A dynamical forcing is composed of geostrophic modes only so that
the nearly balanced motion is excited. We assume that the vertical structure
of the forcing is of the first baroclinic mode, in which the vertical structure
of the horizontal velocity components has one node. A fourth-order linear
hyperviscosity in the horizontal direction and an usual eddy viscosity of the
Laplacian form in the vertical one are adopted for representing small scale
dissipation. The horizontal and vertical eddy viscosity coefficients are set to
$\nu_H = 6.0 \times 10^8$ [m^4/s] and $\nu_V = 2.5 \times 10^{-1}$ [m^2/s] for the higher resolution
model.

3 Results

Figure 1 shows the energy spectra obtained as a function of horizontal
wavenumber. Here the energy spectra are averaged over the vertical column.
The k_H^{-3} and $k_H^{-5/3}$ spectral ranges in the total energy are clearly found in
all the experiments conducted in the present study. As the velocity field is
decomposed into vertical component of vorticity and horizontal divergence
component, kinetic energy can be described as a sum of rotational and divergent energies. While the rotational energy is predominant in the -3 spectrum

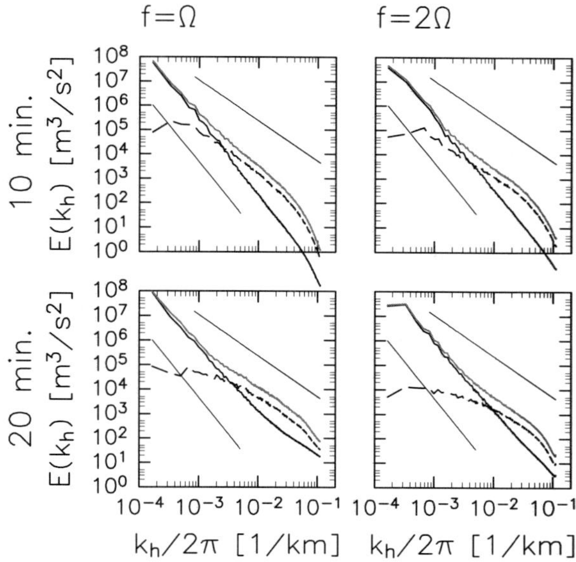

Fig. 1. Energy spectra as a function of horizontal wavenumber. The results for $2\pi/N = 10$ and 20 minutes are illustrated in the upper and lower panels. The solid, dashed and gray lines represent the rotational and divergent parts of kinetic energy and total energy, respectively. Two straight lines in each panel indicate $-5/3$ (upper one) and -3 (lower one) slopes.

regime, the divergent one is predominant in the $-5/3$ range. The spectrum transition can be characterized by the intersection of the rotational and divergent spectra, as is shown in [4]. The rotational energy spectrum has a transition from the -3 to $-5/3$ slopes at $k_H/2\pi \sim 10^{-2}$ for the $2\pi/N = 20$ [min.], $f = 2\Omega$ case. However, this scale is smaller than the transition scale of the total energy spectrum and the magnitude of the rotational energy in the transition scale of the rotational part has minor importance in the experiments. While the energy level of the rotational component does not depend on planetary rotation nor stratification in the scales larger than 100 km in all the experiments, the divergent energy is smaller in the case with faster planetary rotation or weaker stratification. As the result, the transition scale varies by the divergence energy and is qualitatively proportional to the Rossby deformation radius $L_D = NH/f$. This result is consistent with the empirical fact that the horizontal scale in which a geostrophic motion is predominant can be characterized by L_D. However, the scale where the spectral transition occurs could not be diagnosed only by the deformation radius, because the transition scale largely depend on the external forcing, as is reported by [4].

A snapshot of the potential vorticity (PV) and the vertical velocity on a horizontal plane is presented in Fig. 2. The PV is materially conserved and defined as $Q \equiv (f\mathbf{k} + \nabla \times \mathbf{u}) \cdot (\Gamma\mathbf{k} + \nabla\theta')$ ($\Gamma \equiv \mathrm{d}\bar{\theta}/\mathrm{d}z$, $\bar{\theta}$: the temperature

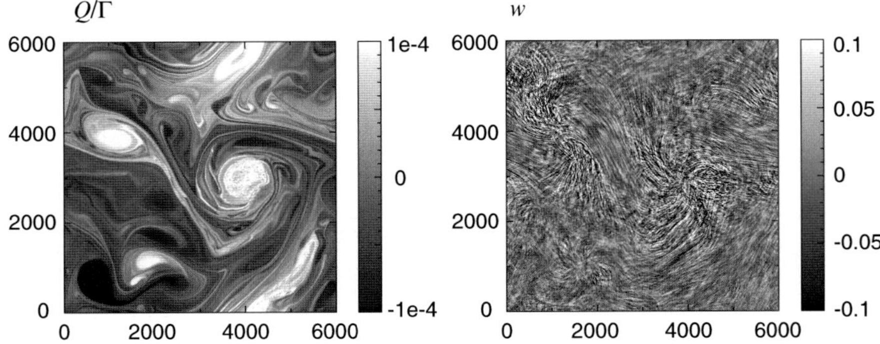

Fig. 2. A snapshot of Q/Γ and the vertical velocity field w at $z = 1250$ [m] in the case of $2\pi/N = 10$ [min.] and $f = 2\Omega$. Note that Q/Γ is simply proportional to the potential vorticity Q, since Γ is constant in the present model.

of the basic state, θ': the temperature disturbance). The PV and vertical velocity fields roughly represent balanced and imbalanced motions, respectively. While the motion associated with the PV is predominant in the large scales, the vertical velocity field has structure of much smaller scale. The characteristic scales of these motions just correspond to the energy spectrum seen in Fig. 1; the rotational and divergent energies predominate in the synoptic and smaller scales, respectively. The PV field evolves slowly and a filament structure elongates around strong vortices. On the other hand, the gravity waves reflected in the vertical velocity field seems to radiate from the strong vortices and propagate to various directions. It should be noted that the vertical motion is generated by only nonlinear interactions between geostrophic modes, since the external geostrophic forcing cannot excite directly a divergent component. Strong magnitude of vertical velocity is remarkably seen in the high PV regions.

Next, we examine kinetic energy transfer in the spectral space. Our principal interest is energy conversion from vortical to divergent motions, because divergent energy converted from vortical one plays an essential role in generating the mesoscale $-5/3$ spectral slope in our experiments. For simplicity, we analyze kinetic energy transfer using decomposition into the vortical and divergent motions, instead of that based on the Craya-Herring frame or eigenframe [3]. As the velocity field is decomposed into vortical and divergent ones, the triad terms appearing in the energy equations can be classified into four types; VVV (composed of three vortical modes), VVD (two vortical and one divergent modes), VDD (one vortical and two divergent modes) and DDD (three divergent modes). Using these notations, the energy equations of the vortical and divergent parts are symbolically written as

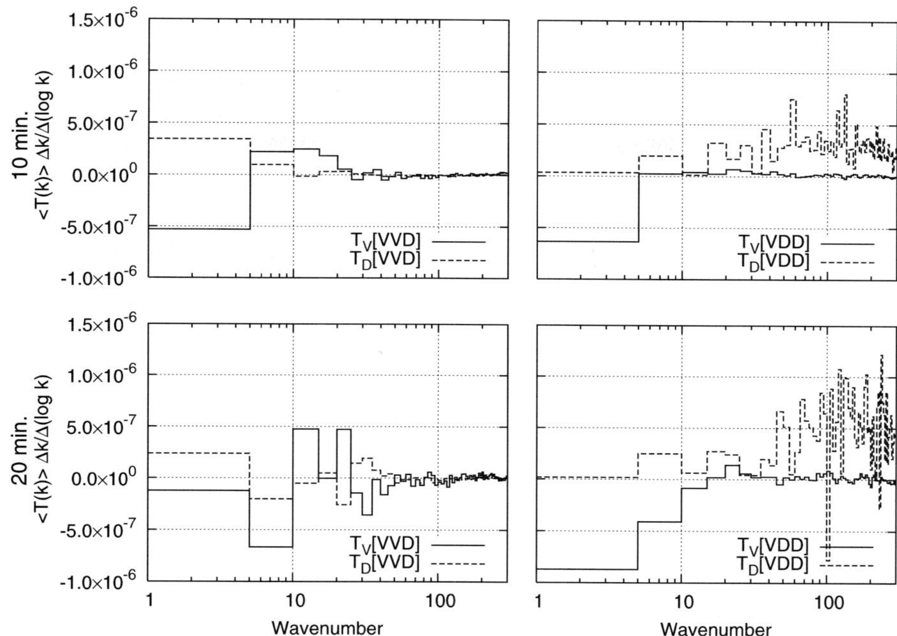

Fig. 3. The energy transfer functions averaged over the vertical column for the $2\pi/N = 10$ [min.] (the upper panels) and 20 [min.] (the bottom panels), $f = 2\Omega$ cases. The left and right panels indicate the energy transfer by the VVD and VDD triads, respectively. The solid and dashed lines shows the energy transfer in the vorticity and divergence energy equations.

$$\frac{\mathrm{d}E_V(k_H,t)}{\mathrm{d}t} = T_V[\mathrm{VVV}] + T_V[\mathrm{VVD}] + T_V[\mathrm{VDD}] + (\text{other terms}),$$

$$\frac{\mathrm{d}E_D(k_H,t)}{\mathrm{d}t} = T_D[\mathrm{VVD}] + T_D[\mathrm{VDD}] + T_D[\mathrm{DDD}] + (\text{other terms}).$$

Here, the E and T are kinetic energy and energy transfer function averaged over the vertical column and the vertical vorticity and horizontal divergence components are denoted by the subscript V and D.

Figure 3 shows energy transfer function of vortical and divergent components. In this figure, the contribution of VVD and VDD triads is presented, since only these triads represent energy exchange between vortical and divergent motions. A mean value over every five wavenumbers is shown and a scaling factor is multiplied so that the area (the integral of transfer function) is correctly represented in the log-linear coordinates. We can see that vortical energy in the lower wavenumber is transferred to divergent energy. However, the scale where the vortical energy is converted to the divergent one is quite different between the VVD and VDD parts of energy transfer function. Namely, energy conversion by VVD triads is confined in the low wavenumber region, while divergent energy in the higher wavenumbers is mainly supplied by VDD triads. Energy transfer to the higher wavenumbers is more remarkable in the case with weaker stratification. [1] showed that only the triads composed of two geostrophic and one ageostrophic modes contribute to energy transfer from geostrophic mode to gravity one, while the triads composed of one geostrophic and two ageostrophic modes are responsible for downscale energy cascade in gravity mode in the Ro, Fr $\rightarrow$ 0 limit. The result that energy transfer to higher wavenumbers in the divergent mode is mainly attributed to the VDD triad is consistent with [1]. However, energy conversion from the vortical to the divergent modes cannot be directly applied to his argument, because potential energy cannot be neglected in the geostrophic part of energy in the scales larger than the Rossby deformation radius, namely in the k_H^{-3} spectral range. The difference between the present V-D decomposition and the QG-AG one cannot be neglected in such the range. A further study for energy transfer involving potential energy would be also necessary.

4 Summary

In the present study, numerical experiments on rotating stratified turbulence are conducted with a high-resolution model to investigate the horizontal energy spectrum observed in the atmosphere. The slope of the energy spectrum excited by the forcing in the synoptic scale is close to -3 in the synoptic scales and $-5/3$ in the mesoscales, as observed in the troposphere. The scale of the spectral transition becomes smaller as the Rossby deformation radius decreases. In the mesoscale $k_H^{-5/3}$ range, the energy of divergent motion is remarkable even if the external forcing excites directly only geostrophic modes. In our experiments the divergent motion is observed to be generated around high PV regions and propagate to various directions. While energy conversion by the VVD triads is confined in the low wavenumber region, divergent energy in the higher wavenumbers is mainly supplied by the VDD triads.

References

1. Bartello P (1995) J Atmos Sci 52:4410–4428
2. Charney JG (1971) J Atmos Sci 28:1087–1095
3. Godeferd FS, Cambon C (1994) Phys of Fluids A 6:2084–2100
4. Kitamura Y, Matsuda Y (2006) Geophys Res Lett 33:L05809, doi:10.1029/ 2005GL024996
5. Lilly DK (1983) J Atmos Sci 40:749–761
6. Lindborg E (2005) Geophys Res Lett 32:L01809, doi:10.1029/2004GL021319
7. Nastrom GD, Gage KS (1985) J Atmos Sci 42:950–960
8. VanZandt TE (1982) Geophys Res Lett 9:575–578
9. Waite ML, Bartello P (2006) J Fluid Mech 546:313–339

Parameter Dependence of Eastward-Westward Asymmetric Jets in Forced Barotropic 2D Turbulence on a β-plane

Shin'ya Murakami[1] and Takahiro Iwayama[2]

Graduate School of Science and Technology, Kobe University, Nada-Ku,
Kobe 657-8501, Japan
[1] `murakami@ahs.scitec.kobe-u.ac.jp`
[2] `iwayama@kobe-u.ac.jp`

Abstract. In barotropic 2D turbulence on a β-plane with high wavenumber forcing, it is well known the existence of alternating zonal jets by many numerical experiments. Moreover, in the case of β to be large, the eastward jets are narrower and faster than the westward jets. This velocity profile corresponds to the saw-tooth vorticity profile. A formation mechanism and parameter dependence of jets have been not well known yet.

In this study, we concentrate our attention on the saw-tooth vorticity profile. We define the asymmetricity of the zonal jets as $r \equiv \overline{\zeta}_{y+}/|\overline{\zeta}_{y-}|$, where $\overline{\zeta}_{y+}$ and $\overline{\zeta}_{y-}$ are the positive and negative mean meridional gradients of zonally averaged relative vorticity, respectively. We examine the β dependence of asymmetricity r by numerical experiments. The asymmetricity r increases as β increases at small β. On the other hand, r decreases as β increases at large β. We interpret this behavior of r as the forcing scale limits the width where vorticity gradient is positive.

Keywords: β-plane turbulence, zonal jets, asymmetry

1 Introduction

The motion of geophysical fluids are approximately two-dimensional (2D) and turbulent. Thus 2D turbulence on a β-plane (or on a rotating sphere) is thought to be a simple model of the large scale motion of geophysical fluids. A pioneer work in this field is done by Rhines [1]. He pointed out that there is a characteristic wavenumber $k_\beta = \sqrt{\beta/(2U_{\mathrm{rms}})}$ in 2D β-plane turbulence, where β is the meridional gradient of the Colioris parameter and U_{rms} is the root-mean-square velocity [1]. Many numerical experiments of forced 2D β-plane turbulence confirmed formation of alternating zonal jets whose width coincides with L/k_β, where L is the meridional size of flow domain [2] [3] [4] [5].

In forced 2D β-plane turbulence, it is well-known that there is zonal asymmetric jets such that the eastward jets are narrower and faster than the westward jets in β being large [2] [3] [4] [6](Fig. 1(a)). This velocity profile

415

Y. Kaneda (ed.), *IUTAM Symposium on Computational Physics and New Perspectives in Turbulence*, 415–420.

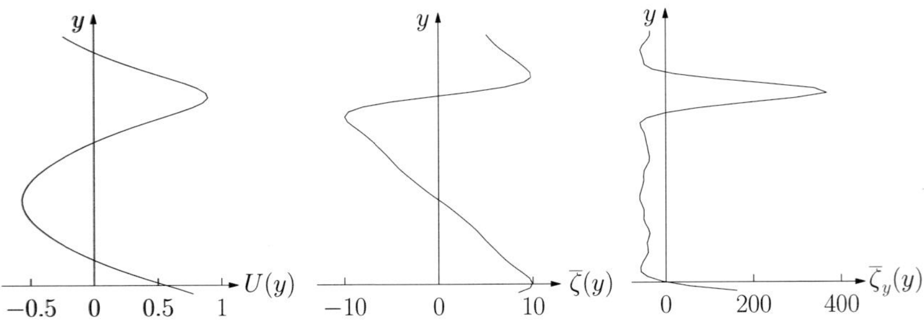

Fig. 1. (a) Zonally averaged eastward-westward velocity profile $U(y)$, (b) zonally averaged relative vorticity $\bar{\zeta}$ and (c) meridional gradient of zonally averaged relative vorticity $\bar{\zeta}_y$ obtained by numerical experiment in the condition of $\beta = 120$, $\lambda_2 = 123$, $r = 4.12$, $k_{\rm j} = 12$. To compare Fig. 2, one set of eastward-westward jets is shown and the horizontal axes are normalized its maximum and minimum values.

corresponds to a peculiar ("saw-tooth") vorticity profile (*e.g.*, Fig. 1(b), (c)). The saw-tooth vorticity profile is thought to be one of universal characteristics of forced 2D β-plane turbulence. Danilov and Gryanik [7] and Danilov and Gurarie [6] showed that the meridional gradient of zonally averaged relative vorticity $\bar{\zeta}_y$ is approximately piecewise-constant, and the region where $\bar{\zeta}_y$ is positive (hereafter we shall call this region as the positively sloping tooth region) is narrower than that where $\bar{\zeta}_y$ is negative (negatively sloping tooth region). They used this characteristic profile of $\bar{\zeta}_y$ to explain simulated one-dimensional energy spectrum that has spiky peak at harmonics of jet wavenumber.

Although they revealed the relation between shapes of the one-dimensional energy spectrum and a characteristic of the zonal jets in physical space, not only the formation of the asymmetric zonal jets but also parameter dependence of the asymmetric zonal jets have not been well known, yet.

In this study, we define the zonal asymmetricity in forced 2D β-plane turbulence, and investigate its β dependence using numerical experiments.

2 Definition of Zonal Asymmetricity

Here, we assume that the zonally averaged vorticities obtained from numerical experiments can be modeled by the saw-tooth profiles as showed in Fig. 2. (Indeed, the existence of saw-tooth vorticity profiles is verified by many numerical experiments.) We denote the total length of positively sloping tooth regions as L_+, and negatively sloping tooth regions as L_-, respectively. Those are defined by

$$L_+ \equiv \int_{\bar{\zeta}_y > 0} dy, \qquad L_- \equiv \int_{\bar{\zeta}_y < 0} dy. \tag{1}$$

We calculate the mean value of the meridional gradient of zonally averaged relative vorticity over the positively sloping tooth regions, $\bar{\zeta}_{y+}$, and that over

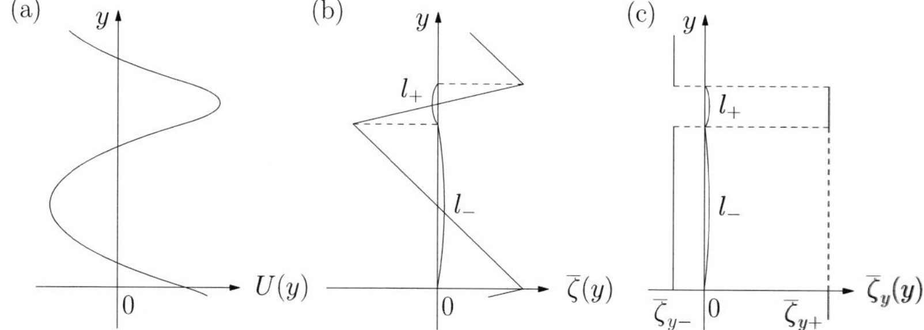

Fig. 2. (a) Zonally averaged eastward-westward velocity profile $U(y)$ which has piecewise-constant vorticity gradient for $r = 4$, (b) zonally averaged relative vorticity $\overline{\zeta}$ and (c) meridional gradient of $\overline{\zeta}$ corresponding to the velocity profile (a).

the negatively sloping tooth regions, $\overline{\zeta}_{y-}$, that is,

$$\overline{\zeta}_{y+} \equiv \frac{1}{L_+} \int_{\overline{\zeta}_y > 0} \overline{\zeta}_y \, \mathrm{d}y, \qquad \overline{\zeta}_{y-} \equiv \frac{1}{L_-} \int_{\overline{\zeta}_y < 0} \overline{\zeta}_y \, \mathrm{d}y. \tag{2}$$

We define a zonal asymmetricity r in terms of $\overline{\zeta}_{y+}$ and $\overline{\zeta}_{y-}$ as

$$r \equiv \frac{\overline{\zeta}_{y+}}{|\overline{\zeta}_{y-}|}. \tag{3}$$

Note that the zonally averaged relative vorticity $\overline{\zeta}$ obeys the following constraint $\int_{L_+ + L_-} \overline{\zeta}_y \, \mathrm{d}y = 0$ in the case of periodic boundary condition. This implies the relation $L_+ \overline{\zeta}_{y+} = L_- |\overline{\zeta}_{y-}|$. With the aid of this relation, the zonal asymmetricity r can be expressed as $r = L_-/L_+$. Moreover, providing that there are eastward-westward jets of k_j pair in the flow region, the width of the positively sloping tooth region of each jet l_+ and that of the negatively sloping tooth region of each jet l_- can be estimated as $l_+ = L_+/k_\mathrm{j}$ and $l_- = L_-/k_\mathrm{j}$, respectively, and the zonal asymmetricity reduces to

$$r = \frac{l_-}{l_+}. \tag{4}$$

3 Governing Equation and Numerical Methods

The vorticity equation for incompressible 2D barotropic fluid on a β-plane is

$$\frac{\partial \zeta}{\partial t} + \frac{\partial \psi}{\partial x} \frac{\partial \zeta}{\partial y} - \frac{\partial \zeta}{\partial x} \frac{\partial \psi}{\partial y} + \beta \frac{\partial \psi}{\partial x} = D\zeta + F, \tag{5}$$

where $\zeta = \nabla^2 \psi$ is the relative vorticity, ψ is the stream function, $D = -\lambda_n(-\nabla^2)^{-n} - \nu_m(-\nabla^2)^m$ is the dissipation operator, and F is a forcing.

We apply the doubly periodic boundary condition. Numerical simulations are performed using a dealiased spectral method with the 2/3 rule and the second order Adams-Bashforth time stepping scheme. The grid size is 512^2, the cutoff wavenumber is $k_{\mathrm{cutoff}} = 170$, and the computational domain is $2\pi \times 2\pi$. Initial conditions are zero vorticities at all points. According to [6], the forcing is Markovian and supplied in the wavenumber range $98 \leq k_{\mathrm{f}} \leq 102$. The integrations continue until both the energy and the enstrophy to be nearly equilibrium. The parameters used in the simulations are as follows: the hypofriction $n = 2$ is adopted with the hypofriction coefficients $\lambda_2 = 50$, 123 and 300, and β varies from 0 to 600. The hyperviscosity coefficient is $\nu_m = 20/k_{\mathrm{cutoff}}^{2m}$ with $m = 4$. The energy input rate ϵ is nearly constant in time, but has weak β dependence, $0.00190 \leq \epsilon \leq 0.00223$.

We examine the β dependence of r, l_+ and l_- for three hypofriction coefficients. These quantities are calculated from time and zonally averaged relative vorticity. The time average is taken over the last 200 unit times of the simulations.

4 Results and Discussion

Fig. 1 shows the zonally averaged zonal velocity profile $U(y)$, zonally averaged vorticity $\overline{\zeta} = -dU/dy$ and the meridional derivative of zonally averaged vorticity $\overline{\zeta}_y = d\overline{\zeta}/dy$ obtained by numerical experiment in the conditions $\beta = 120$, $\lambda_2 = 123$. In contrast, Fig. 2 shows the zonal velocity, the relative vorticity corresponding to the zonal velocity and the meridional gradient of the relative vorticity for $r = 4.0$, that is nearly equal to the value of r in Fig. 1. From Figs. 1 and 2, we can see that the saw-tooth profile well models the zonal jets obtained by the numerical experiments.

Fig. 3(a) shows that the β dependence of zonal asymmetricity r. When $\beta = 0$, the zonal flow is symmetric ($r = 1$). The asymmetricity r increases as β increases in the range $20 \lesssim \beta \lesssim 200$. On the other hand, r gradually decreases as β increases in $\beta \gtrsim 200$. Therefore, there is an upper bound for r.

We show the β dependence of l_+ and l_- in Fig. 3(b). The length l_- gradually decreases as β increases in almost all ranges of β. In contrast, l_+ considerably decreases as β increases and finally reaches forcing length scale $l_{\mathrm{f}} = 2\pi/k_{\mathrm{f}}$. Note that the fact that l_+ is nearly equal to l_{f} is already pointed out by [7], but its β dependence have not been known, yet.

If we accept the behavior of the β dependence of l_+ as shown above, we can interpret the β dependence of l_- using the behavior of l_+ and k_{j} as a function of β. If β becomes larger, the jet wavenumber $k_{\mathrm{j}} \approx k_\beta = \sqrt{\beta/2U_{\mathrm{rms}}}$ also becomes larger,[1] so that the width of one jet $l = l_+ + l_- = 2\pi/k_{\mathrm{j}}$ must decrease. Since l_+ quickly equilibrates l_{f} in $\beta \gtrsim 200$, l_- must decrease as β increase.

---

[1] When β is large, we can estimate the dependence of U_{rms} on β. Firstly, we assume that almost energy dissipates at k_{j}. Secondly, we assume that the Rhines scale $k_\beta = \sqrt{\beta/(2U_{\mathrm{rms}})}$ equals to k_{j}. After easy calculations, we can derive

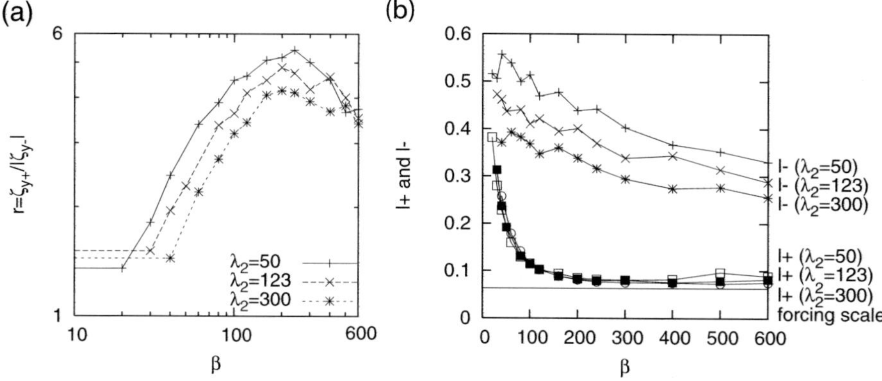

Fig. 3. (a) log-log plot of β dependence of the zonal asymmetricity r. (b) β dependence of the widths of positively and negatively sloping tooth regions of the zonally averaged relative vorticity in each jet.

We interpret the existence of the upper bound of r using the β dependence of l_+ and l_-. Since the decrease rate of l_+ is greater than that of l_- in $20 \lesssim \beta \lesssim 200$, r increases as β increases. In contrast, l_+ is nearly equal to l_f and l_- still decreases as β increases in $\beta \gtrsim 200$, therefore $r = l_-/l_+$ decreases as β increases further. That r has upper bound corresponds to that l_+ has lower bound.

Let us note on a relation between the the zonal velocity profile and the Rayleigh-Kuo stability condition. We check whether the condition is satisfied or not, at every time step of numerical integrations. When the condition $\beta + \overline{\zeta}_y < 0$ was satisfied for the first time, the zonal jets were sufficiently asymmetric ($r \sim 3$ with $\beta = 600$). Figure 8 in [4] also shows that the profile of the zonal jets is not directly related to the Rayleigh-Kuo stability condition. Therefore, there must exist a mechanism except for the barotropic instability to form the asymmetric zonal jets.

5 Summary

We have investigated β dependence of zonally asymmetric zonal jets in forced 2D β-plane turbulence. In particular, we concentrated our attention on the zonally averaged relative vorticity profile $\overline{\zeta}(y)$. To describe $\overline{\zeta}$, we used the fact that $\overline{\zeta}$ has "saw-tooth" profile. Then, we introduced the meridional gradient of $\overline{\zeta}$ averaged over the positively sloping tooth regions, $\overline{\zeta}_{y+}$, and that over the

$$U_{\mathrm{rms}} = \left(\frac{\epsilon}{\lambda_n}\right)^{\frac{1}{n+2}} \left(\frac{\beta}{2}\right)^{\frac{n}{n+2}}. \tag{6}$$

In the case of $n = 2$, U_{rms} is proportional to $\beta^{1/2}$. In our numerical experiments, the dependence of U_{rms} on β is good agreement with the above estimate, especially at large β (figure not shown).

negatively sloping tooth regions, $\overline{\zeta}_{y-}$, and defined the zonal asymmetricity r as $r \equiv \overline{\zeta}_{y+}/|\overline{\zeta}_{y-}|$. We also define l_+ as the width of positively sloping tooth region of $\overline{\zeta}$ and l_- as that of negatively sloping tooth region. The zonal asymmetricity r can be expressed in terms of the ratio l_- to l_+. We investigate the β dependence of r, l_+ and l_- using the results of numerical experiments. The results showed that the zonal flows are zonally symmetric at small β, and the zonal asymmetricity increases as β increases further. When β reaches some critical value $\beta_c \sim 200$, the zonal asymmetricity r reaches its upper bound. The asymmetricity r decreases as β increases further. We interpret the existence of upper bound of r by decomposing r into l_+ and l_-, and interpret β dependence of those. When $\beta \gtrsim \beta_c$, l_+ are approximately equal to the forcing length scale l_f. When l_+ is approximately constant, l_- must decrease, because one jet width l_{jet}, which is evaluated using Rhines wavenumber $k_\beta = \sqrt{\beta/(2U_{\mathrm{rms}})}$ as $l_{\mathrm{jet}} \approx 2\pi/k_\beta$, decreases as β increases. Therefore $r = l_-/l_+$ has an upper bound. The lower bound of l_+ and the value of β_c might depend on the forcing length scale l_f, thus the maximum of zonal asymmetricity r might depend on l_f. Physical explanation on the β dependence of l_+ is left as an unsolved problem. This would be a subject of future study.

Acknowledgement. We have been supported by "The 21st Century COE program of Origin and Evolution of Planetary Systems" in the Ministry of Education, Culture, Sports, Science and Technology of Japan. T.I. is supported by the Grant-in-Aid for Scientific Research No. 1854033.

References

1. Rhines PB (1975) J Fluid Mech 69:417–443
2. Panetta RL (1993) J Atmos Sci 50:2073–2106
3. Vallis GK, Maltrud ME (1993) J Phys Oceanogr 23:1346–1362
4. Chekhlov A, Orszag SA, Sukoriansky S, Galperin B, Staroselsky I (1996) Physica D 98:321–334
5. Danilov S, Gurarie D (2002) Phys Rev E 65:067301-1-067301-3
6. Danilov S, Gryanik VM (2004) J Atmos Sci 61:2283–2295
7. Danilov S, Gurarie D (2004) Phys Fluids 16:2592–2603

Statistics of Quasi-Geostrophic Vortex Patches

Takeshi Miyazaki[1], Li Yingtai[1], Hiroshi Taira[1],
Shintaro Hoshi[1], Naoya Takahashi[1] and Hiroki Matsubara[2]

[1] Dept. Mech. Eng. and Intelligent Systems. Univ. of Electro-Commun., Chofu,
Tokyo 182-8585, Japan
miyazaki@mce.uec.ac.jp
[2] RIKEN, 2-1 Hirosawa, Wako, Saitama 351-0198, Japan
matsubara@riken.jp

Abstract. The statistics of quasi-gestrophic point vortices and spherical vortex patches are investigated theoretically and numerically, in order to understand fundamental aspects of quasi-geostrophic turbulence. The numerical computations are performed using the fast special-purpose computer for molecular dynamics simulations, MDGRAPE-2/3. The most probable distributions are determined based on the maximum entropy theory. The theoretical predictions agree well with the numerical results.

Keywords: quasi-geostrophic, point vortex, vortex patch, maximum entropy theory, molecular dynamics

1 Introduction

Geophysical flows are under strong influence of the buoyancy force associated with stable density stratification and the Coriolis force due to the earth's rotation. The fluid motion is almost confined within a horizontal plane, although different motions are allowed on different horizontal planes. Then, geophysical flows are considered to be two-dimensional at the lowest order of approximation. The next order approximation is the 'quasi-geostrophic approximation', which incorporates the interaction between motions on adjacent horizontal planes. Since isolated vortices persist for a long time in geophysical flows and they dominate the turbulence dynamics, it is of crucial importance to understand the statistics of vortices.

There have been many theoretical and numerical studies on two-dimensional vortex systems. The statistical mechanics of point vortices was investigated by Onsager [1], Joyce & Montgomery [2], Kida [3] and Lundgren & Pointin [4]. Recently, Yatsuyanagi *et al.* [5] performed a very large numerical simulation of 2D point vortices ($N = 6724$), and investigated their statistical properties.

As for quasi-geostrophic vortices, a natural sequence of vortex models from the point model (N degrees of freedom) through the wire-(spheroidal) model ($2N$ degrees of freedom) to the ellipsoidal model ($3N$ degrees of freedom) was

421

Y. Kaneda (ed.), *IUTAM Symposium on Computational Physics and New Perspectives in Turbulence*, 421–426.

constructed. [6, 7] In this paper, we investigate the statistical properties of quasi-geostrophic point vortices and spherical vortex patches both theoretically and numerically. Numerical simulations of N-vortex system ($N = 2000$) in an infinite fluid domain are performed using the fast special-purpose computer for molecular dynamics simulations. The most probable distributions are determined based on the maximum entropy theory. The theoretical predictions agree quite well with the numerical results.

2 Quasi-Geostrophic Approximation

We can introduce a streamfunction $\Psi(x, y, z)$, since the fluid motion in each horizontal plane is two-dimensional:

$$u = \frac{\partial \Psi}{\partial y}, v = -\frac{\partial \Psi}{\partial x}. \tag{1}$$

The time-evolution under the quasi-geostrophic approximation is governed by,

$$\left(\frac{\partial}{\partial t} + \frac{\partial \Psi}{\partial y} \frac{\partial}{\partial x} - \frac{\partial \Psi}{\partial x} \frac{\partial}{\partial y} \right) q = 0, \tag{2}$$

where q denotes the potential vorticity:

$$q = -\Delta \Psi = - \left(\frac{\partial^2}{\partial x^2} + \frac{\partial^2}{\partial y^2} + \frac{\partial^2}{\partial z^2} \right) \Psi. \tag{3}$$

In the point vortex system, we assume that the potential vorticity is concentrated on N points ($\boldsymbol{R}_i, i = 1, 2, \ldots, N$) with the δ-function distribution:

$$q = \sum_{i=1}^{N} \hat{\Gamma}_i \delta(\boldsymbol{r} - \boldsymbol{R}_i). \tag{4}$$

Here N is the number of vortices and $\hat{\Gamma}_i$ denotes the strength of the ith vortex.

3 Statistical Mechanics of Quasi-Geostrophic Point Vortices

3.1 Equations of Motion

The Hamiltonian of an N point vortex system is given as a summation of the interaction energy of $N(N-1)/2$ vortex-pairs:

$$H = \sum_{(i,j)}^{N} H_{mij}, \quad H_{mij} = \frac{\hat{\Gamma}_i \hat{\Gamma}_j}{4\pi |\boldsymbol{R}_i - \boldsymbol{R}_j|}, \tag{5}$$

where H_{mij} is the interaction energy between two point vortices located at $\boldsymbol{R}_i = (X_i, Y_i, Z_i)$, $\boldsymbol{R}_j = (X_j, Y_j, Z_j)$ with strength $\hat{\Gamma}_i$, $\hat{\Gamma}_j$. We have the canonical equations of motion for the $i-$th vortex:

$$\frac{\mathrm{d}X_i}{\mathrm{d}t} = \frac{1}{\hat{\Gamma}_i}\frac{\partial H}{\partial Y_i}, \qquad \frac{\mathrm{d}Y_i}{\mathrm{d}t} = -\frac{1}{\hat{\Gamma}_i}\frac{\partial H}{\partial X_i}. \tag{6}$$

The center of vorticity (P, Q) and the angular momentum I are conserved besides the energy H (Hamiltonian itself). We shift the coordinate origin to the vorticity center and the length scale is normalized using I:

$$I = \sum_{i=1}^{N} \hat{\Gamma}_i(X_i^2 + Y_i^2) = \sum_{i=1}^{N} \hat{\Gamma}_i. \tag{7}$$

3.2 Statistics of Point Vortices

We compute numerically the time evolution of point vortices ($N = 2000$) with strength $\hat{\Gamma}_{1,2,\cdots,N} = 0.5$, located randomly (and uniformly) in a rectangular parallelepiped box ($2.432 \times 2.432 \times 4.864$) initially. We call this case the 'continuous case', below. The total energy is taken to be $E(= H/N^2) = 0.123$, which is the most probable value of 10^6 ensembles produced randomly with the fixed angular momentum I. We investigate the statistics of the equilibrium state, which is attained after $t = 10 \sim 20$, by averaging the numerical datas during $t = 20 \sim 200$.

The equilibrium state is axisymmetric and the probability distribution $F(r, z)$ is a function of the radial coordinate r and the vertical coordinate z. We can see, in Fig. 1, that the equilibrium distribution is almost z-independent for $|z| \leq 1.459$, whereas it becomes more concentrated near the axis as $|z|$ increases ('end effect'). The distribution of the center region is essentially that of a two-dimensional case. [3]

3.3 Maximum Entropy Theory

The equilibrium distribution is determined theoretically, based on the maximum entropy theory, which was applied to the system of two-dimensional point vortices by Kida. [3] We assume that $\hat{N}$ vortices are placed continuously in the vertical range $z_1 \leq z \leq z_2$. The strength of each vortex is taken to be unity ($\hat{\Gamma}_{1,2,\cdots,\hat{N}} = 1$). The energy E is a parameter that determines equilibrium distribution, for the fixed angular momentum $\hat{I}(= I/\hat{N} = 1)$. The number density $n(x, y, z, t)$ is related to the probability distribution function $F(x, y, z, t) = n/\hat{N}$. The vertical distribution of vortices remains unchanged,

$$P(z) = \iint F(x, y, z)dxdy, \tag{8}$$

because each vortex moves only in the horizontal plane where it is located initially.

The equilibrium distribution satisfy the following nonlinear integral equation that is derived based on the maximum entropy theory extended to the quasi-geostrophic flow:

$$1 = \iiint F(x, y, z)\, dx\, dy\, dz \qquad \left(\int_{z_1}^{z_2} P(z)\, dz = 1 \right), \tag{9}$$

$$1 = \iiint (x^2 + y^2) F(x, y, z)\, dx\, dy\, dz \tag{10}$$

$$\log F + 1 + \alpha(z) + \beta(x^2 + y^2) + \frac{\gamma}{4\pi} \iiint \frac{F(\boldsymbol{r}')}{|\boldsymbol{r} - \boldsymbol{r}'|}\, d^3\boldsymbol{r}' = 0, \tag{11}$$

$$\frac{8\pi H}{\hat{N}^2} = \iiiiii \frac{F(\boldsymbol{r}) F(\boldsymbol{r}')}{|\boldsymbol{r} - \boldsymbol{r}'|}\, d^3\boldsymbol{r}\, d^3\boldsymbol{r}'. \tag{12}$$

Here, $\alpha(z)$, β and γ are Lagrange's undetermined constants, corresponding to the invariants $P(z)$, I and E, respectively.

We solve the integral equation by a numerical iteration. The Gaussian integration method with only 'five points' is used in evaluating the vertical integration, which is rather crude but is enough to capture the 'end-effect' correctly. In fact, the obtained equilibrium distribution $F(r, z)$ agrees well with that of the direct numerical simulation (Fig. 1), in which the theoretical results are shown by three thick dotted lines.

4 Statistical Mechanics of Spherical Vortex Patches

The point vortex system is an inviscid system without energy dissipation. In order to incoporate dissipative processes, such as merging of vortices of the same sign, we introduce a spherical vortex patch instead of a point vortex. The initial radius is taken to be $r_s = 0.036$, which is about 30 vortices. The initial strength is $\hat{\Gamma}_{1,2,\dots,N} = 1$. The number of vortex patches in the following simulation is $N = 1000$.

4.1 Merger of Quasi-Geostrophic Spherical Vortices

The motion of vortex centroid is governed by (6) with the Hamiltonian (5), if the vortices are far apart. The total enstrophy S is defined as,

$$S = \sum_{i=1}^{N} \hat{\Gamma}_i, \qquad \hat{\Gamma}_i = \frac{4\pi}{3} r_{si}^3 q. \tag{13}$$

Direct numerical simulations of the quasi-geostrophic equation (based on the CASL-algorithm) indicate that two co-rotating vortices of radii r_i, r_j merge when following two conditions are satisfied, i.e., $a < 1.3(r_i + r_j)$,

$h < r_i + r_j$. Here a and h denote the horizontal and vertical distances between two vortices, respectively.

The energy H and the enstrophy S are dissipated during merger, and a new vortex of radius r_k appears after merger. [8] We introduce a simple rule for such dissipative processes, $F_S(S_i + S_j) = S_k$, $(0 \ll F_S < 1)$. Here, F_S is a constant enstrophy decaying factor. This relation determines the radius r_k of the new vortex, whose centroid coincides with that of the vortices before merger.

4.2 Enstrophy and Energy Dissipation

In CASL-simulations, the energy decay rate was always smaller than the enstrophy decay rate, i.e., $-\Delta S/S > -\Delta H/H$. [7] We adjust the enstophy decaying factor F_S so that the same inequality holds. We find out that this occurs in a narrow range around $F_S = 0.98$. For $F_S = 0.988, r_{s(1,2,\ldots,N)}(0) = 0.036$, we observe power law decays of N, S, E during $3 \leq t < 40$:

$$N(t) \sim t^{-0.277}, \;\; S(t) \sim t^{-0.00636}, \;\; H(t) \sim t^{-0.00407}. \tag{14}$$

The merging process terminates at $t \approx 40$, and no power law behavior is seen after that. For $F_S \leq 0.96$, the energy decay rate is larger than that of the enstrophy, i.e., $-\Delta S/S < -\Delta H/H$. When F_S is nearly unity, the energy increases, which is clearly unphysical. The choice of F_S is crucial for a physically meaningful simulation.

4.3 Statistics of Spherical Vortex Patches

We investigate statistical properties of the vortex patches. The vorticity distribution $\omega(r)$ becomes uniform (Fig. 2), for the case of $F_S = 0.988$. This indicates that a uniform vortex patch is formed after weak dissipative processes, in contrast to the equilibrium distribution is formed in completely inviscid vortex interactions.

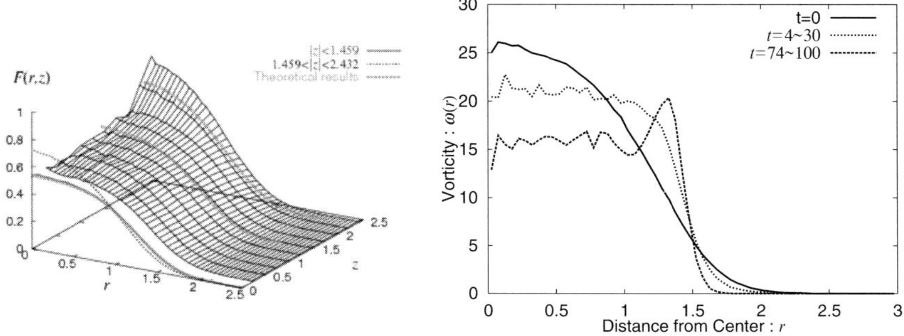

Fig. 1. Equilibrium distribution $F(r, z)$ **Fig. 2.** Vorticity distribution $\omega(r)$: $F_S = 0.988$, $r_s = 0.036$

5 Summary

We have invetigated the statistics of quasi-geostrophic point vortices and spherical vortex patches.

- Point vortices
    - The continuous equilibrium distribution is attained after $t = 10 \sim 20$.
    - The equilibrium distribution of the center region (vertically) is similar to that of two-dimensional point vortices, whereas the distribution near the upper and lower lids suffer from end-effect and concentrates tightly around the axis of symmetry.
    - The predictions based on the maximum entropy theory are in good agreement with the numerical equilibrium distributions.
- Spherical vortex patches
    - For an appropriate enstrophy decay factor, i.e., $F_S = 0.988$, power law decays of N, S, E are observed.
    - The vorticity distribution becomes almost uniform, indicating that weak dissipative processes produce a vortex patch of uniform potential vorticity.

References

1. Onsager L (1949) Nuovo Cimento Suppl 6:279-287
2. Joyce G, Montgomery D (1973) J Plasma Phys 10:107-121
3. Kida S (1975) J Phys Soc Jpn 39(5):1395-1404
4. Lundgren TS, Pointin YB (1997) J Stat Phys 17(5):323-355
5. Yatsuyanagi Y, Kiwamoto Y, Tomita H, Sano MM, Yoshida T, Ebisuzaki T (2005) Phys Rev Lett 94:054502
6. Miyazaki T, Furuichi Y, Takahashi N (2001) J Phys Soc Jpn 70(7):1942-1953
7. Li Y, Taita H, Takahashi N, Miyazaki T (2006) Phys Fluids 18(7):076604
8. Miyazaki T, Asai A, Yamamoto M, Fujishima S (2002) J Phys Soc Jpn 71(11):2687-2699

The Breakdown of a Columnar Vortex with Axial Flow

Naoya Takahashi[1] and Takeshi Miyazaki[2]

[1] Univ.Electro-Communications, Dept. Mechanical Engineering and Intelligent Systems, 1-5-1, Chofugaoka, Chofu, Tokyo 182-8585 JAPAN
`naoya@mce.uec.ac.jp`

[2] Univ.Electro-Communications, Dept. Mechanical Engineering and Intelligent Systems, 1-5-1, Chofugaoka, Chofu, Tokyo 182-8585 JAPAN
`miyazaki@mce.uec.ac.jp`

Abstract. The interaction between a columnar vortex and external turbulence is investigated numerically. The formation of turbulent eddies around the columnar vortex and the vortex-core deformations are studied in detail by visualizing the flow field. In the marginally case with $q = -1.5$, small thin spiral structures are formed inside the vortex core. In the unstable case with $q = -0.45$, the linear unstable modes grow until the columnar vortex completes one turn. Its growth rate agrees with that of the linear analysis of Mayer and Powell. After the vortex completes two turns, a secondary instability is excited which causes the collapse of the columnar q-vortex, after which many fine scale vortices appear spontaneously.

Keywords: direct numerical simulation, columnar vortex, vortex breakdown

1 Introduction

Many vortices in nature, for example tornados, cyclones and the trailing vortex shed from the wing edges of planes etc, have large length scales. On the other hand, we expect that the atmospheric turbulence has a coherent structure down to small length scales. Therefore, we would like to know more about the interaction between vortices of different length scales.

Our ultimate purpose is the same as that of Leibovich and Stewartson [1]: "to determine the role played by hydrodynamic instabilities in highly nonlinear phenomena, such as vortex break down". About this subject, there are a lot of articles. Howard and Gupta [2] investigated a linear vortex stability problem for parallel vortex flows. Batchelor [3] proposed the 'Batchelor vortex' as a model of trailing line vortices. Lessen *et al.* [4] modeled the q-vortex as a simplification of the Batchelor vortex. Leibovich and Stewartson [5] proposed a sufficient condition for the instability of the q-vortex for $m \gg 1$. Khorrami *et al.* [6] studied the instabilities for $m = 0$ and 1. Mayer and Powell [7] investigated the inviscid and viscous stability problem for various azimuthal

427

Y. Kaneda (ed.), *IUTAM Symposium on Computational Physics and New Perspectives in Turbulence*, 427–432.
© 2008 *Springer*.

wave numbers. Based on these papers, we investigate the interaction between a vortex of large length scale (columnar vortex) and that of small length scale (turbulence) using direct numerical simulation (DNS). We are concerned with the evolution of vortical structures and their statistical properties.

In a previous paper [8], we investigated the interaction between the columnar vortex without axial flow and turbulence, and we found an excitation of a vortex wave on the surface of the columnar vortex. It was caused by axisymmetric, elliptical and bending deformations of the vortex core. In small length scale, we found that the spirals are aligned in the azimuthal direction, and that they don't exist uniformly in the axial direction. We also confirmed that the statistical properties agreed qualitatively with Rapid Distortion Theory (RDT) [9]. Additionally the statistical properties agreed quantitatively with RDT only if the amplitude of the initial background turbulence was small.

In this paper, we focus on the interaction between a columnar vortex with axial flow and turbulence.

2 Method

We solve the Navier-Stokes equation for incompressible fluids under periodic boundary conditions with period 4π with a spectral method. The time integration is performed using the fourth order Runge-Kutta-Gill method. The simulations are performed with resolutions of 512^3. The columnar vortex is immersed in an initially isotropic homogeneous turbulence field, which itself is produced numerically by a direct numerical simulation of decaying turbulence [8] with Taylor microscale $\lambda = 0.403$, the integral length scale $L = 1.49$ and Taylor microscale Reynolds Number $R_\lambda = 126$.

As the columnar vortex, we use the q-vortex, which is a model for trailing vortices [4]. The q-vortex in spherical coordinates is defined by

$$(U_r, U_\theta, U_z) = \left(0, \frac{\Gamma_0}{2\pi r} \left\{ 1 - \exp\left(-\frac{r^2}{r_0^2} \right) \right\}, \frac{\Gamma_0}{2\pi r_0 q} \exp\left\{ -\frac{r^2}{r_0^2} \right\} \right), \quad (1)$$

where U_r, U_θ and U_z are the radial, azimuthal and axial components of the velocity field, and q is the swirling parameter. The initial circulation Γ_0 is an arbitrary parameter, so we set the circulation strong enough to dominate the vortex dynamics of the flow field as $\Gamma_0 = 40 r_0^2 \omega_{r.m.s.}$ where $\omega_{r.m.s.}$ is the root mean square of the enstrophy of the initial turbulence. Then the Reynolds number of the columnar vortex Γ_0/ν becomes about $20,000$. We consider two values of q, that is -1.5 (marginally stable case) and -0.45 (unstable case).

3 Results

3.1 Results of the Marginally Stable Case ($q = -1.5$)

Figure 1 shows the time evolution of the isosurface of the enstrophy Ω. Inside the vortex core, we observed that the vortex core was divided into fine

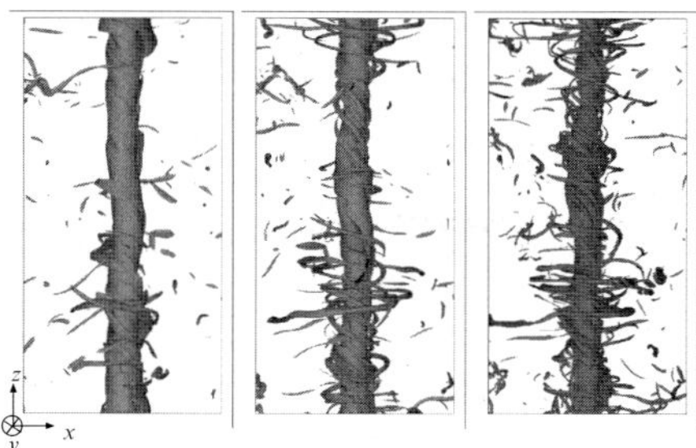

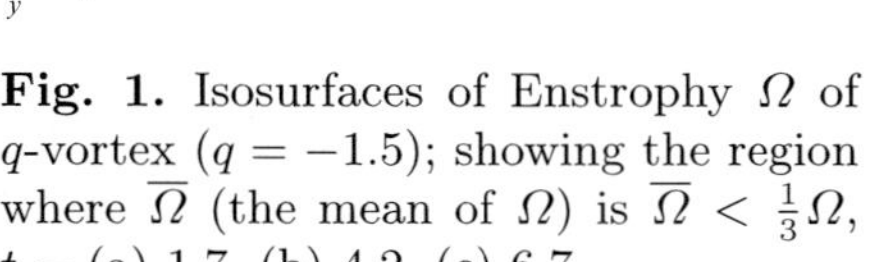

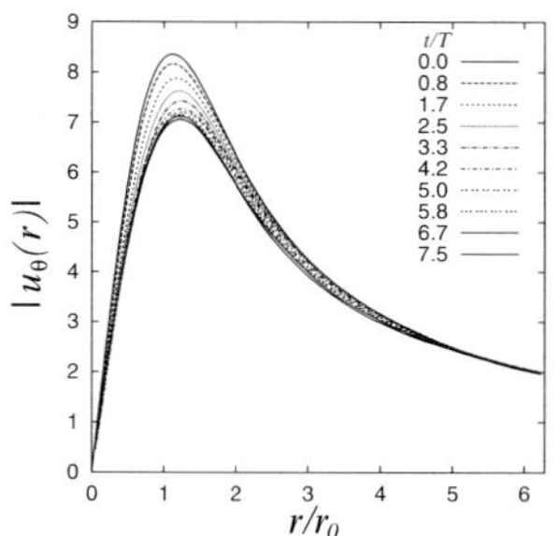

Fig. 1. Isosurfaces of Enstrophy Ω of q-vortex ($q = -1.5$); showing the region where $\overline{\Omega}$ (the mean of Ω) is $\overline{\Omega} < \frac{1}{3}\Omega$, $t =$ (a) 1.7, (b) 4.2, (c) 6.7.

Fig. 2. Time evolution of the radial profiles of azimuthal velocity ($q = -1.50$).

scale structures (Fig. 1(b)). These structures generated around the surface of the core are wound up by the columnar vortex, which stretches and thins the structure. This indicates that the nonlinear interaction is concentrated inside the core. We found that these structures were different from the worm structure [10], and that they were very similar to that of tornado [11]. Simultaneously, we observed the axisymmetrization of many spirals aligned in the azimuthal direction; this is similar to what we observed in the Lamb-Oseen vortex (without axial flow) [8].

After long times ($t/T > 5$), these filaments are thin and elongated enough to be dissipated by viscosity.

In Fig. 2, we show the time evolution of the radial profile of the azimuthal velocity. We confirm that the profiles keep their shape, and that their amplitude decreases with t. After $t \simeq 5$, the decrease of amplitudes becomes slower than that is observed before $t \simeq 5$.

3.2 The Most Unstable Case ($q = -0.45$)

The q vortex with $q = -0.45$ is the case most unstable against the bending disturbance $m = 1$ [7]. Figure 3 shows the time evolution of the vortical structure using enstrophy visualization. At the early stage, we observed the excitation of the linear instability, which causes the deformation of the structure of the columnar vortex. As shown in Fig. 3a, the structure around the columnar vortex has two helices. At the next stage (Fig. 3b), the deformation of the columnar vortex is large, and the helices separate from each other. Simultaneously, we observed that the helices began to subdivide into small length scales on the surface. This implies the occurrence of the secondary instability. These deformations are amplified at the last stage (Fig. 3c). After a while, the columnar vortex breaks down abruptly.

In Fig. 4, we show the time evolution of the radial profile of azimuthal velocity. After $t \simeq 1.3$, we observe that the shape deviates remarkably from

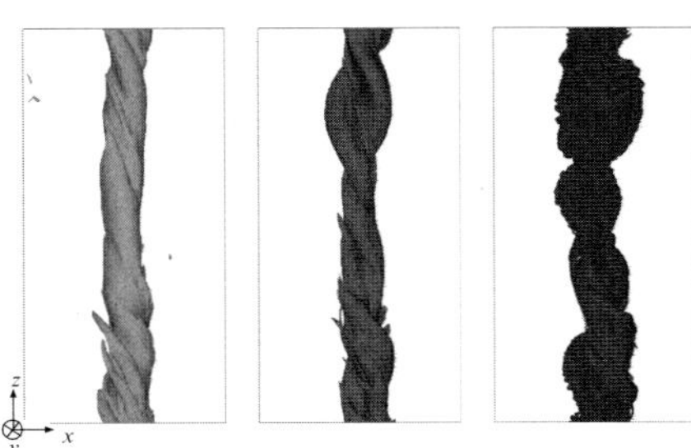

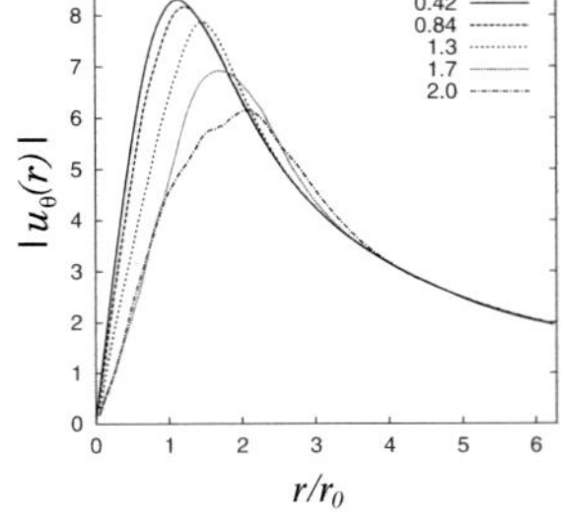

Fig. 3. Isosurfaces of Enstrophy Ω of q-vortex ($q = -0.45$); showing the region $\overline{\Omega}$ (the mean of Ω) is $\overline{\Omega} < \frac{1}{3}\Omega$, $t =$ (a) 0.8, (b)1.3, (c) 1.7 .

Fig. 4. Time evolution of the radial profiles of azimuthal velocity ($q = -0.45$).

their initial profile. Additionally, we observe the overshoots in the profile of the local circulation (Figure is omitted), that induces a Rayleigh type instability [12].

At the nonlinear stage ($t \simeq 2.92$), we could confirm via spectral analysis that the amplitude of the linear unstable mode is about 5% of the mean profile velocity of the axial component. To investigate the effect of the finite-amplitude of the linear unstable mode, we use WKB analysis [13]. Although this method is restricted to the limit of short wave lengths, it has several advantages: It is simple, numerically unproblematic, enables us to consider the time dependent basic flow and further allows us to analyse not only exponential but also algebraic growth. Here, we suppose that the flow is perturbed by a rapidly oscillating localized field $\mathbf{v}(\mathbf{x}, t) = \mathbf{a}(\mathbf{x}, t) \exp\left(\mathrm{i}\Psi(\mathbf{x}, t)/\varepsilon_0\right)$, where ε_0 is a small parameter. Substituting this into the linearized Euler equation, we get the ordinary differential equations

$$\frac{\mathrm{d}\mathbf{x}}{\mathrm{d}t} = \mathbf{V} = \mathbf{U}_q(r) + \varepsilon \Re\left\{\tilde{\mathbf{u}}(r)\right\}, \tag{2}$$

$$\frac{\mathrm{d}\mathbf{k}}{\mathrm{d}t} = -\left(\frac{\partial \mathbf{V}}{\partial \mathbf{x}}\right)^T \mathbf{k}, \quad \frac{\mathrm{d}\mathbf{a}}{\mathrm{d}t} = -\left(\frac{\partial \mathbf{V}}{\partial \mathbf{x}}\right)\mathbf{a} + 2\left(\frac{\partial \mathbf{V}}{\partial \mathbf{x}}\right)\mathbf{a} \cdot \mathbf{k}\frac{\mathbf{k}}{|\mathbf{k}|^2}, \tag{3}$$

where $\mathbf{U}_q$ is the velocity of the q-vortex, $\tilde{\mathbf{u}}$ is the primary instability mode, $\mathbf{a}$ is the amplitude of the secondary perturbation, $\mathbf{k} = \nabla\Psi$ is the vector of the wavenumber and ε is the amplitude of the linear unstable mode. We performed numerical integration of these equations with respect to t with initial conditions $\mathbf{x} = (r\cos\theta, r\sin\theta, z)$, $\mathbf{k} = (\sin\Theta\cos\Phi, \sin\Theta\sin\Phi, \cos\Theta)$ and $\mathbf{a} = \cos\beta(-\sin\Phi, \cos\Phi, 0) + \sin\beta(-\cos\Theta\cos\Phi, -\cos\Theta\sin\Phi, \sin\Theta)$ [13]. The arbitrary parameters θ, z, Θ, Φ and β are sought so as to maximize the corresponding growth rate $\sigma \equiv \log|\mathbf{k} \times \mathbf{a}|/t$. To avoid that the exponential growth of the linear unstable mode dominates in the solution, we evaluated this maximum only for short time ranges $t \leq 5$. At that time, the exponential

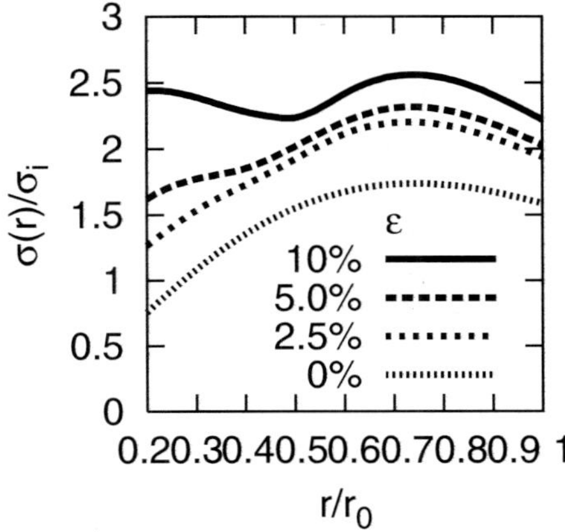

Fig. 5. The radial dependence of the growth rate from the WKB analysis. $\sigma(r)$: corresponding growth rate, defined as $\max_{t \leq 5}(1/t) \log |\mathbf{k} \times \mathbf{a}|$, σ_i: normalization factor, linear growth rate of the q-vortex.

factor $\exp(\omega_i t)$ becomes about 3.3 for the growth rate of the unstable mode ω_i, which is still justifiably small. Hence, we used amplitudes for the linear unstable mode ε of 0 (q-vortex), 2.5, 5.0 and 10%.

Because the σ exhibits periodicity in the axial direction z, we seek the radial dependence of the maximum value of σ with respect to t (≤ 5) for various values of ε (Fig. 5). We confirmed that the growth rate σ is 2–3 times larger than that of the q-vortex if ε_0 is larger, and that σ takes its maximum value around $r \sim 0.7r_0$. This implies that the transient growth induces the breakdown of the blade structure observed in DNS (Fig. 3(c)).

4 Concluding Remarks

We investigated the columnar vortex with axial flow with two different values of the swirling parameter q.

In the marginally stable case with $q = -1.5$, we found that the columnar vortex broke down gradually. In this breakdown process, its vortical structure and statistical properties are similar to that of the Lamb-Oseen vortex. Some different properties are also observed, especially the most characteristic one, small scale structures inside of the vortex core. This vortical structure is similar to that of a tornado.

In the unstable case with $q = -0.45$, we found that the columnar vortex broke down suddenly. For convenience, we divide the time evolution in three stages: the early stage ($t < 2$), the secondary stage ($2 \leq t < \sim 3$) and the nonlinear stage ($t > \sim 3$). At the early stage, it is well known that the unstable mode grows exponentially in time. At the following stage, the unstable mode has a finite amplitude. At the nonlinear stage, the columnar vortex breaks down abruptly. We also investigated this process via WKB analysis of the q-vortex with an unstable mode of finite amplitude. From this analysis emerged a picture of 'transient' growth of the disturbance, that exceeds the exponential growth of the unstable primary mode.

Acknowledgement. This research was partially supported by the Ministry of Education, Science, Sports and Culture, Grant-in-Aid for Young Scientists (B), 17740250, 2005–2006. Computations performed on Numerical Simulator III at Computer Center of Japan Aerospace Exploration Agency (JAXA) are gratefully acknowledged. We are grateful for useful conversation with Prof. H.-G. Matuttis.

References

1. Leibovich S, Kribus A (1990) J Fluid Mech 216:459–504
2. Howard LN, Gupta AS (1962) J Fluid Mech 14:463–476
3. Batchelor GK (1964) J Fluid Mech 20:645–658
4. Lessen M, Singh PJ, Paillet F (1974) J Fluid Mech 63:753–763
5. Leibovich S, Stewartson K (1983) J Fluid Mech 126:335–356
6. Khorrami MR (1991) J Fluid Mech 225:197–212
7. Mayer EW, Powell KG (1992) J Fluid Mech 245:91–114
8. Takahashi N, Ishii H, Miyazaki T (2005) Phys Fluids 17:035105
9. Miyazaki T, Hunt JCR (2000) J Fluid Mech 402:349–378
10. Yamamoto K, Hosokawa I (1988) J Phys Soc Japan 57:1532–1535
11. Katsura J, Conner H (2002) J Natural Disaster Science 24:61–71
12. Govindaraju SP, Saffman PG (1971) Phys Fluids 14:2074–2080
13. Lifschitz A, Hameiri E (1991) Phys Fluids A 3:2644–2651

Evolution of an Elliptical Flow in Weakly Nonlinear Regime

Yuji Hattori[1], Yasuhide Fukumoto[2], and Kaoru Fujimura[3]

[1] Division of Computer Aided Science, Kyushu Institute of Technology, Tobata,
Kitakyushu 804-8550, Japan
`hattori@mns.kyutech.ac.jp`
[2] Graduate School of Mathematics and Space Environment Research Center,
Kyushu University, Fukuoka 812-8581, Japan
`yasuhide@math.kyushu-u.ac.jp`
[3] Department of Applied Mathematics and Physics, Tottori University, Tottori
680-8552, Japan
`kaoru@damp.tottori-u.ac.jp`

Abstract. We study the nonlinear evolution of an elliptical flow by weakly nonlinear analysis. Two sets of amplitude equations are derived for different situations.
First, the weakly nonlinear evolution of helical modes is considered. Nonlinear self-
interaction of the two base Kelvin waves results in cubic nonlinear terms, which
causes saturation of the elliptical instability. Next, the case of triad interaction is
considered. Three Kelvin waves, one of which is a helical mode, form a resonant triad
thanks to freedom of wavenumber shift. As a result three-wave equations augmented
with linear terms are obtained as amplitude equations. They explain the numerical
results on the secondary instability obtained by Kerswell (1999).

Keywords: elliptical flow, elliptical instability, weakly nonlinear analysis,
amplitude equations, secondary instability

1 Introduction

Large-scale coherent vortical structures established in flows are often seen to
destabilize and collapse to small-scale disorder. There are several important
mechanisms for the destabilization among which are the elliptical instability
[1–5] and the curvature instability [6, 7]. Both instabilities are the consequences of parametric resonance of two Kelvin waves mediated by symmetry-
breaking perturbation and well understood in linear regime. For the elliptical
instability [4, 5], this perturbation is an imposed strain. However, the linear
theory is insufficient to explain the whole process.

In this paper, we study the evolution of an elliptical flow by weakly
nonlinear analysis. There are several works on the dynamics of vortices in
weakly nonlinear regime. Knobloch *et al.* [9] derived various types of amplitude equations from the view point of symmetry breaking. Sipp [10] derived

433

Y. Kaneda (ed.), *IUTAM Symposium on Computational Physics and New Perspectives in
Turbulence*, 433–438.

the amplitude equations for disturbance on the Lamb-Oseen vortex by applying weakly nonlinear analysis. In these two works the nonlinear terms are cubic in the amplitudes. In the present paper we derive two types of amplitude equations for different situations, one of which is three-wave interaction augmented with linear terms. It explains numerical results on the secondary instability obtained by Kerswell [8]. The results would be useful not only in analyzing transition to turbulence but also in studying turbulence by low-dimensional dynamical systems.

2 Weakly Nonlinear Analysis

Throughout the paper the flow is assumed to be inviscid and incompressible, though allowance may be made for small viscosity [3].

The base flow U is set to the sum of rigid body rotation and plain strain inside an elliptical cylinder. In a cylindrical coordinate system (r, θ, z), the velocity field (U, V, W) and the pressure P of the base flow is expressed as

$$U = U_0 + \varepsilon U_1,$$
$$U_0 = W_0 = 0, \quad V_0 = r, \quad P_0 = r^2/2 - 1,$$
$$U_1 = r \sin 2\theta, \quad V_1 = r \cos 2\theta, \quad W_1 = 0, \quad P_1 = 0.$$

The boundary of the cylinder is $r = 1 - \varepsilon/2 \cos 2\theta$ at which slip condition is imposed. The total base flow has elliptical streamlines. The magnitude of strain ε is assumed to be small but finite.

Let us expand the total velocity field u_T in both shear strength ε and disturbance amplitude α

$$u_T = U_0 + \varepsilon U_1$$
$$+ \alpha u_{01} + \alpha^2 u_{02} + \alpha^3 u_{03} + \cdots + \varepsilon \alpha u_{11} + \varepsilon \alpha^2 u_{12} + \cdots . \tag{1}$$

Similar expansions are assumed for pressure p_T.

We substitute (1) into the Euler equations

$$\frac{\partial u_T}{\partial t} + u_T \cdot \nabla u_T = -\nabla p_T, \quad \nabla \cdot u_T = 0. \tag{2}$$

At each order $O(\varepsilon^i \alpha^j)$ we obtain linear equations. Depending on the combination of modes in the leading-order disturbance u_{01}, solvability conditions force the amplitude of the modes to vary slowly, which derives amplitude equations [10].

2.1 Kelvin Waves

At $O(\alpha)$ the equation is

$$\frac{\partial u_{01}}{\partial t} + U_0 \cdot \nabla u_{01} + u_{01} \cdot \nabla U_0 + \nabla p_{01} = 0, \quad \nabla \cdot u_{01} = 0. \tag{3}$$

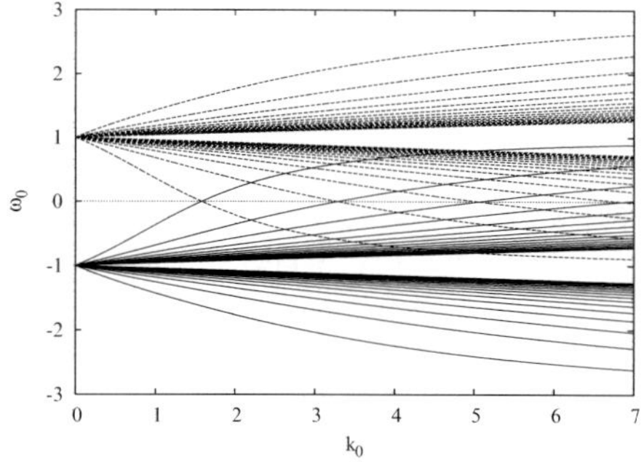

Fig. 1. Dispersion relations of Kelvin waves. $m = \pm 1$.

The leading-order disturbance $\boldsymbol{u}_{01}$ is set to $\boldsymbol{u}_{01} = \sum_m \boldsymbol{u}_{01}^{(m)}(r)\mathrm{e}^{\mathrm{i}(m\theta + k_0 z - \omega_0 t)}$. Then we have the following set of linear ordinary equations for each mode

$$[-\mathrm{i}\omega_0 \mathsf{L} + \mathsf{M}(m, k_0)]\,\boldsymbol{u}_{01}^{(m)} = 0, \tag{4}$$

where L and M are 4×4 matrix operators. Since the above linear equation has no source term, the linear operator $-\mathrm{i}\omega_0 \mathsf{L} + \mathsf{M}(m, k_0)$ should be degenerate so that we have non-trivial solutions. This leads to dispersion relations between ω_0 and k_0 (Fig. 1). The solutions are the well-known Kelvin waves, which are neutrally stable.

2.2 Linear Instability

The set of frequency ω_0 and the wavenumber k_0 is chosen from the cross points of the dispersion curves in Fig. 1 so that two Kelvin waves corresponding to $m = \pm 1$ are excited. We focus on stationary helical modes: $\omega_0 = 0$. Thus the leading-order disturbance is set to $\boldsymbol{u}_{01} = A_+ \boldsymbol{u}_{01}^{(1)}(r)\mathrm{e}^{\mathrm{i}(\theta + k_0 z)} + \overline{A_-}\boldsymbol{u}_{01}^{(-1)}(r)\mathrm{e}^{\mathrm{i}(-\theta + k_0 z)}$ (complex conjugate of A_- is introduced for convenience). At $O(\varepsilon\alpha)$ the equations for $\boldsymbol{u}_{11}^{(\pm 1)}$ are singular; compatibility conditions force the amplitude of the Kelvin waves to vary slowly with time scale ε^{-1} so that the equations become solvable

$$\mathsf{M}(1, k_0)\boldsymbol{u}_{11}^{(1)} = \overline{A_-}\mathsf{N}^+(-1)\boldsymbol{u}_{01}^{(-1)} - \frac{\mathrm{d}A_+}{\mathrm{d}\tau}\mathsf{L}\boldsymbol{u}_{01}^{(1)},$$

$$\mathsf{M}(-1, k_0)\boldsymbol{u}_{11}^{(-1)} = A_+\mathsf{N}^-(1)\boldsymbol{u}_{01}^{(1)} - \frac{\mathrm{d}\overline{A_-}}{\mathrm{d}\tau}\mathsf{L}\boldsymbol{u}_{01}^{(-1)},$$

where $\tau = \varepsilon t$. See [1] for the expression of $\mathsf{N}^+ = \overline{\mathsf{N}^-}$. Finally we arrive at

$$\frac{\mathrm{d}A_+}{\mathrm{d}\tau} = \mathrm{i}a\overline{A_-}, \quad \frac{\mathrm{d}A_-}{\mathrm{d}\tau} = \mathrm{i}a\overline{A_+}. \tag{5}$$

For the present case, a turns out to be real so that the system is unstable with exponential growth rate $|a|$. In general the stability of the above system is determined by the sign of energy of the two Kelvin waves: a pair of positive- and negative-energy waves or of zero-energy waves leads to instability [5].

2.3 Nonlinear Saturation of the Linear Instability

Here we consider nonlinear evolution of helical modes due to self-induced nonlinearity. At $O(\alpha^2)$ the equation is

$$\frac{\partial \boldsymbol{u}_{02}}{\partial t} + \boldsymbol{U}_0 \cdot \boldsymbol{\nabla} \boldsymbol{u}_{02} + \boldsymbol{u}_{02} \cdot \boldsymbol{\nabla} \boldsymbol{U}_0 + \boldsymbol{\nabla} p_{02} = -\boldsymbol{u}_{01} \cdot \boldsymbol{\nabla} \boldsymbol{u}_{01}, \quad \boldsymbol{\nabla} \cdot \boldsymbol{u}_{02} = 0. \quad (6)$$

The solution is written in the form

$$\boldsymbol{u}_{02} = A_+^2 \mathrm{e}^{\mathrm{i}(2\theta + 2k_0 z)} \boldsymbol{u}_{A_+ A_+} + A_-^2 \mathrm{e}^{\mathrm{i}(2\theta - 2k_0 z)} \boldsymbol{u}_{A_- A_-}$$
$$+ A_+ A_- \mathrm{e}^{\mathrm{i}2\theta} \boldsymbol{u}_{A_+ A_-} + A_+ \overline{A_-} \mathrm{e}^{\mathrm{i}2k_0 z} \boldsymbol{u}_{A_+ \overline{A_-}} + \mathrm{c.c.}$$

At $O(\alpha^3)$ the equation is

$$\frac{\partial \boldsymbol{u}_{03}}{\partial t} + \boldsymbol{U}_0 \cdot \boldsymbol{\nabla} \boldsymbol{u}_{03} + \boldsymbol{u}_{03} \cdot \boldsymbol{\nabla} \boldsymbol{U}_0 + \boldsymbol{\nabla} p_{03} = -\boldsymbol{u}_{01} \cdot \boldsymbol{\nabla} \boldsymbol{u}_{02} - \boldsymbol{u}_{02} \cdot \boldsymbol{\nabla} \boldsymbol{u}_{01}, \quad (7)$$

with $\boldsymbol{\nabla} \cdot \boldsymbol{u}_{03} = 0$. The solution is written in the form

$$\boldsymbol{u}_{03} = \left(|A_+|^2 A_+ \boldsymbol{u}_{|A_+|^2 A_+} + |A_-|^2 A_+ \boldsymbol{u}_{|A_-|^2 A_+} \right) \mathrm{e}^{\mathrm{i}(\theta + k_0 z)}$$
$$+ \left(|A_+|^2 A_- \boldsymbol{u}_{|A_+|^2 A_-} + |A_-|^2 A_- \boldsymbol{u}_{|A_-|^2 A_-} \right) \mathrm{e}^{\mathrm{i}(\theta - k_0 z)} + \mathrm{c.c.},$$

where irrelevant terms are omitted. Since the equations for $\boldsymbol{u}_{|A_+|^2 A_+}$ etc. are singular, the amplitudes $A_\pm$ should vary slowly with time scale α^{-2}. Mean flow correction should be also regarded; it arises at $O(\varepsilon\alpha^2)$.

By choosing $\varepsilon = \alpha^2$ we can derive the following set of amplitude equations

$$\frac{\mathrm{d}A_\pm}{\mathrm{d}\tau} = \mathrm{i}a\overline{A_\mp} - \mathrm{i}\left(c|A_\mp|^2 + dC \right) A_\pm, \qquad \frac{\mathrm{d}C}{\mathrm{d}\tau} = \mathrm{i}\left(A_+ A_- - \overline{A_+ A_-} \right), \quad (8)$$

where C is the amplitude of the mean flow correction and the coefficients c and d are determined by the solvability conditions at $O(\alpha^3)$ and $O(\varepsilon\alpha^2)$, respectively. The amplitude equations (8) are in the same form with those for the Lamb-Oseen vortex [10] except that the terms proportional to $|A_+|^2 A_+$ and $|A_-|^2 A_-$ vanish. Saturation of elliptical instability is explained by (8).

2.4 Secondary Instability: the Case of Triad Resonance

Next, we consider nonlinear evolution due to triad resonance. Once helical modes are excited, they can be regarded as perturbation which could induce another instability (secondary instability). Kerswell [8] showed that this is indeed the case for rapidly rotating flow without strain by numerical linear stability analysis; the instability is most likely due to triad resonances of the Kelvin waves.

Although exact resonant triads of neutral Kelvin waves can be hardly realized, the elliptical instability allows $O(\varepsilon)$-margin of wavenumber of the unstable helical modes so that resonance condition is met by some triads

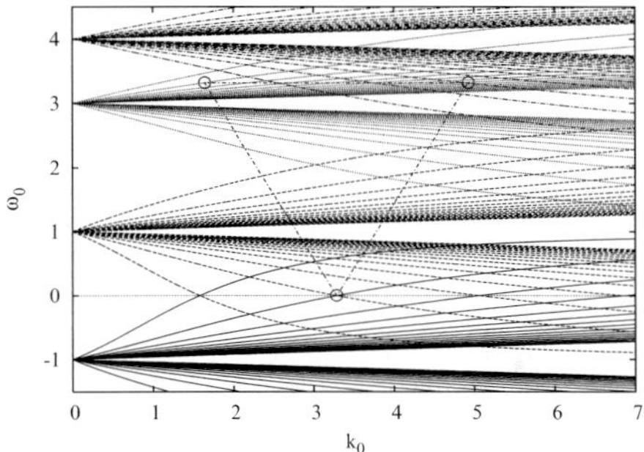

Fig. 2. Dispersion relations of Kelvin waves and triad. $m = \pm 1, 3$ and 4.

of Kelvin waves. In Fig. 2 we show one of these triads: $(m; k, \omega) = (1; k_*, 0)$, $(3; k_*/2, \omega_0)$, $(4; 3k_*/2, \omega_0)$, where $k_* \approx 3.273$ and $\omega_0 \approx 3.32$. The wavenumber of the corresponding neutral helical mode $(m = 1)$ is $k_0 \approx 3.286$. It shifts generally as $k = k_0 + \varepsilon k_1$ for the elliptical flow; the helical mode is destabilized for $|k - k_0| \leq \varepsilon \Delta k$ so that we have $k_* = k_0 + \varepsilon k_1$ for $\varepsilon \geq \varepsilon_c = |k_* - k_0|/\Delta k$.

Let us set

$$\boldsymbol{u}_{01} = A_+ \mathrm{e}^{\mathrm{i}(\theta + k_* z)} + B_+ \mathrm{e}^{\mathrm{i}(3\theta + k_* z/2 - \omega_0 t)} + C_+ \mathrm{e}^{\mathrm{i}(4\theta + 3k_* z/2 - \omega_0 t)}$$
$$+ A_- \mathrm{e}^{\mathrm{i}(\theta - k_* z)} + B_- \mathrm{e}^{\mathrm{i}(3\theta - k_* z/2 - \omega_0 t)} + C_- \mathrm{e}^{\mathrm{i}(4\theta - 3k_* z/2 - \omega_0 t)} + \text{c.c.}$$

Then compatibility conditions arise at $O(\alpha^2)$. By choosing $\varepsilon = \alpha$ we can derive the following set of amplitude equations

$$\frac{\mathrm{d}A_\pm}{\mathrm{d}\tau} = \mathrm{i}\beta_1 \overline{A_\mp} + \mathrm{i}\beta_2 A_\pm + \mathrm{i}\gamma_1 \overline{B_\pm} C_\pm,$$

$$\frac{\mathrm{d}B_\pm}{\mathrm{d}\tau} = \mathrm{i}\gamma_2 \overline{A_\pm} C_\pm, \qquad \frac{\mathrm{d}C_\pm}{\mathrm{d}\tau} = \mathrm{i}\gamma_3 A_\pm B_\pm. \tag{9}$$

The coefficients are evaluated numerically: $\beta_1 = -5.542 \times 10^{-1}, \beta_2 = -2.383 \times 10^{-1}, \gamma_1 = -6.210, \gamma_2 = 1.961, \gamma_3 = -4.264$. Note that the values of β_1 and β_2 change with $O(\varepsilon)$ shift of k.

An example of the evolution of the mode energies is shown in Fig. 3. Initially the energy of helical modes $E_{A\pm} = |A_\pm|^2$ is fueled to large values compared to $E_{B\pm}$ and $E_{C\pm}$. Exponential increase of E_{B+} and E_{C+}, which corresponds to the secondary instability, is observed. The mode energy E_{A+} almost vanishes and E_{B+} and E_{C+} saturate around $t = 23$; then E_{B+} and E_{C+} decrease exponentially. In the exponential increase or decrease, the ratio of mode energies is easily evaluated as $E_{B+}/E_{C+} = |\gamma_2/\gamma_3| \approx 0.46$, which is comparable to the value 0.545 in Kerswell [8]. The difference between the two values is probably due to viscous effects. In our case the flow is assumed to be inviscid, while Kerswell's value 0.545 is obtained for the case of the Ekman number $E = 10^{-4}$; the ratio depends on E when $E \leq 10^{4.5}$ [8].

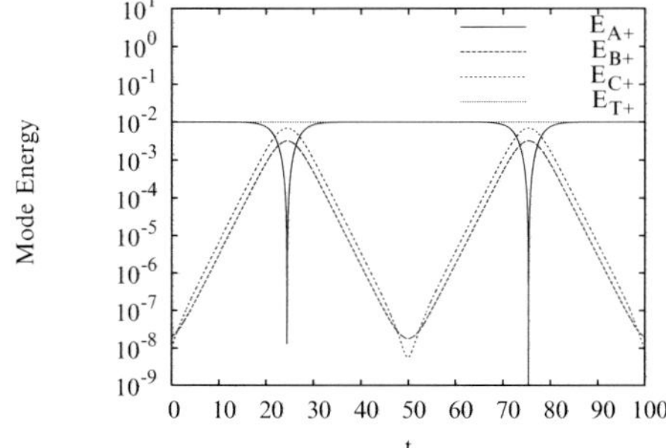

Fig. 3. Evolution of mode energies. Resonant triad. $E_{T+} = E_{A+} + E_{B+} + E_{C+}$.

3 Conclusion

The evolution of an elliptical flow is studied by weakly nonlinear analysis. Two sets of amplitude equations are derived for different situations: the first one can be used to estimate saturation amplitude of destabilized helical modes due to self-induced nonlinear effect; the second one shows that triad interaction is responsible for the secondary instability found numerically by Kerswell [8]. The results would be useful in understanding the nonlinear evolution of elliptical flows. Direct numerical simulation, which is planned as a future work, should assess the applicability of the present analysis and reveal other processes important in the fully nonlinear evolution.

References

1. Moore DW, Saffman PG (1975) Proc R Soc Lond A 346:413–425
2. Tsai C-Y, Widnall SE (1976) J Fluid Mech 73:721–733
3. Eloy C, Le Dizès S (2001) Phys Fluids 13:660–676
4. Kerswell RR (2002) Annu Rev Fluid Mech 34:83–113
5. Fukumoto Y (2003) J Fluid Mech 493:287–318
6. Hattori Y, Fukumoto Y (2003) Phys Fluids 15:3151–3163
7. Fukumoto Y, Hattori Y (2005) J Fluid Mech 526:77–115
8. Kerswell RR (1999) J Fluid Mech 382:283–386
9. Knobloch E, Mahalov A, Marsden JE (1994) Physica D 73:49–81
10. Sipp D (2000) Phys Fluids 12:1715–1729

List of Participants

Abe, Hiroyuki
Japan Aerospace Exploration
Agency
Japan
habe@chofu.jaxa.jp

Araki, Keisuke
Okayama University of Science
Japan
araki@are.ous.ac.jp

Arimitsu, Naoko
Graduate School of EIS
Yokohama National University
Japan
arimitsu@ynu.ac.jp

Arimitsu, Toshihico
Graduate School of Pure
and Applied Sciences
University of Tsukuba
Japan
tarimitsu@
  sakura.cc.tsukuba.ac.jp

Bec, Jérémie
CNRS Observatorie
de la Côte d'Azur
France
jeremie.bec@obs-nice.fr

Brethouwer, Geert
Department of Mechanics
KTH
Sweden
geert@mech.kth.se

Cambon, Claude
LMFA
Ecole Centrale de Lyon
France
claude.cambon@ec-lyon.fr

Chomaz, Jean-Marc
LadHyX École Polytechnique
France
chomaz@ladhyx.polytechnique.fr

Choquin, Jean Philippe
CEA
France

Davidson, Peter A.
Department of Engineering
University of Cambridge
UK
pad3@eng.cam.ac.uk

Doering, Charles R.
Department of Mathematics
and Michigan Center for
Theoretical Physics
University of Michigan
USA
doering@umich.edu

Eckhardt, Bruno
Fachbereich Physik
Philipps-Universität Marburg
Germany
bruno.eckhardt@
  physik.uni-marburg.de

Farge, Marie
Laboratoire de Météorologie
Dynamique
CNRS Ecole Normale Supérieure
France
farge@lmd.ens.fr

Fisher, Robert
University of Chicago
USA
rtfisher@uchicago.edu

Fukudome, Koji
Nagoya Institute of Technology
Japan
18416600@stn.nitech.ac.jp

Fukunishi, Yu
Department of Mechanical Systems
and Design
Tohoku University
Japan
fushi@fluid.mech.tohoku.ac.jp

Godeferd, Fabien
Fluid Mechanics and Acoustics
Laboratory
École Centrale de Lyon
France
Fabien.Godeferd@ec-lyon.fr

Gotoh, Toshiyuki
Department of Engineering Physics
Nagoya Institute of Technology
Japan
gotoh.toshiyuki@nitech.ac.jp

Hagiwara, Yoshimichi
Kyoto Institute of Technology
Japan
yoshi@kit.ac.jp

Hamba, Fujihiro
Institute of Industrial Science
University of Tokyo
Japan
hamba@iis.u-tokyo.ac.jp

Hanazaki, Hideshi
Department of Mechanical
Engineering and Science
Kyoto University
Japan
hanazaki@mech.kyoto-u.ac.jp

Hattori, Yuji
Division of Computer Aided Science
Kyushu Institute of Technology
Japan
hattori@mns.kyutech.ac.jp

Hayashi, Yoshiyuki
Division of Earth and Planetary
Sciences
Hokkaido University
Japan
shosuke@gfd-dennou.org

Herring, Jackson R.
NCAR
USA
herring@ucar.edu

Hoshi, Shintaro
Dept. Mechanical Engineering
and Intelligent Systems
Univ. Electro-Communications
Japan
s.hoshi@miyazaki.mce.uec.ac.jp

Hu, Ning
State Key Laboratory for Turbulence
and Complex System
Peking University
China
h_ning@pku.org.cn

Hunt, Julian C. R.
Department of Space and Climate
Physics
University College London
UK
julian.hunt@cpom.ucl.ac.uk

Ishida, Takaki
Department of Computational
Science and Engineering
Nagoya University
Japan
ishida@fluid.cse.nagoya-u.ac.jp

Ishihara, Takashi
Department of Computational
Science and Engineering
Nagoya University
Japan
ishihara@cse.nagoya-u.ac.jp

Ishii, Katsuya
Information Technology Center
Nagoya University
Japan
ishii@cse.nagoya-u.ac.jp

Ishioka, Keiichi
Division of Earth and Planetary
Sciences
Kyoto University
Japan
ishioka@gfd-dennou.org

Iwamoto, Kaoru
Department of Mechanical
Engineering
Tokyo University of Science
Japan
iwamoto@rs.noda.tus.ac.jp

Iwayama, Takahiro
Graduate School of Science
and Technology
Kobe University
Japan
iwayama@kobe-u.ac.jp

Izawa, Seiichiro
Department of Mechanical Systems
and Design
Tohoku University
Japan
izawa@fluid.mech.tohoku.ac.jp

Jiménez, Javier
School of Aeronautics, Spain
Centre for Turbulence Research
Stanford University, USA
jimenez@torroja.dmt.upm.es

Kambe, Tsutomu
Institute of Dynamical Systems
Japan
kambe@ruby.dti.ne.jp

Kameda, Takatsugu
Department of Mechanical
Engineering
Yamaguchi University
Japan
kameda@yamaguchi-u.ac.jp

Kaneda, Yukio
Department of Computational
Science and Engineering
Nagoya University
Japan
kaneda@cse.nagoya-u.ac.jp

Kareem, W. Abdel
Department of Mechanical Systems
and Design
Tohoku University
Japan
waleed@fluid.mech.tohoku.ac.jp

Kawahara, Genta
Division of Nonlinear Mechanics
Osaka University
Japan
kawahara@me.es.osaka-u.ac.jp

Kimura, Yoshiyuki
Graduate School of Mathematics
Nagoya Univeristy
Japan
kimura@math.nagoya-u.ac.jp

Kitamura, Yuji
Division of Earth and Planetary
Sciences
Kyoto University
Japan
kitamura@kugi.kyoto-u.ac.jp

Kitoh, Osami
Depatment of Engineering Physics,
Electronics and Mechanics
Nagoya Institute of Technology
Japan
kitoh.osami@nitech.ac.jp

Kubo, Takashi
EcoTopia Science Institute
Nagoya University
Japan
t-kubo@nagoya-u.jp

Kuczaj, Arkadiusz K.
University of Twente
The Netherlands
a.k.kuczaj@utwente.nl

Kurien, Susan
Theoretical Division
Los Alamos National Laboratory
USA
skurien@lanl.gov

Matsubara, Masaharu
Department of Mechanical
Systems Engineering
Shinshu University
Japan
mmatsu@shinshu-u.ac.jp

Matsumoto, Takeshi
Division of Physics and Astronomy
Kyoto University
Japan
takeshi@
  kyoryu.scphys.kyoto-u.ac.jp

Matsumura, Ryo
Department of Mechanical and
System Eng.
Kyoto Institute of Technology
Japan
mathyn05@kit.ac.jp

Meneveau, Charles
Department of Mechanical
Engineering and Center
for Environmental and Applied
Fluid Mechanics
Johns Hopkins University
USA
meneveau@jhu.edu

Mininni, Pablo
NCAR
USA
mininni@ucar.edu

Miyazaki, Takeshi
Dept. Mechanical Engineering
and Intelligent Systems
Univ. Electro-Communications
Japan
miyazaki@mce.uec.ac.jp

Mizuno, Yoshinori
Graduate School of Engineering
Nagoya University
Japan
mizuno@fcs.coe.nagoya-u.ac.jp

Mochizuki, Shinsuke
Faculty of Engineering
Yamaguchi University
Japan
shinsuke@yamaguchi-u.ac.jp

Moffatt, H. Keith
Trinity College, Cambridge
UK
h.k.moffatt@damtp.cam.ac.uk

Mouri, Hideaki
Meteorological Research Institute
Japan
hmouri@mri-jma.go.jp

Murakami, Shin'ya
Graduate School of Science
and Technology
Kobe University
Japan
murakami@ahs.scitec.kobe-u.ac.jp

Nagata, Kouji
Department of Mechanical Science
and Engineering
Nagoya University
Japan
nagata@nagoya-u.jp

Nagib, Hassan
Illinois Institute of Technology
USA
nagib@iit.edu

Nakamura, Ikuo
Professor Emeritus
Nagoya University
Japan
nakamura@hm.aitai.ne.jp

Nakano, Tohru
Department of Physics
Chuo University
Japan
nakano@phys.chuo-u.ac.jp

Narasimha, Roddam
Jawaharlal Nehru Centre
for Advanced Scientific Research
India
roddam@caos.iisc.ernet.in

Neophytou, Marina
Department of Civil
and Environmental Engineering
University of Cyprus
Cyprus
neophytou@ucy.ac.cy

Nozawa, Kojiro
Institute of Technology
Shimizu Corporation
Japan
nozawa@shimz.co.jp

Oberlack, Martin
Department of Mechanical
Engineering
Technical University Darmstadt
Germany
oberlack@hyhy.tu-darmstadt.de

Okamoto, Naoya
Department of Computational
Science and Engineering
Nagoya University
Japan
okamoto@fluid.cse.nagoya-u.
 ac.jp

Pouquet, Annick
NCAR
USA
pouquet@ucar.edu

Procaccia, Itamar
Weizmann Institute of Science
Israel
itamar.procaccia@weizmann.ac.il

Rubinstein, Robert
NASA Langley Research Center
USA
r.rubinstein@larc.nasa.gov

Ruppert-Felsot, Jori
Laboratoire de Météorologie
Dynamique
CNRS Ecole Normale Superieure
France
jori@lmd.ens.fr

Sakai, Yasuhiko
Department of Mechanical Science
and Engineering
Nagoya University
Japan
ysakai@mech.nagoya-u.ac.jp

Sasaki, Atsushi
Department of Mechanical
Systems Engineering
Shinshu University
Japan
asasaki@shinshu-u.ac.jp

Schekochihin, Alexander A.
DAMTP
University of Cambridge
UK
as629@damtp.cam.ac.uk

Schneider, Kai
MSNM-CNRS & CMI Université
de Provence
France
kshneid@cmi.univ-mrs.fr

Schumacher, Jörg
Department of Mechanical
Engineering
Technische Universität Ilmenau
Germany
joerg.schumacher@tu-ilmenau.de

Seki, Daisuke
Department of Mechanical
Systems Engineering
Shinshu University
Japan
t06a116@shinshu-u.ac.jp

Sekimoto, Atsushi
Division of Nonlinear Mechanics
Osaka University
Japan
sekimoto@me.es.osaka-u.ac.jp

Shi, Yi Peng
State Key Laboratory for Turbulence
and Complex System
Peking University
China
syp@mech.pku.edu.cn

Shimomura, Yutaka
Keio University
Japan
yutaka@phys-h.keio.ac.jp

Siefert, Malte
Institute of Physics
University of Potsdam
Germany
malte.siefert@gmail.com

Smith, Leslie
Mathematics and Engineering
Physics
University of Wisconsin
USA
lsmith@math.wisc.edu

Sreenivasan, Katepalli R.
International Centre for
Theoretical Physics
Italy
krs@ictp.it

Staplehurst, Philip John
Department of Engineering
University of Cambridge
UK
pjs51@cam.ac.uk

Suematsu, Jun'ichi
Department of Mechanical
Systems Engineering
Shinshu University
Japan
t06c001@shinshu-u.ac.jp

Tagawa, Shinji
Department of Mechanical Science
and Engineering
Nagoya University
Japan
tagawa@sps.mech.nagoya-u.ac.jp

Takahashi, Naoya
Dept. of Mechanical Engineering
and Intelligent Systems
Univ. Electro-Communications
Japan
naoya@mce.uec.ac.jp

Takase, Kazuyuki
Japan Atomoic Energy Agency
Japan
takase.kazuyuki@jaea.go.jp

Tanahashi, Mamoru
Department of Mechanical
and Aerospace Engineering
Tokyo Institute of Technology
Japan
mtanahas@mes.titech.ac.jp

Tatsumi, Tomomasa
Professor Emeritus
Kyoto University
Japan
tatsumi@skyblue.ocn.ne.jp

Toh, Sadayoshi
Division of Physics and Astronomy
Kyoto University
Japan
toh@scphys.kyoto-u.ac.jp

Toschi, Federico
CNR-IAC
Italy
toschi@iac.cnr.it

Tsuji, Yoshiyuki
Department of Energy Engineering
and Science
Nagoya University
Japan
c42406a@nucc.cc.nagoya-u.ac.jp

Uchida, Kenji
Department of Mechanical Science
and Engineering
Nagoya University
Japan
kenji-u@sps.mech.nagoya-u.ac.jp

Ueno, Kazuto
Graduate School of Engineering
Nagoya University
Japan
ueno@fcs.coe.nagoya-u.ac.jp

Umeki, Masayuki
Trinity Industrial Corp.
Japan
m-umeki@trinityind.co.jp

Warhaft, Zellman
Mechanical and Aerospace
Engineering
Cornell Univerity
USA
zw16@cornell.edu

Watanabe, Takashi
Department of Mechanical
Systems Engineering
Shinshu University
Japan
t05a145@amail.shinshu-u.ac.jp

Watanabe, Takeshi
Department of Engineering Physics
Nagoya Institute of Technology
Japan
watanabe@nitech.ac.jp

Wells, John C.
Ritsumeikan University
Japan
jwells@se.ritsumei.ac.jp

Yasue, Toshiya
Department of Mechanical
Systems Engineering
Shinshu University, Japan
t05a141@amail.shinshu-u.ac.jp

Yokoi, Nobumitsu
Institute of Industrial Science
University of Tokyo, Japan
nobyokoi@iis.u-tokyo.ac.jp

Yoshida, Kyo
Graduate School of Pure
and Applied Sciences
University of Tsukuba, Japan
yoshida@sakura.cc.tsukuba.ac.jp

Yoshimatsu, Katsunori
Department of Computational
Science and Engineering
Nagoya University
Japan
yosimatu@fluid.cse.nagoya-u.
 ac.jp

Yoshino, Masato
Department of Mechanical
Systems Engineering
Shinshu University
Japan
masato@shinshu-u.ac.jp

IUTAM Bookseries

1. P. Eberhard (ed.): *IUTAM Symposium on Multiscale Problems in Multibody System Contacts*. 2007   ISBN 978-1-4020-5980-3
2. H. Hu and E. Kreuzer (eds): *IUTAM Symposium on Dynamics and Control of Nonlinear Systems with Uncertainty*. 2007   ISBN 978-1-4020-6331-2
3. P. Wriggers and U. Nackenhorst (eds): *IUTAM Symposium on Computational Methods in Contact Mechanics*. 2007   ISBN 978-4020-6404-3
4. Y. Kaneda (ed.): *IUTAM Symposium on Computational Physics and New Perspectives in Turbulence*. 2007   ISBN 978-4020-6471-5
5. A. Combescure, R. de Borst and T. Belytschko (eds): *IUTAM Symposium on Discretization Methods for Evolving Discontinuities*. 2007   ISBN 978-4020-6529-3
6. A.V. Borisov, V.V. Kozlov, I.S. Mamaev and M.A. Sokolovskiy, M.A. (eds.): *IUTAM Symposium on Hamiltonian Dynamics, Vortex Structures, Turbulence*. 2008   ISBN 978-1-4020-6743-3
7. J.F. Morrison, D.M. Birch and P. Lavoie (eds.): *IUTAM Symposium on Flow Control and MEMS*. 2008   ISBN 978-1-4020-6857-7